THE BACKYARD HOMESTEAD *Guide to Raising* FARM ANIMALS

The mission of Storey Publishing is to serve our customers by publishing practical information that encourages personal independence in harmony with the environment.

Edited by Sarah Guare
Art direction by Dan O. Williams
Book design and text production by MacFadden and Thorpe

Cover illustration and illustrations on pages i-vi by © Michael Austin, Jing and Mike Company
Interior illustration credits appear on page 340

Indexed by Andrea Chesman
Expert review of the rabbit chapter by Bob Bennett, the sheep chapter by Carol Ekarius, the turkey chapter by Eugene Morton, and the cattle chapter by Mary Sherman

Sections of this book previously appeared in the following books by Storey Publishing: *Barnyard in Your Backyard*, edited by Gail Damerow; *Storey's Guide to Keeping Honey Bees*, by Malcolm T. Sanford and Richard E. Bonney; and *Storey's Guide to Raising Pigs*, by Kelly Klober.

Storey Publishing
210 MASS MoCA Way
North Adams, MA 01247
www.storey.com

Printed in the United States by Versa Press
10 9 8 7 6 5 4 3 2 1

Library of Congress Cataloging-in-Publication Data

The backyard homestead guide to raising farm animals / edited by Gail Damerow.
p. cm.
Includes index.
Previously published as: Barnyard in your backyard. North Adams, MA : Storey Pub., c2002.
ISBN 978-1-60342-969-6 (pbk. : alk. paper)
1. Domestic animals. 2. Food animals. I. Damerow, Gail.
SF61.B23 2011
636—dc22

2010051172

THE BACKYARD HOMESTEAD Guide to Raising FARM ANIMALS

Edited by Gail Damerow

Contributing Authors:

Richard E. Bonney, Gail Damerow, Kelly Klober, Darrell L. Salsbury, Malcolm T. Sanford, Nancy Searle, Paula Simmons, Heather Smith Thomas

Storey Publishing

Contents

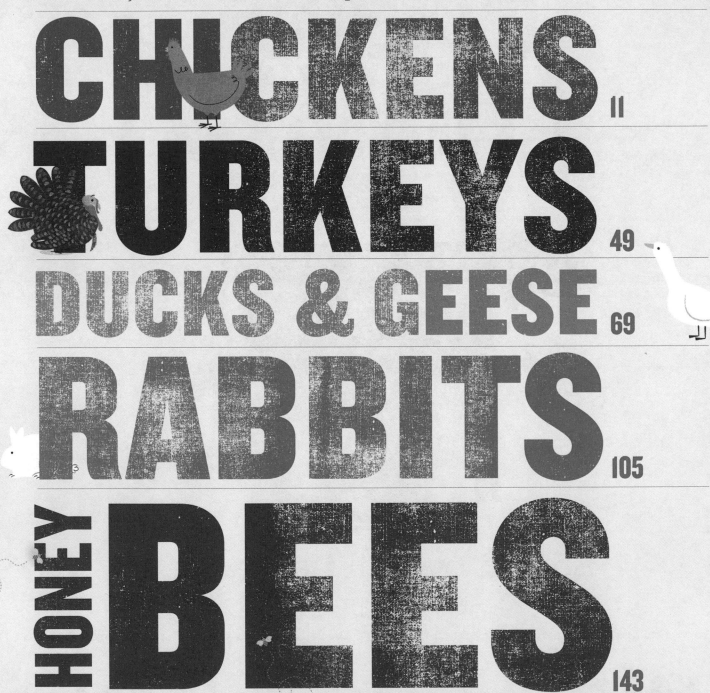

Preface

I did not grow up on a farm. My father was a town boy. My mother was raised in the country, left at her first opportunity, and never looked back. I have fond memories of visiting my grandmother and her big flock of laying hens; of watching an uncle milk his cows; of briefly, while my parents built a new house, living in a rented country house with pigs across the road and a goat next door that surprised and delighted me one noon by poking its head through the kitchen window hoping to snatch a bite of my sandwich. These and similar events made me determined to someday have farm animals of my own.

And so when I came of age I started out with chickens, soon followed by ducks and geese and other poultry. Then came rabbits. Then I had to have goats. In the decades that followed I have, at one time or another, raised nearly every species of farm animal. At first I did it just for fun. Before long I realized I had ready access to eggs, milk, and meat that were fresher and better tasting than anything available at the store. I have been grateful for these resources during the several economic crises that have occurred over the years, and especially today in the face of growing concern about the safety of industrially produced food.

I didn't have any background in farming or raising livestock, and by the time I was ready to seek their advice no farmers were left in my family. I gleaned most of my start-up information from books, and the rest I learned simply by rolling up my sleeves and jumping into what has become a life-long and ongoing learning experience. My point is that you need not have grown up on a farm to raise farm animals. This book is offered as your first step toward enjoying the rewarding experience of keeping livestock of your own and to assure you — you can do it!

Gail Damerow

Introducing Backyard Farm Animals

Food security. The term means different things to different people. One definition involves having a reliable source of basic foods and not having to worry about going hungry. Another requires the food to be of sufficient quantity and quality to meet your dietary needs and satisfy your food preferences. Still other definitions specify that the food be nutritious, safe, and healthful. And some definitions incorporate the concepts of local self-sufficiency and environmental sustainability.

Taken together, these various definitions point in one direction: Grow your own. And unless you are a vegetarian, that means raising livestock. As a lot of people are learning, you don't need to live on a farm to raise food animals. A pair of rabbits in the carport or on the back porch will provide a year-round supply of meat while taking up hardly any space at all. A beehive or two will give you healthful honey while pollinating your garden. A few hens will provide you with fresh eggs while living happily in one corner of the garden. When I started out with livestock, I lived on approximately one acre on which I raised a variety of rabbits, chickens, turkeys, ducks, and geese, along with a big garden and a small orchard. The garden and orchard benefited from the manure produced by the animals, and the animals benefited from surplus produce gleaned from the garden. Best of all, my family enjoyed food security of the highest order.

Today I live on a farm, where we have space to raise more food animals. We keep dairy goats for their delicious milk, and on occasion use surplus milk to raise a calf for awesomely tasty homegrown burgers or to raise piglets that fill our freezer with succulent pork. But you don't have to live on a farm, or even on one acre of land, to produce nutritious, safe, healthful eggs, milk, and meat. This book is here to point the way.

Why Raise Food Animals?

Aside from supplying wholesome eggs, milk, meat, and honey, keeping farm animals has another health benefit. Observing the sight and sounds of the animals in your yard offers a refreshing, stress-reducing change from the scurry of modern life. Provided you leave the cell phone behind, doing barnyard chores is a great way to take time out for relaxation and quiet enjoyment. Since livestock must be cared for daily, they pull you away from your indoor activities and force you to get out for a little exercise and fresh air. I often spend long days in the office and look forward to doing evening chores that not only let me stretch my muscles but also help clear my head.

Educational value is another important reason for raising backyard livestock. Too many kids today believe eggs come in plastic cartons and milk comes from a cardboard box. Keeping animals for food helps kids learn, hands on, the basics of producing their food. As any farm family will avow, helping to care for animals at home is a wonderful way for children to learn responsibility, patience, dedication, and compassion. And you could find no better way than raising barnyard animals for children to learn about the natural processes of procreation, birth, and death. As a wonderful bonus, children who grow up around farm animals are generally healthier than children who grow up isolated from the land.

A big question livestock novices frequently ask is, "Can I save money by growing my own ________ (fill in the blank)?" One thing is for certain: It's not about money. If you take into consideration the cost of building facilities and acquiring stock, as well as feeding and otherwise maintaining your animals, and especially if you include the amount of time you put into the endeavor, you cannot produce your own meat, milk, eggs, and honey as cheaply as you can purchase them from a big-time industrialized producer. If, on the other hand, you are buying quality products from small-scale producers, you can come pretty close to matching or beating their price.

So each chapter of this book includes a discussion on whether or not you can save money by growing your own, taking into consideration mainly the cost of feed, which is the major expense of keeping livestock once your animals have been acquired and their housing established. Even if your chosen endeavor turns out not to be a spectacular financial success, if you take into consideration all the advantages of growing your own — including the satisfaction of putting healthful, wholesome food on the table and the fun you have doing it — you should be able to produce your own food at a reasonable price.

Accentuate the Positive

The marvelous rewards that come from raising your own food animals don't come without a price, however, and here I'm not referring to money. For one thing, barnyard animals require constant care, day after day, week after week. No matter what else may be going on in your life on a particular day, or how tired you are at day's end, you must make time to take care of your livestock. Although daily chores don't involve a lot of time or hard work, they are an important responsibility. If you have trouble finding a substitute caretaker for times when you must be away, you may soon feel tied down by your animals.

But that's mainly a matter of attitude. Many's the time I have felt like not doing barnyard chores, only to have my spirits lifted when I got to the barn and was greeted by animals eagerly awaiting my arrival. Sometimes when I turn down an invitation from a friend or relative because I can't leave my animals for extended periods of time, I'm exhorted to "get rid of all those animals" so I can "be free." Anyone who could say such a thing doesn't have livestock of their own; my friends who do have barnyard animals would never consider suggesting anything of the sort. Keeping livestock gives us freedom of a different kind — freedom to eat what we choose, produced by animals raised the way we choose to raise them.

Other downsides that can be part and parcel of keeping backyard livestock include the need to deal with manure, odor, noise, flies, and complaining neighbors. All of these potentially negative factors can be handily dealt with through proper management. I consider manure to be a bonus rather than a disadvantage because I am a gardener, and manure makes outstanding compost that grows a bounty of scrumptious vegetables. My barnyard animals provide me with a constant supply of manure. Odor-free composting techniques are covered in such books as *The Complete Compost Gardening Guide,* by Barbara Pleasant and Deborah L. Martin. If you are not a gardener, surely you know someone who is who would be delighted to have a source of free, natural fertilizer and more than likely would be happy to clean out your barn to get it.

Properly dealing with manure automatically solves the problems of odor and flies, which leaves us next to consider noise. Barnyard noise is particularly problematic because not everyone considers it a problem. When I hear my neighbor's cow bellow, I know her calf is being weaned or the cow is ready to be rebred. I once had a neighbor who, when she heard a neighbor's cow bellow, became so alarmed she called in a vet at her own expense. Now that can get pretty annoying if you are the cow's owner.

Even if your chosen endeavor turns out not to be a spectacular financial success, if you take into consideration all the advantages of growing your own you should be able to produce your own food at a reasonable price.

A crowing rooster is another noisemaker that not everyone considers to be a problem. I enjoy hearing the sound of a cock crow, but I no longer have to worry about neighbors complaining about it. That's one of the reasons I moved from one acre to a farm, while some of my chicken-keeping friends have been in constant battles, sometimes ending up in court, over their crowing roosters. Laying hens don't make nearly as much racket as roosters, although the occasional cranky neighbor may take exception to their cackling. If barnyard noise is a potential problem, consider silent animals, such as honey bees, rabbits, or Muscovy ducks. The latter are sometimes called quackless ducks because their sound is so muted it can be heard only at close range.

Dealing with "The Day"

Unless your interest is in fresh eggs or honey, raising food animals means you must be prepared for the eventuality that one day animals will be butchered. Even a dairy animal involves meat production, since in order to give milk the female must give birth, giving you an annual crop of young ones to deal with. Butchering an animal you raised yourself can be traumatic if you, or especially your child, have become attached to the animal. How well I remember the rabbits our family had when I was little. I had thought they were my pets until the day I came home from school and found them hanging from the basement rafters to be skinned. I can't tell you how betrayed I felt. I eventually got over it, and today rabbit is one of my favorite meats. But as a child, I would have appreciated knowing the rabbits our family was raising, that I had spent so much time playing with, were destined for the dinner plate. With a tactful approach, no child is too young to learn.

The cardinal rule among those of us who raise animals for meat is to never name one. Well, that's not quite realistic, especially when you have more than one and need to differentiate between them. But at least avoid giving them affectionate petlike names. Instead use either numbers or names that serve as a reminder of the animal's purpose in life, such as Finger Lickin', Hambone, or Sir Loin.

Dealing with butchering involves not only overcoming the emotional aspects, but also following the prescribed procedures that result in safe, tender, tasty meat. Educate yourself by reading a book such as *Basic Butchering of Livestock & Game,* by John J. Mettler, Jr., and if the process sounds like something you'd rather not get involved with, find out ahead of time if you can count on someone else to do it for you. That someone might be a friend or neighbor raising similar livestock, or perhaps a professional slaughterhouse.

Not all slaughterhouses accept all kinds of livestock. Some take only poultry, whereas others take only larger animals. Even a custom butcher who handles larger stock might have a seasonal schedule: for example, taking in only game animals during the hunting season. When you find a slaughterhouse you plan to use, seek endorsements from past customers. We once had a pig butchered by a shop we had not used before, and they included far too much fat in the ground sausage (which you can understand when you realize that such a shop charges by the pound). As a result, 1 pound of sausage cooked down to less than ½ pound of meat, and — since fat does not keep as well as lean meat — a lot of the sausage went rancid in the freezer before we could use it up.

Perhaps you don't want to get involved in raising meat at all. Consider that right from the start. If you want a cow or goat for milk, the animal will lactate only as a result of giving birth, so you will have one or more babies to deal with in the future. If you don't raise them for meat, what you will do with them? If you raise chickens or other fowl for eggs, one of the birds may eventually steal off into some secluded place to lay her eggs and hatch a batch of chicks, thus greatly increasing your backyard population. The offspring might be considered a bonus if raised for meat, a burden if not.

One of the big advantages to raising livestock solely for meat is that the project can be short-term. A batch of broiler chickens, for example, can be raised and butchered all within 8 weeks' time. A lamb can be ready to turn into chops in six or seven months. These short time frames give you a chance to decide whether you like raising livestock at all. If the answer is yes, you then have the choice of doing another short-term meat project in the future or engaging in a long-term project involving breeding your own animals.

The Value of Networking

Once you've decided which animals to keep, educate yourself further about what's involved. Read not only this book, which provides an overview of each type of food animal, but also some of the books mentioned in the Resources section on page 333, which offer more in-depth details on each specific breed. Subscribe to a periodical dealing with your chosen breed. Network with others who raise the breed by joining a local club, if one exists, and regional or national breed clubs. Visit the fair in your county, and perhaps in surrounding counties, to meet people who have the breed that interests you.

A super place to gather information is at a 4-H show, where the kids involved are well educated about their animals and eager to share their knowledge. Nothing pleases children more than the opportunity to show an adult how smart they are. The people you meet during your networking will become invaluable when you have questions about such things as how to harvest honey, milk a cow, or trim a goat's hooves.

Finding Stock

The same places that offer good networking opportunities are also excellent sources for locating livestock to purchase. Avoid purchasing stock at an auction or sale barn, where animals are constantly coming and going. You will have no idea where your animal came from, and you can't tell how healthy or unhealthy it may have been to start with or what kind of diseases it may have been exposed to along the way. The last thing you want is for your first livestock experience to turn into a fiasco involving multiple expensive visits with the veterinarian, administering medications to a reluctant animal, and in the end possibly losing the animal despite your best efforts.

If possible, buy animals from someone who lives nearby. Livestock purchased close to your home already will be adapted to your area, and you will have someone to turn to if you need help later on. When you buy from a local breeder, you can see for yourself whether the animals come from a clean, healthful environment and whether the breeding population has the proper conformation. If you are buying a female breeding animal — a cow, ewe, sow, or doe (goat or rabbit) — the seller may have a male animal to which you could breed her when the time comes. Keeping a rabbit buck for

Raising a rare breed for food may seem contradictory, but doing so supports breeders and encourages them to perpetuate that breed.

breeding is no big deal, but keeping a bull, ram, boar, or goat buck just to breed one or two females is neither safe nor cost-effective.

An excellent place to find local sellers of livestock is the farm store. Many farm stores maintain a bulletin board where breeders may advertise livestock for sale, and the clerks can tell you who buys feed for the species you are seeking. The county Extension office is another possible source of information, although some agencies are more active and knowledgeable than others. Larger livestock operations might advertise in the Yellow Pages of your phone book, in the newspaper classified ads, or in the freebie shopper newspapers that abound in every community. The farm store and Extension office can also tell you if your area has a club or other interest group dedicated to your breed. Also check with the national association that promotes your chosen breed or species, most of which maintain a membership list that is available to the public. Some organizations publish their membership list on websites to help you locate members nearest you.

If you are interested in a less common breed, contact the American Livestock Breeds Conservancy or Rare Breeds Canada for their periodically updated list of breeders. Raising a rare breed for food may seem contradictory, but doing so supports breeders and encourages them to perpetuate that breed; if nobody wants these animals, nobody will continue producing them. Besides, taste tests prove time and again that rare breeds are often the best tasting, primarily because they have not been selectively bred for rapid growth.

Getting the Animal Home

If you have difficulty finding what you want locally, cast your net a little farther afield. When purchasing animals from a distance, try to travel to the seller's location to view the breeder stock and pick up your purchase. No matter how carefully animals are transported, shipping always involves certain risks.

We have occasionally purchased a calf from a dairy in the next county and transported it home in our pickup camper, and we never had a problem until one extremely hot day. During the 45-minute drive home we stopped to offer the calf some water. It was too frightened to drink, so we decided the better plan was to get home fast and get it off the hot truck. By the time we arrived home the calf was nearly prostrate from heat and dehydration. After a good hosing down with cold water and several gallons of Gatorade, the calf was fine, but the incident gave us quite a scare. Since then if we have to transport livestock in the summer, we do so in the cool hours of early morning or late evening.

If you cannot pick up your purchase in person but must arrange to have it shipped, have a clear written understanding with the seller regarding who bears the risk if the animal gets sick or dies. The stress of long-distance travel compromises an animal's immune system, risking infection during travel or on arrival at its new home.

Preparing a Home

Before bringing home your first animal, have everything ready for it. A little advance preparation will smooth the way.

Ensure family support. Check with all members of your family to see how they feel about having livestock in your backyard. It's always best to have everyone's full support, especially when you may need a substitute to do your daily chores whenever you must be away. If not all members are involved in maintaining the livestock, strife can result when the uninterested members feel the others spend too much time at the barn, yet they share in the bounty. By contrast, relations in families in which everyone is involved in some phase of animal care are usually harmonious. In our family, my husband and I normally do chores together; we each have certain responsibilities, but each of us pitches in for the other when need be. We enjoy our time together walking to and from the barn, but at the barn, we devote our full attention to the animals.

Establish caretaking responsibilities. Establish a caretaking schedule and decide who in your family will do what chores daily, weekly, monthly, and seasonally. Children, for instance, can be in charge of the daily routines of feeding, milking, and gathering eggs; these simple tasks will help them learn about responsibility. Adults or older teenagers should probably be involved in the less frequent but more difficult tasks, such as vaccinating, cleaning stalls, or attending births.

Check zoning regulations. Every area has a slightly different set of zoning laws, which may prohibit you from keeping certain species, limit the number of each species you may keep, regulate the distance animal housing must be from nearby human dwellings or your property line, or restrict the use of electric fencing.

I saw firsthand how zoning works on my little one-acre farmstead, which was rezoned after I moved there. Although my poultry activities were grandfathered in — meaning the authorities could not make me get rid of the birds I already had — I was not allowed to increase the population. Now the nature of raising chickens is that after the spring hatch you have more, and as the year progresses and you butcher some, you have fewer. Complying with the new law meant I would not be able to hatch and raise young chickens for meat. I managed to prevail as long as I lived there, but not without hassles from neighbors and occasional visits from the authorities. If you plan to raise livestock on property you have yet to purchase, check not only existing laws but also proposed changes. If existing zoning laws are not livestock friendly, several websites explain how to get the laws changed.

Prepare facilities. Once you learn of any zoning regulations that will influence where on your property you may keep animals, prepare their housing. Most animals require all-weather housing. If your area has particularly hot days or cold days, take those extremes into consideration right from the start, or you may never get around to providing proper housing. If you are starting out with babies, remember they will grow; make sure your facilities are big enough to handle them when they mature. If you wish to breed your stock to raise future babies, chances are pretty good you'll want to keep one or more of the babies, so allow space for expansion. Since things have a way of taking longer than expected, have your facilities ready and waiting *before* you bring home your first animal. Provide adequate feed and water stations.

Lay in a supply of feed. Unlike wild animals, which are adept at balancing their own nutritional needs, domestic animals rely on us to furnish all the nutrients they need. The best choice for a beginner is to use bagged feed from the farm store. If you are concerned about what's in the ration you can opt for an all-natural formula. If you prefer certified organic feed, expect to pay 50 to 100 percent more. After you become knowledgeable about your chosen species' habits and dietary needs, you will be in a better position to develop an alternative ration, should you so desire. Meanwhile, if the bagged feed you will be using is different from what the animal has been eating, purchase some of its usual feed from the animal's seller. Gradually mix in greater quantities of new feed with the old to avoid an abrupt change that can cause digestive upset in an animal already stressed from the move.

Install sturdy fencing. Secure the livestock area with a stout fence that not only keeps in your livestock but also keeps out predators. When most people hear the word *predator*, they think of wild animals such as foxes, raccoons, or coyotes. But the number one predators of domestic livestock are dogs. Our neighborhood was once terrorized by a dog that killed countless chickens (including some of mine), a calf, a couple of sheep, and dozens of 4-H rabbits. When the animal-control officer finally caught and euthanized the dog, the owner was furious that her children had been deprived of their beloved pet. Sometimes a predator dog is not the neighbor's but your own. I've heard many a tale of dogs that got along well with poultry, and even guarded them, then for reasons only the dog could know eventually went on a rampage and killed the birds.

Most livestock books recommend farm fencing of one sort or another, which securely confines stock and excludes predators, but may not be legal in more populated areas. The type of fencing you use must be acceptable in your area, both legally and aesthetically. To make your fence animal safe as well as publicly acceptable, you may have to fudge a bit by camouflaging farm fence to look like something else from the outside; for instance, having small-mesh woven wire on the inside with attractive post and rail board fencing on the outside. Attractive fencing that blends well with the neighborhood landscape is more likely to be acceptable to neighbors.

Inform your neighbors. Let your neighbors know about your plan to raise livestock. Explain that you are taking great pains to keep your animals from getting into other people's yards and to keep other people's animals out of your yard. Describe what you are doing to maintain clean housing and minimize odors and flies. By letting the neighbors in on your plans, you are less likely to hear complaints from them later. You might even get them involved by asking for their input and advice. Perhaps they'd be willing to help out, for instance when you go on vacation, in exchange for fresh eggs from your chickens, fresh milk from your cow or goat, or barnyard compost for their garden. Who knows — you might pique their interest enough that they'll want backyard farm animals of their own.

How Many Animals Can You Keep?

The following illustrations show some of the possibilities for the number of animals that can be kept in an average yard. A quarter-acre lot, planned out well and intensively maintained, can provide milk, meat, honey, and eggs for a small family. Adding another quarter acre allows you to inexpensively raise steers for beef. These examples show what can be done in a given amount of space, but remember: the less living space your animals have, the more time you will have to spend cleaning and maintaining their quarters.

A Homestead on One-Tenth of an Acre

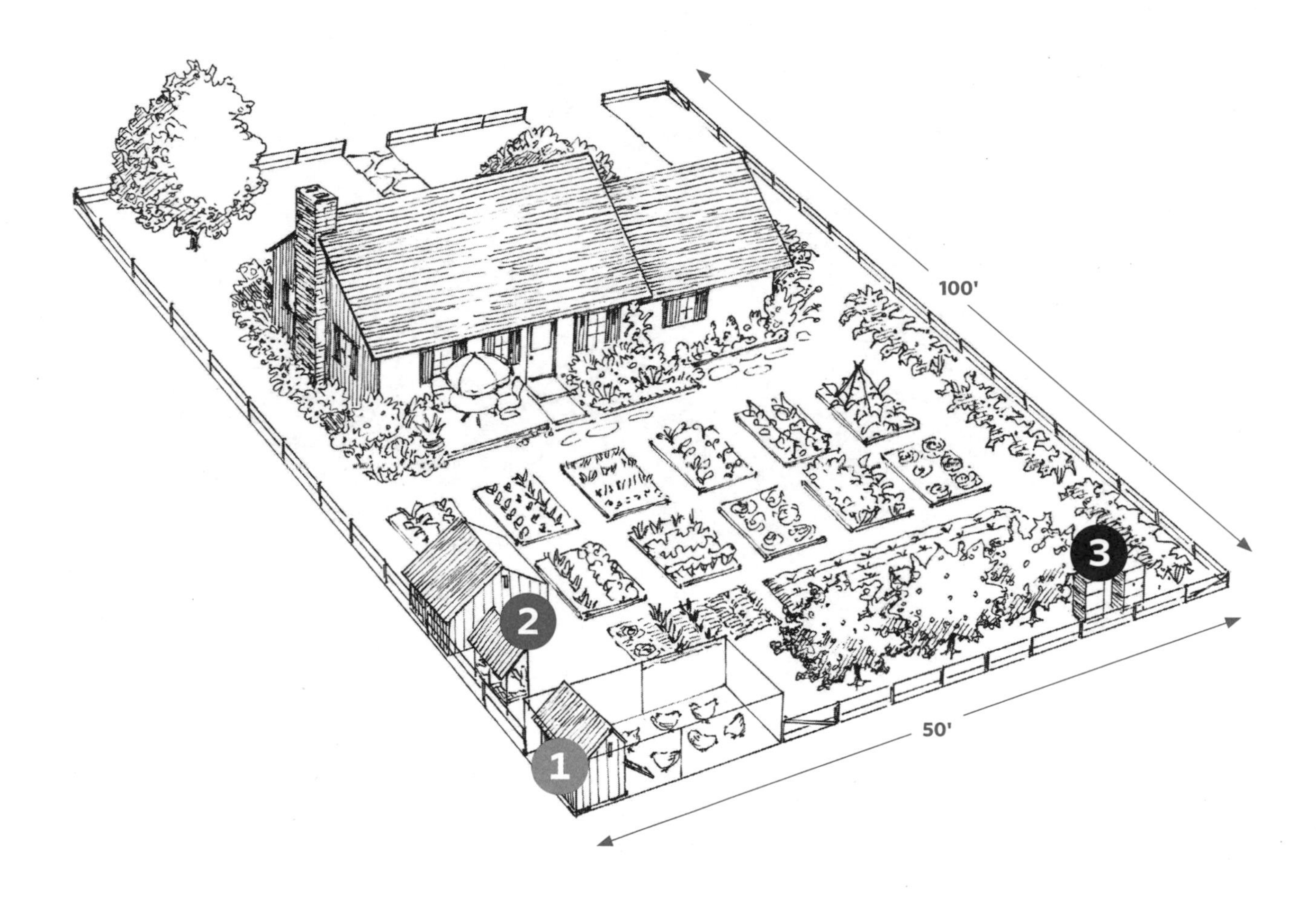

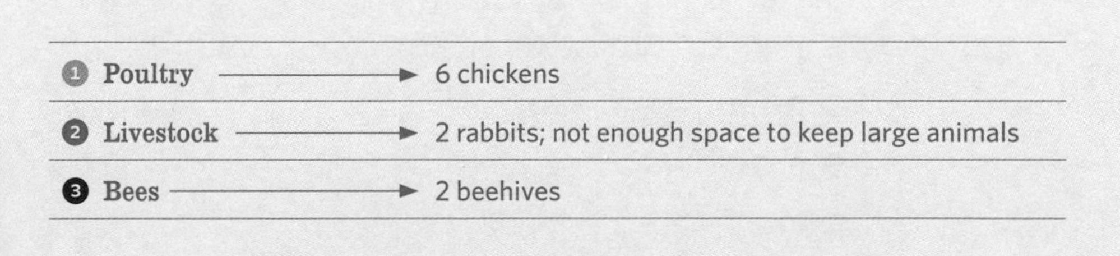

❶ **Poultry** →	6 chickens
❷ **Livestock** →	2 rabbits; not enough space to keep large animals
❸ **Bees** →	2 beehives

A Homestead on a Quarter Acre

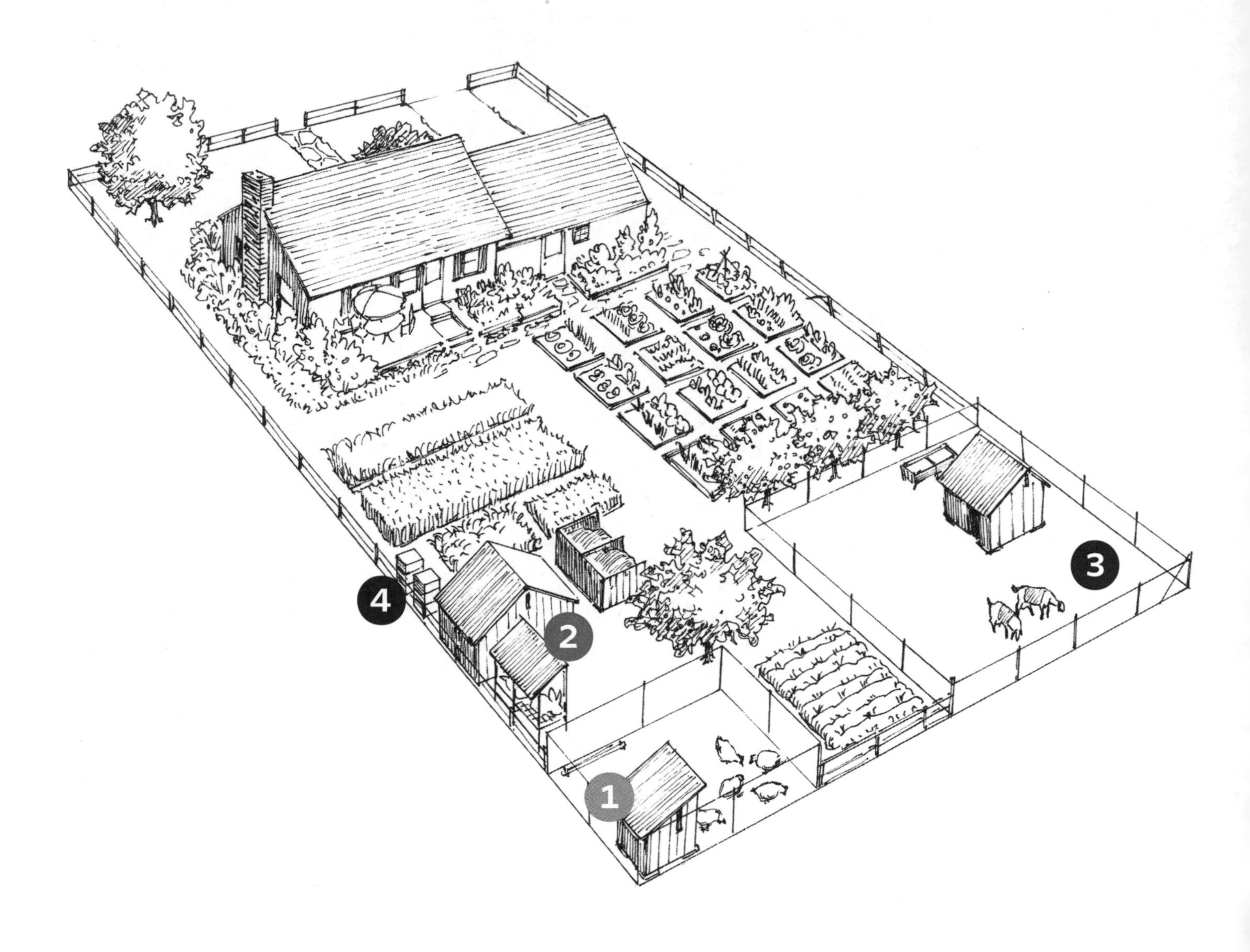

❶ **Poultry** ⟶	6 layers or 24 broilers or 3 turkeys
❷ **Rabbits** ⟶	3 rabbits
❸ **Goats** ⟶	2 dairy goats
❹ **Bees** ⟶	2 beehives

A Homestead on Half an Acre

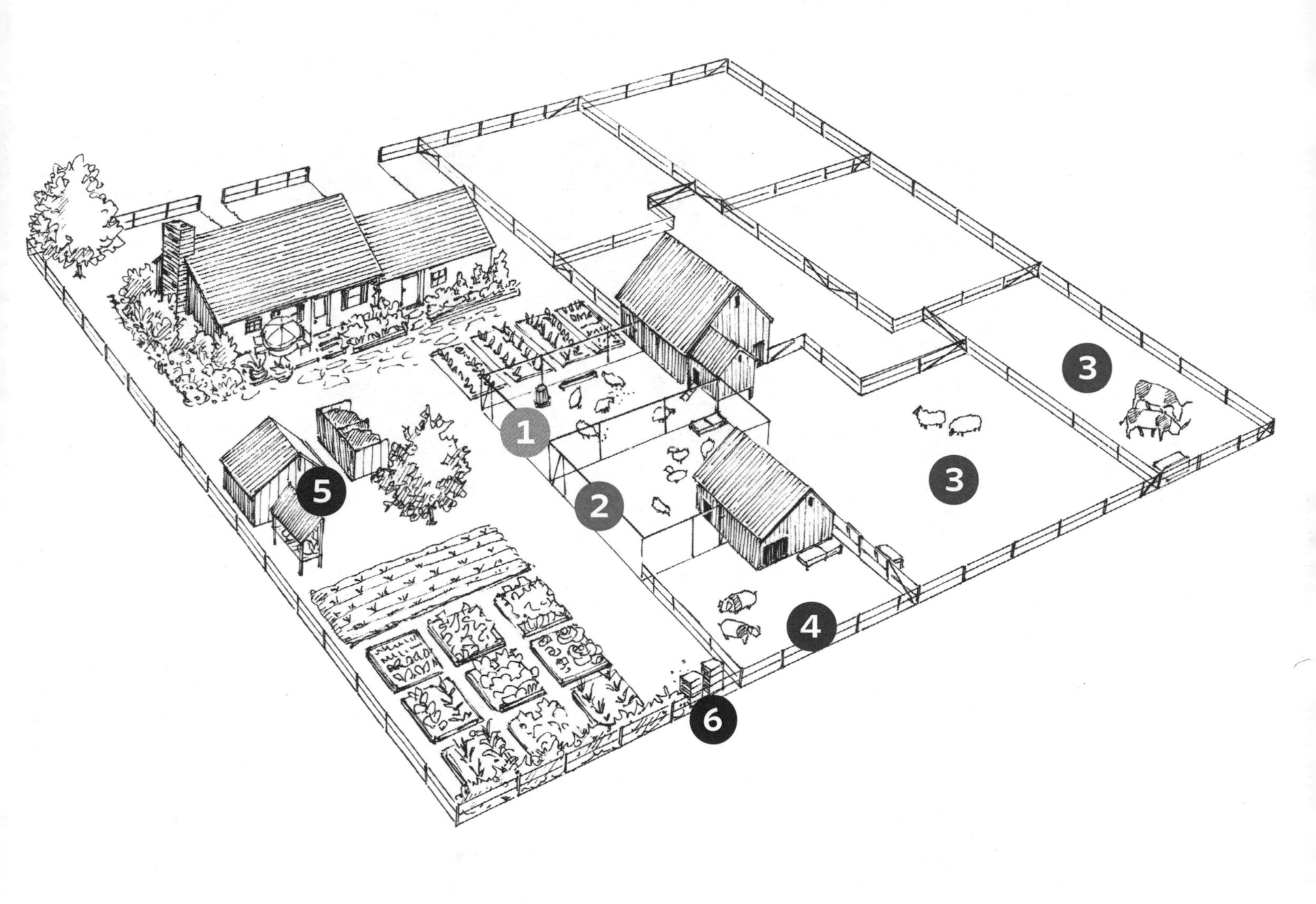

1	**Poultry**	6 layers or 24 broilers or 3 turkeys
2	**Waterfowl**	6 ducks or 3 geese
3	**Livestock**	1 cow with 1 or 2 calves or 2 beef steers and 2 dairy goats or 2 beef steers and 2 lambs
4	**Pigs**	2 pigs
5	**Rabbits**	3 rabbits
6	**Bees**	2 beehives

Calcium
Grit

CHAPTER 1

Chickens

If you've never raised livestock before, chickens make a great place to start. They're easy to raise, they don't need a lot of space, and they don't cost a lot of money to buy or to feed. Everything you learn about feeding, housing, and caring for your chickens will help you later if you decide to raise some other kind of animal.

All chickens belong to the genus *Gallus*, the Latin word for cock, or rooster. The English naturalist Charles Darwin traced all breeds of domestic chicken back tens of thousands of years to the wild red jungle fowl of Southeast Asia (*Gallus gallus*). These fowl look something like today's brown Leghorns, only smaller. Recent evidence indicates that gray jungle fowl (*Gallus sonneratii*) may also have been involved in the domestic chickens' lineage.

Wild jungle fowl are homebodies, preferring to live and forage in one place as long as possible. This trait made taming wild fowl an easy task. All people had to do was provide a suitable place for the chickens to live and make sure they got plenty to eat. As a reward, their keepers had ready access to fresh eggs and meat.

Early chickens didn't lay many eggs, though, and they made pitifully scrawny meat birds. Over time, chicken keepers selected breeders from those that laid best or grew fastest and developed the most muscle. Different chicken breeders made their selections based on different criteria, and thus came about today's many different breeds. The Romans called household chickens *Gallus domesticus*, a term scientists still use.

In 1868, Darwin took inventory of the world's chicken population and found only 13 breeds. Now we have many times that number. Most of today's breeds were developed during the twentieth century, when chickens became the most popular domestic food animal.

Getting Started

How much money you spend to get started depends on such factors as the kind of chickens you want and how common they are in your area, how simple or elaborate their housing will be, and whether you already have facilities you can use or modify. Chickens must be protected from wind and harsh weather, but the housing need not be fancy. If your yard isn't fenced, you'll need to put one up.

In deciding where to put your chicken yard, consider whether crowing may bother your neighbors. Male chickens — called roosters, or cocks — are well known for their inclination to crow at dawn. Cocks occasionally crow during the day, and if two cocks are within hearing distance, they will periodically engage in an impromptu crowing contest. A rooster rarely crows during the dark of night, unless he is disturbed by a sound or a light.

If keeping a rooster is illegal in your area, or the sound of crowing might cause a problem, consider keeping hens without a rooster. Although the rare persnickety neighbor may complain about hen sounds, the loudest noise a hen makes is a brief cackle upon leaving the nest after she's laid an egg. Contented hens sing to themselves by making a happy sound that only a grouch could object to. Without a rooster, hens will still lay eggs. The rooster's function is not to make hens lay eggs but to fertilize the eggs so they can develop into chicks. Without a rooster, you won't be able to hatch the eggs your hens lay.

Caring for a home flock takes only a few minutes each day to provide feed and water and to collect eggs.

Comparing Benefits and Drawbacks

Raising chickens has some downsides. One is the dust they stir up, which can get pretty unpleasant if they are housed in an outbuilding where equipment is stored. Another is their propensity to scratch the ground and dig dust holes, which can be a big problem if they get into a bed of newly planted seedlings. Chickens also produce plenty of droppings that, if not properly managed, will smell bad and attract flies.

Until you raise your own chickens, it may be hard to believe that people become attached to their chickens and have difficulty letting them go when it's time to butcher meat birds or replace old layers with younger, more efficient hens. The only alternative, though, is to run a retirement home for chickens, which gets pretty expensive, and the birds will still get old and die eventually. You'll have to come to grips with the loss.

For many people, the upside of raising chickens far outweighs the downside. Chickens provide wholesome eggs and meat for your family, and you can take pride in knowing that the flock that puts food on your table lives under pleasant conditions.

Caring for a home flock takes only a few minutes each day to provide feed and water and to collect eggs. In hot or cold weather, these jobs must be done twice daily, seven days a week. If you raise chickens for meat, the project will be finished in two to three months. If you raise hens for eggs, you must care for them year-round. As long as you keep in mind that your flock relies on you for its survival, raising chickens is a breeze.

Check First to See If You're Allergic

Before you set up a chicken farm, make sure that you and your family are not allergic to chicken dander. You can find this out ahead of time by visiting a poultry show at your county fair or spending a few hours helping care for someone else's chickens. If you discover you have an allergic reaction, you will have avoided the expense and heartache of setting up a flock you immediately have to get rid of.

Choosing the Right Breed

No one knows for certain how many breeds of chicken can be found throughout the world. Some breeds that once existed have become extinct, new ones have been developed, and forgotten ones have been rediscovered. Only a fraction of the breeds known throughout the world are found in North America.

The *American Standard of Perfection*, published by the American Poultry Association, contains descriptions and pictures of the many breeds and varieties officially recognized by that organization. These breeds are organized according to whether they are large chickens or bantam (miniature chickens suitable for exhibition or as pets), and each group is divided by class. Large chickens are classified according to their place of origin: American, Asiatic, English, Mediterranean, Continental, and Other (including Oriental). Each class is further broken down into breeds and varieties. Chickens of the same breed all have the same general type, or conformation. A chicken that looks similar to the ideal for its breed, as depicted in the *Standard*, is true to type, or typey.

Most chickens have white or yellow skin. If you are raising chickens for meat, the skin color may make a difference to you. People of Asian cultures tend to prefer chickens with dark skin, Europeans prefer white skin, and most Americans prefer yellow skin.

Some breeds come in more than one plumage color; each color constitutes a variety. The colors may be plain — such as red, white, or blue — or they may have a pattern, such as spangled, penciled, or barred. Two varieties of the Plymouth Rock are white and barred.

Another feature that can distinguish one variety from another is the style of comb. Most breeds sport the classic single comb, with its series of sawtooth zigzags. The Sicilian Buttercup, by contrast, has two rows of points that meet at the front and back, giving the comb a flowerlike look. Other comb styles are carnation, cushion, pea, rose, and strawberry. Rose comb and single comb are two varieties of Leghorn. Comb style becomes important in areas of weather extremes: Chickens with small combs that lie close to the head are less likely to suffer frostbite than chickens with large combs.

Some breeds come in more than one plumage color; each color constitutes a variety.

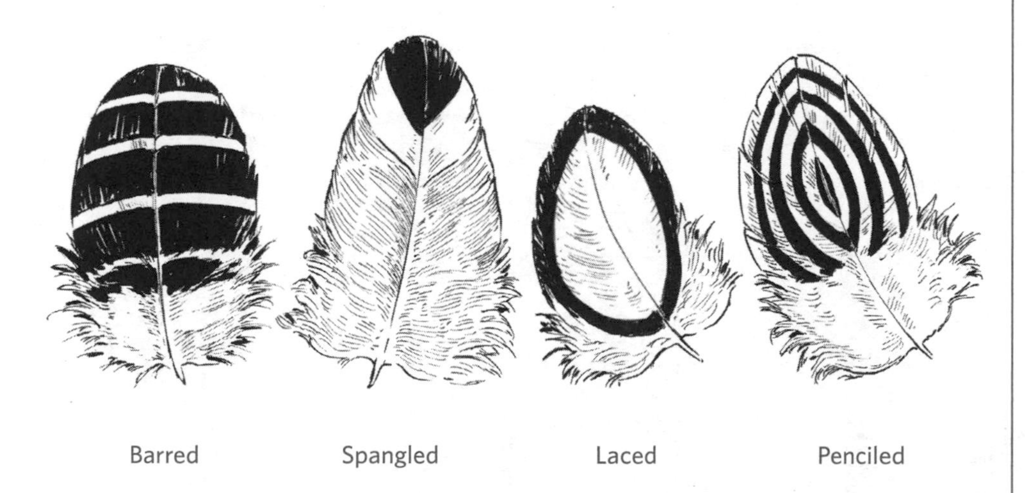

Barred Spangled Laced Penciled

With all these different possibilities, how do you choose the breed and variety that is best for you? Narrow your choices by deciding what you want your chickens to do for you:

- Do you primarily want eggs?
- Do you want to raise your own chicken meat?
- Do you want both meat and eggs?
- Do you want to help preserve an endangered breed?

Comb Styles

Buttercup comb

Carnation comb

Cushion comb

Pea comb

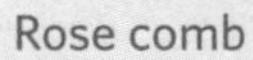
Rose comb

Rose comb (spiked)

Single comb

Strawberry comb

V comb

Walnut comb

Quick Guide to Chicken Breeds

Breed	Eggs	Meat	Foraging	Temperament	Climate
Ameraucana	good	good		docile	hardy
Ancona	best		best	excitable	hardy
Andalusian	good		best	excitable	heat tolerant **
Appenzeller	good		best	excitable	cold hardy
Araucana	good			docile	hardy
Aseel		good		docile*	all
Australorp	good	good		docile	cold hardy
Barnvelder	good	good	good	docile	hardy
Brahma		good	good	docile	cold hardy
Buckeye	good	good	good	docile*	cold hardy
Campine	good		good	docile	**
Catalana	good	good	good	excitable	heat tolerant **
Chantecler	good	good	good	excitable	cold hardy
Cochin		good	good	docile	cold hardy
Cornish		best		docile*	temperate
Delaware	good	best	good	docile	hardy
Dominique	good	good	good	docile	hardy
Dorking	good	good	good	docile	**
Faverolle	good	good		docile*	hardy
Fayoumi	best		best	excitable	heat tolerant **
Hamburg	good		good	excitable	temperate
Holland	good	good	good	docile	**
Houdan	good	good	best	docile	temperate
Java	good	good	good	docile	cold hardy
Jersey Giant		good		docile	cold hardy
La Fleche	good	good	good	excitable	hardy
Lakenvelder	good		good	excitable	**
Langshan	good	good	good	docile	cold hardy
Leghorn	best		good	excitable	heat tolerant **
Malay		good	good	excitable	heat tolerant
Maran	good	good	good	excitable	**
Minorca	best	good	best	excitable	heat tolerant **
Naked Neck	good	good	good	docile	temperate
New Hampshire	good	good	good	docile	**
Norwegian Jaerhon	best		good	excitable	cold hardy **
Orloff		good		docile	cold hardy
Orpington	good	best	good	docile	cold hardy **
Penedesenca	good	good	best	excitable	**
Plymouth Rock	good	good	good	docile	**
Polish	good			docile	temperate
Redcap	good	good	best	excitable	hardy
Rhode Island Red	good	good	good	docile*	**
Rhode Island White	good	good	good	docile	hardy
Shamo	good	best		docile*	temperate
Sussex	good	good	good	docile	**
Welsummer	good	good	best	docile	cold hardy **
Wyandotte	good	good	good	docile*	cold hardy

* breeds known to have occasional aggressive individuals

** breeds having cocks with large combs that are prone to frostbite

Egg Breeds

Ancona

Leghorn

Minorca

Meat Breeds

Australorp

Cornish

Orpington

Rock-Cornish cross

Dual-Purpose Breeds

New Hampshire

Plymouth Rock

Rhode Island Red

Egg Breeds

All hens lay eggs, but some breeds lay more eggs than others. The best laying hen will yield about 24 dozen eggs per year. The best layers are smallish breeds that produce white-shell eggs. These breeds originated near the Mediterranean Sea, hence their classification as Mediterranean. Examples are Minorca, Ancona, and Leghorn, respectively named after the Spanish island of Minorca and the Italian seaport towns of Ancona and Leghorn (Livorno).

Leghorn is the breed used commercially to produce white eggs for supermarkets. Leghorns are inherently nervous, flighty birds that are unlikely to calm down unless you spend a lot of time taming them. The most efficient layers are crosses between breeds or strains within a breed. The strains used to create commercial layers are often kept secret, and the best production strains are not available to the general public.

While most layers produce white-shell eggs, some lay brown-shell eggs. Brown egg layers are calmer than Leghorns, and therefore most people find them more fun to raise. Brown egg layers are usually dual-purpose breeds or hybrids.

The laying ability of all hens is greatest during their first year. Each year thereafter, the eggs increase slightly in size and decrease in number. At some point you will have to decide if you are getting enough eggs to justify feeding and caring for your hens. Laying hens that don't lay well enough to earn their keep are called spent hens, and usually end up as stewing hens. Most people who keep hens for eggs replace them with young pullets every two to three years.

Meat Breeds

Good layers are scrawny, because they put all their energy into making eggs instead of meat. If you want chickens mainly to have homegrown meat, raise a meat breed. For meat purposes, most people prefer to raise white-feathered breeds, because they look cleaner than dark-feathered birds after plucking. The best breeds for meat grow plump fast. The longer a chicken takes to get big enough to butcher, the more it eats. The more it eats, the more it costs to feed. A slow growing broiler therefore costs more per pound than a faster grower.

Most meat breeds are in the English class, which includes Australorp, Orpington, and Cornish. Of these, the most popular is Cornish, which originated in Cornwall, England. The fastest growing broilers result from a cross between Cornish and New Hampshire or Plymouth Rock. The Rock-Cornish cross is the most popular meat hybrid. Those 1-pound (0.5 kg) Cornish hens you see in the supermarket are actually 4-week-old Rock-Cornish crosses and may not be hens but cockerels (young cocks).

When raising chickens on pasture became popular, growers quickly learned that commercial strain Cornish cross chickens are not active foragers, so to raise pastured poultry, they turned to a cross between the slower growing heritage Cornish and other old-time breeds. Most of these broilers do not have white plumage like industrial broilers, hence some of the trade names for similar hybrids in the United States include a reference to color: Black Broiler, Red Broiler, and Silver Cross to name a few.

A Rock-Cornish eats just 2 pounds (1 kg) of feed for each pound of weight it gains. By comparison, a layer eats three times as much to gain the same weight. You can see, then, why it doesn't make much sense to raise a laying breed for meat or a meat breed for eggs. If you want both eggs and meat, you could keep a flock of layers and raise an occasional batch of fryers, or you could raise a dual-purpose breed.

Strains

A strain consists of related chickens of the same breed and variety that have been selectively bred for emphasis on specific traits, making the strain readily identifiable to the trained eye. Most strains are named after their breeder, although sometimes a breeder will develop more than one strain and give each a code name.

Two strains may be so different from each other that generalized comparisons about their breed become difficult. Indeed, sometimes more variation occurs among strains than among breeds. To further complicate matters, the farther a strain gets from its source, the more it may change, because the new breeder may not share the vision of the original breeder. When a strain is taken up by several new breeders, chances are good that each will selectively breed in a different direction until the original strain is barely recognizable.

Dual-Purpose Breeds

Dual-purpose breeds kept for both meat and eggs don't lay quite as well as a laying breed and aren't quite as fast growing as a meat breed, but they lay better than a meat breed and grow plumper faster than a laying breed. Most dual-purpose chickens are classified as American because they originated in the United States. They have familiar names like Rhode Island Red, Plymouth Rock, Delaware, and New Hampshire. All American breeds lay brown-shell eggs.

Chantecler

Dominique

Some hybrids make good dual-purpose chickens. One is the Black Sex Link, a cross between a Rhode Island Red rooster and a Barred Plymouth Rock hen. Another is the Red Sex Link, a cross between a Rhode Island cock and a White Leghorn hen. (A sex link is a hybrid whose chicks can be sexed by their color or feather growth.) Red Sex Links lay better than Black Sex Links, but their eggs are smaller and dressed birds weigh nearly 1 pound (0.5 kg) less.

Each hatchery has its favorite hybrid. Although hybrids are generally more efficient at producing meat and eggs than a pure breed, chicks hatched from their eggs will not be the same as the parent stock. If in the future you want more of the same, you will have to buy new chicks from the hatchery. One of the advantages of keeping a nonhybrid is that when you are ready for replacement layers you can hatch your own. Since about half the chicks will be cockerels, you can raise the surplus cockerels for meat.

Endangered Breeds

Many dual-purpose breeds once commonly found on farmsteads are now endangered. Because these chickens have not been bred for factory-like production, they've retained their ability to survive harsh conditions, desire to forage, and resistance to disease.

The two organizations American Livestock Breeds Conservancy and Rare Breeds Canada keep track of breeds and varieties they believe are in particular danger of becoming extinct and encourage breeders to engage in poultry conservation. Among the most endangered breeds is the Dominique, sometimes incorrectly called "Dominecker," the oldest American breed. A few years ago, it almost disappeared, but it is now coming back thanks to the efforts of conservation breeders. Canada's oldest breed, the Chantecler, experienced a similar turn of fate.

Making the Purchase

After settling on a breed, variety, and strain, your next decision is whether to purchase newly hatched chicks or grown chickens. Starting with chicks is less expensive than buying the same number of mature birds, and the chicks will grow up knowing you are their keeper. Some breeds and hybrids are sold sexed, meaning you know when you buy them how many are cockerels and how many are pullets (young hens). Sex-link hybrids are easy to sex, because the pullets are a different color from the cockerels. Most breeds must be sexed by examining their private parts, which requires experience and skill to avoid injuring the young birds. Unsexed chicks are sold straight run or as hatched, in which case about half will be cockerels and half pullets. If you want to raise laying hens and no roosters, sexed pullets are the best option. If you want to raise a batch of broilers for the freezer, all cockerels will grow faster and larger.

Starting with grown laying hens carries certain risks, among them the greater likelihood of buying diseased

Starting a Breeder Flock

If you're happy with hybrid layers or buying an occasional batch of broilers to raise, you don't need a breeder flock. But if you wish to hatch eggs and raise chicks in the future, you'll need to maintain a breeder flock. Chicken keepers who hatch eggs include homesteaders who enjoy producing their own meat and eggs and preservationists who work with an endangered breed.

To have consistent results in your chicks year after year, your breeding flock must include chickens only of a single breed. Your results will be inconsistent if you hatch eggs from a flock containing different breeds or commercially crossbred chickens. The birds in your breeding flock should be healthy and free of deformities. They should be reasonably true to type, meaning that each bird is of the correct size, shape, and color for its breed.

You can hatch eggs in two ways: Let a hen hatch them for you or hatch them in a mechanical incubator. Letting a hen handle the hatching is called natural incubation. If you hatch eggs in a mechanical device, the process is called mechanical or artificial incubation. For in-depth details on breeder flock management and egg hatching, see *Storey's Guide to Raising Chickens* by Gail Damerow.

or spent hens. To make sure you are getting young hens, look for legs that are smooth and clean, and breastbones that are soft and flexible. Advantages to purchasing grown chickens are that you can easily tell the roosters from the hens, and you won't have to wait as long to start enjoying fried chicken or fresh eggs.

Where you buy your chickens depends on what kind you want. If you want a production breed or hybrid, get chicks from a commercial hatchery. If none is nearby, deal with a reputable firm that ships by mail. You may find chicks at a local feed store, although chances are you won't be able to learn much about what they are or where they came from. Avoid bargain chicks that come free with your first purchase of a sack of feed; they are likely to be excess cockerels of a laying breed.

Dual-purpose breeds are sold by hatcheries, as well as by individuals who advertise in local newspapers. Visiting the seller lets you see what the flock looks like and the conditions under which the birds are raised.

In making your final selection, the main thing to look for is good health. Chicks should be bright-eyed and perky. If they come by mail, open the box in front of the mail carrier, in case something has gone wrong and you need to file a claim. In grown chickens, signs of good health include the following:

- Feathers that are smooth and shiny, not dull or ruffled
- Eyes that are bright, not watery or sunken
- Legs that are smooth and clean, not rough and dirty
- Combs that are full and bright, not shrunken and dull
- Soundless breathing; no coughing, sneezing, or rattling sounds

When you visit a seller, whistle as you approach the flock. The chickens will pause to listen, letting you easily hear any unusual breathing sounds. Before taking home your selection, look under each chicken's wing and around the vent under its tail to make sure it isn't crawling with body parasites.

How many chickens to get depends, again, on your purpose in keeping them. If you are interested in egg production, determine how many eggs you want per day, divide that number by two, and multiply by three. If you want, for example, six eggs per day, you'll need at least nine hens. Since hens don't lay at the same rate all year long, sometimes you'll have more eggs than you can use, and other times you'll have too few. It's the nature of the game. With meat birds the calculation is a bit simpler: Each chicken equals one fryer. You might raise one batch a year, based on how much chicken your family eats in a year, or you might avoid having a whole lot of butchering to do at one time by dividing up the total into several smaller batches and raising them sequentially.

When you order chicks by mail, open the box in front of the mail carrier to verify any claim you may have for losses. Introduce the chicks to household pets to let them know the chicks are yours and shouldn't be touched.

Should I Buy Chicks or Adult Birds?

Type	Pros	Cons
Chicks	• less expensive • chicks know you	• sexing of other than sex links not 100% accurate • you have to wait for eggs and meat
Mature birds	• know the sex • don't have to wait for meat or eggs	• more expensive • greater likelihood of buying birds with problems

How Long Do Chickens Live?

A chicken may live 10 to 15 years. Few chickens live out their full, natural lives. Chickens raised for meat have a short life of only 8 to 12 weeks. Chickens raised for eggs or as breeders are usually kept for 2 or 3 years, until their productivity and fertility decline. A chicken kept in a protective environment may survive as long as 25 years, although geriatric hens lay precious few eggs.

Raising Chicks

When a hen hatches out a brood, she keeps them in the nest until she feels they are ready to venture out into the world. Even after they leave the nest, she keeps them warm and helps them find food. A mother hen gathers her brood under her wings if it rains or she senses danger. She squawks and puffs up, making herself look as big as possible to chase away any dog, cat, or human that might come near.

When you raise chicks yourself, you become their mother hen. No, you needn't keep your chicks in your pocket or puff yourself up and squawk if something threatens them, but you do need to make sure they are well fed, warm, and safe.

The Brooder Box

Chicks need a warm, dry, draft-free place where they are protected from dogs, cats, and other animals. Such a place is called a brooder. The simplest brooder is a sturdy cardboard box big enough to hold a feeder and drinker, and still have space so the chicks can move around.

Fasten a piece of chicken wire or hardware cloth to the top of the box so air can get in but pets can't. Place a piece of cardboard or some newspapers over part of the top if necessary to keep out drafts.

At one end of the box, hang a lightbulb in a reflector; you can purchase this setup at a farm store or hardware store. The heat from the light will keep the chicks warm. To increase the warmth level, lower the light or increase the wattage; to decrease the warmth level, reduce the wattage, raise the light, or use a larger box.

Line the bottom of the box with several layers of newspaper topped with paper toweling. The paper toweling gives the chicks better footing than newspaper, which they would find slippery. After at least a week, when the chicks are walking and eating well, you can use shredded paper, wood shavings (but not cedar, which is toxic), sand, or other bedding at the bottom of the brooder. Bedding absorbs droppings and helps keep the chicks warm and dry but should not be used until they learn to eat chick feed instead of bits of bedding. Each day, sprinkle a little clean bedding over the old bedding.

You can tell whether chicks are comfortable in the brooder by the way they act:

- If they are too warm, they will pant and move as far away from the light as they can, crowding into corners and possibly smothering one another.
- If they are not warm enough, they will complain loudly, crowd under the light, sleep in a pile, and possibly smother one another.
- When they are just right, they will move freely around the brooder, make contented sounds, and sleep nicely spread out. The brooder box is the right size if the chicks don't crowd near the feeder and drinker while sleeping.

The advantage to using cardboard boxes is that they are disposable, so you don't have to worry about disinfecting and storing them for reuse. A popular brooder used by chicken newbies is a plastic storage tote, which works okay provided the paper and bedding are cleaned out and replaced often to prevent an accumulation of moisture. Moisture leads to moldiness, and moldiness leads to brooder pneumonia.

If you plan to raise a lot of chicks in the future, you may eventually wish to

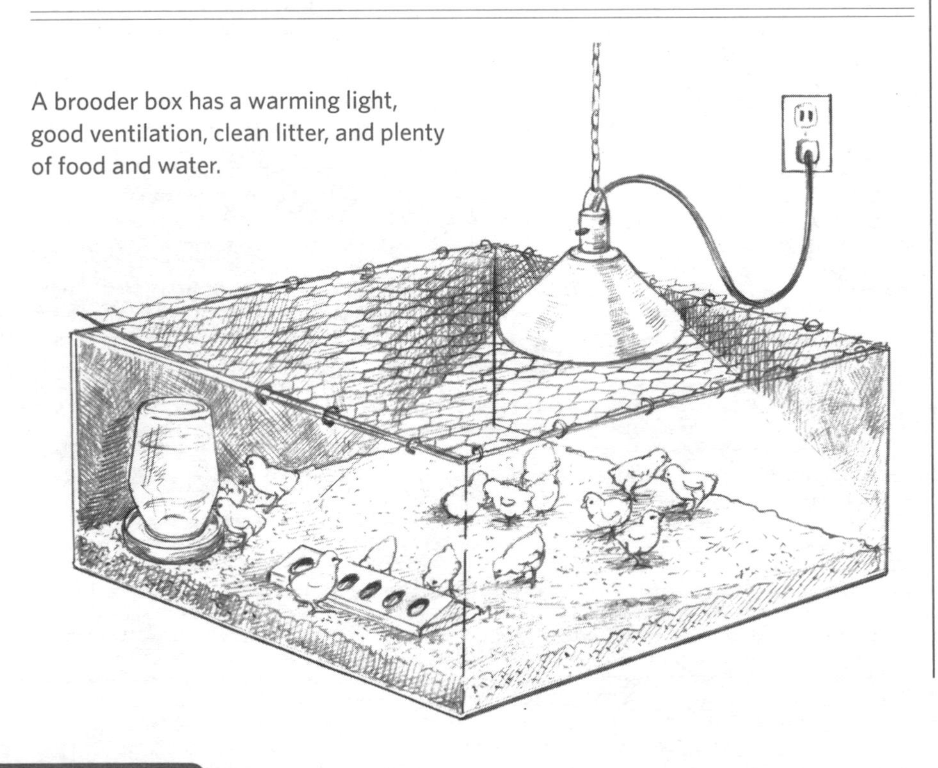

A brooder box has a warming light, good ventilation, clean litter, and plenty of food and water.

Chicken Development

Development stage	Age in weeks
Chicks feather out	4-6
Cockerels crow	6-8
Pullets start laying	20-24

purchase or build a permanent brooder. But it's a good idea to start with cardboard boxes, which will give you an idea of the features your brooder will need and its appropriate size given the number of chicks you brood at a time and the growth rate of your chosen breed. Whatever style of brooder you use, if your chicks are kept warm, dry, and away from drafts and predators, fed properly, and always have clean water, chances are good they will thrive and grow.

Feeding Chicks

A chick should drink its first water as soon as possible, and the water should be warmed to room temperature. If you set out the water before you go to pick up your chicks, it will be a comfortable temperature by the time you get back. Dip the beak of each chick into the water and make sure it swallows before you release it into the brooder.

Clean, fresh water must be available at all times thereafter. The easiest way to water chicks is to use a 1-quart (1 L) glass jar fitted with a chick-watering basin from the farm store or poultry supply catalog. Fill the jar with water, place the basin on top, and flip the jar over. Every time a chick takes a drink, water flows out of the jar into the basin. These devices are designed to keep chicks from getting the water dirty by walking in it and from drowning by falling into the water. As the chicks get older and need more water and a larger basin, switch to the 1-gallon (4 L) size.

Feed them a starter ration purchased from the farm store. Starter

How to Read the Brooder Temperature

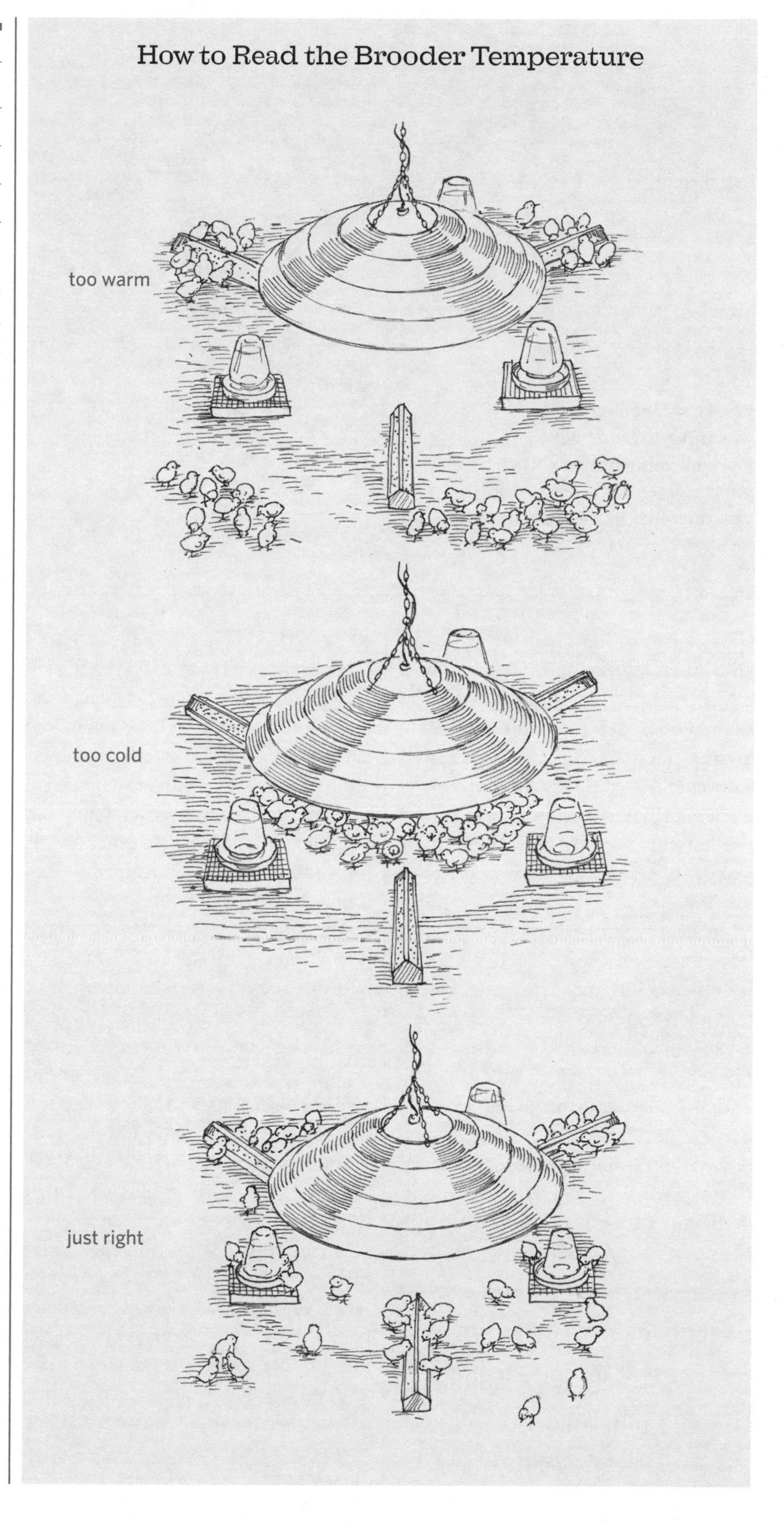

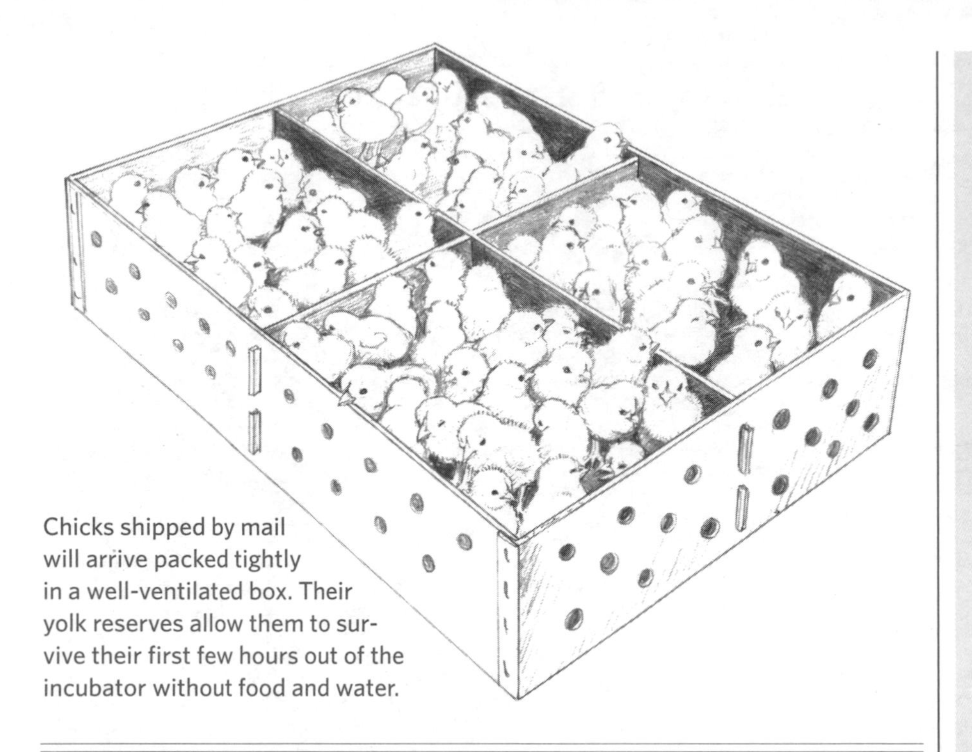

Chicks shipped by mail will arrive packed tightly in a well-ventilated box. Their yolk reserves allow them to survive their first few hours out of the incubator without food and water.

is higher in protein and lower in calcium than layer ration, which should never be fed to young chickens. Some brands of starter are medicated, some are not. Medicated ration is designed to prevent coccidiosis. If your chicks were vaccinated against coccidiosis at the hatchery, do not feed them medicated starter as it will neutralize the vaccine.

In areas where chickens are big business, farm stores sell a variety of rations for chicks. You may find starter ration for newly hatched chicks and grower ration for older chicks. You may find one kind of grower ration for meat birds and another kind for layers. In most parts of the United States, though, you will find only one all-purpose starter or starter-grower.

If you run out of starter, or you forget to pick some up and you have chicks to feed, you can make an emergency starter ration by cracking scratch grains in the blender or, if you have no scratch, by running a little uncooked oatmeal through the blender and mixing it 50/50 with cornmeal. Grains are high in calories and low in the protein, vitamins, and minerals a chick needs for good growth and health, so don't routinely use this mixture.

Depending on their breed, chicks will be fully feathered by the time they are 4 to 6 weeks old.

Initially, sprinkle a little feed on the paper towels lining the brooder box and put the rest in a shallow dish or tray. Once the chicks learn where to find chow, switch to a feeder designed especially for chicks. If the feeder has a cover with slots in it, allow one slot per chick. If the feeder is a trough type without a cover, allow 1 inch (2.5 cm) of trough length per chick, or half that if the chicks can eat from both sides.

To minimize waste, fill the feeder only two-thirds full. The top of the feeder should be as high as the chicks' backs. Raise the feeder as the chicks grow, either by using a hanging feeder with an adjustable chain or by putting wooden blocks under the feeder. When your chicks outgrow their baby feeder, switch to a chicken-sized model.

Feeding Layers versus Broilers

If you are raising layers, when they reach 18 weeks of age, gradually mix layer ration into the starter until they are completely switched over to adult ration by the time they start laying. Of the laying breeds, each pullet will eat about 25 pounds (11 kg) of feed by laying age at 20 weeks. Dual-purpose breeds take a bit longer to reach laying age and eat about 27 pounds (12 kg) of feed.

If you raise chicks for eggs, you want them to grow slowly so they are fully developed by the time they start to lay. If you raise broilers, on the other hand, you want them to grow as fast as possible to keep them nice and tender until they get big enough to butcher. The younger the chicken, the more tender it will be. Feed meat chicks often to stimulate their appetite. The more they eat, the faster they grow.

As They Grow

Chicks start growing feathers on their wings within a day or two of hatching. Depending on their breed, they will be fully feathered by the time they are 4 to 6 weeks old. By then, they need at least 1 square foot (30 sq cm) of space each and no longer require artificial heat. They are ready to be moved out of the brooder and into the chicken house.

As the chicks grow, be sure to provide more space and less heat. The rule of thumb is to start them at 95°F (35°C) and reduce the heat 5 degrees a week until ambient temperature is reached. Chicks that are kept too warm or crowded start pecking at one another, causing serious wounds that lead to cannibalism. Cannibalism is a learned habit

From Chicks to Chickens

As your chicks grow, you will soon be able to tell the cockerels from the pullets. At 3 to 8 weeks of age, depending on the breed, they will develop reddened combs and wattles. Cockerels have larger, more brightly colored combs and wattles than pullets.

Cockerels will develop spurs on their legs. The older a cock gets, the longer his spurs grow. Hens have no spurs, or tiny spurs, or little round knobs in place of spurs. Game hens are one exception — they may have spurs as long as 1½ inches (4 cm).

In some breeds, the cock's feathers are a different color from the hen's. In most breeds, the cock's hackle and saddle feathers are pointed, whereas a hen's feathers are rounded. The cock also develops long, sweeping tail feathers.

And, of course, the cockerels are the ones that crow. Their first attempts will sound pretty funny, but soon enough they'll get the hang of it. When the cockerels start chasing the pullets, it's time to separate them. Select the best cockerels for breeding and fatten the rest for butchering.

Pullets start laying at 20 to 24 weeks, depending on the breed. The first eggs will be small and probably laid on the floor. After a few weeks, you should find regular-size eggs in the nests.

that usually starts about the time feathers start growing on the lower back. Once cannibalism starts, it's difficult to stop. Preventive measures include increasing the available space, reducing heat and light, and using a red lightbulb that minimizes the attractiveness of blood and emerging blood-filled feathers.

Relieving boredom is another way to prevent cannibalism. Chicks brooded on a wire floor peck each other more readily than chicks brooded on a solid floor covered with litter. The bedding gives chicks an opportunity to engage in normal chicken behavior of pecking, scratching, and dust bathing — things chicks instinctively do when just days old. Lacking the opportunity to engage in these activities, they instead peck each other's feathers and toes.

Perches also help relieve boredom by giving chicks a place to play. Given a chance, chicks will practice perching when only a few days old. Pretty soon, they'll be playing perching games. A chick may jump onto the perch and then jump off the other side, scaring the dickens out of the other chicks. Or one chick may follow another onto the perch, causing the first to lose its balance and hop down. After a few weeks, some chicks will roost on the perch overnight. By the time they are 4 or 5 weeks old, all of the chicks will roost. Allow 4 inches (10 cm) of roosting space per chick.

Don't use a perch, however, if you're raising broilers. A perch causes heavy meat birds to get blisters on their breasts. Blisters and calluses may also occur on heavy birds housed on a wire floor or on packed, damp litter. If you raise broilers, forget the perches and freshen the bedding every day.

Chick Health Issues

If you keep your chicks warm, dry, and away from drafts and predators, feed them properly, and make sure they always have clean water, chances are they'll thrive and grow. Here are three common conditions that affect chicks: pasting, coccidiosis, and brooder pneumonia.

Pasting. Pasting occurs when droppings stick to a chick's rear end and clog the vent opening. Moisten the wad of dried droppings with warm water and gently pick it off, taking care not to tear the chick's tender skin. To prevent pasting, make sure your chicks don't get a chill. If pasting persists, mix a little cornmeal or ground-up raw oatmeal with their starter; or better yet, seek a brand of starter that does not invite pasting. By the time your chicks are 1 week old, pasting should no longer be a problem.

Coccidiosis. Coccidiosis is a parasitic infection that causes droppings to be loose, watery, and sometimes bloody. Chicks raised in the cool weather of early spring are unlikely to get this disease unless they live in filthy conditions or are forced to drink dirty water. Coccidiosis occurs more often during warm humid weather, when the parasites naturally flourish. To prevent this disease, keep drinking water free of droppings and scrub the waterer every time you refill it. Keep the brooder bedding clean and dry by immediately replacing dirty or wet bedding.

As a first-time chicken owner, you would be wise to avoid coccidiosis by either having your chicks vaccinated or feeding them a medicated starter, which contains a coccidiostat. *Do not* feed medicated starter to chicks that have been vaccinated, as it will neutralize the vaccine, which is simply a low-level dose of the coccidia chicks are most likely to encounter in their environment. Medicated starter is designed as a preventive measure and won't help chicks that already have coccidiosis. Treating the disease requires stronger medication available from a farm store, poultry supplier, or veterinarian.

Brooder pneumonia. This fungal infection of the lungs causes affected chicks to have difficulty breathing, or they may just sicken and die. To prevent brooder pneumonia, make sure the feed and bedding are always free from moisture, since the combination of moisture and brooder warmth creates an ideal environment for fungal growth.

Raising Broilers

Of all the forms of livestock, chickens put meat on your table with the least amount of time and effort. In a matter of weeks your chicken-keeping chores are over and your freezer is full of tasty, healthful poultry. If, on the other hand, you keep a dual-purpose flock, the availability of poultry meat can be ongoing as you butcher surplus cockerels and spent hens throughout the year.

Meat Classes

Chicken meat can be divided into five basic classes.

Rock-Cornish game hen, Cornish game hen. Not a game bird at all and not necessarily a hen, but an immature (usually 5 to 6 weeks old) Cornish or Cornish cross chicken of no more than 2 pounds (1 kg) dressed weight. This single-serving chicken is typically stuffed and roasted whole.

Broiler, fryer. A tender chicken, usually less than 13 weeks of age, that has soft, pliable, smooth-textured skin and a flexible breastbone. A broiler/fryer is tender enough to be cooked by any method.

Roaster, roasting chicken. A tender chicken, usually between 3 and 5 months of age, that has soft, pliable, smooth-textured skin and a breastbone that is somewhat less flexible than that of a broiler. A roaster is usually stuffed, roasted whole, and sliced for serving.

Hen, fowl, baking chicken, stewing chicken. A mature hen, usually older than 10 months of age, with meat that is less tender than that of a roaster and a nonflexible breastbone. This chicken must be cooked by a moist method such as stewing, braising, or pressure cooking. Stewing hens are generally laying hens that are no longer economically productive.

Cock, rooster. A mature male chicken with coarse skin, a hardened breastbone tip, and tough, dark meat. Such a chicken is generally not fit to eat, although with long, moist cooking may be made tender enough to chew. A better plan is to butcher surplus cockerels at the fryer stage, which is about the time they begin to crow.

Managing Meat Birds

The trick to growing tender, tasty chickens is to feed them for good growth and ensure they live a healthful life with a minimum of stress. Chickens that don't have a lot of room for activity will grow more quickly, and their meat will be more tender but less flavorful when compared to chickens that have more opportunity to be active. Methods for raising meat birds fall into three basic categories.

Indoor confinement involves housing chickens indoors on litter and bringing them everything they eat. This method works best where space is limited. The goal of confinement is to get the most meat for the least cost by efficiently converting feed into meat. The standard feed conversion ratio is 2 to 1 — each bird averages 2 pounds (1 kg) of feed for every 1 pound (0.5 kg) of weight gain. To get a feed conversion ratio that high, you must raise industrial strain Cornish-cross hybrids, which have been developed for their distinct ability to eat and grow. Efficient feed conversion means allowing birds only enough space to get to feeders and drinkers, and no more. If you don't like the idea of imitating this industrial production method, give your meat birds more room than the minimums shown in the accompanying chart, but be prepared to feed them a bit longer than the usual eight weeks or less to grow them to target weight.

Range confinement, like indoor confinement, involves keeping broilers in a building, but this building is portable, is kept on pasture, and is moved daily. Range confinement reduces feed costs, especially if you move housing first thing each day to encourage hungry birds to forage for an hour before feeding them their morning ration. On the other hand, you need enough good pasture (or unsprayed lawn) to move the shelter daily, and you must do so each day without fail. As they reach harvest size, the birds will graze faster and deposit a higher concentration of droppings, making it necessary to move them at least twice a day. Chickens raised by this method take longer to reach target weight than do those confined indoors.

Minimum Space for Confined Meat Birds

Age in weeks	Floor space per chick (sq. ft. /sq. cm)
0–2	0.5/465
2–8	1/929
8+	2–3/1,858–2,787

Breeds by Skin Color

Yellow
Aseel, Barnvelder, Brahma, Buckeye, Chantecler, Cochin, Cornish, Delaware, Dominique, Holland, Java, Jersey Giant, Langshan, Malay, Naked Neck, New Hampshire, Orloff, Plymouth Rock, Rhode Island, Welsummer, Wyandotte

White
Ameraucana, Australorp, Dorking, Faverolle, Houdan, La Fleche, Maran, Minorca, Orpington, Penedesenca, Redcap, Sussex

Pinkish
Catalana

Black
Silkie

Free range lets chickens freely come and go from their range shelter. This method requires more land than either form of confinement, because you need enough space for both a shelter and a pasture for grazing (and trampling), multiplied several times to allow for fresh forage. Figure at least one-quarter acre for 100 birds. Industrial Cornish cross chickens are unsuitable for this method because they are not active foragers. A better option is one of the many heritage Cornish cross strains, sometimes called "*Label Rouge*" chickens after the French organically grown hybrids they are modeled on. This method requires less labor than range confinement (because you don't have to move the shelter daily) but more labor than indoor confinement (because you do have to move the shelter occasionally). Allowing the chickens to exercise creates darker, firmer, more flavorful meat but also causes them to eat more and grow more slowly — they don't reach target weight until about 13 weeks. A typical feed conversion ratio for pastured poultry is 4 to 1 — each bird averages 4 pounds (2 kg) of feed for every 1 pound (0.5 kg) of weight gain.

A combination management plan is to raise chickens in confinement for eight weeks, butcher some as fryers, and range feed the rest for another four to five weeks until they reach roaster size. This plan stretches butchering over a longer period of time, making the job more manageable if you grow lots of meat birds but have few helpers when the time comes to harvest them.

Feeding Meat Birds

Young birds convert feed into meat more efficiently than older ones. The most economical meat is a broiler or fryer weighing 2½ to 3½ pounds (1.25 to 1.5 kg). Raising roasters that weigh 4 to 6 pounds (2 to 3 kg) costs more per pound.

A confinement-fed broiler eats about 2 pounds (1 kg) of starter for every pound of weight it gains. If you raise an efficient meat breed or hybrid to 3½ pounds (1.5 kg), each will eat 7 pounds (3.25 kg) of starter by butchering time at 7 to 8 weeks. If you raise a dual-purpose breed, your broilers won't grow as rapidly. Depending on their breed, by the time they weigh 3½ pounds (1.5 kg), they may eat twice as much as a specialized meat breed. Pastured broilers are typically raised to 5 or 6 pounds (2.5 to 3 kg), during which time they may eat as much as 24 pounds (11 kg) of ration.

The older a chicken is, the less efficiently it converts feed into meat and the costlier it becomes to raise. The conversion ratio starts out below 1 in newly hatched chicks and reaches 2 to 1 at about the fifth or sixth week. During the seventh or eighth week, the cumulative, or average, ratio reaches 2 to 1 — the point of diminishing return. From then on, the cumulative ratio has nowhere to go but up. Yet, even though the most economical meat comes from younger, lighter birds, most folks prefer meatier broilers or fryers, which of course cost more per pound to raise.

One way to assess the readiness of a meat bird for butchering is to determine if its shape more closely resembles a rectangle (left), which is preferred, or a triangle.

Can you save money growing your own broilers? That depends on the price of feed, how much the cost can be reduced by pasturing, how efficiently your broilers grow, and the weight at which they are harvested. If you are satisfied with cheap, industrially produced supermarket chicken, all bets are off; growing your own compares in value, if not in price, to broilers purchased from alternative sources. Using averages (and for simplicity's sake, ignoring the cost of housing, water, electricity, and so forth), do the math: By the time a pastured broiler reaches the live weight of 5 pounds it will have eaten about 20 pounds of feed. At 75 percent of live weight, a 5-pound chicken dresses out to approximately

Water Is Essential

Regardless of your management method, provide free access to fresh water at all times. Chickens that don't get enough to drink eat less and therefore grow more slowly.

3¾ pounds. Your break-even cost can be calculated by comparing the purchase price of a broiler with the cost of starter/grower ration. At current rates, a pastured broiler sells for approximately $6.50 per pound, or about $24.00 for a 3¾ pounder. The current rate for one brand of all-natural starter/grower ration is about 25 cents per pound, or about $5.00 for 20 pounds, which is approximately one-fifth the value of the meat. Even if you pay $3.50 for each broiler chick, and 50 to 100 percent extra for certified organic feed, it's still a pretty good deal.

Avoiding Drug Residue

Starter ration comes in medicated and nonmedicated formulas. The medicated version contains a coccidiostat to prevent coccidiosis, an intestinal disease that interferes with nutrient absorption and drastically reduces the growth rate of infected birds. If you choose to start your broiler flock on a medicated ration, you must find a nonmedicated feed to use during the drug's withdrawal period, which represents the minimum number of days that must pass from the time drug use stops until the drug residue dissipates from the birds' bodies to a level deemed acceptable for human consumption. If the label does not specify a withdrawal period, ask your feed dealer to look it up for you in his spec book.

If you choose to use nonmedicated feed throughout the growing period, you'll have to be careful to prevent coccidiosis. This disease is especially problematic in areas where conditions are warm and humid. Keep litter clean and dry for indoor birds, move range-fed birds frequently to prevent a build-up of droppings, and keep drinking water free of droppings. If, despite your best efforts, your chickens should require a coccidiostat or any other medication, observe the specified withdrawal period. For many drugs, the withdrawal period is 30 days.

An alternative for entirely avoiding the issue of drug residue is to have the hatchery vaccinate your chicks. The vaccine stimulates a natural immunity by introducing a low-level infection of the species of coccidia the chicks are most likely to encounter in their environment. Many hatcheries offer the option of having chicks vaccinated, which produces lifetime protection against coccidiosis, provided the chicks are never fed medicated rations, which would neutralize the vaccine.

Butchering

As butchering time draws near, seek out a fellow backyard chicken keeper willing to show you how to clean your broilers, or learn the procedure from a comprehensive chicken-raising book such as *Storey's Guide to Raising Chickens* by Gail Damerow. If you prefer not to butcher your own chickens, look for a custom slaughterhouse in your area that handles chickens.

Egg Production

A pullet starts laying when she is about 22 weeks old. Her first eggs are quite small, but by 30 weeks old, her eggs should be of normal size. Each egg takes about 25 hours to develop, causing a hen to lay about an hour later each day. As laying and sleeping time converge, the hen will skip a day, then start a new cycle.

Some hens take more time than normal between eggs and therefore lay fewer eggs per cycle than a hen that lays every 25 hours. Conversely, some hens lay closer to every 24 hours and so lay more eggs per cycle. The laying cycle of individual hens therefore varies, and may be as little as 12 days. Production hens are bred to have the shortest possible interval between eggs so they will lay as many eggs as possible per cycle. The best heavy-breed hens lay about 40 eggs in a cycle; a Leghorn lays closer to 80 eggs.

Can you save money by keeping your own hens? That depends on the price of feed, how much the cost can be reduced by judiciously feeding kitchen or garden scraps, how well your hens lay, and the price you're paying for eggs now. If you are satisfied with cheap, industrially produced supermarket eggs, all bets are off; growing your own compares in value, if not in price, to eggs from alternative sources. Using averages (and for simplicity's sake, ignoring the cost of housing, water, electricity, and so forth), do the math: A hen eats 2 pounds of rations per week or about 100 pounds per year, during which she lays maybe 240, or 20 dozen, eggs. The approximate break-even cost can be calculated by multiplying how much you are paying per dozen eggs times 20 dozen, and comparing that figure to what you would pay for 100 pounds of lay ration. At current rates, 1 dozen organic eggs sells for $4.00 or more per dozen. That works out to $80.00 per year for 20 dozen eggs. The current rate for one brand of all-natural lay ration is about $26.00 per 100 pounds, or approximately one-third the value of the eggs. Even if you buy the chick for $3.50, or a laying hen for $15.00, and pay 50 to 100 percent more to purchase certified organic feed, you're still money ahead.

The Life of a Layer

A good laying hen produces about 20 dozen eggs in her first year. At 18 months of age, she stops laying and goes into a molt, during which her old feathers gradually fall out and are replaced with new ones. Chickens molt once a year, usually in the fall, and the process generally takes 2 to 3 months. Because a hen needs all her energy to grow replacement feathers during the molt, she lays few eggs or none at all. Once her new feathers are in, she looks sleek and shiny, and she begins laying again.

After her first molt, a hen lays larger but fewer eggs. During her second year, she will lay 16 to 18 dozen eggs. Some hens may lay more, others fewer. Exactly how many eggs a hen lays depends on many factors, including breed and strain, how well the flock is managed, and the weather. Hens lay best when the temperature is between 45°F and 80°F (7°–27°C). When the weather is much colder or much warmer, hens lay fewer eggs than usual. In warm weather, hens lay smaller eggs with thinner shells.

All hens stop laying in winter, not because the weather is cold but because winter days have fewer daylight hours than summer days. When the number of daylight hours falls below 14, hens stop laying. If your henhouse is wired for electricity, you can keep your hens laying year-round by installing a 60-watt lightbulb. Use the light in combination with daylight hours to provide at least 14 hours of light each day.

Because hens, like humans, need 6 to 8 hours of rest each day, don't leave the light on 24 hours a day. If you tend to be forgetful about turning the light on and off, plug it into a timer switch from an electrical supply or hardware store. If you use a timer, remember to adjust it occasionally as daylight hours change, as well as any time the power goes out and throws off your lighting schedule.

If you raise chickens primarily for eggs, you have the same concern as commercial producers — a time will come when the cost of feeding your hens is greater than the value of the eggs they lay. For this reason, commercial producers rarely keep hens more than two years. To keep those eggs rolling in, buy or hatch a batch of chicks every year or two. As soon as the replacement pullets start laying, the older ones become stewing hens. Until that time, you can improve egg production by watching out for lazy layers.

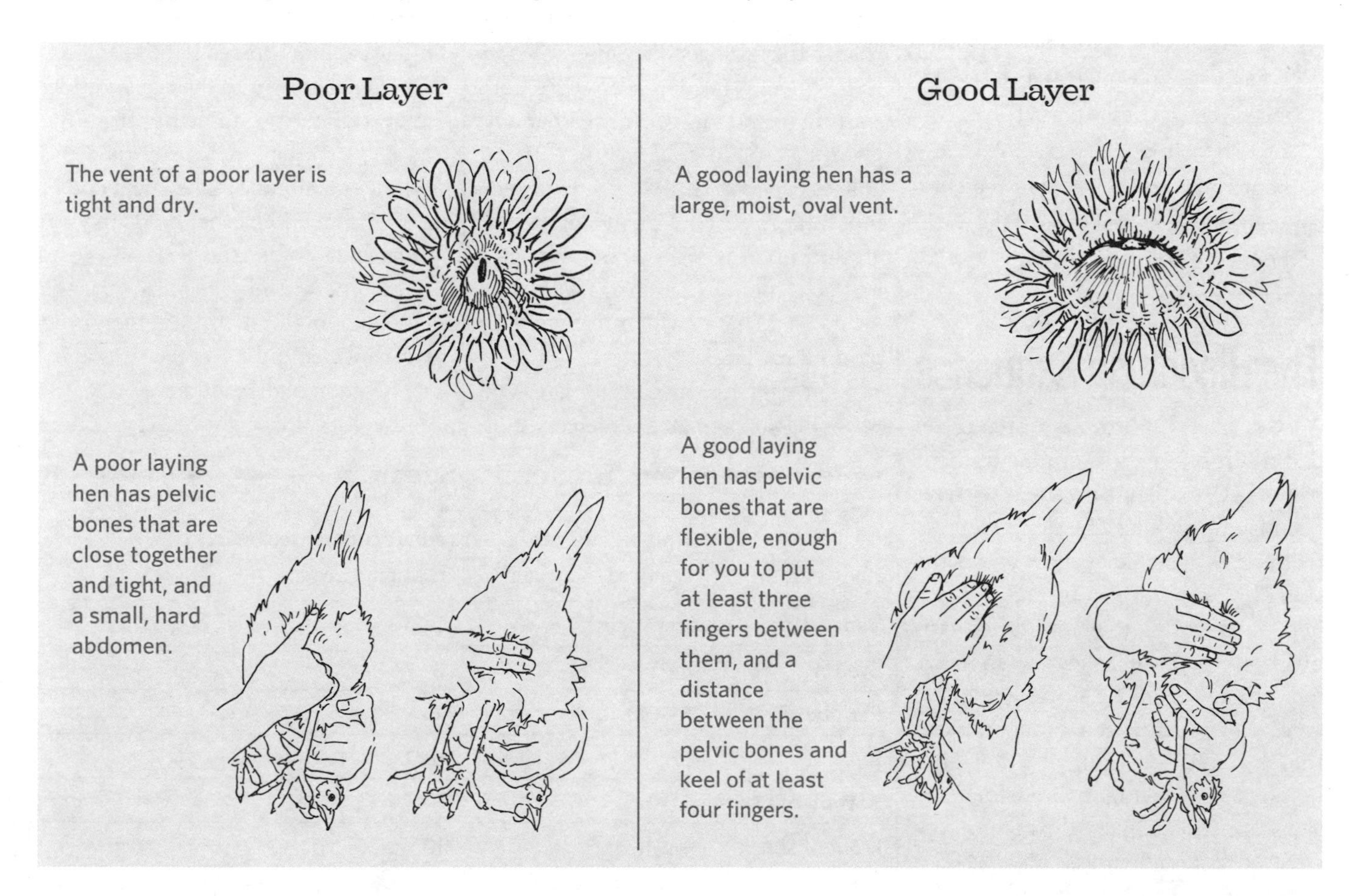

Poor Layer
The vent of a poor layer is tight and dry.
A poor laying hen has pelvic bones that are close together and tight, and a small, hard abdomen.
Good Layer
A good laying hen has a large, moist, oval vent.
A good laying hen has pelvic bones that are flexible, enough for you to put at least three fingers between them, and a distance between the pelvic bones and keel of at least four fingers.

Layers versus Lazy Hens

To improve your flock's overall laying average, cull and slaughter the lazy hens. The hens you cull can be used for stewing or making chicken soup. When your flock reaches peak production at about 30 weeks of age, you can easily tell by looking at your hens and by handling them which ones are candidates for culling:

- Look at their combs and wattles. Lazy layers have smaller and paler combs and wattles than good layers.
- Pick up each hen and look at her vent. A good layer has a large, moist vent. A lazy layer has a tight, dry vent.
- Place your hand on the hen's abdomen. It should feel round, soft, and pliable, not small and hard.
- With your fingers, find the hen's two pubic bones — located between her breastbone and her vent. In a good layer, you can easily press two or three fingers between the pubic bones and three fingers between the keel and the pubic bones. If the pubic bones are close and tight, the hen is not a good layer.

The Bleaching Sequence

If you raise a yellow-skin breed, you can sort out the less productive hens by the color of their skin after they have been laying awhile. The same pigment that makes egg yolks yellow colors the skin of yellow-skinned breeds. When a hen starts laying, the skin of her various body parts bleaches out in a certain order. When she stops laying, the color returns in reverse order. You can therefore tell how long a yellow-skin hen has been laying, or how long ago she stopped laying, by the color of the exposed skin on her beak and legs.

Collecting and Storing Eggs

An egg is at its best quality the moment it is laid, after which its quality gradually declines. Properly collecting and storing eggs slow that decline.

Collect eggs often so they won't get dirty or cracked. Pullets sometimes lay their first few eggs on the floor. A floor egg is usually soiled and sometimes gets trampled and cracked. Well-managed pullets soon figure out what the nests are for. If they continue laying on the floor, perhaps the nests are not easily accessible to them or not enough nests are available for the number of pullets in the flock.

Eggs also get dirty when a hen with soiled or muddy feet enters the nest. Eggs crack when two hens try to lay in the same nest or when a hen accidentally kicks an egg as she leaves the nest. The more often you collect eggs, the less chance they will get dirty or cracked.

Frequent collection keeps eggs from starting to spoil in warm weather and from freezing in cold weather. Try to collect eggs at least twice a day. Since most eggs are laid in the morning, around noon is a good time for your first collection. Discard eggs with dirty or cracked shells, which may contain harmful bacteria.

Store eggs in clean cartons, large end up so the yolk remains centered within the white. Store them in the refrigerator as soon as possible after they are laid.

The egg rack on a refrigerator door is not a good place to store eggs. Every time you open the refrigerator, eggs on the door get blasted with warm air. When you shut the door, the eggs get jarred. The best place to keep eggs is on the lowest shelf of the refrigerator, where the temperature is coldest. Raw eggs in a carton on the lowest shelf keep well for four weeks.

Egg Anatomy

An egg consists of many parts. The first thing you see is the shell. If you look closely, you can see tiny pores, which allow oxygen and carbon dioxide to pass through the shell. The big end has more pores than the little end, causing more air to be trapped at the big end.

The shell is lined with a leathery outer membrane. Within the outer membrane is an inner membrane. These two membranes help keep bacteria from getting into the egg and slow the evaporation of moisture from the egg. Between the outer membrane and the inner membrane at the larger end of the egg is an air cell that holds oxygen for the chick to breathe.

The inner membrane surrounds the egg white, or albumen. Albumen is 88 percent water and 11 percent protein. One function of the albumen is to cushion the yolk floating in it. The yolk is

Bleaching Sequence

Body part	Number of eggs required to bleach	Approximate weeks to lay that many eggs
Vent	0–10	1–2
Eye ring	8–12	2–2½
Earlobe	8–10	2½–3
Beak	35	5–8
Bottom of feet	50–60	8
Front of shank	90–100	10

made up of fats, carbohydrates, proteins, vitamins, and minerals that feed the growing embryo.

On two sides of the yolk are cords, or chalazae, that keep the yolk floating within the albumen. When you crack open an egg, the chalazae break and recoil against the yolk. They then look like white lumps on opposite sides of the yolk.

On top of the yolk is a round, whitish spot called the germinal disk or blastodisc. If the egg is infertile, the blastodisc is irregular in shape. In a fertile egg, this spot is called the blastoderm. A blastoderm looks like a set of tiny rings, one inside the other. When an egg is incubated, the blastoderm develops into an embryo that becomes a baby chick. If you break an egg into a dish and examine the yolk, you can tell whether the egg is fertile by whether it has an irregularly shaped blastodisc or a perfectly round blastoderm.

Determining Freshness

Sometimes you'll find eggs in a place you haven't looked before, so you can't tell how long they've been there. One way to determine whether an egg is fresh is to put it in cold water. A fresh egg sinks, because it contains little air. As time goes by, moisture evaporates through the shell, creating an air space at the large end of the egg. The older the egg, the larger the air space. If the air space is big enough to make the egg float, the egg is too old to eat.

Another way to tell if an egg is fresh is to use a light to examine the air space, the yolk, and the white. This examination is called candling, since it was once done using candles. These days an electric light is used, but the process is still called candling. A penlight with a strong beam is ideal for candling.

Candle eggs in a dark room. Grasp each egg by its small end and hold it at a slant, large end against the light, so you can see its contents through the shell. In a fresh egg, the air space is

Double Yolkers and No Yolkers

A double-yolk egg may be laid by a pullet whose production cycle is not yet well synchronized, or occasionally by a heavy-breed hen, often as an inherited trait. Sometimes an egg contains more than two yolks. The greatest number of yolks on record is nine in one egg.

A small egg without a yolk most often occurs as a pullet's first effort, produced before her reproductive system is fully geared up. In a mature hen, a no yolker may occur if a bit of reproductive tissue breaks away, stimulating the egg-producing glands to treat it like a yolk and wrap it in egg white, membranes, and a shell as it travels through the oviduct. A no yolker laid by a mature hen may therefore have a small particle of grayish tissue contained within.

Fresh egg (AA quality) ⅛" (3.1 mm) or less

Aging egg (A quality) up to 3/16" (4.7 mm)

Old egg (B quality) over 3/16" (4.7 mm)

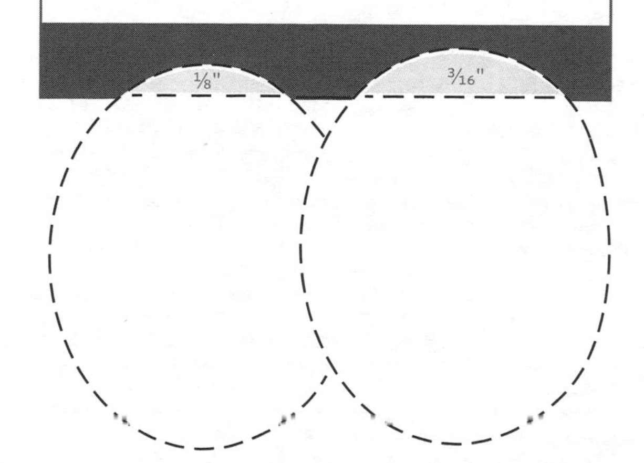

How to Gauge Freshness

Photocopy this gauge at left, paste it onto a piece of cardboard, and use it as a guide when candling an egg to determine freshness (shown below). Remember: In a fresh egg, the air space is no more than ⅛" deep.

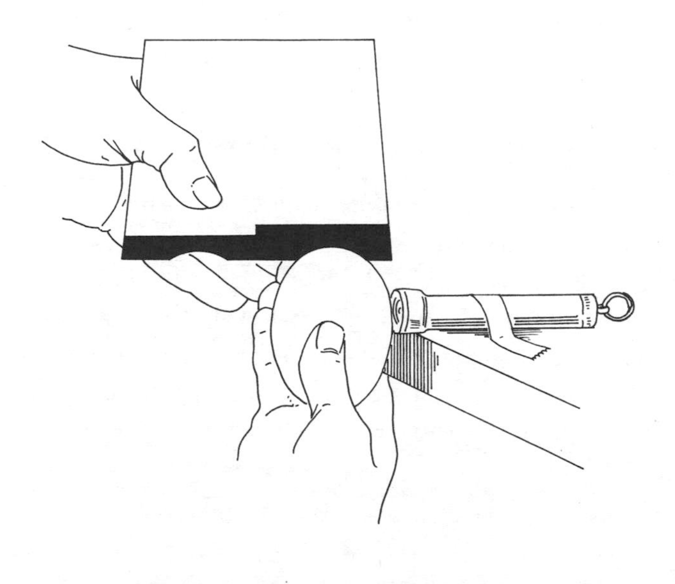

no more than ⅛ inch (3.1 mm) deep. The yolk is a barely visible shadow that hardly moves when you give the egg a quick twist. In an old or stale egg, the air space is large and sometimes irregular in shape, and the yolk is a plainly visible shadow that moves freely when you give the egg a twist.

Occasionally you may see a small, dark spot near the yolk or floating in the white. This spot is a bit of blood or flesh that got into the egg while it was being formed. Although these spots are harmless, they are not appetizing.

If you aren't sure what you are seeing when you candle an egg, break the egg into a dish and examine it. Soon you will be able to correlate what you see through the shell with what you see in the dish.

When you break a fresh egg into a dish, the white is compact and firmly holds up the yolk. In an aging egg, the white is runny and the yolk flattens out. Compare one of your homegrown eggs with a store-bought egg — it's easy to tell which is fresher.

Egg Aging

An egg stored at room temperature ages more in 1 day than an egg stored in the refrigerator ages in 1 week.

Shell Color

Each hen's eggs have a specific shell color. All her eggs might be white, light brown, dark brown, speckled, blue, or green. The color of the shell has nothing to do with the nutritional value of the contents.

Hens of the Mediterranean breeds lay white-shell eggs. Since Mediterraneans are the most efficient layers, they are preferred by commercial egg producers. Many consumers prefer eggs with white shells because that's what they're used to seeing.

Hens of the American breeds lay brown-shell eggs. Since the Americans are dual purpose, they're popular in backyard flocks. Some consumers prefer brown eggs because they look homegrown. Brown eggs come in every shade from a dark reddish color to a light tan that appears almost pink. The darkest eggs of all come from Barnvelders, Marans, Penedesencas, and Welsummers, although none of these breeds lays as well as an American or Mediterranean hen.

Blue-shelled eggs are laid by a South American breed called Araucana and its relative, the Ameraucana. Green-shelled eggs are laid by hens bred from a cross between Araucana or Ameraucana and a brown-egg breed; such a cross is called an Easter Egger.

An old egg floats.
A fresh egg sinks.
An old yolk flattens.
A fresh yolk stands up.

Nutritional Value

Eggs have been called the perfect food. One egg contains almost all the nutrients necessary for life. The only essential nutrient it lacks is vitamin C. Most of an egg's fat, and all the cholesterol, is in the yolk. To reduce the cholesterol in an egg recipe use two egg whites instead of one whole egg for half the eggs in the recipe. If the recipe calls for four eggs, for example, use two whole eggs plus four egg whites.

To eliminate cholesterol in a recipe for cakes, cookies, or muffins, substitute 2 egg whites and 1 teaspoon of vegetable oil for each whole egg in the recipe. In a recipe calling for 2 eggs, for example, use 4 egg whites plus 2 teaspoons of vegetable oil. If the recipe already has oil in it, you can omit the extra 2 teaspoons.

Housing a Flock

Chickens are territorial. They rarely roam far during the day, and they come back at night to roost in a familiar place. In ancient times, that place would have been a tree. Even some modern chickens prefer to roost in trees, but trees give them no protection from inclement weather or from owls and other predators. For that reason, chickens should be provided with protective housing and encouraged to sleep there at night.

The type of housing you provide will be influenced to a great extent by suitable existing facilities, available space, and the amount of time you wish to devote to maintenance. Portable housing is ideal for chickens, because they are periodically moved to fresh, clean ground, which may be in the garden or out in a pasture; but such a system works only if you have a large enough yard and are willing to take time to do the moving. An alternative is to divide the area around a stationary building into several separate yards and rotate the chickens among them to give vegetation a chance to regrow.

Because of the keeper's time and space constraints, the usual backyard setup involves a small stationary coop with an attached fenced yard. In short order, the chickens eat all the vegetation in the yard, which then develops a hard-packed barren surface. Provided such a yard is clean and dry, the chief disadvantage to such a situation is that the chickens have no access to forage and all their feed must be brought to them.

Coop Location and Design

The ideal spot for a chicken shelter is on a hill or slope that offers good drainage in rainy weather. The shelter might be a remodeled existing structure, such as a playhouse the children have outgrown, an unused toolshed or other outbuilding, or a camper shell from a pickup truck. You may prefer to buy a ready-made coop or to design and build your coop from scratch, making it as plain or as fancy as your heart desires. If you live in a mild climate, or if you raise an annual batch of broilers that won't be around all year, you need only a rudimentary shelter as protection from wind and rain. In a harsh climate where chickens are kept year-round, housing must be insulated and heated to keep combs and wattles from freezing.

The more room chickens have, the happier and healthier they are. Crowding leads to stress that can cause chickens to eat each other's feathers or flesh. Total living space includes the coop and yard combined; the larger the yard, the less critical a large shelter becomes.

The usual backyard setup involves a small stationary coop with an attached fenced yard.

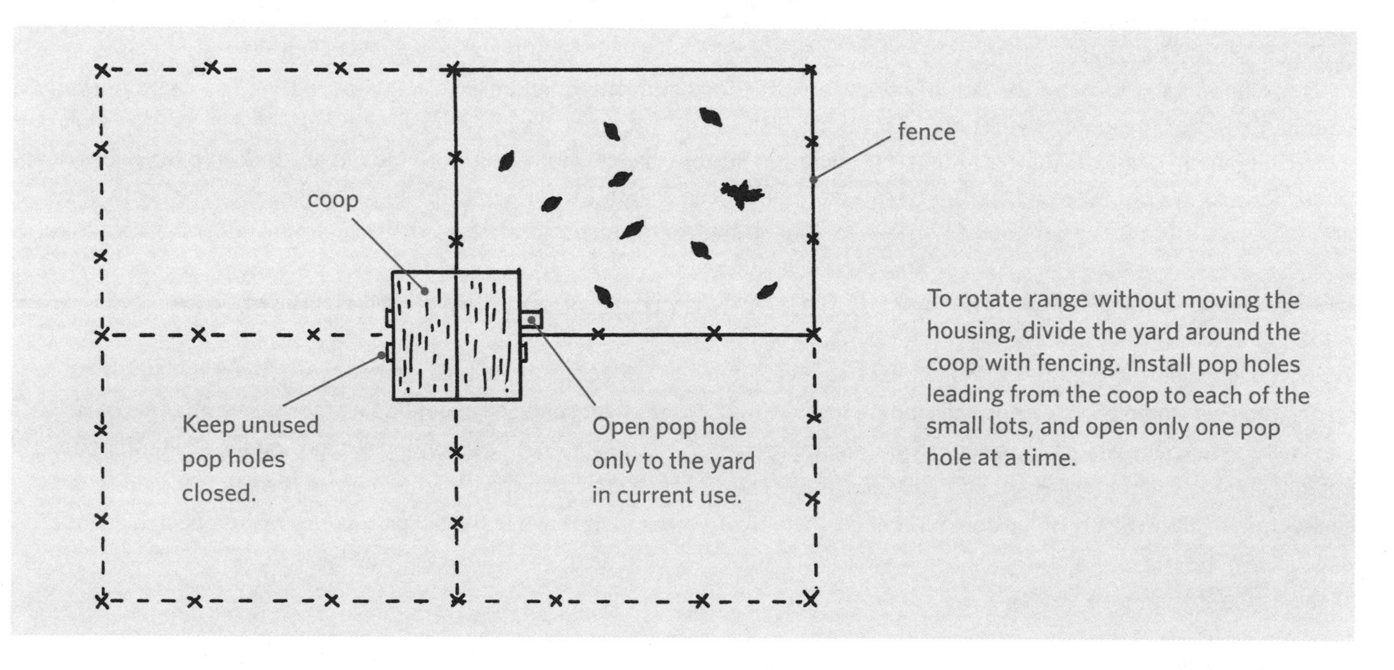
fence
coop
Keep unused pop holes closed.
Open pop hole only to the yard in current use.

To rotate range without moving the housing, divide the yard around the coop with fencing. Install pop holes leading from the coop to each of the small lots, and open only one pop hole at a time.

This three-level hutch includes everything chickens need to be safe and would comfortably suit 4 chickens of a heavy breed, 5 to 6 of a light breed.

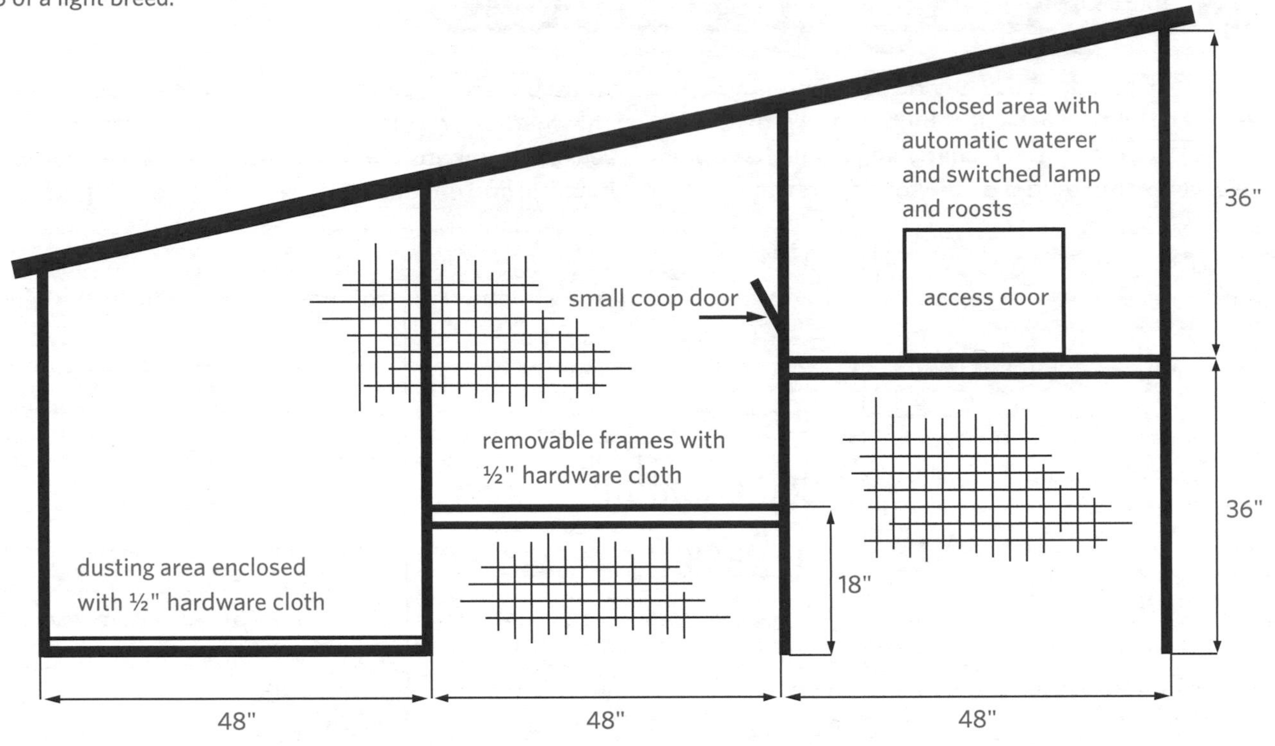

The minimum living space per chicken is generally quoted as being 4 square feet (1.25 sq m) per heavy breed and 3 square feet (1 sq m) per light breed where the chickens are free to go outside; 10 square feet (3 sq m) per heavy breed and 7½ square feet (2.25 sq m) per light breed where the chickens are confined within the coop.

These figures work pretty well for a sizable flock. For instance, if you have two dozen laying hens, the figures indicate that a coop connected to an outdoor run would be 72 square feet (22 sq m), or about 8½ feet (2.5 m) on a side. That's a pretty good size. But if you scale down to, say, two laying hens, a 6-square-feet (2 sq m) coop would be only 3 feet by 2 feet (1 m by 0.5 m), which isn't much better than keeping them in a cage. In such a case, increasing the living space, or at least making the coop split level, will go a long way toward increasing the chickens' comfort level.

The coop should have a door you can close and latch at night to protect your flock from roaming predators, and a few screened openings to provide ventilation when the door is shut. If you raise layers, you will need electricity so your hens have 14 hours of light during winter, when the days are short. Be sure the wiring is properly installed; hire an electrician if you are not experienced with wiring. Running an extension cord from the house is unsafe.

An alternative to a stationary coop and run is a portable shelter that can be moved around on pasture or a lawn that has not been sprayed with any toxins. It's a good option for a short-term

Shelter-Moving Issues

When moving a portable shelter without a floor, take care that no chickens get trampled. Chickens can learn to move with the shelter, especially when they see fresh grass coming up. But initially they will try to stay where they were, and as the shelter moves, they will bunch up and may get crushed. Until they catch on to the idea of moving, have a helper shoo them ahead as the shelter moves along.

Chickens that range outside their shelter during the day may get confused about finding their home after it's been moved. You can help out by not moving the shelter far outside the previous range, by rounding up stragglers that insist on bedding down at the old place, or by using a portable fence to confine them close to the new location.

project of raising broilers, but for long-term chicken keeping, it requires plenty of land and labor. Leave the portable shelter in one place too long, and the vegetation will be destroyed and the bare spots will grow up to weeds. On the other hand, if the shelter is moved around a garden plot, the chickens will eat weeds and weed seeds, and fertilize the soil for future planting. But you'll need a place to keep the chickens when the vegetables are growing.

Portable shelters have the advantage of being less expensive than permanent housing, are not taxed as property improvements in some areas, and are frequently moved to give chickens healthful ground and fresh forage. To aid in moving them, portable shelters come in four basic styles:

- With handles, to be moved by hand
- On skids, to be moved by hand if light enough; otherwise by ATV, truck, tractor, or draft animal
- On wheels without an axle, to be moved by hand; the wheels may be incorporated into the shelter design or on a separate dolly
- On wheels with an axle, to be moved by ATV, truck, trailer, or draft animal

Whatever design you choose, the portable shelter must be so easy to move that you won't be discouraged from moving it as often as necessary. A shelter that is movable by one person is more likely to get moved in a timely fashion than one requiring one or more helpers who may not always be around when you need them.

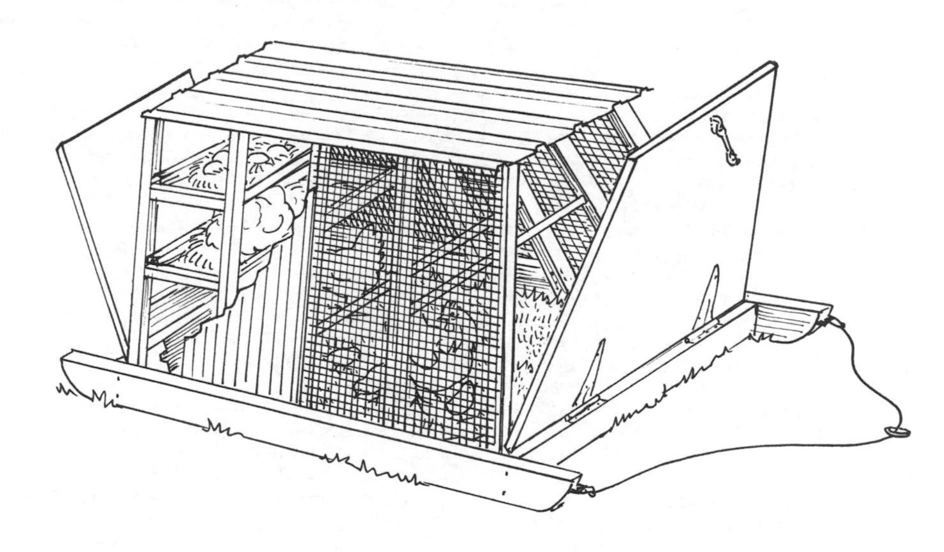

This portable shelter on skids offers all the comforts of a permanently fixed shelter.

A ladder-style perch allows chickens to hop up to a high roosting spot without much trouble.

The Perch

A chicken's natural inclination is to roost in a tree at night. To satisfy this instinct and encourage chickens to roost indoors, proper chicken housing is fitted with a perch. An ideal chicken perch is about 2 inches (5 cm) in diameter. Allow 8 inches (20 cm) of perching space for each chicken.

An old unused ladder makes a dandy perch. You can buy dowels from the hardware store, or use new 2-by-2 lumber with the corners rounded off so the chickens can wrap their toes around it. Two things not to use as a perch are plastic or metal pipe, both of which are too smooth for chickens to maintain a good grip with their toes.

A platform perch requires more space than a ladder-style perch, but minimizes jostling by chickens that want to roost on the highest rung.

If your coop isn't big enough to have one long perch, install two or more shorter ones, spacing them 18 inches (46 cm) apart and at least 18 inches from the wall. Place the lowest perch 2 feet (60 cm) off the ground and the second one 12 inches (30 cm) higher than the first one. Your chickens will hop onto the first perch and use it as a step to get up to the second one, and so on until they reach the top. Attach the perches securely so they won't rotate or sag.

If you're raising broilers for the short term, you won't need to provide a roost. Perching can cause heavy meat birds to get blisters on their breasts from pressing against the roost.

When a hen wants to lay an egg, she looks for a private place where she feels the egg will be safe.

Nests

When a hen wants to lay an egg, she looks for a private place where she feels the egg will be safe. If you don't supply nests, your hens will lay on the floor, where their eggs will become soiled or cracked. Once chickens get a taste of a broken egg, they'll deliberately peck open fresh eggs. The bad habit of egg eating is hard to stop.

Furnish one nest per four hens. You can buy ready-made nests from a store or catalog, make your own, or use found items. A small pet carrier or a covered cat litter box, for instance, makes a good nest. A reasonable nest size is 14 inches wide by 14 inches high by 12 inches deep (35 cm wide by 35 cm high by 30 cm deep).

Place nests on the floor until your pullets get used to laying in them, then firmly attach them to the wall 1½ to 2 feet (46 to 61 cm) off the floor. A two-tier system also works, with a lower tier for pullets and a higher tier that the hens will prefer as they mature. Install a rail in front of each tier so a hen can check out a nest before entering. If one nest is occupied, she will wander down the rail until she finds a vacant nest.

Nail a 4-inch- (10 cm) wide board across the front of the nests to keep the eggs from rolling out and to hold in nesting material. A handy feature is to have a door at the back of the nests that lets you clean out and replace nesting material and collect eggs from outside the coop.

When pullets first start to lay, leaving an egg in each nest helps teach them what the nests are for. You don't have to use real eggs; use plastic or wooden ones from a poultry supply catalog or hobby shop. Golf balls are perfect for this purpose.

Litter

Nesting material, also called bedding or litter, keeps eggs clean. Litter may consist of wood shavings, rice hulls, peanut hulls, chopped straw, soft hay, ground-up corncobs, shredded paper, dry leaves, or any other soft, absorbent material. Place 3 to 4 inches (7.5 to 10 cm) of litter in each nest.

Occasionally an egg will break, and sometimes a hen will leave her droppings in a nest, creating quite a mess. A handy way to clean things up is to keep a supply of squares cut from a cardboard box to line your nests. When you clean out a nest and remove the old cardboard, the nest floor underneath it will be clean. Put down a new piece of cardboard before adding fresh nesting material.

A thick layer of the same litter spread on the floor of your coop will help keep your chickens clean and healthy. If you don't use litter, manure will collect on the floor, particularly beneath the perches, and will smell unpleasant and attract flies. When you use litter, your chickens will scratch in it and stir in the manure. The result makes excellent garden compost.

Start with a layer of litter at least 4 inches (10 cm) deep. If the litter around the doorway or under the perch gets packed down, break it up with a shovel and rake. If any litter gets wet from a leaky waterer or poor drainage, remove the wet patch and replace it with dry litter. Be sure to fix the problem that caused the litter to get wet.

A chicken coop that smells like manure or has the pungent odor of ammonia is mismanaged. These problems are easily avoided by keeping litter dry, adding fresh litter as needed to absorb droppings, and periodically removing the old litter and replacing it with a fresh batch.

The Chicken Fence

Chickens enjoy being in sunlight and fresh air. A fence keeps them from straying and protects them from dogs and other predators. The fence should be at least 4 feet high (1.25 m), or higher if you keep one of the lightweight breeds that tend to fly. Even chickens that don't fly as adults may stretch their wings while they are younger, lighter, and enjoying the exuberance of youth.

The best kind of fence for chickens is a wire mesh fence with small openings. Chicken wire, also called poultry netting, has 1-inch- (2.5 cm) wide openings woven in a honeycomb pattern. It is often used as chicken fencing but does not hold up over the years, and raccoons have no trouble tearing holes in it.

A more durable fence is yard-and-garden wire with 1-inch (2.5 cm) spaces toward the bottom and wider spaces toward the top. The small openings at the bottom keep chickens from slipping out and small predators from getting in. Another, more expensive option is chain link.

To further protect your chickens from predators, string an electrified wire along the top and another along the outside bottom of your fence.

A plastic storage crate, with a lip added at the front bottom to hold in nesting material, makes a handy nest that offers good ventilation in warm weather to prevent layers from getting overheated. It can be placed on the floor for pullets then raised onto a sturdy shelf for hens.

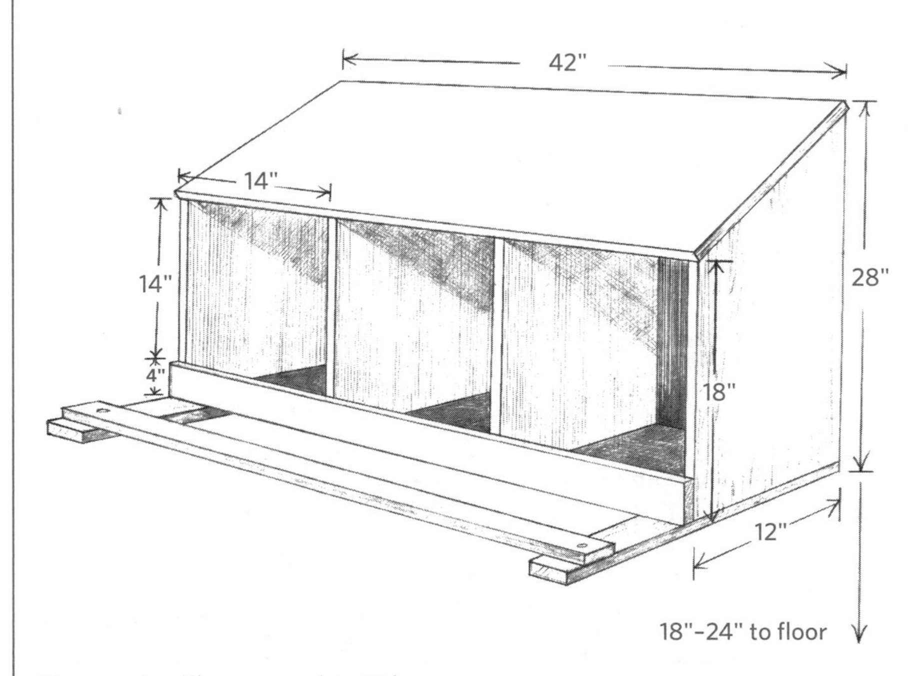

Three nests will accommodate 12 hens.

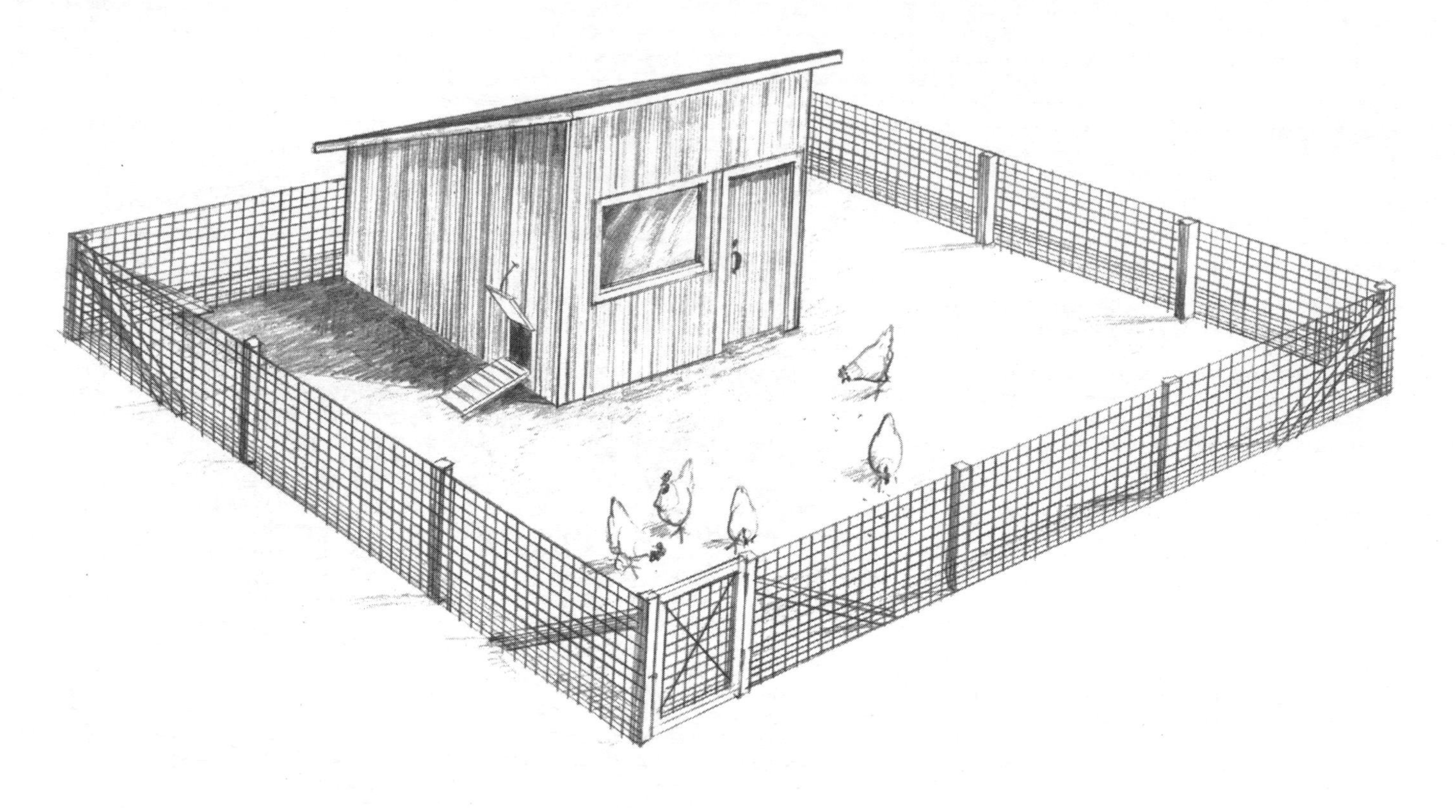

Woven-wire fencing is ideal for chicken yards; it is sturdy enough to protect against predators, is finely meshed to keep chickens from slipping outside, and offers a great view of the chicken yard culture.

Check first to make sure local regulations allow electric fence in your area.

Newly hatched chicks can slip through 1-inch openings. If you plan to let your hens hatch baby chicks that will run in the yard, get a roll of 12-inch- (30 cm) wide aviary netting, which looks like chicken wire but has ½-inch (13 mm) openings. Fasten the aviary netting tightly along the bottom of your fence.

Keeping House

How often a chicken coop needs to be cleaned depends on how big it is in relation to the number of chickens it houses, how well kept the litter is, and the prevalence of disease. A large coop housing a properly maintained and healthy flock may not need to be cleaned more often than once a year; a small coop may need to be cleaned out once or twice a week. You can tell when the coop needs cleaning — if the smell and consistency of the litter start to become unpleasant to you, it's certainly unpleasant for the chickens.

The best time to clean a coop is on a warm, dry day. Wear a dust mask, or tie a bandanna over your nose and mouth, so you won't inhale the fine dust you'll be stirring up. First, remove all fixtures, feeders, waterers, perches, and nests. Shovel out the used litter. With a hoe, scrape droppings from the perches, walls, and nests. Use an old broom to brush dust and cobwebs from the walls, especially in corners and cracks. If a source of electricity is nearby, this job is better handled with a shop vacuum.

Mix 1 tablespoon of chlorine bleach per 1 gallon of boiling water. Use an old broom to scrub the inside of the coop with the bleach-water. Leave the doors and windows open so the coop will dry fast. While the building is drying, scrub all the items you removed with fresh bleach-water and leave them in the sun to dry. When the structure and fixtures are dry, replace the fixtures and add fresh litter and nesting material.

While you are at it, check around outside and pick up any junk that may be lying around, including feed-bag strings and bits of broken glass that your chickens might be tempted to eat. Remove any scrap lumber or rolls of fencing that provide hiding places for rodents and snakes, and get rid of old tires and other items that can accumulate water where pesky insects might breed.

Feeding Chickens

Wild jungle fowl, the ancestors of modern chickens, met their nutritional needs by consuming a variety of plants and insects. Given enough room to roam, some of today's breeds remain active foragers. In most backyard situations, though, chickens are confined to a small area and must be furnished a balanced diet.

Chicken Feed

The easiest way to give your chickens a good start toward a nutritionally balanced diet is to buy chicken feed, or poultry ration, at a farm store. The basic formula for mature chickens is a layer ration containing 16 percent protein plus a lot of other important nutrients. Layer ration comes in crumbled or pelleted form. Less feed will be wasted if you use pellets. Since two-thirds of the cost of keeping chickens goes into feed, the less you waste, the lower your cost.

Premixed Formula Issues

The trouble with most commercially prepared rations is that they include soybean meal for protein, despite the fact that soy meal may contain solvent residues left after the oil has been extracted and, further, has inherent properties that inhibit the absorption of other nutrients.

Another issue of concern with any premixed formula is that it contains a fixed amount of nutrients, so when it is the only ration offered, each chicken may get more of some nutrients than it needs while trying to satisfy its requirement for others. Chickens that are fed a variety of feedstuffs, and especially those that actively forage, can pick and choose what they need to obtain a balanced diet.

Each chicken eats about 2 pounds (1 kg) of ration per week. All chickens eat less in summer than in winter, when they need extra energy to stay warm.

Layer ration comes in 25- or 50-pound (11.3–22.7 kg) sacks. It's a good idea to keep a little extra on hand so you won't run out, but don't stock up too far ahead. Chicken feed goes stale, especially in warm weather, causing the nutritional value to decrease. Buy only as much as your flock will consume within two or three weeks.

Chickens love cracked corn or a mixture of grains called scratch, which should be viewed as chicken candy. You can tame a flock by throwing down a handful of scratch whenever you visit your chickens; pretty soon, they will learn to come to you. In cold weather, a little scratch fed late in the day helps keep chickens warm overnight. But scratch is low in vitamins, minerals, and protein. Too much scratch radically reduces total protein intake. In the diet of laying hens, insufficient protein reduces egg production and the hatchability of incubated eggs, and makes hens fat and unhealthy. Scratch should therefore be fed sparingly, if at all.

After opening a sack of scratch or layer ration, store the feed in a trash container with a tight-fitting lid. A 10-gallon (4 L) container holds 50 pounds (23 kg) of feed. The closed container helps prevent feed from getting stale and keeps out mice. Place the container in a cool, dry area, out of the sun. Use up all the feed in the container before opening another sack.

If you buy more feed at a time than will fit into the container, store unopened bags away from moisture and off the floor or ground. Wooden or plastic pallets are ideal for this purpose. Avoid attracting rodents by keeping your feed storage area swept clean. To prevent mice from nibbling holes in your feed sacks and inviting their cousins to dinner, reduce the rodent population by frequently setting traps.

Reducing Feed Costs

You can reduce the cost of buying feed by treating your chickens to leftover table scraps and garden produce. Chickens love tomatoes, lettuce, apple parings, bits of toast, and other tasty treats from the kitchen. Take care, though, to avoid strong-tasting foods such as onions, garlic, or fish, which will give an off flavor to eggs and meat. Don't feed chickens anything that has spoiled or rotted. And avoid feeding raw potato peelings, which are not digestible. If you wish to feed potato peelings, cook them first. Do not rely on table scraps as your flock's sole source of food. Hens fed scraps for more than 10 percent of their diet lay fewer eggs.

A better way to reduce the cost of feed is to let your chickens roam on a lawn or in a pasture for part of the day. By eating plants, seeds, and insects, they will balance their diet and eat less of the expensive commercial stuff. Take care not to put your chickens on grass or around buildings that have been sprayed with toxins.

How much you save on commercial rations by allowing your chickens to forage depends on the quantity and quality of the forage, which varies

Insoluble versus Soluble Grit

When grains and fibrous vegetation pass through a chicken's gizzard, muscular action breaks them up by grinding them together with small pebbles and large grains of sand the chicken has swallowed. In short, grit serves as a chicken's teeth. Eventually the grit, too, gets ground up and passes through the digestive system, furnishing minor amounts of trace minerals. Known as inert or insoluble grit, the most common commercially available source is granite.

Some calcium supplements, such as oyster shell, can serve double duty as grinding grit and therefore are sometimes called calcium grit or mineral grit. Because calcium grit gets ground up much more quickly than inert grit, and serves as a time-release source of minerals, most notably calcium carbonate, it is also called soluble grit. This type of grit is sometimes given to laying hens as their sole source of grit, although the hens may be getting enough calcium from other sources and may prefer an inert form of grit. And if the flock includes young chickens and/or mature males, they can easily overdose on the calcium. Therefore soluble grit should not be offered as the sole source of grit.

seasonally. Some layer flocks can survive entirely on high quality forage. Broilers, on the other hand, would grow much more slowly and in the long run may end up costing more per pound.

Supplements

Depending on how your chickens are managed, they may need supplemental grit, calcium, phosphorus, or salt. Flaxseed is an additional supplement that may be used to boost the omega-3 content of eggs.

Grit is not needed by chickens that eat only commercially prepared rations since their saliva is sufficient to soften the feed. But foraging chickens need grit to grind up plant matter. Yarded or pastured birds pick up natural grit from the soil, although they may not get enough. Granite grit, available from any farm store carrying poultry rations, should be offered in a separate hopper and available at all times.

Calcium is needed by laying hens to keep eggshells strong. The amount of calcium a hen needs varies with her age, diet, and state of health; older hens, for instance, need more calcium than younger hens. Hens on pasture obtain some amount of calcium naturally, but illness can cause a calcium imbalance. In warm weather, when all chickens eat less, the calcium in a hen's ration may not be enough to meet her needs, and a hen that gets too little calcium lays thin-shelled eggs. On the other hand, a hen that eats extra ration in an attempt to replenish calcium gets fat and becomes a poor layer. Eggshells consist primarily of calcium carbonate, the same material found in oyster shells, aragonite, and limestone. All laying hens should have access to a separate hopper full of crushed oyster shells, ground aragonite, or chipped limestone (not dolomitic limestone, which can be detrimental to egg production).

Phosphorus and calcium are interrelated — a hen's body needs one to metabolize the other. Range-fed hens obtain some phosphorus and calcium by eating beetles and other hard-shelled bugs, but they may not get enough. To balance the calcium supplement, offer phosphorus in the form of defluorinated rock phosphate in a separate hopper and make it available at all times.

Salt is needed by all chickens but only in tiny amounts. Commercially prepared rations contain all the salt a flock needs. Foraging chickens that eat primarily plants and grain may need a salt supplement. Salt deficiency causes hens to lay fewer, smaller eggs and can cause any chicken to become cannibalistic. Loose salt (not rock salt) should always be available to range-fed chickens in a separate hopper. Iodized salt is suitable, although either a trace-mineral salt mix or kelp is a better source because it also supplies many other necessary minerals in addition to salt.

Flaxseed can be fed to hens to boost the omega-3 content of their eggs. Omega-3s are polyunsaturated fatty acids required for human growth, development, and good vision. Pastured hens lay eggs with yolks that are high in omega-3s. Although the total fat content remains the same as if the hens were not on pasture, the percentage of polyunsaturated fat increases. As an alternative to pasture, omega-3 can be boosted in eggs by adjusting the hens' diet to include 10 percent flaxseed. Besides being high in omega-3s, flaxseed is high in protein, as well as a large number of vitamins and minerals, but feeding too much can cause eggs to taste fishy.

Feeders

Eating is the main activity of chickens. They like to peck a little here and a little there, eating all day long, which works fine for chickens that can forage. Confined chickens should be fed free choice so they can eat whenever they want to. Free-choice feeding requires a large enough feeder to ensure that your flock won't run out of rations before the next feeding.

Feeders come in many designs. Good ones share these important features:

- They are designed to discourage chickens from sitting on top and messing in the feed. Not all feeders come with lids. If your feeder lacks a lid, fashion one from the lid of a 5-gallon bucket by notching out two sides to fit under the feeder's handle. Or cut the bottom from a plastic 1-gallon jug and hang it where it will dangle over the feeder to discourage roosting.
- They are designed to prevent beaking out, a habit chickens have of using their beaks to scoop feed onto the ground, where it is wasted. Beaking out is discouraged by a feeder with edges that roll or bend inward.
- They are easy to clean. A multipart feeder that comes apart for cleaning is preferable to one that is welded or glued together.
- Their height can be adjusted to suit the chickens' size, so the feed remains at the height of the chickens' backs. A hanging feeder works well because the chain from which it hangs can be shortened to raise the feeder as the chickens grow.

Solid-bottom rabbit feeders make dandy hoppers for supplements, offered on a free choice basis so each chicken can balance its nutritional needs.

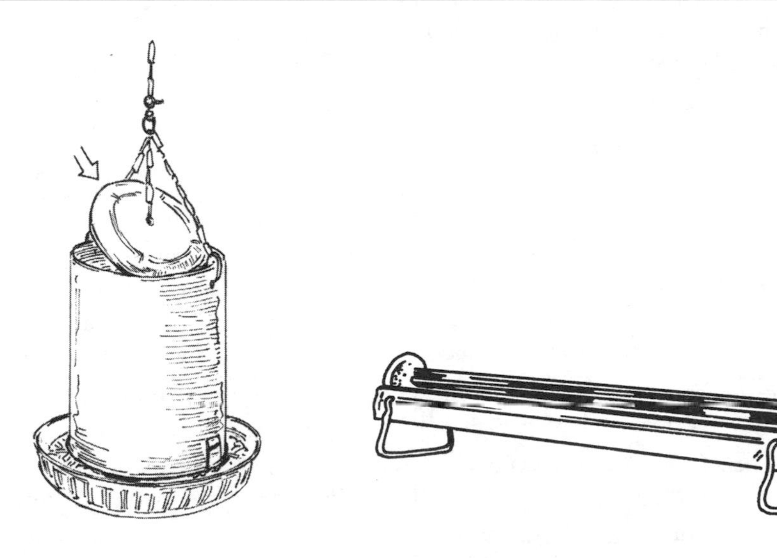

The height of this hanging tube-type feeder is adjusted by raising or lowering the chain, and the plastic jug bottom dangling over the top discourages chickens from roosting.

This trough has adjustable legs to increase or decrease the trough's height, as well as an antiroosting reel that spins and dumps any bird that tries to hop up on top of it.

A typical hanging feeder consists of a bucket-shaped container, open at both ends, with a shallow dish attached at the bottom and a handle at the top from which the feeder is hung. Feed is poured in at the top, and as the chickens eat from around the dish, gravity causes more feed to replace that which has been eaten.

As a rule of thumb, allow 1½ inches (4 cm) of space per bird around a tube feeder or 1 inch (2.5 cm) of space along a trough feeder. Another rule of thumb for free-choice feeding is to furnish enough feeder space so at least one-third of your chickens can eat at the same time. You can't go wrong by providing more than the minimum feeder space, but you can go wrong by providing less.

If your flock includes more than one cock, provide one feeder per cock and place the feeders at least 10 feet (3 m) apart. Each cock will claim a feeder and entice some of the hens to join him, reducing competition between the males and thus reducing fights.

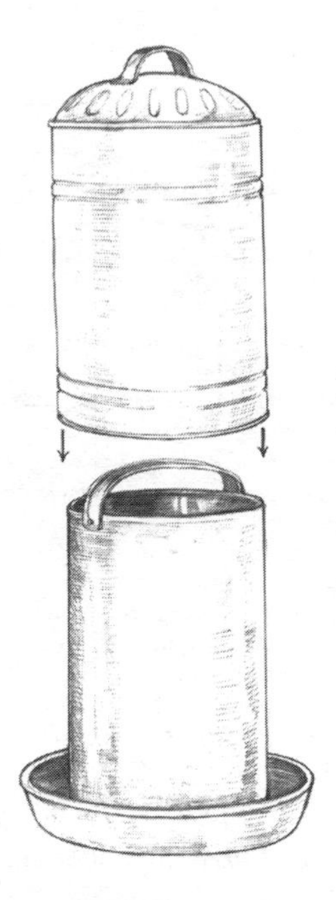

To fill a metal waterer: Fill the inner container with water then slide the outer shell over it.

To fill a plastic waterer: Fill the container with water, screw on the base, and flip it over.

Water

The most important part of your flock's diet is water. Chickens must be able to drink whenever they desire. A chicken can't drink much at once, so it has to drink often. Depending on the weather and on the chicken's size, each bird will drink 1 to 2 cups (225 to 475 mL) of water per day.

When the temperature gets above 80°F (27°C), chickens may drink two to four times more than usual. Provide fresh cool water several times a day to encourage drinking.

In cold weather, chickens will drink less than normal, but they still need water. To make sure they drink enough when the weather is cold enough to freeze water, provide warm water several times a day. If your coop has electricity, use a water-warming device from the farm store to keep the drinking water from freezing.

Like feeders, waterers come in many designs. The best waterer holds enough that your flock doesn't run out before you have a chance to supply more. It keeps the water clean by not allowing chickens to step in it or to roost over it.

The worst kind of waterer is a rain puddle. Chickens will walk in it, mess in it, drink from it, and get sick. Even if you provide plenty of clean drinking water, your chickens will drink from puddles if they can. Avoid puddles by filling areas of standing water with dirt, sand, or gravel. Since chickens love to dig dust holes in the yard, and dust holes become puddles in wet weather, filling puddles in the chicken yard is a never-ending job.

Automatic waterers are wonderful. Every time a chicken takes a sip, fresh clean water flows in. The water is always clean, and the waterer is always full. All you have to do is check every day to make sure nothing has clogged the waterer and clean out the water bowl or trough once a week. The farther your waterers are from the feeders, the less feed your chickens will get in their water and the less often you will have to clean the waterers. Farm stores and poultry supply catalogs carry automatic waterers. They are not expensive, but piping water to your coop may be.

The least expensive kind of waterer is made of plastic and holds 1 gallon (4 L). Fill the container, screw on the base, and turn the waterer over. Each time a chicken drinks, water runs out of the container through a little hole in the base. Plastic waterers don't hold up well, especially when left outside in the sun or in freezing weather, so be prepared to replace them often.

A bit more expensive is a metal waterer consisting of an inner container and an outer shell. The inner container has a basin at the bottom. When you fill the inner container and slip on the outer shell, the shell presses against a clip that lets water flow into the basin. This style comes in various sizes starting at 2 gallons (7.5 L). They eventually rust out, but sometimes you can delay this by patching holes with epoxy.

In the summer, locate waterers in the shade to keep it cool. In winter, put water in the sun to help prevent freezing. If your flock includes more than one cock, provide one waterer per cock, just as with feeders, creating a separate territory for each male.

Provide drainage beneath each waterer to prevent standing puddles in the event of spillage. A bed of sand or gravel beneath the waterer will improve drainage. Make a small platform by nailing together a wooden frame, 42 inches wide by 42 inches long by 12 inches high (107 cm wide by 107 cm wide by 30 cm). Staple hardware cloth to the top. Set the platform on the sand or gravel, and place the waterer on top of the platform. Any spills will be absorbed into the sand or gravel. Your chickens will stay healthier because they won't be able to drink dirty water from the ground.

Handling Chickens

Calm chickens are a joy to be around. Chickens that squawk and fly up every time you come near can be annoying to humans, not to mention being stressful for the chickens. Part of flightly behavior is genetic — some breeds are more easily frightened than others — but all chickens can be taught to be calm.

Remain calm when you are around your chickens. Whenever you approach them, whistle or sing so they'll hear you coming and are less likely to be startled. Walk slowly among them, talking or singing softly. Spend at least five minutes a day with your flock, not just feeding and watering, but standing still and watching. The more time you spend, the less easily frightened your chickens will be.

Peck Order

Chickens are social animals. They are happiest when they are with others of their own kind. In every group of chickens, one emerges as the leader. Scientists call this "establishing the pecking order." The peck order reduces stress by ensuring that every chicken knows how to relate to every other chicken. Most of the time, roosters are higher in rank than hens. If a hen is higher in rank than a cock (in other words, the cock is "henpecked"), it is usually because the cock is young and inexperienced or old and weak.

Chicks start establishing their place in the peck order when they are only three weeks old. After a couple of weeks, most of the fighting stops, because the peck order has been established. If a new chicken is introduced, the fighting starts again until the new chicken has reached its position in the peck order.

When two chickens meet, the first thing they do is attempt to establish the peck order with respect to one another. The farther a chicken is from home, however, the more timid it becomes. The home flock therefore has an advantage over any newcomer, which will get badly beaten up if you simply turn it loose.

Instead, keep the new chicken apart from the flock for a few days. After the new chicken becomes used to you, start bringing in your other chickens, one at a time. Introduce the lowest ranking ones first. They are usually the youngest or the oldest birds. After the new chicken has met all the other chickens one by one, put her with the whole flock. She should get along fine, because she has already established her place in the peck order. She will probably rank somewhere in the middle.

If you are introducing several new birds, you can help ease the transition by placing a fence between the old flock and the new for a few days. When you remove the separating fence, leave each group its own feed and watering station. If both groups include a mature cock, you can expect the cocks to fight, even through the fence.

Aggression

Most roosters are not mean, but they will defend their flock from a threat. However, they can't always tell when a threat is real. Something as simple as floppy shoelaces may look like a real threat to a rooster. If you always wear jeans when you tend your flock and one day you wear overalls or a rain suit, the floppy legs might be perceived as a threat. Sometimes a rooster will feel threatened by any man, or any woman, or any child, or any person but the one who normally brings feed and water.

When a rooster gets his hackles up, watch out — he's threatening to attack.

Sparring is normal when chicks are young; once the peck order has been established, fighting should subside.

A rooster may learn to be mean if he gets teased by a person or another animal. Children sometimes poke sticks at chickens and dogs enjoy chasing them, in both cases scaring the chickens out of their meager wits. Keep dogs and misbehaved children away from your flock, or teach them how to behave properly.

Sometimes a rooster gets the notion that you are a big chicken he must outrank in the peck order. He may move toward you sideways, with his head down, pretending he is looking at something on the ground. The term cockeyed comes from this ability of a cock to look someplace else while he is really watching you. He may not actually attack, but he may threaten by rushing at you or bumping against your leg. Try to make friends with him by picking him up and rubbing his wattles.

If a rooster does attack, pay attention to what you are wearing or carrying, especially anything different from the usual. Maybe he doesn't like your rubber boots, flip-flop sandals, or bare legs, or the way the new feed bucket swings in your hand. If you are doing or wearing something different, avoid it next time and see if it makes a difference.

Head Shaking

A chicken that shakes its head from side to side is telling you that you have frightened it by moving too fast or being too loud.

When a rooster gets his hackles up — that is, raises the feathers on his neck to make himself look big and fierce — he means serious business. Occasionally a cock remains mean no matter what you do. Such a rooster is big trouble. Get rid of him.

Chickens and Other Animals

Chickens get along well with other pets and livestock. You can pasture them with sheep, goats, cows, or horses, but if the chickens have access to roosting on the manger or in the hay storage area, they will foul the hay for the other livestock.

Keeping chickens with ducks or geese is not a good idea unless you have a large yard. Waterfowl like it wet, and chickens need to stay dry to be healthy. Chickens shouldn't be kept with pigs or turkeys, either, because of the possibility of exchanging diseases.

Chickens get along with dogs and cats, provided the pets are properly trained. The best way to train a pet is to introduce it to grown chickens as a puppy or kitten, to which chickens look big, scary, and something to stay away from. Adult cats need to be watched around baby chicks, and all dogs must be watched around chickens. Discourage your dog from playing with your chickens, as a dog tends to play rougher than chickens can withstand. Many a chicken has been killed by a family dog that didn't mean it.

Banding

When all the chickens in a flock look alike, it's a good idea to band them so you can tell them apart. Thin spiral leg bands may be used for whole flock identification. If, for example, you want to keep track of your laying hens according to the year they were hatched, use a different color band each year.

Numbered leg bands help you identify chickens in your flock.

To keep track of individual birds, such as for mating purposes, use numbered leg bands. Leg bands are sold through poultry supply catalogs and at some feed stores. Be sure to purchase the right size band for your breed:

- Number 9 (9/16") fits hens of the light breeds, such as Leghorn and Ancona
- Number 11 (11/16") fits cocks of the light breeds and most dual-purpose chickens, such as Wyandotte and Plymouth Rock
- Number 12 (12/16") fits cocks of the heavy breeds, such as Jersey Giant and Cornish

Catch and Carry

The easiest way to catch a chicken is to go out after dark, when the chicken is asleep, and pick it off its perch. If you need to catch a chicken during the day, you'll be happy if yours are tame enough for you to walk right over and pick them up. A chicken that isn't tame will stay just out of reach or run away from you. In that case, try to trap it in a corner, which is easier if you have help. Once you start to catch a chicken, don't give up or you'll only teach the bird to evade you in the future.

If you will have occasion to frequently catch chickens during the day, invest in a crook or a poultry net. The crook lets you snare a leg. With a net, you can snag the whole bird, if you're quick. Both devices are available from poultry suppliers.

After you have caught a chicken, hold it for a moment until it calms down. Stroke its neck and wattles to let it know you mean no harm. Cradle the chicken in one arm while you hold its legs with the other hand. If the chicken is frightened and tries to get away, carry its weight on the arm with the hand holding the legs, and with the other hand hold the wings over its back. Do not carry a chicken by its wings alone, as doing so may cause damage to the bird as well as to you (from flailing claws and spurs). Covering a bird's eyes by laying a handkerchief over its head usually helps calm it down.

Chickens are less stressed when carried one at a time and held with the head upward. But in the event you have to carry more than one at a time, turn them upside down and hold on to both legs. Most of the time a chicken held by both legs with its head downward will stop struggling. Be sure to hold on to both legs; otherwise the chicken may churn in an attempt to get away, possibly injuring itself or you. Carry only as many chickens in one hand as you can comfortably hang on to but never more than four at a time.

To catch a chicken with a hook, slip the hook around one leg and pull toward you.

Transporting

Chickens can be safely transported in anything from a paper bag to a pet carrier, as long as the following conditions are met:

- The chicken can't get out
- The container is not so big that the bird can hurt itself flying in an attempt to get out
- The container is not so small that the bird can't stand up and move
- The container has no sharp edges or other injurious protrusions
- The chicken has access to drinking water; in long-distance transit at least provide water during occasional stops
- The container is sufficiently ventilated to allow the chicken to breathe
- The chicken is protected from drafts, cold, and rain

A car trunk is not suitable for transporting chickens, as it may accumulate lethal carbon monoxide fumes. A stock rack or wire cage on an open pickup bed is not suitable, either, unless some form of wind protection is provided. A pickup with a topper may be suitable if the topper is not airtight and the truck is never parked where the chickens are left suffering in the hot sun. Chickens will be safe and comfortable traveling inside a car or truck that humans are comfortable riding in, but don't forget to protect the floor and seats from droppings with a tarp, some feed sacks, or several layers of newspaper.

Chicken Health

Chickens can suffer many different health problems from many different sources. Most disorders, though, are preventable through good management. A well-maintained backyard flock, derived from healthy stock, should require little by way of preventive medicine or drug treatment.

If you regularly introduce new birds, you will likely encounter more problems than if you maintain a closed flock with no travel and no new additions. Consult your veterinarian or Extension agent for advice.

The longer you keep a chicken, the more likely it is to develop some disease, which is one reason commercial growers and many experienced backyard flock owners won't keep chickens for more than a year or two — as soon as this year's replacement flock matures, last year's flock is out the door. Given a little extra care, though, your chickens can remain safe and healthy for many years.

How Diseases Spread

Disease-causing organisms are always present in the environment, but they may not cause problems unless a flock is stressed or kept in unclean conditions. Diseases are carried through the air, soil, and water. They may be spread through contact with other chickens or other animals, especially rodents and wild birds. They may be carried on the clothing, particularly the shoes, of the person who tends the flock.

Wild birds can spread diseases by flying from one chicken flock to another, looking for handouts. If you live in an area where chickens are common, netting over your chicken yard will keep out freeloading birds.

Visiting other chicken yards is a good way to bring back diseases by way of manure clinging to your shoes. After such a visit, clean your shoes thoroughly before visiting your own flock.

One chicken can get a disease from another, even if both birds appear to be healthy. After your flock is established, avoid introducing new chickens. Every time you bring home a new bird, you run the risk of bringing some disease with it. If you do acquire a new chicken, house it apart from the rest of your flock for at least two weeks, until you are certain the bird is healthy.

Chickens have a greater chance of remaining healthy if you take the following measures:

- Scrub out waterers and refill them with clean fresh water daily
- Avoid feeding old or moldy rations
- Clean out the coop at least once a year; promptly remove wet litter and accumulated piles of droppings between annual cleanings
- Provide enough space at feeders and waterers so the lowest chickens in the peck order don't get pushed away
- Vaccinate as recommended for your area by your county Extension agent, state poultry specialist, or veterinarian
- Maintain a stress-free environment, and train your chickens to be calm

Signs of Illness

Once you become familiar with how a healthy chicken looks and acts, you can easily detect illness by noticing changes. Each time you enter the coop, stand quietly for a few moments until your chickens get used to your presence and go on about their business. Then look for anything unusual.

Sound. Chickens in a healthy flock make pleasant, melodious sounds. Sick chickens may sneeze, gulp, or make whistling or rattling sounds, especially at night.

Smell. Notice how your chicken house usually smells. Any change in odor is a bad sign.

Appearance. A well chicken has a bright, full, waxy comb, shiny feathers,

Feather Loss

Chickens lose feathers for various reasons:

Molting. Chickens naturally lose and renew their plumage annually

Picking. Chicks pick newly emerging blood-filled feathers from one another, a form of cannibalism that is preventable through proper management

Mating. A cock's claws can strip the feathers from a hen's back; this feather loss is avoidable by rotating breeding cocks or housing cocks away from hens part of the time

Parasites. Lice and mites cause itching that results in chickens pulling out their own feathers

A healthy chicken looks perky and alert, with its head and tail held high.

and bright, shiny eyes. A sick chicken's feathers may look dull, and its comb may shrink or change color. Its eyes may get dull and sunken, or swell shut. Sticky tears may ooze from the corners of its eyes. Its nostrils may drip or become caked.

Droppings. A chicken's droppings are normally brownish or grayish with a white cap. The droppings of a sick chicken may be white, green, yellow, or bloody, or become loose. Occasional foamy droppings, however, are normal.

Behavior. A healthy chicken looks perky and alert, with its head and tail held high. A sick chicken hangs its head or hunches down, sometimes ruffling its feathers to get warm. It may drink more than usual, eat or drink less than usual, or lay fewer eggs. A mature bird may lose weight. A young bird may stop growing.

Dead chickens are, of course, one sign of disease, but don't jump to hasty conclusions if one chicken dies. Just like among humans, an individual chicken may die suddenly from any number of causes. Naturally you'd be upset if you found a dead chicken, but it's not a serious issue unless more chickens die or your flock shows other signs of disease.

Providing Treatment

Properly treating a sick chicken requires knowing what disease it has. Unfortunately, many chicken diseases mimic one another. If you do not know exactly what disease you are dealing with, administering the wrong medication can make things worse. Seek help from an experienced poultry person in your area, your county Extension agent, or your state poultry specialist, or consult a comprehensive chicken health manual such as *The Chicken Health Handbook* by Gail Damerow.

If one of your chickens appears sick, immediately isolate it well away from the rest of your flock. To avoid spreading disease, tend your well chickens before feeding and watering the sick one. For the same reason, always tend chicks or growing chickens before taking care of mature ones, even when they all appear to be perfectly healthy.

It sounds harsh, but the best way to treat many diseases is to do away with the sick bird and burn or deeply bury the body. Usually by the time you notice that a chicken isn't well, it's too sick to be cured. By getting rid of it, you can keep its disease from spreading to the rest of your chickens. Even if you do cure the bird, it may remain a carrier and continue to infect other chickens in your flock, and it is almost certain not to fully recover its reproductive capabilities. Eliminating diseased chickens from your breeding flock helps make future generations more disease resistant.

Lice and Mites

Lice and mites may be brought to your flock by wild birds, rodents, and new chickens. They may be carried on used feeders, waterers, nests, and other equipment; if you recycle used equipment, scrub and disinfect it before putting it to use. Lice and mites bite or chew a chicken's skin and suck its blood, and a serious infestation can result in death. The habit chickens have of dusting themselves in dry soil, which can be infuriating if they do it in your flower bed, helps keep their bodies free of lice and mites.

Periodically check your chickens for lice and mites. Examine them at night, using a flashlight to look between the feathers around the head, under the wings, and around the vent. Also look carefully at the scales along the shanks. You won't need to check more than a few chickens, since these

Raised scales on a chicken's legs indicate leg mites.

Salmonella

Salmonella is a bacterial disease that can affect poultry. Eggs and meat contaminated by salmonella bacteria pose a significant human health risk. Infected birds may have diarrhea and obvious signs of illness, often leading to death. Carrier birds are those with no signs of illness that pass bacteria to their eggs, to other birds, and to small children who handle the chickens and then stick their fingers in their mouths. Be sure to purchase stock from a salmonella-free flock.

parasites spread rapidly from one chicken to another.

Lice leave strings of tiny light-colored eggs or clumps that look like miniature grains of rice clinging to feathers. You are unlikely to see the lice themselves, since they move and hide quickly, but you may see scabs they leave on the skin.

Body mites are tiny red or light-brown insects that look like spiders crawling on the skin at night. During the day, they inhabit perches and nests. A setting hen on the nest is an easy target for mites.

If lice or body mites get into your flock, dust all chickens with an insecticidal powder. Use only products approved for chickens; these products are available through farm stores and poultry suppliers. Thoroughly clean out your coop and sprinkle insecticidal powder into all the cracks and crevices. Make sure your chickens have a place where they can bathe in loose dirt or fine sand.

Leg mites get under the scales on a chicken's shanks, causing them to be raised instead of lying smoothly. A serious infestation is painful and causes the chicken to walk stiff-legged. To control leg mites, once a month coat perches and the legs of all your chickens with vegetable oil.

Internal Parasites

Different kinds of internal parasites occur in different areas of the United States. The two most common are coccidia and worms.

Coccidia are everywhere, but a properly managed flock develops a natural immunity to them. When coccidia get out of hand, they cause coccidiosis. Although many different animals are affected by coccidiosis, the coccidia that infect chickens do not infect other kinds of animals, and vice versa. Coccidiosis usually affects chicks, but adult chickens can also get it, especially when the weather is hot and humid. The first sign is loose droppings, sometimes tinged with blood. A medication for treating coccidiosis, called a coccidiostat and sold through farm stores and poultry supply catalogs, must be used to treat the whole flock at once. Many hatcheries offer to vaccinate chicks with an anticoccidial vaccine, which confers a lifetime immunity (provided the chicks are never fed medicated starter, which neutralizes the vaccine).

Worms in chickens are similar in nature to those affecting dogs and cats. A chicken gets roundworms by picking up worm eggs as it pecks for food on the ground. It gets tapeworms by eating an infected earthworm, grasshopper, housefly, ant, snail, or slug. Confined chickens are more likely to have roundworms, whereas foraging chickens are more likely to have tapeworms. Signs of both kinds of worm are droopiness, decreased laying, and weight loss, or in young birds, slow or no weight gain. Loose droppings or diarrhea are also possible. Sometimes you can see worms in the droppings.

To reduce the chances of your chickens getting worms, periodically let your chickens into a rested yard, prevent puddles from forming in the yard, and keep the coop floor covered with clean, dry litter. Move pastured chickens often, then mow the previous pasture plot if necessary to open it up to fresh air and sunlight. Discourage wild birds and rodents from visiting, and worm any new chicken you bring into your flock.

Parasites in the environment are killed naturally by drying in the sun and, even more effectively, by being frozen in the winter. If your climate is mild, or if you have a particularly mild winter, you may need to initiate aggressive parasite control.

Fecal Analysis

To find out for sure whether your chickens have worms, scoop some fresh feces into a plastic bag, seal it, and take it to your vet for analysis. If your chickens have worms, the vet can tell you what kind they are and recommend an appropriate treatment. If your chickens do not need to be wormed, you will have saved yourself the cost of unnecessary medication, as well as saving your chickens the stress of being unnecessarily medicated.

Cancer

Only a few forms of cancer occur in chickens, and unfortunately they appear with some frequency. The cancer viruses that infect chickens do not affect humans.

Marek's disease is caused by a herpes virus. It causes a cancer of the circulating immune cells. Affected birds are usually less than 4 months old, and they often develop signs of weakness and leg paralysis. Some birds may just sicken and die. Vaccination is an effective preventive measure.

Lymphoid leucosis is a similar cancer that occurs in birds more than 4 months old and is caused by a different virus. Affected birds often sicken and die. Pale combs, indicative of anemia, may be seen. There is no vaccine for lymphoid leucosis, but some chickens are genetically resistant.

Cancers of the ovary and uterus occur in older hens. Any hen more than 2 years old can be affected. Birds with cancer of the reproductive tract may lay misshapen eggs or no eggs at all. They will gradually lose condition and develop a prominent breastbone and an enlarging soft, fluid-filled abdomen.

Winter Care

A chicken stays warm by ruffling its feathers to trap warm air next to its body. A cold draft blowing through the feathers removes the warm air and causes chilling. Check for drafts by quietly standing in your coop awhile, and then squatting to detect cold air moving at chicken level. Take any necessary measures to reduce coop draftiness.

Unlike a hen, a cock doesn't sleep with his head tucked under his wing. In cold weather, his comb may therefore freeze during the night, becoming quite painful and perhaps reducing his fertility. Insulation helps prevent frozen combs, as does mounting a small electric pet heater (such as a sealed infrared panel made by Infratherm) above the perch. To make sure your chickens won't get too warm, plug the heater into a thermostatic control that kicks on when the temperature falls below 35°F (2°C) and kicks off above 35°F (2°C).

Summer Care

To keep themselves cool in warm weather, chickens pant. A mature chicken starts breathing through its mouth when the temperature reaches 85°F (30°C), a chick when the temperature reaches 100°F (38°C). As the temperature rises, chickens breathe more rapidly and spread their wings away from their bodies. Both signs are indications of heat stress. At a temperature of 105°F (41°C) or above, chickens may die.

To prevent deaths in hot weather, ensure continuous access to fresh water that remains as cool as possible, either by putting waterers in the shade or by periodically furnishing cool water. Since chickens drink less when their water isn't pure, avoid putting medications in drinking water during hot weather. Make sure your chickens can get into shade without crowding together. In a dry climate, you can cool off your chickens by spraying them lightly with a hose. In a humid climate, spraying doesn't help, because the air is already too full of moisture to allow evaporation.

During hot weather, hens lay smaller eggs with thinner shells, a problem you can mitigate somewhat by using a layer ration with a higher than usual protein level. If you live in a hot climate, your farm store may carry high-protein ration to help layers through the summer months.

CHAPTER 2

Turkeys

Compared to other species of livestock, little is known about how the turkey was first domesticated. What is known is that the wild turkey is indigenous to the Americas and was kept by Native Americans centuries before the arrival of Europeans. Various tribes ate turkey eggs, as well as turkey meat both fresh and dried as jerky.

Spanish explorers of the early sixteenth century brought turkeys to Europe, where the birds were selectively bred for a couple of centuries before being brought back to America by early settlers. From those birds sprang the supermarket turkey we know today.

Turkeys are, in many ways, similar to chickens, and the best way to get started raising turkeys is to first raise some chickens. Even though turkeys are much bigger than chickens, they can be quite a bit touchier. Baby turkeys cost about seven times more than baby chicks, yet chicks are more forgiving of common mistakes made by poultry novices. And you can raise a batch of broiler chicks in about half the time needed to raise turkeys to harvest weight. So it makes sense to practice poultry husbandry on chickens and then use your newfound knowledge to try your hand at raising turkeys, especially since both require much the same sort of facilities.

Choosing a Breed

Many turkey breeds are available to choose from, not all of which are desirable for home meat production. In selecting the best breed for you, consider some basic criteria: how big they get, how fast they grow, how they will look in your yard or on your table, and how self-sufficient and hardy they are.

1. Size matters. A dressed turkey must conveniently fit into your refrigerator while it is aging, and into your oven while it is roasting. If you plan to harvest all your turkeys at once, you also need to have enough freezer space to store them all; if your home freezer isn't big enough, you will have to rent a locker. The other option is to harvest each turkey as you are ready to roast it, which stretches out the amount of time you have to feed your turkeys (therefore the cost of raising them). And the older they get the less tender they become.

2. Color considerations. The majority of turkeys raised for meat have white feathers, because white pinfeathers are less visible on a roasted bird. Dark or black pinfeathers on a roasted turkey on the platter look awfully unappetizing. Some of the nonwhite breeds have light-colored pinfeathers that don't show much. On the other hand, if you plan to breed turkeys to raise their young for meat, you want birds you find aesthetically pleasing. Turkeys are large, imposing birds that will be highly visible in your yard.

Another color consideration is the amount of white meat a turkey produces. The broad breasted strains have been selectively bred to have a large chest, because consumers have been conditioned to prefer lots of white meat. However, white meat is not as succulent or flavorful as dark meat.

The heritage breeds don't grow as large as the broad breasted strains, and therefore are more suited for a small family.

3. Heritage breed versus commercial strain. Industrially produced turkeys are selectively bred broad breasted strains originally developed from standard breeds. The standard breeds are those that were accepted into the *American Poultry Association Standard of Perfection* in the late eighteenth century. Now called heritage breeds, they include (among others) the Black, Bourbon Red, Bronze, Narragansett, and Royal Palm. Raising a heritage breed for meat helps support the turkey breeders who choose to maintain the genetic lines of these minor breeds. But that's not the only reason why a heritage breed can be ideal for family meat production.

The heritage breeds don't grow as large as the broad breasted strains, therefore are more suited for a small family with limited refrigerator and oven space. On the other hand, if you're into white meat, the breast of a heritage breed is about half the size of that of a broad breasted strain.

The heritage breeds are better foragers and have greater disease resistance than industrial strains, making them more suitable for raising on pasture. On the other hand, they grow at a much slower rate, finishing in 6 to 7 months where the broad breasted strains are ready to harvest in 4 to 5 months. A heritage turkey meat project therefore requires an extra two months of labor, but the reward is juicier, more flavorful meat.

Talking Turkey

A baby turkey of either gender is a poult. A young male turkey under 1 year of age is sometimes called a jake, and a young female is a jenny. A mature female is a hen. A mature male is a tom or gobbler.

Taxonomically, turkeys are the largest members of the order Galliformes, comprising chickenlike birds. Indeed, a turkey is often described as a large chickenlike bird with a fan-shape tail. Along with chickens, turkeys are in the family Phasianidae, which includes pheasants and peafowl. Their genus, *Meleagris*, includes *Meleagris ocellata*, or the ocellated turkey of southeastern Mexico, northern Belize, and Guatemala, and *Meleagris gallopavo*, the wild turkey of North America from which today's domesticated turkeys were developed.

Finally, should you decide to keep your best tom and a few of your hens so you can raise poults of your own, be aware that the broad breasted strains cannot mate naturally because they are too awkward due to their enormous weight and outsize chests. Propagating a broad breasted strain requires artificial insemination.

Parts of a Turkey

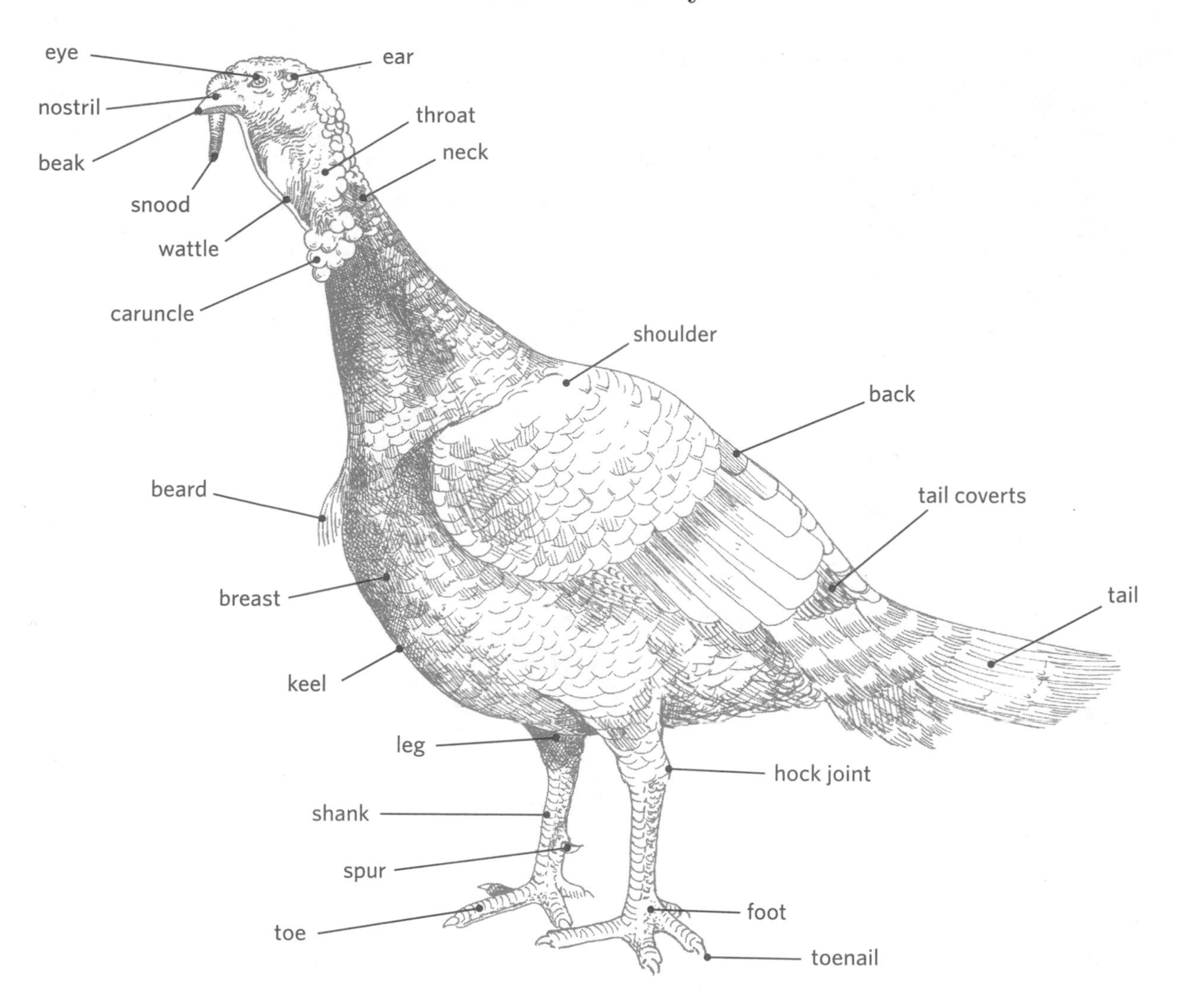

Quick Guide to Turkey Breeds

Breed	Weight, young* m/f in lbs. (kg)	Weight, old* m/f in lbs. (kg)	Temperament	Color	Status
Black	23 (10.5)/14 (6.5)	33 (15)/18 (8)	calm	black	uncommon
Bourbon Red	25 (11.5)/16 (7.5)	35 (16)/20 (9)	variable	chestnut and white	common
Broad Breasted Bronze	30 (13.5)/18 (8)	45 (20.5)/25 (11.5)	placid	dark bronze	uncommon
Broad Breasted White	30 (13.5)/18 (8)	45 (20.5)/25 (11.5)	placid	white	common
Bronze	25 (11.5)/15 (7)	30 (13.5)/19 (8.5)	calm	coppery bronze	uncommon
Midget White	13 (6)/8 (3.5)	18 (8)/10 (4.5)	friendly	white	rare
Narragansett	23 (10.5)/14 (6.5)	28 (12.5)/16 (7.5)	calm	black and gray	rare
Royal Palm	16 (7.5)/10 (4.5)	20 (9)/12 (5.5)	docile	white and black	uncommon

* Some strains may not reach these weights due to dietary factors and how actively the strain has been selectively bred for growth.

Breeds

Of the many turkey breeds available, some are bred primarily for exhibition while others are extremely rare and costly. For the turkey raising novice, the following breeds are worth considering.

Black. Originating in Mexico, the Black turkey was brought to Europe by early explorers, then brought to the United States by early settlers. Although it has black feathers, it remained a popular meat bird for centuries because of its calm disposition, rapid growth, and early maturation.

Bourbon Red. One of the older breeds developed in the United States, the Bourbon Red is named after Kentucky's Bourbon County. This handsome heavy breasted bird has chestnut-colored feathers accented in white- and light-colored pinfeathers. The Bourbon Red is an active forager and is good for pest control.

Broad Breasted White. Selectively bred by industrial producers for fast growth and a large breast, the Broad Breasted White accounts for some 90 percent of all turkeys grown for meat. Although it is suitable for family meat production, it cannot be used to breed future turkeys because its rapid growth leads to heart problems, and its heavy weight results in lameness and an inability to mate naturally.

Broad Breasted Bronze. Once the main commercial turkey until the Broad Breasted White was developed, the Broad Breasted Bronze is the hardier of the two, but its dark multicolored feathers with a coppery hue result in a less clean-looking dressed bird. Like the Broad Breasted White, this heavy turkey is too clumsy to breed naturally, and artificial insemination is required in order to raise future poults.

Narragansett

Royal Palm

Tom or Hen?

When the tom turkey opens out his fantail and struts his stuff, he looks absolutely beautiful, but close up he's as strange looking as they come. The tom has no feathers on his head and neck and his face is covered with fleshy caruncles that vary in color from pink to pale purplish blue to nearly white. A loose growth called a snood hangs from the top of his beak, a fleshy pouch called a wattle dangles beneath his beak, and a beard grows from his chest. To make things more interesting, when the tom gets excited his snood gets long and floppy and his caruncles turn a brilliant red.

The hen looks much like the tom, but is smaller and less colorful. She may or may not have a beard growing from her chest, and she may occasionally spread her tail and briefly strut like a tom, although her tiny snood won't grow long and floppy and her face doesn't light up like the tom's.

Close up, a tom looks downright weird.

Bronze. Ancestor to the Broad Breasted Bronze, the Bronze turkey was developed by early American settlers by crossing Black turkeys brought from Europe with local wild turkeys. The resulting birds have dark pinfeathers. The Bronze turkey looks similar to the Broad Breasted Bronze, but it is smaller and able to reproduce naturally.

Midget White. Developed in Massachusetts as a small turkey for backyard production, the Midget White is the smallest of the turkey breeds and an ideal size for a small family. With its broad breast, the Midget White looks like a miniature Broad Breasted White, but its much lighter weight gives it the ability to fly really well — an important consideration when designing facilities intended to keep turkeys in.

Narragansett. Developed in Rhode Island by early settlers as a cross between Black turkeys brought from Europe and local wild turkeys, the Narragansett is similar in color to the Bronze but has a steel-gray tint in contrast to the bronze's copper cast. It is known as an excellent forager with a pleasant disposition, good egg production and brooding instincts, and early maturation.

Royal Palm. Developed in Florida, the Royal Palm turkey appears to be basically white until the tom opens his tail to display stunning bands of black feathers. Having never been selectively bred for fast growth and heavy muscling, Royal Palms are considered primarily ornamental, but they are a nice size for a small family. They are active foragers and good flyers, an important consideration in designing facilities for them.

The Midget White is an ideal size for a small family.

Making the Purchase

Turkeys can be tricky for a novice to raise. Because a turkey is curious about things, it tends to get itself into trouble, such as wedging its body into a tight spot and not having any inclination to back out. Young turkeys tend to pile on top of one another when cold or frightened and those on the bottom can be smothered. And sometimes a poult dies without apparent cause, which is more common with poults that have endured the stress of being shipped by mail compared to those hatched locally, and more so among broad breasted strains than among standard breeds.

Broad breasted strains run a few dollars less per poult than standard breeds, but don't let that be your main consideration in making your purchase. Because the broad breasted strains don't have as strong an immune system as the standard breeds, their lower price is somewhat offset by greater losses.

For these reasons, start your first turkey project with small numbers, but not too small. Anywhere between a dozen and two dozen poults should be few enough for you to keep track of and learn from, but not so few you are likely to lose them all before Thanksgiving rolls around.

Most hatcheries sell poults as straight run, meaning the number of males and females is the same ratio as whatever hatches. Some hatcheries offer sexed poults in the broad breasted strains, so you could purchase only males or only females. Since toms and hens grow at a slightly different rate, with sexed poults all the turkeys will be ready to harvest at the same time — an advantage for commercial production but not necessarily desirable for turkeys raised at home.

The broad breasted strains are ready to harvest in 4 to 5 months, so if you start your poults in June or July they will be ready to dress in time for Thanksgiving and Christmas. The heritage breeds are ready to harvest in 6 to 7 months, so start them in April or May. You don't have to harvest all your turkeys at the same time, although the longer you keep them the more expensive they become to feed.

Almost any hatchery that sells chicks sells poults as well. Place your order early in the year, especially if you want one of the less common breeds, to ensure the hatchery doesn't sell out. Ordering well in advance also gives you plenty of time to get your brooding facility set up.

Raising Poults

Prepare your brooding area well before the poults are scheduled to arrive. Turn on the heater a day ahead, so the bedding and floor have time to warm up and not draw heat away from the new arrivals. With the brooder warmed up, check for drafts and take measures to eliminate any you find. Fill the drinker well ahead, so the water will be brooder temperature for the poults' first drink.

The Brooder

Poults need to be brooded for five to six weeks before they are big enough to be put out on pasture. For the first week or so the poults don't need much room, but as they grow they need more space to move around. Providing sufficient space reduces stress, helping prevent the poults from becoming frustrated and picking on each other. An initial brooding space of 5 feet by 5 feet (1.5 × 1.5 m) is adequate for two dozen poults. The brooding area must be secure from all predators, including dogs, cats, rats, weasels, and snakes.

For warmth and comfort, cover the brooder floor with a thick layer of bedding. Good bedding materials include shredded paper, well-dried grass clippings, rice hulls, and pine shavings (but not cedar shavings, which are toxic to poults). Until the poults are eating well, cover the bedding with paper toweling so they aren't tempted to fill up on bits of bedding.

Brooder Bedding Benefits

Good bedding benefits brooding poults in many ways.

- It provides a soft place for the poults to rest.
- By insulating the floor, it keeps the brooder toasty warm.
- It absorbs moisture from droppings and spilled drinking water.
- It keeps poults busy by giving them something to scratch in.
- By getting mixed with droppings it prevents manure caking.
- It allows poults to engage in the natural activity of dusting.

Brooding Space per 24 Poults

Age in weeks	Minimum space in feet
0–1	5 × 5 (1.5 × 1.5 m)
1–3	8 × 8 (2.5 × 2.5 m)
3–6	10 × 10 (3.0 × 3.0 m)

Until they grow a full set of feathers, your poults will need heat. An infrared lamp in a reflector is commonly used to provide heat for small-scale brooders, although an infrared pet heater panel (such as those made by Infratherm) is superior because it is not a fire hazard, can't burn out and leave your poults without heat, and when they start perching they can't accidentally knock the panel down.

Since, unlike an infrared lamp, a heater panel does not emit light, if you use that style heater you will need to light the brooding area so the poults can find feed and water. For the first week leave the light on all the time. After a week, change the bulb to a low wattage (dim) light and turn it off for at least one hour every 24 to accustom the poults to darkness so they won't panic in a power failure.

As a general rule, start with the heat source about 18 inches (46 cm) above the brooder floor and raise it about 2 inches (5 cm) each week until it is 24 inches (61 cm) above the floor. The temperature at poult level should start at 95°F (35°C) and be reduced 5°F (2.7°C) per week until the brooder temperature is 70°F (21°C) or the same as ambient temperature. When the weather is warm, or the poults are well feathered, they no longer need heat. You can judge their comfort level the same way as you would for chicks, illustrated on page 21.

At about 3 weeks of age (4 to 5 weeks for broad breasted strains) your poults will want to perch on anything above the floor — typically on top of the feeder, drinker, or heater. At this point you might install a 2-inch (5-cm) round roost, 12 inches (30 cm) off the floor, to give them something suitable to perch on. Allow 6 inches (15 cm) of roosting space per turkey.

When poults reach the age of three weeks they appreciate a small homemade perch like this one.

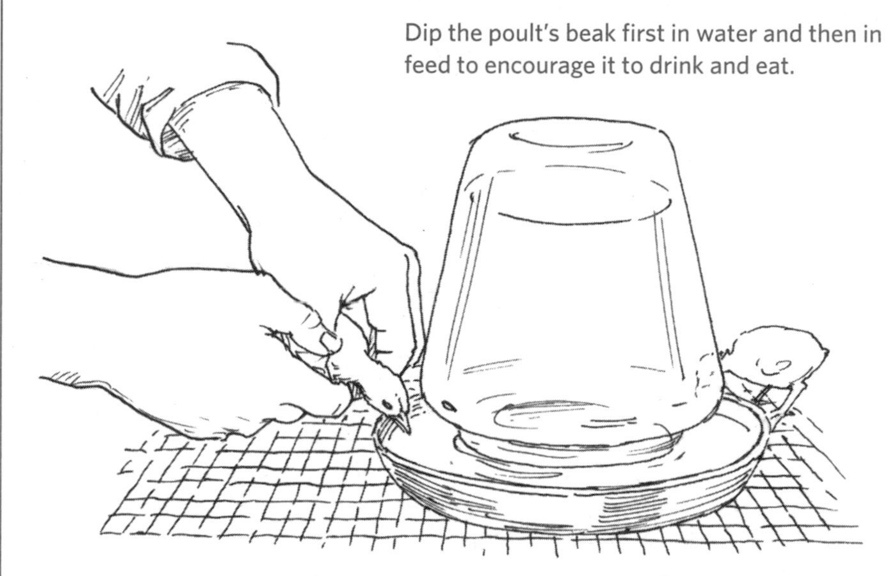
Dip the poult's beak first in water and then in feed to encourage it to drink and eat.

Feeders

Turkey poults often need help finding feed and water, so for the first few days keep an eye on them to make sure they all learn to eat. As soon as your poults arrive, give each one a drink of water. Gently pick up the poults one by one and dip their beaks into the drinker. When a poult tips back its head to let the beakful of water slide down its throat, you know it has had a drink. You can then dip its beak into the feeder to show it where to eat, then tuck it under the heater.

Initially sprinkling feed onto a piece of cardboard, a cookie sheet, or the paper towels covering the brooder floor gives the poults something to peck at, and once they start pecking the scattered feed they are likely to look for more feed to peck. You may need to scatter feed several times before they are all eating from the feeders. To attract the poults to the feeders, put a few shiny or brightly colored marbles into the feed, cover the feed with mashed hard-boiled egg, or sprinkle uncooked oatmeal over the feed. When they are all eating from the feeder, remove the cardboard or cookie sheet.

For the first week or so, a chick-size feeder works fine for poults. Later they'll need something larger, which could be a commercially purchased trough feeder, a homemade wooden trough, or a hanging tube feeder. Furnish enough feeder space so at least half the turkeys can eat at the same time. For 24 poults you'll need 6 linear feet (1.8 m) of trough space. If the trough is attached to the wall, so they can eat from only one side, you'll need troughs totaling 6 feet; if they can eat from both sides, you'll need 3 total feet (0.9 m), such as two 18-inch (46 cm) long troughs. With hanging tube feeders, you'll need 16 inches (41 cm) total diameter, such as one 16-inch or two 8-inch (20 cm) diameter feeders. Less feed will be spilled if the feeders are adjusted to remain about the height of the poults' backs as they grow.

Feed your poults free choice, meaning feed should be available to them at all times. To reduce feed waste by scratching and beaking out (using the beak to scoop feed onto the floor), never fill a trough feeder more than half full, and refill it often. Scratching is not possible with a tube feeder, and a proper tube rim has a rolled-in lip to prevent beaking out.

At first your poults will peck a little feed here and there and won't seem to eat much. By the time they reach 2 weeks of age they could be eating as much as half a pound of feed per bird per week. As they approach harvest weight they may eat as much as a pound (450 g) of feed per bird per day.

Turkey Feed

The feed you put into the feeders will depend in good part on what is available locally. In an area where turkeys are commonly raised, turkey starter, grower, developer, and finisher rations may be available. In other areas you'll need to find reasonable substitutes.

Turkey starter ration, fed for the first seven weeks or so, contains 26 to 30 percent protein. Game bird starter is suitable for turkeys. Another option is to use chick starter (usually 20 to 24 percent protein) and add 4 pounds (1.8 kg) of fish meal (averaging 65 percent protein) per 25-pound (11.3 kg) bag of ration. With this option, when the poults are about 4 weeks old taper the fish meal down to 2 pounds (0.9 kg) per 25-pound bag of ration.

Turkey grower or developer ration, fed from about 7 to 14 weeks, contains 16 to 24 percent protein. Where turkey ration is not available, you can feed the same ration that would be fed to chickens raised as broilers. This ration can be continued until the birds reach harvest weight. Or, where regular turkey rations are available, when the poults reach 14 weeks of age you can switch to turkey finisher, containing 14 to 16 percent protein, and continue until the birds reach harvest weight. If the turkeys are putting on too much fat toward the end, substitute whole oats for 20 percent of their ration (see page 65 on how to assess the fat cover, or finish).

Commercially prepared rations come in the form of pellets or crumbles. Pellets are made from a ground-up mixture that has been compressed. Crumbles

Rodent Control

Rodents can eat an amazing amount of feed, vastly increasing the cost of raising turkeys. Keeping rodents away is not easy, as they are attracted by the ready availability of feed. Store rations in a container with a tight lid, such as a clean trash can, and sweep up any spilled feed. At the first sign of rodents — droppings, tunnel openings, holes chewed in unopened feed sacks — set out traps. Poisoning is another option, but must be done with great care to avoid poisoning the turkeys.

Three Styles of Feeders

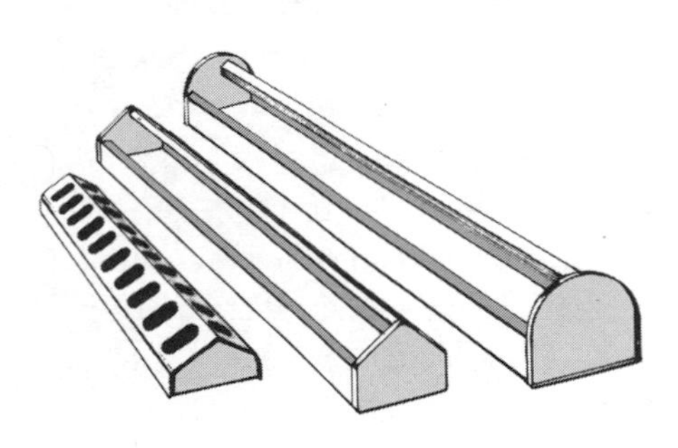
Ready-made trough feeders come in various lengths and diameters.

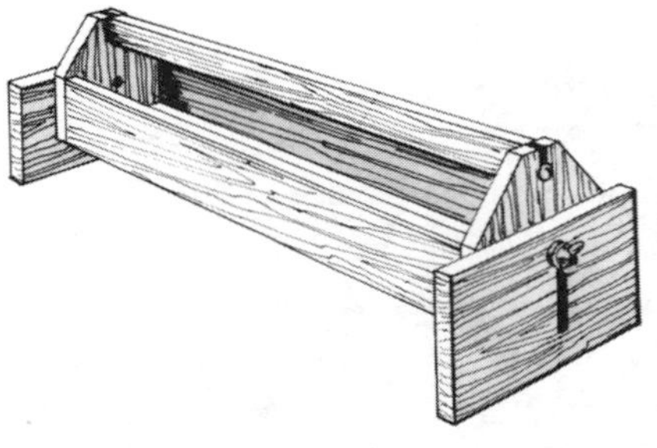
Homemade wooden trough feeder with feet at both ends to adjust the height as the poults grow.

A hanging tube feeder can be easily raised by means of a chain to keep it at the level of the turkeys' backs as they grow, thus minimizing wasted feed.

What to Feed

Age in weeks	Ration	Protein
0-7	turkey or game bird starter	26-30%
7-14	turkey grower or developer	16-24%
14+	turkey grower, developer, or finisher	14-16%

Skin Color

Turkeys naturally have white skin. When turkeys are fed corn or greens (including foraging on pasture) their skin may become yellowish.

are crushed pellets that are fed to poults not yet big enough to swallow whole pellets. Starter ration comes in crumble form. Once your turkeys pass the starter ration stage at about 7 weeks of age, they are big enough to handle pellets.

Whenever you change from one ration to another, make the change gradually to avoid digestive upset. Begin by adding a little of the new ration to the one you have been feeding. Each day, for the course of about a week, add a little more until the changeover is complete.

Green Feed

Turkeys love greens, and that goes for poults, too. But until they reach about 6 weeks of age, poults are not ready to forage for themselves. Before then they would appreciate a daily salad consisting of greens finely chopped with a salad chopper or a pair of scissors.

Suitable greens include such things as lettuce, chard, grass clippings from a lawn that has not been sprayed, and tender alfalfa or clover leaves. Another option is to sprout alfalfa or other seeds for them. Just be sure anything you feed your poults is fresh, tender, and chopped small enough for them to eat.

As soon as you introduce greens to your poults, they also need access to clean sand or granite grit to help digest the roughage. For information on granite grit, see page 38.

Water

Turkeys must have access to fresh, clean water at all times. Provide enough drinkers so at least one-fourth of your turkeys can drink at the same time. When your poults first arrive, mixing a water-soluble vitamin-mineral electrolyte into their drinking water will help reduce shipping stress. Most hatcheries offer an electrolyte mix and will ship it in the carton with the poults.

Drinkers are made of either plastic or galvanized steel. Plastic eventually cracks, and steel eventually rusts. You can extend the life of your waterers by keeping plastic out of the sun and keeping metal off the ground.

Drinkers come in various sizes from 1 quart (1 L) to several gallons. A 1-gallon (4 L) drinker is sufficient for two dozen poults during their first month. After that you can add more drinkers or use larger ones. A comfortable size for most people to carry is a 2-gallon (7.5 L) model, which weighs 16 pounds (7 kg) when full of water. By the time two dozen turkeys are fully grown you would need three 2-gallon drinkers. A good idea is to always keep an extra drinker on hand in case one springs a leak.

Waterers can be set on the floor or hung from a chain. The advantage to using hanging drinkers is that they can easily be raised as your turkeys grow. The disadvantage is that they tend to spill, especially when poults bump into the drinker or try to perch on top and cause the drinker to swing from the chain. Placing bricks, blocks of wood, or concrete blocks underneath to stabilize the drinker will keep it from swinging, while the chain keeps the drinker from being knocked off the blocks.

The chief advantage to a floor drinker is that it can be moved around more easily. But floor models tend to get cluttered with bedding, which can be minimized by putting each drinker on a hardware cloth platform built over a sturdy wooden frame. Using platforms

Water Requirements per 24 Turkeys

Age in weeks	Approximate amount needed*
0-4	½-1 gallon (2-4 L)
4-8	1-2½ gallons (4-9.5 L)
8-12	2-3 gallons (7.5-11.5 L)
16+	3-5 gallons (11.5-19 L)

*Water needs vary with rations fed, stress, and ambient temperature.

A drinker hanging from a chain can be easily raised as turkeys grow.

for metal drinkers keeps them off the ground and thus delays rusting. Be sure the platform doesn't place the water too high for young turkeys to reach.

In hot weather, reduce stress by frequently providing cool water to encourage drinking. Check waterers often to ensure your turkeys never run out; for every 20°F (11°C) increase in temperature, the turkeys' water needs can as much as double.

At least once a week, or sooner if algae begins to appear, scrub drinkers with a dishwashing brush and detergent, then sanitize them with a solution of chlorine bleach (1 teaspoon [5 mL] of bleach per gallon [3.8 L] of water). Before refilling the drinkers with water, rinse them well to remove all traces of bleach, which can be fatal to poults.

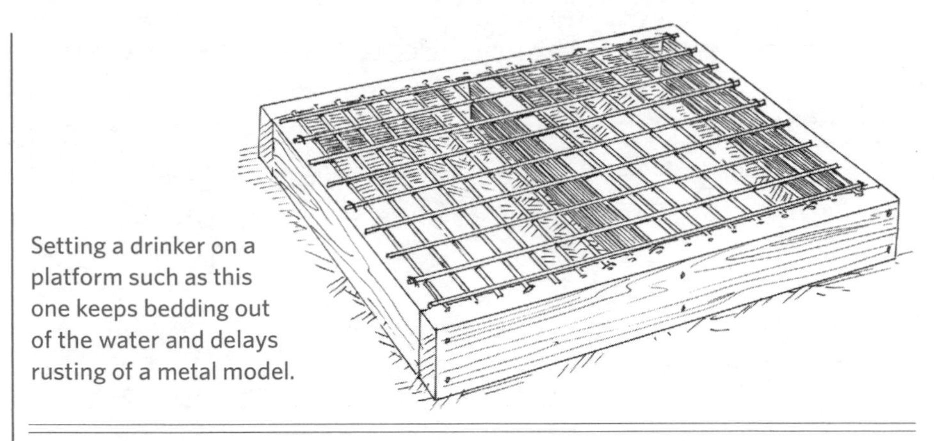

Setting a drinker on a platform such as this one keeps bedding out of the water and delays rusting of a metal model.

As They Grow

Depending on the weather, turkeys can go outside when they are about 6 weeks old. When the temperature reaches 70°F (21°C) or above, and the weather is not inclement, the birds can spend a little time outdoors each day. At 8 weeks of age, they may be allowed to come and go as they please.

Management Systems

You can use one of several different management systems to raise a few turkeys for meat for the family. Which system works best for you will depend in large part on how much land you have available.

Confinement

Turkeys can be kept indoors for their entire lives. The facility should be large enough so each turkey has at least 6 square feet (0.5 sq m) of living space. The floor should be covered with at least 4 inches (10 cm) of clean, dry, absorbent, dust-free bedding and the litter must be refreshed as often as necessary to eliminate caked manure and wet spots. The building must be well ventilated but not drafty, and it must have enough light for the turkeys to find the drinkers and feed troughs. Under this system, all feed is brought to the turkeys, and in most cases they are fed commercial rations the entire time.

Confinement is most common where predators are numerous, where the weather is always rainy or otherwise bad, or where insufficient space is available to provide a yard or pasture. This system is not ideal for the turkeys, as they never get to enjoy fresh air and sunshine, and if they become bored they may pick on each other. Confinement is most often used for broad breasted strains, which grow rapidly and may be harvested in as little as 16 weeks. It is much less suitable for heritage breeds, which are more active and can take as long as 6 months to reach harvest weight.

Yarding

A more typical backyard option is to allow turkeys access to a yard during the day. The indoor facility is much the same as for confinement, although it needn't be quite as large. The same building that was used for brooding the poults can be used to provide shelter for yarded turkeys. One-quarter acre, or about 150 feet by 75 feet (46 m × 23 m), is sufficient for two dozen turkeys. The yard should be protected with woven-wire field fence or chain link.

Yarding has a huge advantage over confinement in that it gives the turkeys an opportunity to spend time in the fresh air and sunshine, and gives them more room to move around and engage in the natural activities of pecking, scratching in the soil, and dust bathing. On the other hand, turkeys that spend more than a couple of months in a yard will soon eat or trample all the vegetation, and the yard will become packed solid (or, in rainy weather, will become a sea of muck).

Pasturing

Raising turkeys on pasture involves providing enough land for the turkeys to forage over that they do not destroy

Prevent Flying

The flying breeds like to perch on fences and gates, often coming down on the wrong side and then running up and down the fence trying to get back in with the rest of the gang. Even a fence as high as 6 feet (1.8 m) will not keep in some breeds.

Stringing several strands of smooth wire or a single electrified wire along the top of the fence serves as a deterrent to roosting. A net cover over the entire yard, where feasible, will keep turkeys where they belong. Clipping the flight feathers of one wing puts a bird off balance so it can't fly but lasts only until new feathers grow in. Clipping off the first joint of one wing of newly hatched poults is a drastic but surefire method of permanently preventing flight.

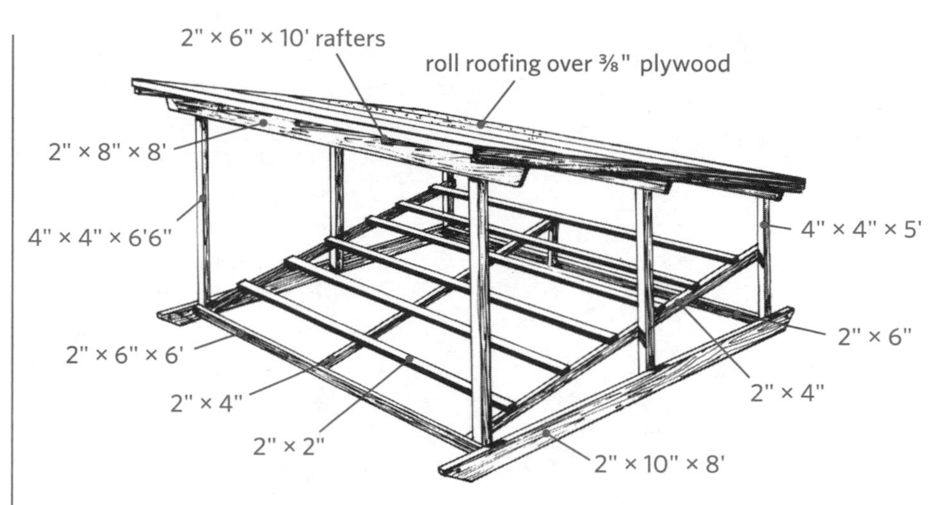

Portable range shelter on skids.

the vegetation, which requires periodically moving the turkeys to new ground. This system is more natural, and therefore healthier, than confinement or yarding, but it requires more land. Two dozen turkeys need at least one-half acre (2,025 sq m) of good pasture, such as alfalfa or ladino clover. By foraging for insects and greens, the turkeys will eat as much as 25 percent less of the commercial ration.

One method of pasturing turkeys is basically an extension of yarding, except that it involves multiple yards. Situate the shelter in the center of the pasture and divide the pasture into paddocks, or pens, that radiate out from the shelter like spokes of a wheel and let the turkeys into each pen sequentially. When one pen has been grazed down, or if bare spots appear, move them to the next pen.

How long a flock takes to graze down a given area depends on a number of factors, including the size of the flock, the kind and condition of the pasture, temperature, and rainfall. When sun, rain, and warm weather combine to help plants grow quickly, your flock can graze a given area for two weeks or more. In cool, hot, or dry weather, when plants grow slowly, the same number of turkeys may graze down the same area to nothing in a matter of days. The best you can do is keep a watchful eye on the vegetation and move the turkeys when they have grazed the plants down to 1 inch (2.5 cm), or when bare spots begin to appear.

Meanwhile, vegetation in other pens must not be allowed to grow taller than about 5 inches (12.5 cm), so during times of rapid vegetative growth you will have to mow unused pens. Bear this fact in mind when installing fences so you can avoid odd corners where your mower cannot reach; otherwise you'll have a lot of extra work removing weeds that grow out of control in the unreachable corners. For a typical rotation plan see page 31.

A more healthful method of pasturing, but one that involves more work, is to use a portable range shelter and move it to fresh ground at least once a week. Between times, move the feeder and drinker every day to spread droppings and reduce trampling of the

Pasture Confinement

A pasturing alternative that gets wide press is to confine poults within a range shelter that is moved daily. The shelter provides 4 square feet (0.4 sq m) per turkey, or 10 feet by 10 feet (3 × 3 m) for two dozen birds (12 feet by 12 feet [3.5 × 3.5 m] if they are broad breasted). Since the poults never leave the shelter, feeders and drinkers must be contained within the shelter. Each morning without fail the shelter must be moved to new ground, as the turkeys will voraciously eat down the fresh vegetation during the day and foul the ground overnight.

Pasture confinement was originally designed for raising broilers, which take about eight weeks to grow to harvest weight. Turkeys, on the other hand, take four to six months to grow out, vastly increasing the climatic variety the poults must endure, the amount of land required, and the amount of work involved in continually moving the pasture shelter. More information can be found on the Internet by doing a keyword search for "pastured poultry" and for "pastured turkeys."

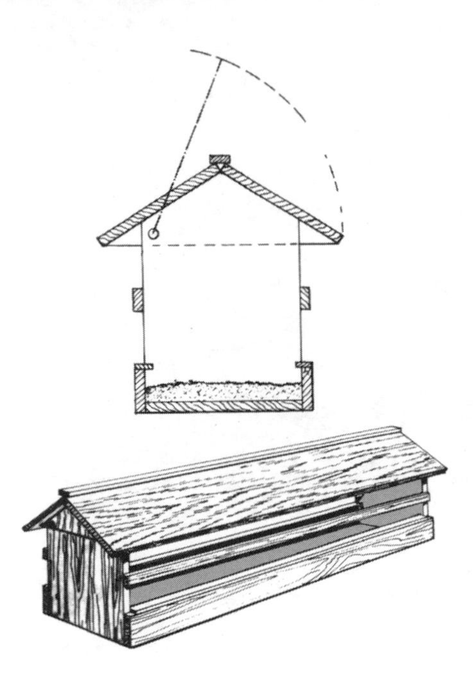

A homemade range trough feeder with a hinged roof to keep out rain.

vegetation. During times of rapid vegetative growth, keep the pasture mowed to no taller than 5 inches (13 cm).

The range shelter should provide at least 2 square feet (0.2 sq m) per turkey, or about 6 feet by 8 feet (1.8 × 2.5 m) for two dozen birds. Roosts must be sturdily built to support the aggregate weight of the growing turkeys. Allow a minimum of 15 inches (38 cm) of roosting space for large breeds and 12 inches (30 cm) for smaller breeds. The roosts can be made from 2 × 4s, with the edges rounded, oriented narrow side up and spaced 18 inches (46 cm) apart if they are slanted ladder fashion, or 24 inches (61 cm) apart and 15 to 30 inches (38 to 76 cm) off the ground if the roosts are arranged platform-style.

Turkeys can be pastured from the age of 8 weeks until they reach harvest weight. Pasturing is more suitable for the active heritage breeds than for the more lethargic broad breasted strains, although people have successfully raised both types on pasture.

Handling Turkeys

Growing turkeys have sharp claws that can easily wound your flesh, and powerful wings that can cause bruising or even a black eye if the turkey flails at you. Catching and carrying a turkey must therefore be done with caution.

If you need to move a group of turkeys, sometimes the easiest method is to drive them. Unlike chickens, turkeys can be herded like cattle, which is handy when the time comes to move them from the brooder to pasture, for instance, or when you want to load them to take to the slaughterhouse. To control and guide the turkeys, a stick comes in handy. It should be several feet long and have a piece of survey tape or a strip of cloth tied to one end to get their attention.

The turkeys will travel together as a group. Guide them with the stick in the direction you want them to go. Move slowly and don't chase them. If they get rushed or otherwise frightened — such as being pushed too rapidly toward something they are unfamiliar with — they may panic and stampede, causing them to crash into a fence or a building and trample or smother each other.

If you need to catch a turkey, you will cause less confusion and incur less chance of injury to yourself or the turkey if you do it in a darkened area. The best time to catch turkeys is at night, after they have gone to roost. Having a helper comes in handy, to briefly turn on a flashlight while you locate the turkey you want to catch.

For a young or small turkey, grasp both legs in one hand and cradle the breast with the other arm. For an older or heavier turkey, take one leg in each hand, then transfer both legs to one hand and clasp the bird against you with its rear pressed against your body, then reach around and get hold of the wing on the far side, where it attaches to the turkey's body. With the legs and wings immobile, the turkey is less apt to struggle.

Poult Health

The list of diseases turkeys are susceptible to is long enough to discourage anyone from trying to raise a turkey. But if you start with healthy poults, keep them in a clean environment, and feed them a healthful, nutritious diet with plenty of fresh, clean drinking water, you shouldn't experience a disease outbreak.

A dead poult is, of course, a sign of disease, but don't panic if one bird dies. Poults need time to build up their immune systems, and until they are about 3 months old they can be pretty vulnerable. The average mortality rate for young turkeys raised commercially is about 15 percent, and some growers do much worse. Naturally you'll be upset when you find a dead bird, but it's not a major issue unless several die at once or show other signs of disease. Should the latter occur, isolate the sick poults and seek help from someone experienced with turkeys, an avian veterinarian, or your county Extension agent. The most common health concerns among backyard poults are blackhead, cannibalism, coccidiosis, and predation.

Signs of Illness

When you are familiar with how a healthy poult looks and acts, you can easily detect illness by noticing changes. Each time you visit your poults, stand quietly for a few moments and watch for anything unusual.

Sound. Healthy poults make pleasant, melodious sounds. Sick birds may sneeze or make other abnormal sounds.

Smell. Notice how your poults normally smell. Any change in odor is a bad sign.

Appearance. A healthy poult looks sharp and perky. A sick poult may become droopy and inactive, pull in its neck, appear sleepy, and let its feathers hang loosely.

Droppings. The droppings of a tom differ from those of a hen. The tom's droppings are cylindrical in shape, blunt at the ends, and arranged like a J, L, question mark, or sometimes a straight line with white urates (the equivalent of turkey urine) at one end. The hen's droppings are slightly smaller in diameter and fall into a roundish pile of loops or spirals with white urates on top or to one side. For both genders, the droppings are firm, although occasional thick, sticky droppings are also normal. Abnormal droppings may be soft, watery, or bloody.

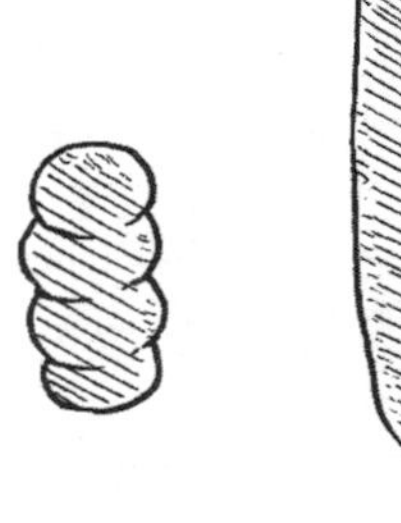

A tom's droppings are cylindrical and often J-shaped (right); a hen's droppings make a rounder pile (left).

Behavior. Notice the amount of feed your poults normally eat and the amount of water they drink. Unhealthy poults may drink more than usual, eat or drink less than usual, and may stop growing.

Blackhead

Histomoniasis is a serious disease of turkeys caused by *Histomonas meleagridis* protozoa, which live in most poultry environments, except where the soil is dry, loose, and sandy. The disease is commonly known as blackhead, which is misleading because the head of an affected bird does not always darken.

The protozoa lodge in blind intestinal pouches or ceca, enter the bloodstream, and eventually migrate to the liver. Early signs of the disease include droopiness, diarrhea, yellow droppings, and weight loss. Sometimes reduction of oxygen in the blood causes the skin to appear dark blue or black, leading to the name blackhead. Damaged tissue in the ceca may result in bloody droppings that might be mistaken for coccidiosis. Turkeys die from either liver failure or a bacterial infection following damage caused by the protozoa; sometimes dead turkeys are the first sign of this disease.

The protozoa are carried by the cecal worm (*Heterakis gallinarum*), the most common parasitic worm in North American poultry. The infection is spread by cecal worms and their eggs that are expelled in droppings and then picked up by a foraging turkey. Earthworms, too, may eat infected cecal worm eggs, and then infect a turkey that eats the earthworms.

Protozoa that lack the protection of a cecal worm egg quickly die, but given such protection can survive in the environment for years. Once a flock becomes infected, blackhead is difficult to eliminate from the flock or the land. Prevention is therefore a preferable plan, and starts by helping the poults develop a strong immune system; see "Strengthening Immunity" on page 63.

A sure way to prevent blackhead is to keep young turkeys away from older turkeys and chickens, and avoid raising turkeys on land where any poultry have lived for at least three years. The time period can be shortened to two years if the land is rototilled and gardened before being returned to pasture. During the rearing period, pasture rotation helps prevent a buildup of infectious organisms, as does periodically moving feeders, drinkers, and roosts.

Cannibalism

Poults that pull each other's feathers or peck each other's flesh are unhappy, mismanaged birds. Cannibalistic behavior is more common among commercial turkey strains raised in industrial confinement than among standard breeds kept in a typical backyard setting. Conditions that encourage cannibalism are overcrowding, keeping the brooder temperature too high, and letting feeders go empty.

Poults raised on slats or wire, or in confinement, are more likely to pick on

How Long Does a Turkey Live?

The vast majority of domestic turkeys live only a few months before they land on a serving platter. The industrially bred toms and hens that produce those turkeys are slaughtered before the age of 2. Standard, or heritage, turkeys raised on pasture have a longer productive life; toms typically breed for 3 to 5 years, hens for 5 to 7 years. The average lifespan of a domesticated turkey is 10 years, although some may live to the ripe old age of 15.

Turkeys and Chickens Together?

Conventional wisdom says you should never keep turkeys and chickens together because turkeys are susceptible to blackhead, a disease with devastating consequences. However, lots of backyarders raise chickens and turkeys together without a problem, and with some benefits.

Newly hatched poults tend to get off to a slow start, but chicks are somewhat quicker on the uptake. Poults brooded with chicks learn to eat and drink more readily by following the chicks.

Broody chickens are often used to hatch turkey eggs. A medium-size chicken can handle about half a dozen turkey eggs. Soon after the eggs hatch, most broody hens will accept six or so additional poults slipped in with the ones she hatched.

Chickens raised with turkeys acquire a sort of immunity to Marek's disease. Turkeys carry a related, although harmless, virus that keeps the Marek's virus from causing tumors in chickens.

On the downside, young turkeys need considerably more protein than baby chicks. And as they grow, chickens are attracted to a tom turkey's tail display, and may follow the tom, picking feathers from his rear end. If not stopped — by separating the offending chicken(s) from the tom — the situation can turn bloody.

And, of course, there's the blackhead issue. Although chickens commonly carry the blackhead protozoa without being infected, turkeys — especially young ones — are highly susceptible and can easily get infected by chickens. Since an outbreak of blackhead requires the presence of both histomonads and cecal worms, the danger of blackhead can be minimized by not bringing in new poultry that may bring trouble with them and, where cecal worms are already present, by regular deworming to reduce the cecal worm population.

A chicken hen will brood at least twice as many poults as she can hatch.

each other than poults in a yard or on pasture, where boredom is less likely. Cannibalism can be deterred by giving the turkeys something to do: Provide a bin of loose soil for dust bathing; set out a bale of hay for them to hop up on and peck at; suspend a flake of hay or a fresh head of cabbage or lettuce where they must reach up to peck it; hang a few shiny pie tins so they can amuse themselves pecking.

Most importantly, give your turkeys plenty of room to grow. At the first sign of pecking, increase their living space, preferably by moving them to different housing, or, best of all, putting them on pasture if they are old enough and the weather is nice enough.

Coccidiosis

Like chicks, poults are susceptible to coccidiosis — an infection caused by parasitic protozoa — but the protozoa that affect chickens are not the same as those affecting turkeys, and vice versa. So chickens cannot transmit coccidiosis to turkeys, and turkeys cannot transmit it to chickens.

Coccidiosis infects poults before they have had a chance to build up an immunity, which occurs through gradual exposure to protozoa in the environment as the poults grow. Too-rapid exposure, by way of infected droppings, results in an outbreak of coccidiosis. Affected poults sit hunched up with ruffled feathers and their heads pulled back, may have bloody diarrhea, and, left untreated, may die. Poults that have been infected and then treated likely will not grow well.

Poults housed on a slat or wire floor, where their droppings can fall through, are less likely to be infected than turkeys kept on litter or pasture, where droppings may accumulate. However, raising turkeys on a slat or wire floor does not allow gradual exposure to the protozoa and therefore may only delay an outbreak of coccidiosis; furthermore, it can also lead to other problems including lameness and cannibalism.

The best preventive measure is good management. Keep feeders and drinkers free of droppings, manage litter to prevent damp spots and manure caking, and move pastured poults often enough that they don't have to forage in accumulated droppings.

As a first-time turkey owner, you would be wise to avoid coccidiosis by feeding your poults a medicated turkey starter, which contains a coccidiostat. Medicated starter is a preventive measure and won't help birds that already have coccidiosis. Treating the disease requires stronger medication available from a farm store, poultry supplier, or veterinarian.

Strengthening Immunity

You can help your poults develop strong immune systems using a probiotic and a vitamin/mineral mix for their first three months of life. The simplest option is to use a product such as Vita-Pro-B Concentrate containing vitamins, minerals, electrolytes, and a probiotic all in one, and mixing it into the drinking water according to directions.

Another form of probiotic is a live culture yogurt, which can be purchased at the market or made at home using a yogurt maker and starter culture. For two dozen poults, evenly stir ½ cup (120 mL) of yogurt into 1 quart (1 L) of starter crumbles. Feed this mix first thing in the morning, and when it's gone refill the feeder with plain crumbles for the rest of the day.

If you use yogurt as your probiotic, you'll also need a water-soluble vitamin/mineral mix — available from a poultry supplier, farm store, and some hatcheries — to dissolve in your poults' drinking water. Broiler Booster brand is formulated for turkeys and other meat birds. If you opt for a standard poultry mix, double the recommended dose for the first month, then reduce to the regular amount.

Predation

Predators abound that like to eat turkey as much as you do. The type of predators most likely to visit your turkeys depends on where you live, how populated the area is, and how secure your turkeys are. Predators can be loosely divided into three groups, as defined by the turkeys' age and the system under which they are managed.

Predators affecting young, confined poults. They include rats, snakes, and weasels. These predators are the most difficult to control because they can squeeze through a hole so small you might not notice it. Watch for small gaps in housing and holes or tunnels coming through the floor, particularly in corners and along the walls.

Predators affecting yarded poults. They are likely to fly in — eagles, hawks, and owls — and therefore cannot be deterred by the stoutest fence. Since owls hunt at night, you can protect your turkeys by closing them in at night. Eagles and hawks, however, hunt in the daytime. They can be kept away from yarded poults by covering the yard with netting or by criss-crossing the yard with wire filament and hanging something that spins or flutters when a raptor flies near; old CDs and DVDs are ideal for this purpose, but strips of fabric such as old bedsheets also work.

When a raptor comes to land in the chicken yard and stirs up a breeze with its wings, CDs and DVDs will spin (or fabric strips will flutter) and scare it away.

Predators affecting pastured turkeys, including both raptors and four-legged hunters such as coyotes, dogs, foxes, and raccoons. A strong perimeter fence, preferably electric, is a good start toward keeping them out, and a second fence within the perimeter fence is even better. Since most four-legged predators prowl at night, enclosing your turkeys in a well-built range shelter overnight will protect them should a predator get through the fence(s). Solar operated Nite Guard flashing lights are also effective when used as directed by the manufacturer.

Once a predator gets a taste of turkey, keeping it away from the flock becomes more difficult. So keep an eye out for predators and their signs (scat, tracks, and sometimes scent), patrol your turkey facility often looking for security breaches, and take corrective measures as necessary. Information on predators common in your area, and their signs, is available from your local wildlife office and county Extension office.

Helpful guides to identifying predators include *Scats and Tracks of North America* by James C. Halfpenny, PhD, and *The Encyclopedia of Tracks and Scats* by Len McDougall.

Butchering

In 16 to 20 weeks for broad breasted strains, and 24 to 28 weeks for standard breeds, your turkeys will be ready to harvest. The industrial strains have a high meat-to-bone ratio, finish with little fat, and are relatively free of pinfeathers. The standard breeds are a little less meaty and are harvested when nearly mature.

At the age of 22 weeks they will begin storing a layer of fat, which by nature is intended to help them survive the lean winter months. The flavor is in the fat, which is one reason heritage breeds are tastier than industrial strains. The fat layer under the skin also makes heritage breeds self-basting while they roast.

Tracking Growth

By weighing a few turkeys every four weeks, you can track their growth rate and determine if they are on target. Remember that heritage breeds gain weight more slowly than industrial strains, and a turkey on pasture will gain weight more slowly than the same turkey kept in confinement.

The table on page 65 indicates typical weight gains for industrial strains. Establishing a similar table for heritage breeds would be difficult because they have not been selectively bred for uniformity of growth and size. However, by tracking your turkeys' weights you will determine if they are gaining at a steady and satisfactory rate.

Weighing your turkeys also helps you determine when they are approaching harvest weight, depending on whether you prefer a smaller turkey or a larger turkey. You might butcher some at a lighter weight and let the rest grow, for instance harvesting smaller turkeys for Thanksgiving and growing larger ones for Christmas. Be aware, however, that the feed conversion rate goes down as the weight goes up, so the more your turkeys weigh at harvest, the more they cost per pound to raise.

Keep in mind that a turkey's weight is not all meat. A turkey that weighs 20 pounds at the time of butchering, for example, will not yield 20 pounds of meat. Broad breasted hybrid turkeys dress out to approximately 80 percent of their live weight. Heritage breeds dress out to approximately 75 percent of their live weight. For both, hens typically dress to a slightly lower percentage of live weight than toms.

Feed Conversion

In a highly controlled industrial confinement setting, the average feed conversion rate for broad-breasted hybrids is 2.5 to 1, meaning for every 2.5 pounds of feed a turkey eats, it gains 1 pound of weight. If you raise a broad breasted strain at home, you can't hope to achieve much better than 3 or 3¼ to 1. So a turkey that weighs 16 pounds at harvest will have consumed approximately 50 pounds of feed.

Feed conversion rates start out low and go up as a turkey ages. The closer the turkey comes to reaching its mature weight, the higher its feed conversion rate. Most industrially raised hens are harvested at about 16 weeks of age at the weight of about 16 pounds (7 kg) live or 12 pounds (5.5 kg) dressed.

Versatile Meat

Although turkey is traditionally served at Thanksgiving and Christmas, there's no reason you can't enjoy your homegrown turkeys year-round. The versatile meat can be ground and served like hamburger, made into sausage, sliced as cold cuts, chopped into stir-fry, or served any way you might prepare chicken or pork.

Average Growth Rate of Industrial Turkey Strains

Age in weeks	Tom in lbs. (kg)	Hen in lbs. (kg)
4	2½ (1.1)	2 (0.9)
8	10 (4.5)	6¼ (2.8)
12	17 (7.7)	11 (5)
16	24 (10.9)	16 (7.3)
20	30 (13.6)	20 (9.1)

Toms are generally harvested at about 18 weeks of age at a weight of about 28 pounds (13 kg) live or 22 pounds (10 kg) dressed. If you prefer a smaller turkey, either gender can be butchered at the live weight of as little as 10 pounds (4.5 kg) and still yield a respectable amount of meat.

The feed conversion rate for standard, or heritage, breeds varies considerably with breed and strain. Furthermore, heritage breeds are generally raised on pasture, which automatically increases their feed conversion rate, which can range anywhere from 4.75 to 1, to more than 6 to 1. On the other hand, an industrial strain raised under the same system will have a similar conversion rate. The bottom line is that industrial and heritage turkeys raised on pasture will eat approximately the same amount of feed to reach approximately the same weight, because the hybrid eats more but the heritage eats longer. At a feed conversion rate of 5 to 1, a 16 pound (7 kg) turkey will consume 80 pounds (36.5 kg) of ration.

Determining Butchering Time

In addition to a turkey's weight, two other factors determine when the turkey is ready for harvest — minimal pinfeathers and adequate fat cover, or finish. Exactly when those three things come together depends on the turkeys' breed and strain, as well as a variety of management factors. As your turkeys grow close to the weight at which you wish to harvest them, start keeping an eye on their pinfeathers and finish.

Pinfeathers are immature feathers emerging from the skin. After a poult acquires its first set of feathers, it almost immediately begins to molt, or gradually lose the old feathers as new ones grow in. Generally by the age of 10 weeks or so this molt is complete enough that pinfeathers are not a problem. However, the poult goes through another molt, which overlaps with the previous molt and lasts until the bird reaches about 14 weeks of age. To top things off, a poult then goes through a partial winter molt. With each molt, whenever an old feather falls out, a pinfeather appears in its place.

Pinfeathers are difficult to remove, and unless they are white they appear as dark spots all over the turkey's skin. When the turkey is roasted whole with the skin on, the pinfeathers can look decidedly unappetizing. By checking the skin beneath the main feathers of the back and breast, you can determine whether or not the turkey is loaded with pinfeathers. Of course, if you plan to skin the turkey before it's cooked or served, pinfeathers are irrelevant.

Finish refers to a thin layer of fat beneath a turkey's skin. A turkey without good finish will cook up dry and somewhat tasteless, since fat is where the flavor is. To assess finish, locate the thinly feathered area on the turkey's breast, about halfway between the breastbone and the wing attachment.

If necessary, pull a couple of feathers so you can get a good view of the skin. A turkey with little fat will appear to have papery-thin, almost transparent skin that looks reddish or bluish purple, because the muscle shows through the skin. A turkey with a good layer of fat covering the muscle will have creamy white or yellowish skin.

Gently pinch the skin between your thumb and forefinger. The pinched skin will be thin on a turkey with little fat, and quite thick on a properly fattened turkey. If you are growing your turkeys on the large side, toward harvesttime they may put on more fat that is desirable. If your turkeys seem to be putting on too much fat, as determined by the pinch test, adjust their diet by substituting whole oats for 20 percent of the ration.

Homegrown versus Store Bought

The meat of a wild turkey is almost all dark and has an intense flavor. The meat of a supermarket turkey is mostly white and lacks flavor. A standard turkey raised on pasture lies somewhere between the two — it has more white meat than a wild turkey and better flavor than a supermarket turkey. Why is that?

A turkey uses breast muscles for flying and leg muscles for walking. Wild turkeys actively fly as well as walk while foraging. Industrially raised turkeys don't fly and in fact can barely walk — their legs are too thin and weak for their heavy bodies. Heritage turkeys raised on pasture spend a lot of time foraging and, given the opportunity, may fly. Active muscles of the legs and breast require oxygen, and oxygen is carried in blood cells. The more active the muscles, the more blood they need, and the more blood they need, the darker the meat.

Muscles get energy from fat stored within the muscle cells. The more the muscles are exercised, the more energy they need, so the more fat they store. A wild turkey that exercises its breast and leg muscles equally has more fat in its muscles than an industrial turkey that never flies and scarcely walks. Since flavor is in the fat, the meat of a wild turkey has more flavor than a supermarket turkey. Similarly, heritage turkeys are more active than supermarket turkeys and therefore more flavorful, but they are not as active as wild turkeys and therefore not as intensely flavored.

Can you save money growing your own turkeys? That depends on a lot of factors, including whether you raise broad breasted or heritage poults, whether you feed them in confinement or on pasture, and whether you are satisfied buying supermarket turkeys or are paying a premium for pastured turkeys. Let's look at some facts. The cost of purchasing heritage poults is about 60 percent higher than for broad breasted poults. On the other hand, the survival rate of heritage poults can be quite a bit higher than for broad breasted poults.

Using averages (and for simplicity's sake, ignoring the cost of housing, water, electricity, and so forth) do the math: Let's say you purchase two dozen heritage poults for $8.00 each and 80 percent of them survive; your effective cost per poult is $10.10. You raise each poult to 16 pounds live weight, or 12 pounds dressed, during which time it will have consumed about 80 pounds of ration. At the current rate of 30 cents per pound, the total cost of feed per turkey comes to $24. Added to the $10.10 initial cost it comes to $30.10 for 12 pounds of turkey, or about $2.50 per pound. The price of supermarket turkeys is about $1.00 per pound; the price of pastured turkeys ranges upward from $4.00 to $6.00 per pound. So your homegrown turkeys will cost more than twice as much as supermarket turkeys, but will be as flavorful as commercially grown pastured turkeys at a cost savings of at least $1.50 per pound.

In some areas, custom slaughterhouses will handle the entire process of butchering, dressing, wrapping, and freezing your homegrown turkeys for you. For step-by-step details on how to slaughter, dress, and store your turkeys yourself see *Storey's Guide to Raising Turkeys* by Leonard S. Mercia.

Breeding Turkeys

So now that you've had so much fun raising your first turkeys, you're thinking of keeping your best tom and a few of your hens to raise your own poults next year and are wondering what's involved. First off, if your turkeys are broad breasted, don't even think about it. They are too heavy and awkward to breed naturally, so they must be bred by artificial insemination. But if you opted for one of the heritage breeds, you're in business.

Here are the chief advantages of raising your own: They are beautiful in the yard and fun to have around, and with proper management the poults generally have a better survival rate than poults purchased by mail order. Some disadvantages: you'll be feeding mature turkeys year-round (see chart at right to estimate how much they'll eat); the hens don't lay many eggs (expect somewhere between 35 and 100, depending on the breed and strain); mature toms can get aggressive; not all hens have the instinct to brood, and among those that do, not all make good mothers.

Turkeys have not been selectively bred for their mothering instinct, so over time the natural inclination of wild turkeys to raise young has been bred out of our domestic turkeys. Among the breeds that retain the strongest mothering instinct are the Black, Bourbon Red, and Narragansett. For others, and even for some individuals among these breeds, you may need an incubator to be successful in hatching poults. Also, once a hen starts setting she stops laying, so to maximize the number of poults you get each spring you'll need an incubator.

If you like a challenge, you might try to breed the mothering instinct back into your heritage hens by actively encouraging them to hatch their own eggs. The more hens you have, and the more places you provide for them to nest, the better your chances are of success. One tom can handle between two and six hens.

Approximate Feed per Week per Breeder

Standard turkey breed size	Hen in lbs. (kg)	Tom in lbs. (kg)
Large	4 (1.8)	8½ (3.9)
Small	3½ (1.6)	5 (2.3)

Housing Breeders

Breeding turkeys need more space than brooded poults. The shelter should provide at least 8 square feet (0.75 sq m) per bird for a large breed

Turkey Eggs

A turkey hen lays best during her first year, after which she lays about 20 percent fewer eggs each year. A turkey egg is about twice the size of a chicken egg, and has a huge golden orange yolk. The shell is a pale creamy tan color speckled with dark brown spots. Fertile eggs take 28 days to hatch, which is one week longer than for chicken eggs. You can eat unfertile eggs, just as you would chicken eggs.

or 6 square feet (0.5 sq m) per bird for a smaller breed. The shelter will need nests for the hens to lay their eggs in. Figure on one nest for up to four hens. A nest that measures 18 inches (46 cm) wide by 24 inches (61 cm) deep by 24 inches high is a suitable size for all breeds. Outdoor space should provide at least 5 square feet (0.5 sq m) for yarded turkeys or 150 square feet (14 sq m) for pastured turkeys. For details on managing a breeding flock of turkeys see *Storey's Guide to Raising Turkeys* by Leonard S. Mercia.

Turkey Behavior

Turkeys, like chickens, enjoy dust bathing to condition their feathers and rid themselves of external parasites. Poults go through the motions even when they are brooded on slats or wire without access to a dusting wallow. To take an honest-to-goodness dust bath, the turkey finds a spot of loose soil or litter with which to cover its body, then flaps its wings and kicks its feet to work it through the feathers before shaking it off. In addition to dust bathing, turkeys like to sunbathe, and one can look quite dead lying on the ground with a wing and a leg stretched out to catch the rays.

A tom turkey looks pretty much like a hen until he fans out his tail, puffs out his chest, gobbles loudly, and struts his stuff. While he's all puffed up, the strutting tom will glide in a semicircle, and when he's really wound up will make two sounds in succession that are interpreted as "chum" and "humm." On rare occasions a hen will strut, too, especially when she feels threatened while raising a brood of poults.

Some breeding toms can be aggressive, but don't mistake curiosity for aggression. A big turkey sidling up to you to find out what you're up to, or following you to see if you've brought a treat, can be intimidating if you aren't acquainted with the bird in question. A tom that challenges or attacks you has not been taught from a young age that you are the dominant turkey. At the first sign of a problem, be bold, stand up to the turkey, and drive him off rather than running away, but don't strike the turkey or you'll only provoke a fight.

Turkeys make a lot of interesting sounds — at least 20 different words have been identified in turkey talk — the most widely recognized of which is the tom's gobble. Hens never gobble, which is why toms are sometimes called gobblers. You can almost always get a tom to gobble by imitating a gobble. Turkey gobbling is similar in function to a rooster's crowing, the purpose being twofold: to attract hens and put competing toms on notice. Strutting, too, is intended to assert dominance and attract hens. A tom will generally strut in the presence of a female, but if no female is handy he might strut for any inanimate object that catches his fancy.

Why Is a Turkey a Turkey?

If the turkey is indigenous to North America, and not Turkey, why is it called a turkey? All sorts of explanations have been theorized. One is that the name sounds like the call a turkey makes when frightened: "turk, turk, turk, turk." Another is that a Native American word for the bird is *firkee*, which early settlers heard as turkey.

Yet another explanation is that Christopher Columbus, on his expeditionary voyage to discover a northwest passage between India and China, saw wild turkeys when he landed in America. Because the male turkey, like a male peacock, displays his colorful tail to attract a mate, and because peacocks are indigenous to India, Columbus called turkeys by the Indian word *tuka*, meaning peacock.

Probably the most likely explanation is that it was a simple case of mistaken identity. Early settlers thought the turkey was a type of guinea fowl, which in those days was called a turkey cock because the guinea was introduced into Europe by way of Turkey. Even today people who are unfamiliar with guineas often believe they are some kind of turkey. So, although eventually it became apparent that guineas and turkeys are only distantly related, the die was cast and the turkey remained a turkey.

CHAPTER 3

Ducks & Geese

Keeping ducks and geese is a relatively simple proposition. They require little by way of housing; in a temperate climate, a fence to protect them from wildlife and marauding neighborhood pets and to keep them from waddling far afield will suffice. They prefer to forage for much of their own food. They are resistant to parasites and diseases. In short, they are the easiest to raise of all domestic poultry.

So why doesn't everyone have waterfowl? Well, for one thing, they like to have at least a small pond to splash in to help them stay clean. Their quacking and honking can get annoying, especially to neighbors. And, while ducks are basically gentle, geese can be decidedly aggressive. But any downside becomes irrelevant if your purpose is to raise a few ducks or geese for roasting or sausage making. Besides, many keepers of ducks or geese enjoy listening to the sounds their birds make, and aggressive geese make terrific watch birds.

Waterfowl Families

The word *waterfowl* collectively refers to birds in the Anatidae family — ducks, geese, and swans. The Anatidae family comprises more than 140 species, all of which are web-footed swimming birds that have a row of toothlike serrations along the edges of their bills (which allows them to strain food out of water).

Waterfowl fall into two basic subfamilies. On one side are geese, along with swans and whistling ducks. These fowl share many common characteristics. Males and females have the same color pattern and a similar voice, molt annually, and mate permanently. The males in this group help incubate and care for the offspring. Being large birds, geese and swans have a hard time hiding while on the nest; thus, to protect themselves, they are instinctively aggressive. They nest after laying only a few eggs, and their hatchlings mature slowly. This group includes the 14 species of goose, which are classified into two groups. One of those groups includes the two species from which domestic geese were developed — the swan goose, which gave us African and Chinese geese, and the graylag, which gave us all other domestic breeds.

The second basic subfamily of waterfowl includes all ducks except the whistlers. The shared characteristics of these ducks are that males and females differ in color pattern and voice, molt twice a year, and mate for a single season. When the female starts nesting, the male moves off by himself. These ducks are smaller than the birds in the first subfamily and can hide themselves more easily. They hatch more eggs at a time, and their hatchlings mature more quickly. This subfamily includes three groups: diving ducks, dabbling or surface-feeding ducks, and perching or wood ducks. The Muscovy was derived from the perching ducks. All other domestic ducks were derived from the Mallard, which is one of the 11 species in the dabbling group.

Waterfowl Family Tree

- Waterfowl (Anatidae)
  - Swans
  - Geese
    - Anser
      - Greylag
        - All domestic geese except Chinese and African
      - Swan goose
        - Chinese and African geese
    - Branta
  - Whistling ducks
  - Other ducks
    - Dabbling
      - Anas
        - Mallard
          - All domestic ducks except Muscovy
    - Diving
    - Perching
      - Aix (wood ducks)
        - Muscovy

Terminology

Talking about waterfowl can get a bit confusing. A male duck is a drake, but a female duck is a duck. A male goose is a gander, but a female goose is a goose. So a drake is a duck, but a duck isn't always a drake, and a gander is a goose, but a goose isn't always a gander. Got it? Thanks to this confusion, a female duck or goose is typically called a hen.

Dealing with these fowl in groups is much simpler: A bunch of ducks is a bevy, and a gang of geese is a gaggle.

A hatchling is a duck or goose fresh out of the egg. If the hatchling is a duck, it is a duckling; if it is a goose, it is a gosling. A hatchling that survives the first few critical days of life and begins growing feathers is called started. When it goes into its first molt (shedding of feathers), it's called junior or green. Until the bird reaches 1 year of age, it's young; after that, it's old.

Waterfowl Traits

Ducks spend a lot of time in water, nibbling at plants, bugs, and other shoreline inhabitants. If you let them wander in your vegetable garden or flower bed, they will help control garden pests, although you have to take care they don't run out of other things to eat and start in on your pea vines or pansies.

Muscovies in particular relish slugs, snails, and other crawly things. In fact, the San Francisco area once had a rent-a-duck service that loaned out Muscovies to local gardeners. Ducks also enjoy chasing flies, in the process offering not only fly control but also endless entertainment. Ducks keep mosquitoes from getting beyond the larval stage, but unfortunately tadpoles suffer the same fate.

All ducks lay eggs, although some breeds lay more than others. Khaki Campbells are particularly known for their laying ability, and their eggs make wonderful baked goods. Any breed can be raised for meat, and putting your excess ducks into the freezer both keeps meat on your family table and keeps down the population at the pond.

Some breeds like to fly around, and occasionally one will fly off into the sunset, but such wanderlust can readily be controlled by clipping a wing or better yet, if your duck yard is small enough, covering it. The primary downside to ducks is their eternal quacking. In a neighborhood where the noise could become a nuisance, the answer is to keep Muscovies, also known as quackless ducks. Among the other breeds, the male makes little sound, but the female quacks loudly, and each bevy seems to appoint one particularly loud spokesduck to make all the announcements.

Geese also make a racket with their honking. Usually they holler only with good reason, but a less observant human might not detect that reason: for instance, a cat or weasel the geese have spotted slinking along the fence line. Besides announcing intruders, geese have a tendency to run them off. A lot of people are more afraid of geese than of dogs, probably because they are less familiar with geese and feel intimidated by their flat-footed body charge, indignant feather ruffling, and snakelike hissing. Even as they fend off intruders, geese can become attached to their owners and are less likely to charge the family dog or cat than roaming pets and wildlife.

Geese are active grazers, preferring to glean much of their own sustenance from growing vegetation. They are often

Parts of a duck (or goose)

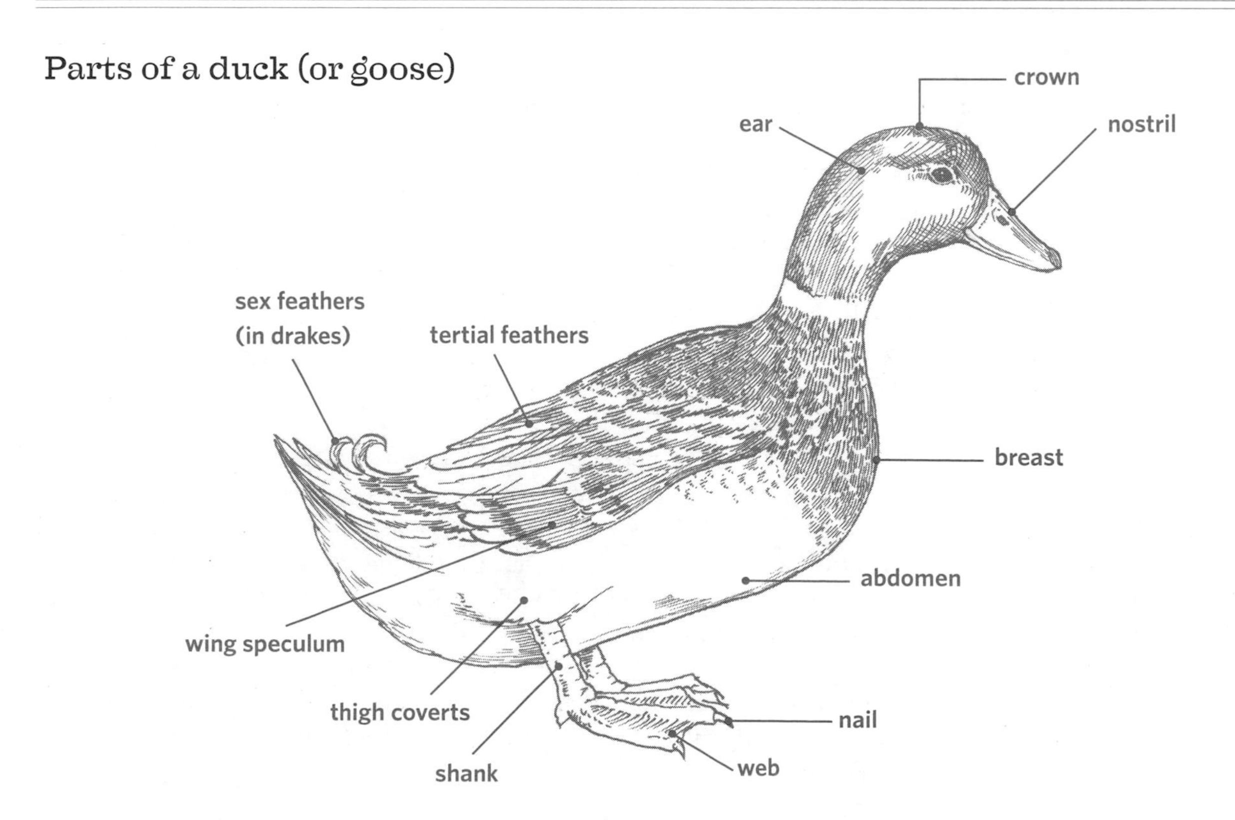

used as economical weeders for certain commercial crops; farmers take advantage of their propensity to favor tender shoots over established plants.

Geese lay enough eggs to reproduce, but no breed lays as many eggs as the laying breeds of duck. Goose meat, however, is plentiful and delicious. Goose is a traditional holiday meal, and when roasted correctly, the meat defies its reputation for being greasy. Goose fat rendered in the roasting makes terrific shortening for baking (leaving your guests wondering what your secret ingredient might be) and in the old days was used as a flavorful replacement for butter on bread. The feathers and down from plucked geese can be saved up and made into comforters, pillows, and warm vests.

Ducks and geese get along well together and can be kept in the same area. In fact, since domestic ducks don't fly well or at all, and some are too heavy to even walk fast, and since geese tend to be aggressive toward trespassers, ducks enjoy some measure of protection from predators when kept together with geese.

Ducks can make a highly effective pest patrol in the garden, as long as you take care to keep them from tender plants and low-growing fruit.

Choosing the Right Breed

Numerous breeds of duck and goose can be found worldwide, and more are created all the time. Others, however, are becoming endangered or extinct. Only a few of the developed breeds are commonly found in the United States. Other breeds with much lower populations are kept by fanciers or conservation breeders. Your purpose in keeping waterfowl will to some extent determine which breed is right for you.

Duck Breeds

Some breeds of duck have been genetically selected for their outstanding laying habits, but unless this ability is maintained through continued selective breeding, the laying potential of a particular flock may decrease over time. For this reason, not all populations of a breed known for laying are equally up to the task. Some sources offer hybrid layers that are bred for efficient and consistent egg production, but their offspring will not retain the same characteristics. Since laying stops when nesting begins, the trade-off among laying breeds is that they generally do not have strong nesting instincts. Although some ducks lay eggs with pale green shells, most of the outstanding layers produce white-shell eggs.

Any breed can be raised for meat, but some breeds and hybrids have been developed to grow faster and larger than others while consuming less feed. The larger, meatier breeds do not lay as well as the midsize dual-purpose (egg and meat) breeds or the smaller breeds kept primarily for eggs. Breeds with white or light-colored feathers pluck cleaner, although the breeds with more colorful plumage are less visible to predators.

Ancona. Originating in England, the Ancona is a hardy dual-purpose breed that lays well and grows rapidly. It is an excellent forager, and because it is active, its meat tends to be leaner than that of less active breeds. The Ancona comes in several color varieties with a patternless patchy plumage color that is often compared to the appearance of Holstein cattle.

Ancona

Appleyard. Originating in England, the Appleyard is a dual-purpose breed that lays well, grows rapidly, and forages actively, producing somewhat lean meat. Its color is similar to that of the wild Mallard, only a little lighter. Thanks to its silvery hue, this breed is sometimes called Silver Appleyard.

Aylesbury. Originating in England, the Aylesbury is among the largest duck breeds raised primarily for meat, and also among the breeds that put on the most fat. Ducklings can grow to about 5 pounds (2.25 kg) in only seven to nine weeks. This breed is not as hardy as other meat breeds and is not an active forager. Unlike most other ducks, which have yellow skin, the Aylesbury has white skin underneath its white plumage.

Buff (Orpington). Originating in England, the Buff duck is a dual-purpose breed, being a relatively good layer and rapid grower. Its light-brown plumage color is generally known as fawn brown. In England this breed is called the Orpington, with buff being but one of its color varieties.

Campbell. Originating in England, the Campbell is the most prolific layer of all

Quick Guide to Duck Breeds

Breed	Eggs/year	Egg shell color	Live weight in lbs. (kg) *male/female*	Foraging	Temperament	Status
Ancona	210–280	white, tan, blue	6 (2.75)/6½ (3)	best	calm	uncommon
Appleyard	200–265	white	8 (3.5)/6 (2.75)	good	calm	common
Aylesbury	35–125	white, pale green	10 (4.5)/9 (4)	good	calm	common
Buff	150–220	white, grayish	8 (3.5)/7 (3)	good	calm	uncommon
Campbell	250–340	white	4½ (2)/4 (1.75)	best	excitable	uncommon
Cayuga	100–150	black to white	8 (3.5)/7 (3)	best	docile	uncommon
Magpie	220–290	green, blue	6 (2.75)/5½ (2.5)	best	excitable	rare
Muscovy	50–125	waxy white	12 (5.5)/7 (3)	best	f: docile m: aggressive	common
Pekin	100–180	white	10 (4.5)/8 (3.5)	fair	docile	common
Rouen	125–160	green	8 (3.5)/7 (3)	good	docile	common
Runner	150–300	white	4½ (2)/4 (1.75)	best	excitable	common
Saxony	190–240	white	9 (4)/8 (3.5)	good	docile	very rare
Swedish	100–150	white	8 (3.5)/7 (3)	good	docile	common
Welsh Harlequin	250–330	white	6 (2.75)/5 (2.5)	good	docile	rare

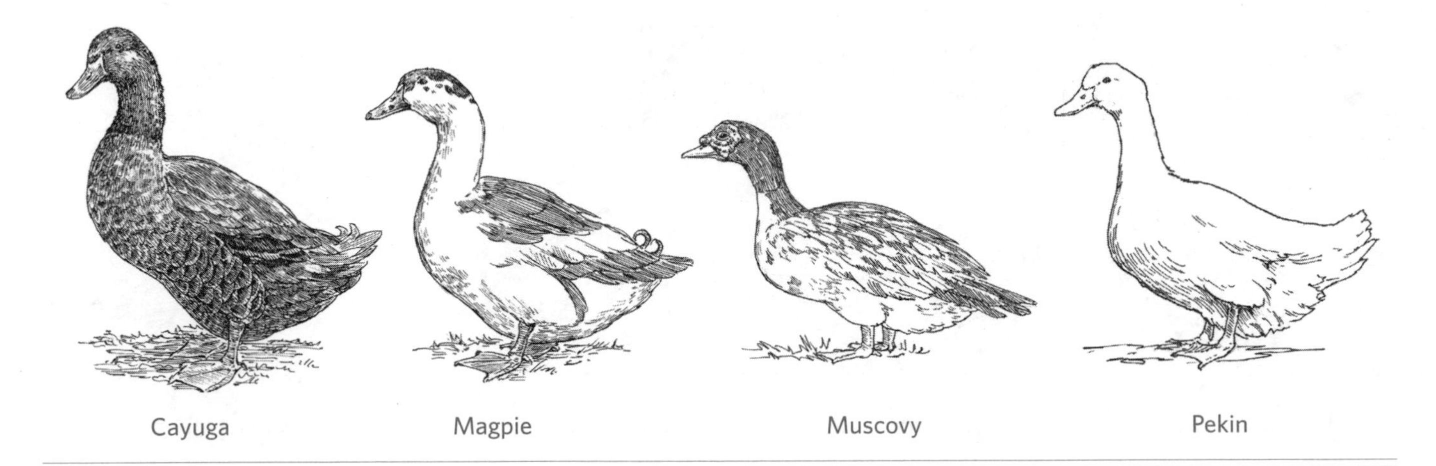

duck breeds. It is an active forager that is adaptable to a wide range of climates. Of the various color varieties, khaki is the most common in North America. This small-bodied duck, if slaughtered when young and tender, makes a nice meal for two.

Cayuga. Originating in New York, the Cayuga is a dual-purpose breed. It is an active forager and among the hardiest of domestic ducks. Eggs laid by young hens start out with a black film that disappears as the hens age, and eventually they lay white eggs. Because of the Cayuga's greenish black plumage, ducks don't pluck clean and instead are sometimes skinned.

Magpie. Originating in Wales, Magpies are among the better laying lightweight breeds. They are active foragers, and the lean meat of a young, tender duck is enough to serve two or three diners. The Magpie plumage is primarily white, with dark to black patches on the head and back.

Muscovy. Native to Mexico, and Central and South America, the Muscovy doesn't look like any other domestic duck and, indeed, is only distantly related to the others. Muscovies are sometimes called quackless ducks because, in contrast to the loud quacking of other female ducks, the Muscovy female speaks in a musical whimper, although she can make a louder sound if startled or frightened. The male's sound is a soft hiss. Aging drakes take on a distinctive musky odor, giving this breed its other nickname of musk duck. Both the males and the females wear a red mask over the bridge of their beak that features lumpy warts, called caruncles.

Muscovies are arboreal, preferring to roost in trees and nest in wide forks or hollow trunks. In confinement, they like to perch at the top of a fence, but don't always come back down on the right side. To get a good grip while perching on high, these ducks have sharply clawed toes. If you have to carry a Muscovy, try to keep it from paddling its legs so those sharp claws don't slice your skin. Aging males can be aggressive, usually toward other male ducks but occasionally toward their keeper. With their powerful wings and sharp claws they can be pretty dangerous.

The male Muscovy matures to be the largest of the meat ducks. Although the variety with white plumage is more suitable for meat, since it has a cleaner appearance when plucked, the original color is an iridescent greenish black with white patches on the wings. Throughout the years about a dozen additional colors have been developed. Both sexes have an enormous appetite for insects, slugs, snails, and baby rodents. With their massive bodies and large flat feet, though, they can be destructive to garden seedlings.

Muscovy meat differs in flavor and texture from that of other ducks, in part because it is much leaner. It has been compared to veal, beef, and (when smoked) ham.

Even though Muscovies are a different species from other domestic ducks, they will interbreed, and the resulting offspring are mules — sterile hybrids that cannot reproduce. A cross between a Muscovy drake and a Pekin duck produces meat birds called Moulards, which have the large breast of the Muscovy with less fat than the Pekin.

Because of concerns about feral Muscovies in the environment, the United States Fish and Wildlife Service requires captive-bred Muscovies to be physically marked by one of four prescribed methods — toe removal, pinioning, banding, tattooing — and prohibits the release of Muscovies into the wild. For information on current regulations, check with your local Fish and Wildlife office, which can be located through the national service's website.

Pekin. Developed in China, Pekins are fast growing big ducks that tend to pile on the fat. Because they have snow-white feathers, they appear clean when plucked; their white pinfeathers don't show as much as the pinfeathers of

Rouen
Runner
Saxony
Swedish
Welsh Harlequin

colored breeds. Strains of giant or jumbo Pekins, which grow bigger and meatier than is typical for the breed, have been developed for commercial meat production. If you order duck in a restaurant, or buy a duck at the meat counter, it will most likely be a Pekin. The hens tend to be noisier than other breeds, which may be an important factor if you plan to raise your ducks to maturity.

Rouen. Developed in France, the Rouen looks like an overstuffed Mallard and comes in two distinct types. The production type is a reasonably decent layer and good forager; the exhibition type grows to about 2 pounds (1 kg) more and is less active. Rouens are slow growing, taking half a year or more to reach harvest weight, and their dark feathers make them difficult to pluck clean.

Runner. Developed in Scotland from stock originating in the East Indies, these ducks are put to work gleaning snails and waste grain from rice paddies. Runners have an upright stance that allows them to move around with greater agility than other breeds; they are called runners because they prefer to run rather than waddle. They are good layers and excellent foragers, with small bodies and meat that tends to be somewhat like wild duck. Known also as Indian Runners, they come in more color varieties than any other breed.

Saxony. Developed in Germany, the Saxony is a large dual-purpose breed that lays well, actively forages, and easily adapts to a wide range of climates. The Saxony is a slow grower, yet its meat is not especially fatty. The drake's plumage is similar to that of a Mallard; the duck is a buff color with silvery blue wings.

Swedish. Coming from Pomerania — an area once occupied by Sweden but that has since been divided between Poland and Germany — the Swedish duck is a dual-purpose breed that lays decently well and is similar in body type to a Pekin. It comes in several color varieties, the best known of which is blue with a white bib, giving rise to its common name of Blue Swedish. It is one of the hardiest breeds, an extremely active forager that does not do well in confinement, and a slow grower.

Welsh Harlequin. Developed in Wales, the Harlequin is a lightweight dual-purpose breed, excellent layer, and active forager, although it lacks a strong sense of self-preservation that makes it vulnerable to predators and inclement weather. Its body type is similar to that of a Campbell. The plumage of the hen is creamy white with dark brown spots along the back. The drake's color pattern is somewhat similar to, though much lighter than, a Mallard.

Goose Breeds

Most domestic goose breeds have been developed for meat, although some are bred with emphasis on other attributes. The Sebastopol, for instance, has long, curly feathers that look like a misguided perm, while the diminutive Shetland was bred to thrive in a harsh environment. Nearly every breed has a tufted version, meaning the goose has a puff of feathers growing upright on top of its head. No breed lays as prolifically as a duck, and although a single goose egg makes a formidable omelet, the eggs are more often used for hatching or for creating craft items such as decorative jewelry boxes.

The main criteria when selecting a breed to grow for meat is the size relative to the number of people you typically feed and the color of the plumage. The darker breeds, especially Toulouse, are more difficult to pluck clean due to their dark pinfeathers. For meat grown as naturally and as economically as possible, ability to forage becomes important. All breeds forage to some extent, although if your geese will be garden weeders, you may want to avoid the soil compaction that typically occurs with the heavier breeds.

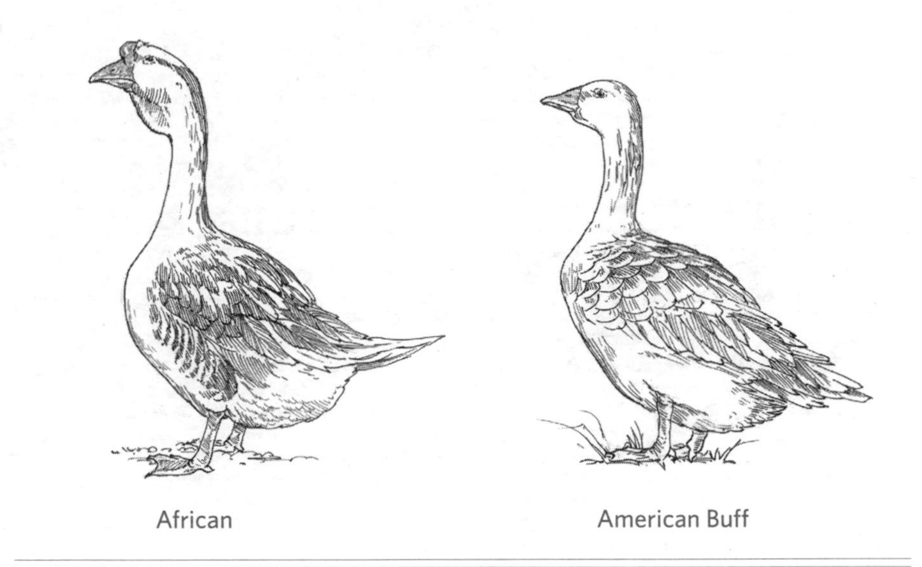

African   American Buff

African. The origin of the African goose is unknown; it is most likely related to the Chinese goose. The African is a graceful breed with a knob on top of its head and a dewlap under its chin. The brown variety, with its black knob and bill, and brown stripe down the back of its neck, is more common than the white variety with orange knob and bill. Because the knob is easily frostbitten, Africans must be sheltered in cold weather. This breed is among the most talkative and also among the calmest, making it easy to confine. Africans, like Chinese, tend to have leaner meat than other breeds, and the young ganders grow fast — reaching 18 pounds (8 kg) in as many weeks.

American Buff. Developed in North America for commercial meat production but remaining quite rare, the American Buff is a pale brown goose

Quick Guide to Goose Breeds

Breed	Eggs/year	Live Weight in lbs. (kg) *male/female*	Foraging	Temperament	Status
African	35–45	22 (10)/18 (8)	best	gentle	common
American Buff	25–35	18 (8)/16 (7)	good	docile	rare
American Tufted Buff	35–50	15 (6.75)/13 (5.75)	good	calm	uncommon
Chinese	30–50	12 (5.5)/10 (4.5)	best	usually calm	common
Embden	15–35	25 (11.5)/20 (9)	good	calm	common
Pilgrim	20–45	14 (6.5)/12 (5.5)	good	docile	rare
Pomeranian	15–35	17 (7.7)/14 (6.5)	best	*	rare
Roman	25–35	12 (5.5)/10 (4.5)	good	docile	rare
Sebastopol	25–35	14 (6.5)/12 (5.5)	good	*	uncommon
Shetland	15–30	10 (4.5)/7 (3)	best	feisty	very rare
Toulouse	25–50	20 (9)/18 (8)	good	calm	common
Toulouse, dewlap	20–30	26 (11.75)/20 (9)	poor	docile	common

* More than most breeds, individuals may be either docile or aggressive.

Chinese Embden Pilgrim Pomeranian

with brown eyes. This breed is known for being docile, friendly, and affectionate. The American Tufted Buff is a separate breed (developed by crossing American Buff with Tufted Roman), but similar except for having a bunch of feathers sprouting from the top of its head. The Tufted is hardier and somewhat more prolific than the American Buff. Both breeds are active, curious, and relatively quiet.

Chinese. Originating in China, the Chinese goose is similar in appearance to the African but lacks the dewlap. It comes in both white and brown, with the brown variety having a larger knob than the white. Like the African, Chinese need protective winter shelter to prevent frostbitten knobs. This breed is the one most commonly used as weeder geese. Because they are both active and small, they do a good job of seeking out emerging weeds while inflicting little damage on established crops. Thanks to their light weight and strong wings, they can readily fly over an inadequate fence. In contrast to the typical goose honk, this breed (like the African) emits a higher pitched "doink" that can be piercing if the bird is upset or irritated. Chinese geese are prolific layers that produce a high rate of fertile eggs even when breeding on land rather than on water. Like African geese, the young grow relatively fast and have lean meat.

Embden. Originating in Germany, the Embden is the most common goose to raise for meat because of its fast growth, large size, and white feathers. Hatchlings are gray and can be sexed with some degree of accuracy, as the males tend to be lighter in color than the females. Their blue eyes, tall and erect stance, and a proud bearing give these geese an air of intelligence. Although they are not as prolific at laying as some other breeds, the eggs are the largest, weighing 6 ounces (170 g) on average.

Pilgrim. Originating in the United States, the Pilgrim is slightly larger than the Chinese and one of the few domestic breeds that can be autosexed — the male hatchling is yellow and grows into white plumage, while the female hatchling is olive-gray and grows into gray plumage similar to the Toulouse but with a white face. Due to their light weight, Pilgrims will often fly over a fence if attracted to something on the other side. The Pilgrim is a quiet breed and more docile than most other geese.

Pomeranian. Originating in northern Germany, the Pomeranian is a chunky breed with plumage that may be all buff, all gray, all white, or saddleback (white with a buff or gray head, back, and flanks). This breed is winter hardy and an excellent forager starting at a young age, when goslings need plenty of quality greens to thrive. More than most, the Pomeranian's temperament is variable and can range from benign to belligerent.

How Long Do Waterfowl Live?

Few waterfowl live out their full, natural lives. Too many get taken by predators. Those raised for meat have a short life of as little as 6 weeks for ducks or up to 6 months for geese. Ducks raised for eggs or as breeders are usually kept for 3 to 5 years, until their productivity and fertility decline. But when properly fed and protected from harm, a duck may live as long as 20 years, while a goose may survive to the ripe old age of 40.

Roman   Sebastopol   Shetland   Toulouse

Roman. Coming from Italy, the Roman is a small, white goose that may be smooth headed or tufted — having a stylish clump of upright feathers at the top of the head. The Roman is similar in size to the Chinese, making it equally ideal as a meat bird for smaller families, although the Roman's short neck and back make it somewhat more compact than the Chinese. This breed is known for being docile and friendly.

Sebastopol. Arising from the Black Sea area of southeastern Europe, the Sebastopol's claim to fame is its long, flexible feathers that curl and drape, giving the goose a rumpled look. The looseness of the feathers makes this goose less able to shed rain in wet weather or stay warm in cold weather. Varieties come in white, gray, or buff plumage, which requires bathing water to maintain its good condition. Lacking proper wing feathers, the Sebastopol cannot fly well.

Shetland. Coming from Scotland, Shetland geese are exceptional foragers that, given ample access to quality greens, can basically feed themselves. Like Pilgrims, they are autosexing — the gander is mostly white, while the goose is a gray saddleback (white with a gray head, back, and flanks). The Shetland is the smallest, lightest-weight domestic breed with powerful wings that result in a dandy ability to fly. These tough little geese have a reputation for being feisty, but given time and patience, they can be gentle and friendly.

Toulouse. Originating in France, the Toulouse comes in two distinct types. The production Toulouse is the common gray barnyard goose; the Dewlap, or Giant, Toulouse gains weight more rapidly, puts on more fat, and matures to a much more massive size, especially when bred for exhibition. The dewlap consists of a fold of skin hanging beneath the bill, growing more pendulous as the goose grows older. In contrast to the more active production Toulouse, the Dewlap Toulouse is less inclined to stray far from the feed trough and puts on more fat, which when rendered lends a wonderful flavor to baked goods.

Making the Purchase

If you can't decide whether to raise ducks or geese, why not get both? Ducks and geese are compatible and can be housed together in the same facilities, as long as sufficient space is provided. Like most birds, ducks and geese are social animals that don't like to be alone, so you will need at least two. If you want to raise young in the future, they must be a pair. If you plan to keep your waterfowl into maturity but don't expect to breed them, you can get by with all females. They will lay eggs, and may try to nest, even if no male is around to fertilize the eggs. Among ducks, the male is quieter than the female (except among Muscovies, which are both quiet); among geese, the male tends to be louder than the female.

If you are looking for a breeding pair, getting grown birds has the advantage that you can see what you're getting, but they may be expensive and will not necessarily recognize you as a friend. If you cannot find the breed you want locally, shipping mature birds is both expensive and a hassle. Ducklings and goslings are much cheaper. If you want breeders to produce future ducklings and goslings, getting a batch of straight-run hatchlings lets them grow up in familiar surroundings and learn to recognize you as their keeper. As they grow, you can pick out the best for future breeding and put the rest in the freezer. When you grow meat from hatchlings, you will know exactly what they have been fed throughout their lives.

Hatchlings are available from more sources than are adult waterfowl. One place to look is the local farm store. Either the farm store or your county

Purchasing Options

Type of purchase	Pros	Cons
Hatchlings	Inexpensive; will learn to recognize keeper; available from many sources	Still a risk that you don't quite know what you're getting
Adults	Know health and sex of birds	May be expensive and aggressive toward you

Extension office can tell you if any hatcheries or waterfowl farms are in your vicinity. The larger operations might advertise in the Yellow Pages of the phone book, newspaper classified ads, or freebie newspapers. The farm store and Extension office can also tell you of any local waterfowl or poultry clubs in your area. Attending a poultry show at the county fair is another way to connect with sellers, as well as with like-minded folks who are willing to share their experiences and expertise.

If you have difficulty finding the breed you want locally, seek out a mail-order source, many of which advertise in farm and country magazines and on the Internet. If you desire one of the less common breeds, you might find a seller through the American Livestock Breeds Conservancy, the Society for the Preservation of Poultry Antiquities, or Rare Breeds Canada. Hatchlings shipped on their first day of life usually survive the trip, although you'll be required to purchase a minimum number so they can keep each other warm. When the shipment arrives, open the box immediately and inspect the ducklings or goslings in front of the postal carrier, so you can make a claim in case any have died in transit.

When purchasing adult waterfowl, seek birds that are still young. Even though ducks and geese can live for many years, their fertility and laying ability decline with age. Ducks are at their prime age for breeding from 1 to 3 years of age, and geese from 2 to 5 years of age. Indications that a bird is still young are a pliable upper bill, a flexible breastbone, and a flexible windpipe. Another sign is blunt tail feathers, where the once-attached down has broken off. After a bird molts into adult plumage, it acquires the rounded tail feathers of an adult, and once it reaches full maturity, you'll have a hard time determining its age and will have to take the seller's word.

Before taking your birds home, make a final check for problems, such as crooked backs or breastbones. In geese, also check for nonsymmetrical pouches between the legs. Look for signs of good health: alertness, clear eyes, proper degree of plumpness, and normal activity level.

Ducklings & Goslings

Like other barnyard poultry, waterfowl hatchlings are precocial, meaning that soon after they hatch they are out and about, exploring their surroundings. Their downy coats offer some protection from the elements, but if they have no mother duck or goose to shelter them from cold and rain (not to mention fending off predators), they must be housed in a brooder until they are old enough to manage on their own.

The Brooder

A brooder is a protected place that provides a growing bird with safety, warmth, food, and water. A home brooder can be made from a large, sturdy cardboard box. Line the box with newspapers then a layer of paper towels for the first week, to provide sure footing. A sheet of small-mesh wire fastened across the top of the box keeps out rodents, household cats, and other predators. An empty feed sack or a few sheets of newspaper covering the mesh guards against drafts.

The typical source of brooder warmth is a lightbulb or infrared heat lamp screwed into a metal reflector, but a hot glass bulb can shatter if splashed with cold water by playful hatchlings. A safer alternative, and a good investment if you plan to raise young waterfowl in the future, is a sealed infrared pet heater, such as the panels made by Infratherm. Hang the heater by a chain that allows you to raise the panel as your waterfowl grow. Be prepared to provide heat until the birds feather out. The rules of thumb are as follows.

Start the brooder temperature at 90°F (32°C). Decrease it five degrees per week until you get down to 70°F (21°C). If the days are warm and nights are cool, you may need to provide heat only at night after the first couple of weeks. Once the hatchlings have feathered out, at about 6 to 8 weeks, they won't be bothered by a temperature as low as 50°F (10°C).

Spraddle Leg

A hatchling does not have strong legs. If it is forced to walk on a slick surface in the incubator, brooder, or shipping box, its feet can slide out from under and splay in two directions. To keep the feet together, bind the legs loosely with a piece of soft yarn tied in a figure eight or a rubber band knotted at the center. Provide a rougher surface, such as paper towels or hardware cloth, to ensure firm footing. Watch closely to make sure the hobble does not get too tight, and remove it after a couple of days as the legs toughen up enough to hold the bird.

Duckling with spraddle leg

Leg binding

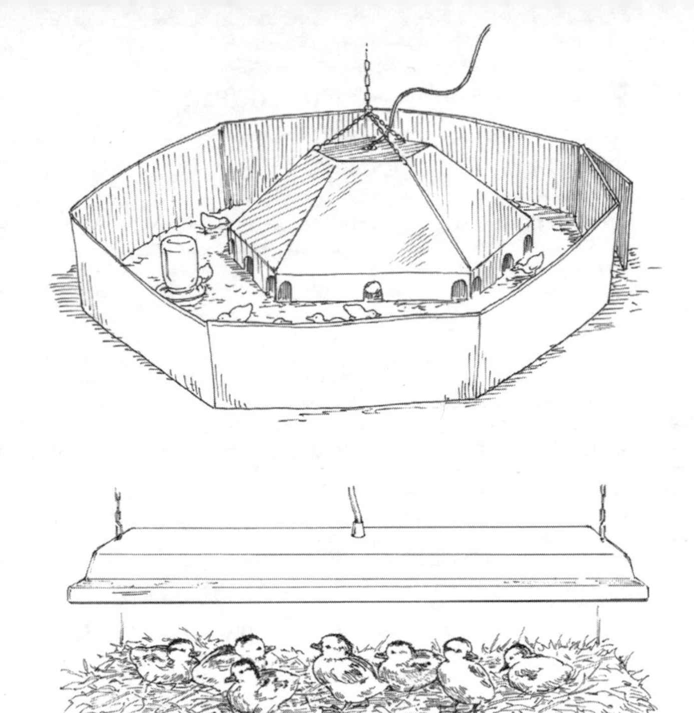

When hung from a chain, a hover brooder — shown here with draft guard — can be raised as ducklings grow in size.

A sealed infrared panel pet heater won't shatter like a lightbulb or infrared heat lamp might if splashed with water by playful baby waterfowl.

To gauge brooder temperature, you don't need a thermometer. Just watch how your ducklings or goslings react to the source of heat. If they remain normally active, they are comfortable. If they huddle under the heater and peep loudly, they are cold. If they spread as far as possible away from the heat and pant, they are hot. (See the illustrations on page 21 for details.)

As they grow, the hatchlings will naturally generate more body heat and rely less on artificial heat. The more rapidly you reduce the heat you provide (by raising the heater panel or, if using bulbs, reducing the wattage), the more rapidly your waterfowl will feather out and become less dependent on the electric heat you are using.

For the first day or two the hatchlings will spend a lot of time resting and require little more than about ¼ square foot (8 sq cm) per duckling and twice that per gosling. As soon as they become active they will need three times the space up to 2 weeks of age, when their space needs will double. Their needed space will double again by the third or fourth week and double again from the fifth or sixth week, or until the weather is warm enough for them to spend time outdoors. The sooner they can get outside, the easier it will be to keep their brooder space clean and dry.

Good sanitation throughout the brooding period means dealing with all the moisture and fluid droppings waterfowl generate. For the first few days while the hatchlings are on paper towels, freshen the box by adding a new layer of towels, then periodically roll them all up and start over. As the young waterfowl grow, they will generate increasingly greater quantities of wet mess that can't be handled with paper. At that point, you must move them to an area covered with at least 3 inches (7.5 cm) of absorbent bedding, such as shredded paper or wood shavings, and replace any bedding that becomes soaked.

Water

Baby ducks and geese love to play in water. This attraction leads to two serious problems: They can quickly make a mess of even the best-kept brooder, and they can just as quickly make an unsanitary mess of their drinking water. An open source of drinking water, such as a pan or bowl, is therefore unsuitable for brooding waterfowl. For the first two weeks of their life, a satisfactory container is a circular chick waterer that screws onto a water-filled jar.

No Swimming Allowed

Raised naturally, ducklings and goslings will happily swim with their parents. The older birds will control how long the babies stay in the water and when they must come out. Left on their own, however, ducklings and goslings may play in the water long after they should come out, becoming soaked, chilled, and possibly ill (or worse). Artificially brooded waterfowl should not have unsupervised access to water deep enough to swim in.

This homemade waterer is made with a wire cylinder in an open pan.

As the birds grow and require enough water in which to submerge their heads, a deeper trough becomes more suitable. The trough must not be easy for the birds to tip by stepping onto the rim, should be placed over a drain that channels overflow and spills away from the brooding area, and should be covered with a wire grate to prevent swimming. If your ducklings and goslings can swim or walk in the water, they will leave droppings in the drinking water that can lead to disease. The grate or wire mesh must be big enough to allow them to get their heads through for a drink.

If, in a pinch, you need to use an open container, you can float a piece of clean untreated lumber on the surface. Cut the wood to fit the container's shape but make it slightly smaller, so your birds can drink from around the edges. Or place a short cylinder of fine-mesh wire in the center of the container, leaving space between the cylinder and the rim to allow a good drink.

Feeding

Waterfowl hatchlings should have free-choice access to feed at all times. Feed them just enough twice a day to last until the next feeding. Since their appetites will grow at an alarming rate, you will need to constantly adjust the amount supplied at each feeding. Allowing waterfowl to go hungry will result in a frantic frenzy when they finally get something to eat, turning your cute downy ducklings and goslings into bug-eyed mini-monsters encrusted with caked-on feed.

Clean out the feeders frequently, taking care not to leave old, stale feed in the corners. Initially you can use feeders with little round holes designed for baby chicks. Watch carefully as your ducklings and goslings grow. When their heads get close to being too big to fit through the holes, switch to troughs with wire guards that prevent the birds from walking in the feed.

Ducklings and goslings need water to wash down dry feed. They fill their mouths with feed, waddle to the water for a drink to wash it down, then waddle back for another mouthful. Before long, a trail of dribbled feed marks the path between the feeder and the drinker. If the drinker is close to the feeder, the feed will soon be a wet mess and the water will turn to sludge.

To avoid these problems, you can moisten the feed with water or skim milk, but you will have to offer less feed at a time and therefore provide feed more often. Left too long, moistened mash clumps together, discouraging your waterfowl from eating. Furthermore, in the warmth of the brooder, it may turn sour or moldy, causing illness if it is eaten.

Finding a commercial starter ration formulated for baby waterfowl can be a challenge. A good substitute for the first few days is mashed hard-cooked egg. Chick starter crumble can then be fed, provided it is not medicated and you fortify it with livestock grade brewer's yeast (3 pounds [1.5 kg] brewer's yeast per 25 pound [11.5 kg] bag starter) to prevent niacin deficiency. Grower ration is generally high in protein to promote fast growth in meat birds. If that is not your goal, reduce the protein level by supplementing the ration with high-fiber feed, such as chopped lettuce or succulent grass. The natural diet of goslings consists almost entirely of freshly sprouting grass. If you have access to a grassy area that has not been sprayed with chemicals, on sunny days you can move your ducklings and goslings outside to graze in a small portable enclosure that provides protection from predators and cold wind. Alternatively, chop lettuce and other tender greens, or sprout alfalfa or similar seed, for your babies. Small amounts, floated in the drinking water, will quickly disappear.

When warm weather allows ducklings and goslings to spend most of their time grazing, you can cut their

Excess protein in the diet and insufficient exercise may result in twisted wing.

commercial feed back to the amount they can clean up in 15 minutes, twice a day. When they reach the age of 1 month, you can feed only once a day, preferably at night, so they'll be hungry in the morning and take greater advantage of the natural forage.

Letting young waterfowl graze helps prevent twisted wing, a condition in which the flight feathers of one or both wings angle away from the body like an airplane's wings. This problem, also known as slipped wing, occurs because the flight feathers grow faster than the underlying wing structure. The heavy feathers pull the wing down, causing it to twist outward. When the bird matures, one or both of its wings remain awkwardly bent outward instead of gracefully folding against its body. Meat birds don't live long enough for this clumsy appearance to present a problem, but it is unsightly in mature waterfowl. You can easily prevent twisted wing by avoiding excess protein.

Sexing Ducklings and Goslings

Although baby ducks all sound the same, as they grow, the drakes lose their voice. Soon the drakes of Mallard-derived breeds develop conspicuous drake feathers that curl up and forward at the end of the tail. Among Muscovies, the drakes are about twice the size of the hens.

If you must separate the hens from the drakes while they are young, the only way to do it is by vent sexing, as described on page 102. Among geese that cannot be sexed by a difference in color between the males and the females (autosexing), vent sexing is the only accurate way to distinguish sex, even once they're mature. Vent sexing young waterfowl takes a considerable amount of practice to be accurate without injuring a delicate bird. If you don't have experience in vent sexing, find a professional to help you learn.

Probiotics

Probiotics are beneficial bacteria that are naturally present in the digestive tract of animals. These beneficial bacteria increase in numbers as the animal matures and, when the body is in proper balance, fight off disease-causing bacteria. By adding a probiotic formula designed for poultry to the drinking water of ducklings and goslings during their first week of life, you can give them a good start.

Growing Up

Ducklings and goslings are incredibly hardy and, when properly cared for, are not particularly susceptible to disease. Their health is easily maintained by providing a clean, dry environment, adequate heat and ventilation, proper nutrition, fresh feed, and clean water. Then stand back and watch them grow.

Ducklings must be brooded until they are about 4 weeks old and goslings until about 6 weeks. If the weather is mild by then, they can be moved outdoors, but they still need shelter from wind, rain, and hot sun. In stormy weather, gather them up and bring them inside until they are fully feathered, which will be around 12 weeks for Mallard-derived ducklings and 16 weeks for Muscovies and goslings.

When you first turn them outdoors, keep an eye on them. Ducklings and goslings have an uncanny ability to find escape routes, get stuck in odd nooks, or get separated from their feed and water. If their yard is large or they are being turned out with mature ducks and geese, confine them to a small part of the yard for the first few days, enlarging their space only after they have had time to get their bearings.

28 Great Backyard Breeds

Many livestock breeds are suitable for the backyard. Which breed is best for you will depend in part on your climate, your space constraints, and your economic resources. For example: If you live in the north where winters are harsh, you might consider the rugged Highland cow, which was developed in the tough terrain of the Scottish highlands. If you live in the south where summers are particularly sultry, you might choose a more heat-tolerant breed, such as the Miniature Jersey.

In the following pages, we showcase a number of wonderful breeds for the small-scale backyard raiser. Consider this your starting point for exploring all of the unique breeds available to you.

SEE ALL 28 BREEDS ⟶

Wyandotte chicken

Chickens

Chickens come in a wide variety of breeds, colors, and feather patterns. We've chosen a couple of hardy dual-purpose breeds that will keep you in eggs for a few years and ably fill your stew pot when production declines, as well as a laying breed that produces unusually colored eggs.

Araucana A South American breed that lays blue eggs, the Araucana is unique. When crossed with any other breed, its offspring, also known as "Easter Eggs" chicks, will produce greenish blue eggs, and occasionally green, pink, or yellow ones.

Plymouth Rock A dual-purpose breed, the Plymouth Rock is kept for both meat and eggs. It is a very hardy breed, it has a docile disposition, and hens make good mothers. Birds can be raised in confinement or free-range.

Wyandotte The Wyandotte is a good laying and meat bird. Because of its relatively short back, a dressed Wyandotte is nearly proportional to a Cornish-Rock bird found at the store. It is very hardy, and it is generally docile.

Bourbon Red turkey

Turkeys

Three things to consider when choosing a turkey variety are size (it has to fit in the oven), color (dark pinfeathers on a roasted bird can be off-putting), and whether the bird is a heritage or industrial variety. We've highlighted three heritage breeds that are a good size for a family.

Bourbon Red One of the older breeds developed in the United States, the Bourbon Red has some desirable traits: it is a good forager (nice for pest control), and it has a relatively heavy breast, light-colored pinfeathers for a clean body when picked, and richly flavored meat.

Bronze Though ancestor to the Broad Breasted Bronze (a large bird that was once the main commercial turkey), this turkey is smaller and able to reproduce naturally. It has bronze feathers, which, when seen in the light, have high iridescence in shades of copper and blue, and dark pinfeathers.

Royal Palm Though developed as an ornamental bird, the Royal Palm is a good size for a small family's dinner table. When displayed, the banded markings on its tail feathers are breathtaking. It is a good flyer and forager, so prepare strong fences.

Muscovy duck

Ducks

Most duck breeds can be kept for both meat and eggs, though some, such as the Muscovy, are best for meat, and others, such as the Runner, are best for eggs. Most good layers produce white-shell eggs and breeds with white or light-colored feathers pluck cleaner.

Muscovy A very large bird, the domestic Muscovy is raised for meat that is known for being less greasy than the meat of some Mallard-derived breeds. As a voracious omnivore and active forager, it is great for pest control. It is a quiet and personable duck.

Pekin Developed in China, the Pekin is a fast-growing bird with light feathers. It is an excellent meat bird. A wonderful pet, it is gregarious, talkative, and a good layer (many lay upwards of 150 eggs per year).

Runner Developed in Scotland from stock originating in the East Indies, Runners prefer to run rather than waddle. In Indonesia, they glean snails and waste grain from rice paddies. These energetic and prolific layers make great pets. They come in more color varieties than any other breed.

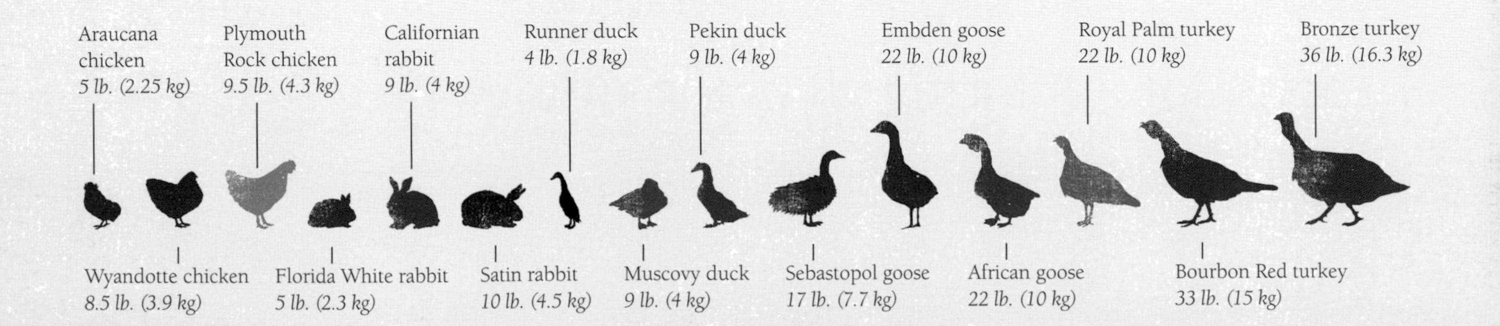

Dexter cow
African goose
Spotted pig
Highland cow
Bourbon Red turkey
Cheviot sheep
Tamworth pig
Californian rabbit

Highland cow

Dairy Cows & Beef Cattle

We've chosen to highlight a dairy breed, a dual-purpose breed, and a meat breed. Miniature Jerseys and Dexter cattle are smaller than standard-size cattle and good for small homesteads, while the medium-size shaggy Highland will yield a fair amount of tasty meat.

Dexter The Dexter is the smallest of the cattle breeds; the average cow weighs less than 750 lb., the average bull less than 1,000 lb. Used for both milk and meat, it is a good milker and produces extremely tasty cuts of succulent beef. The Dexter is gentle, fertile, and a great mother.

Highland Originating in the remote Highlands of Scotland, the Highland is a very tough breed that copes well with exceedingly harsh weather. It yields lean yet tasty meat and has sturdy feet and legs, which help account for its longevity. Its long horns and shaggy coat make it an eye-catcher.

Miniature Jersey The Jersey is one of the oldest dairy breeds. Original Jerseys were of miniature size, but over time Americans bred the cows to become larger, creating the standard-size Jersey. Gentle, people-oriented cows, Miniature Jerseys produce 2 to 4 gallons of sweet milk that contains 5 to 6 percent butterfat.

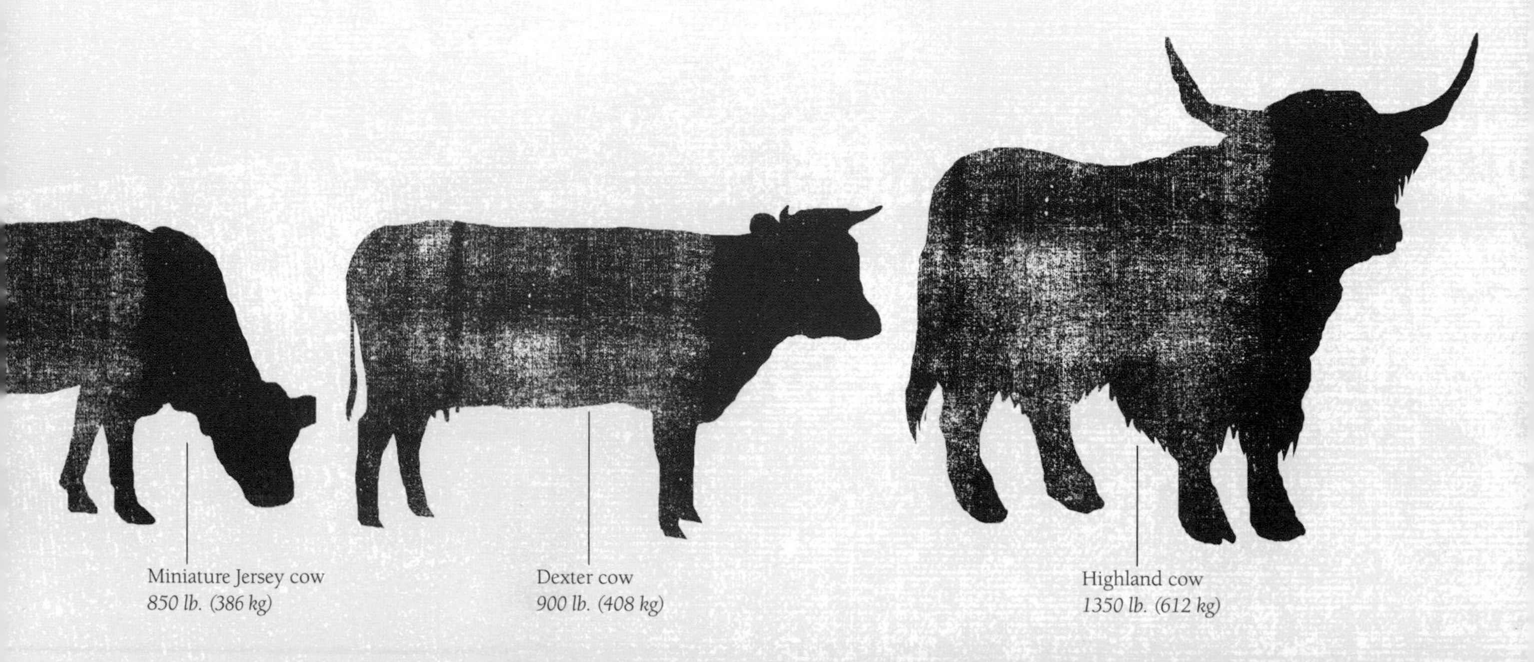
Miniature Jersey cow *850 lb. (386 kg)*

Dexter cow *900 lb. (408 kg)*

Highland cow *1350 lb. (612 kg)*

Wyandotte chicken
Honey Bee
Runner duck
Nigerian Dwarf goat
Muscovy duck
Berkshire pig
Sebastopol goose
Plymouth Rock chicken
Embden goose
Nubian goat

Nigerian Dwarf goat

Goats

All of these dairy goat breeds are very friendly and will produce great-tasting milk. If your milk needs are minimal and you don't want to make cheese or yogurt, the Nigerian Dwarf is the best choice of the three.

LaMancha Developed in California, the medium- to large-size LaMancha is the calmest and friendliest of the dairy breeds and comes in any color. LaManchas are easy to recognize because they have only small ears (elf ears) or no visible ears at all (gopher ears). This unique feature doesn't hurt their hearing one bit.

Nigerian Dwarf A miniature dairy goat developed in West Africa, the Nigeria Dwarf produces less milk and meat than a full-sized breed but is ideal for a small family, if the doe comes from milking, not pet, stock. The breed is known for a docile temperament, good hardiness, and longevity. It breeds throughout the year.

Nubian A large, friendly goat, the Nubian is the most energetic, active, and talkative of the dairy breeds. It will breed throughout the year, and its extremely rich milk makes it popular among cheese and ice cream makers.

Cheviot sheep

Sheep

All of these breeds will produce tasty meat, give you some nice wool, and even do an admirable job of mowing your lawn. The Cheviot is the smallest of the three, the Dorset the largest.

Cheviot A good general-purpose sheep, the medium-size Cheviot will mow your lawn, give you strong resilient wool, and fill your larder with lots of meat. A very hardy breed, it can handle harsh winter weather. It is an active sheep that does well grazing.

Dorset One of the earliest breeds of British sheep, the medium-size Dorset is one of the best choices for a first sheep for both wool and meat. Its lightweight fleece is excellent for handspinning, and it has a large, muscular body and gains weight fast. The Dorset is one of the few breeds that can lamb in late summer or fall (as compared to spring and early summer).

Scottish Blackface A small to medium-size dual-purpose breed for both meat and fiber, the Scottish Blackface's meat is lean and flavorful and its wool is strong and durable without having excessive crimp. It is well suited to cool, damp conditions and produces admirably on sparse forage.

Berkshire pig

Pigs

The Berkshire and the Spotted are heritage breeds, meaning they have a long association with North American farms. The Tamworth is an endangered breed. All yield excellent-tasting meat.

Berkshire An exceptional meat hog, the Berkshire is a well-muscled breed known for giving a high yield of lean, superb-flavored meat. Pork from pure- and high-percentage-bred Berkshires has finishing qualities akin to marbling in high-grade beef. The breed performs very well outdoors, especially when grazing on pasture.

Spotted An excellent hog for small-scale production on pasture or in hoop houses, the Spotted efficiently produces tasty meat. The breed also has good mothering abilities.

Tamworth The hardiest and strongest forager of the swine breeds, the Tamworth produces pork that is lean yet moist and well flavored. Many consider it to have the best-quality bacon of any breed. Quite social and intelligent, it requires stout pens and fences, as it is harder to contain than most pigs.

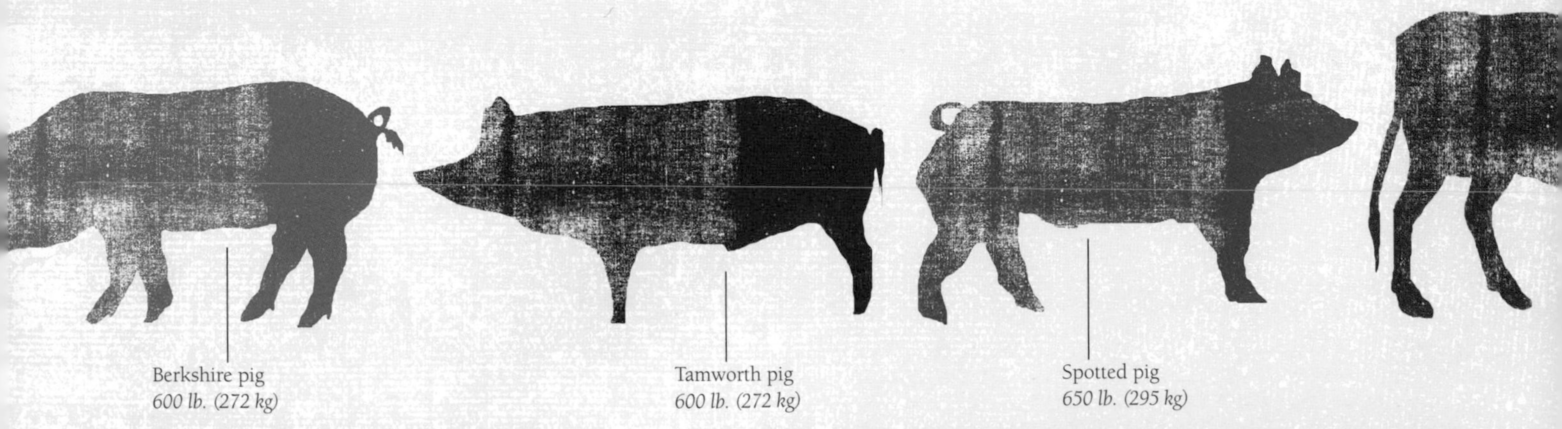

Berkshire pig
600 lb. (272 kg)

Tamworth pig
600 lb. (272 kg)

Spotted pig
650 lb. (295 kg)

Sebastopol goose

Geese

Most geese breeds are raised solely for meat. These breeds are good foragers or fast growers, or both, making them good economical choices. If your goose will be employed to weed your garden, you may want to avoid a heavy breed that will compact the soil.

African An extremely calm breed, the African is considered the gentlest of domestic geese and easy to confine. It is very talkative, though not extremely loud, and it produces a good carcass with tasty, lean meat. The knob on the top of its head is easily frostbitten, so it must be sheltered in cold weather.

Embden Due to its fast growth, large size, and white feathers, the Embden is the most popular goose to raise for meat. An old breed that originated in Germany, it is the tallest of the geese breeds and lays the largest eggs (average weight 6 ounces).

Sebastopol The Sebastopol is known for its long (up to four times longer than other breeds), curly feathers. The looseness of its feathers means that it has trouble staying warm. Lacking proper wing feathers, it cannot fly well.

Florida White rabbit

Rabbits

These rabbit breeds convert feed to meat at a very efficient rate, making them good, economical meat breeds. They also yield an ideal 4-pound fryer and dress out with little waste.

Californian The second most popular meat rabbit (after the New Zealand), the medium-size Californian is a plump breed with full hips and a meaty saddle, ribs, and shoulders.

Florida White A compact but meaty breed, the Florida White is a good choice for a small family. It dresses out better than any other breed. At twelve weeks you can get as much edible meat from this breed as you can from a breed that's twice as big at eight weeks.

Satin An excellent meat rabbit, the medium-size Satin is also known for its long silky fur that comes in just about any color you'd like, making it a dual-purpose breed for fiber and meat. The does are great mothers and the kits grow well with one of the best meat-to-bone ratios.

Honey Bee

Bees

Two popular subspecies of the western honey bee are Carniolan bees and Italian bees. Both are gentle and productive, but their differences may make you choose one over another.

Honey Bee Italians are adaptable to most climates, from cool to subtropical, but don't cope well with hard winters and cool, wet springs. Carniolans fly at cooler temperatures than Italians, making them ideal for northern climes and areas where the early spring nectar flow is strong. Carniolans require more careful management than Italians.

Miniature Jersey cow
Satin rabbit
Araucana chicken
Pekin duck
Bronze turkey
Scottish Blackface sheep
Dorset sheep
Florida White rabbit
Royal Palm turkey
LaMancha goat

Ducks & Geese for Meat

Some breeds of waterfowl have been developed especially for efficient meat production, but any breed is good to eat. If you purchase a duck or goose at the grocery store or butcher shop, the duck would most likely be a Pekin and the goose an Embden. These white-feathered breeds appear cleaner when plucked, because their white pinfeathers don't show as much as the pinfeathers of darker waterfowl.

Feed Conversion

A typical feed conversion rate for waterfowl raised for meat is 3 pounds (1.5 kg), meaning each bird gains 1 pound (0.5 kg) of weight for every 3 pounds (1.5 kg) of feed it eats up to about 8 weeks old. The older a bird gets, the higher the feed conversion rate, creating a trade-off between economical meat and more of it. To improve the conversion rate, commercially raised waterfowl are encouraged to eat continuously by being kept under lights 24 hours, having their feed troughs topped off every few hours, and being confined to a limited area to discourage them from burning off calories. In addition to growing fast, these ducks put on a lot of fat.

Pasture-raised waterfowl take longer to grow to similar weight, and more of that weight is meat instead of fat. During that time, the feed conversion rate for ducks will go up. On the other hand, geese that have plenty of forage for grazing will have a reduced need for commercial rations. Once goslings have been growing well on starter for about a month, they will continue to grow on good grazing for about four months (although they'll grow faster if given supplemental grain), and then finished on grain or finisher ration for the last month before slaughter.

Can you save money by raising your own ducks and geese? That depends on the price of feed, how much the cost can be reduced by providing good forage, and the price you're paying for meat now. Using averages (and for simplicity's sake, ignoring the cost of housing, water, electricity, and so forth), do the math: By the time a Pekin reaches the live weight of 7 pounds (3 kg) it will have eaten about 20 pounds (9 kg) of feed. At 75 percent of live weight, a 7-pound (3 kg) duckling dresses out to approximately 5¼ pounds (2.25 kg). Your break-even cost can be calculated by comparing the purchase price of a dressed Pekin of similar weight with the cost of starter/grower ration. At current prices, a dressed duckling sells for anywhere from $15 to $35. The current rate for one brand of all-natural starter/grower ration is about 25 cents per pound, or about $5 for 20 pounds, which is one-third the value of a $15 duck. Even after adding $4 for each duckling, that's a pretty good deal. But if you choose to pay 50 to 100 percent more to purchase certified organic feed, you come close to breaking even. Raising geese is a much better deal, provided they have access to high-quality forage, because they can live almost entirely by grazing for the better part of the growing period.

Feathering Means Butchering Time

Weight gain is only one important factor in determining when a duck or goose is ready for butchering. The other is the stage of the molt. All those feathers that allow ducks and geese to swim comfortably in cold water take a long time to remove, and plucking is considerably more difficult if some of those feathers are only partially grown. Plucking is less time-consuming, and the result is more appealing, if a duck or goose is in full feather.

Soon after a duck or goose acquires its first full set of feathers, it begins molting into adult plumage and won't be back in full feather for another two months or so. You'll know the optimal time to pluck a duckling or gosling has passed when the feathers around its neck start falling out. From then on, the bird won't grow as rapidly as it has been and for ducks the feed conversion rate goes up. After this point, feeding ducks for meat purposes becomes more costly, the meat becomes tougher, and the meat of Muscovy males takes on an unpleasant musky flavor. Geese, on the other hand, are traditionally roasted during the holidays, so are generally butchered in the fall after their second molt, when they are about 6 months old.

Most ducks reach full feather when they are between 7 and 10 weeks old. Muscovies reach full feather at about 16 to 20 weeks. A duck or goose is in full feather and ready for slaughter when all of these signs are present:

- Its flight feathers have grown to their full length and reach the tail
- Its plumage is bright and hard looking, and feels smooth when stroked
- You see no pinfeathers when you ruffle the feathers against the grain
- It has no downy patches along the breastbone or around the vent

Dressed Weight

A duck or goose loses 25 to 30 percent of its live weight after the feathers, feet, head, and entrails have been removed. Heavier breeds lose a smaller percentage than lighter breeds. The breast makes up about 20 percent of the total meat weight; skin and fat make up about 30 percent.

When all of these signs are right, a duck or goose is ready to be butchered. If you are doing your own butchering for the first time, have an experienced person guide you, or refer to a comprehensive book such as *Storey's Guide to Raising Ducks* by Dave Holderread (butchering geese is similar to ducks, the main difference being they are much larger). If you don't wish to do your own butchering, you might find a custom slaughterer willing to handle ducks and geese. Alternatively, a fellow backyard waterfowl keeper or a hunter might be willing to kill and pluck your ducks or geese for a small fee. If not having to kill your own waterfowl is important to you, determine before you start whether or not someone in your area will do it for you.

Storing the Meat

Freshly butchered duck or goose must be aged in the refrigerator for 12 to 24 hours before being cooked or frozen; otherwise, it will be tough. If you will not be serving the duck or goose within the next three days, freeze it after the aging period until you're ready to cook it. To avoid freezer burn, use freezer storage bags. Most ducks will fit in the 1-gallon (3.5 L) size. A Muscovy female should fit in the 2-gallon size. Geese and Muscovy males should fit in the 5-gallon (19 L) size. Remove as much air from the bag as you can by pressing it out with your hands or by using a homesteaders' vacuum device designed for that purpose. Properly sealed and stored at a temperature of 0°F (–18°C) or below, duck and goose meat can be kept frozen for six months with no loss of quality. To thaw a frozen goose or duck before roasting, keep it in the refrigerator for two hours per pound (0.5 kg).

A whole duck or goose takes up a lot of freezer space. Unless you intend to stuff and roast the bird, you can save space by halving or quartering it, or filleting the breasts and cutting up the rest. Muscovy breast makes an exceptionally fine cut that is the most like red meat of any waterfowl. After removing the fillets, you might package the hindquarters for roasting or barbecuing and boil the rest for soup. A great use for excess Muscovy and goose meat is making sausage, using the same recipes that are intended for pork. Small amounts can easily be made into sausage patties, whereas larger amounts can be stuffed into links.

Roasted to Perfection

Contrary to common belief, a properly prepared, homegrown duck or goose should not be greasy. Although ducks and geese have a lot of fat, the meat itself is pretty lean. All the fat is either just under the skin or near one of the two openings, where it can easily be pulled away by hand before the bird is roasted.

Proper roasting begins by putting the meat on a rack to keep it out of the pan drippings while the bird cooks. Remove any fat from the cavity and neck openings, and stuff the bird or rub salt inside. Rub the skin with a fresh lemon, then sprinkle it with salt. Pierce the skin all over with a meat fork, knife tip, or skewer, taking care not to pierce into the meat. Your goal in piercing is to give the fat a way out through the skin as it melts during roasting. This melted fat will baste the bird as it drips off, so no other basting is required. Do not cover the duck or goose with foil during roasting as you would a turkey.

Slow roasting keeps the meat moist. Roast a whole duck at 250°F (120°C) for 3 hours with the breast side down, then for another 45 minutes with the breast up. Roast a goose at 325°F (160°C) for 1½ hours with the breast side down, plus 1½ hours breast up, then increase the temperature to 400°F (205°C) for another 15 minutes to crisp the skin.

Since not all birds are the same size and not all ovens work the same way, keep an eye on things the first time around, to avoid overcooking your meat, which makes it tough and stringy. Once you settle on the correct time range for birds of the size you raise, you can roast by the clock in the future.

When the meat is done, it should be just cooked through and still juicy. You can tell it's done when the leg joints move freely, a knife stabbed into a joint releases juices that flow pink but not bloody, and the meat itself is just barely pinkish. To keep the meat nice and moist, before you carve the bird, let it stand at room temperature for 10 to 15 minutes to lock in the juices. During this time, residual heat will cook away any remaining pink. Duck or goose does not have light and dark meat like a chicken or turkey; rather, it is all succulent dark meat.

Cooking Methods

The most suitable method of cooking a duck or goose depends on its tenderness, which in turn depends on its age. Fast dry-heat methods, such as roasting, broiling, frying, and barbecuing, are suitable for young, tender birds; slow moist-heat cooking methods, such as pressure cooking or making a fricassee, soup, or stew, are required for older, tougher birds.

Duck & Goose Eggs

The better layers among ducks, kept under optimal conditions, may begin producing eggs as early as 18 weeks of age and continue laying nearly year-round. Breeds, and strains within breeds, that have been developed for meat or ornamental purposes do not lay nearly as well as those that have been selectively bred for their laying ability. The best layers stop production only briefly during the fall molt.

Most geese hit their stride in their second year of life and peak at 5, although they may lay for up to 10 years. Muscovies may lay for six or more years. Mallard-derived ducks normally lay well for three years, and some strains continue to lay efficiently for up to five years. If your purpose in keeping a bevy is egg production, replace older ducks before they lose efficiency.

Most domestic ducks and geese start laying early in the spring of their first year. How well they lay and the length of their laying season depends on the following factors:

- Breed
- Strain (subgroup within the breed)
- Age
- Diet
- Overall health
- Weather
- Daylight hours

Waterfowl tend to lay in the early morning hours. Ducks generally lay one egg every day, geese one every other day. The less productive breeds will lay a consecutive batch of eggs, called a clutch, and then take up to two weeks off before starting another clutch. If you leave a clutch in the nest, a duck or goose may start setting and stop laying for the season.

Egg size depends in part on the breed and the strain, but mostly on the size of the bird. Larger birds lay larger eggs. Smaller, yolkless eggs commonly signal the beginning and end of a bird's laying season, but they may also be laid during hot weather, especially if drinking water is in short supply.

Geese lay white eggs. Ducks may lay white, greenish, or bluish eggs.

Feeding Layers

As the laying season approaches, ducks and geese require more nutrition. About three weeks before you expect them to start laying, gradually switch from maintenance ration to a layer ration that contains 16 to 20 percent protein. In warmer weather, fowl need more protein to keep eggs from getting smaller. If your farm store doesn't carry layer ration for ducks, layer ration formulated for chickens works well.

For sound, thick eggshells, provide calcium continuously throughout the laying season. Ducks and geese that have access to forage usually get plenty of calcium from the shells of bugs and snails they eat. Still, it's a good idea to provide them with a free-choice hopperful of crushed oyster shells, ground aragonite, or chipped limestone (not dolomitic limestone, which can be detrimental to egg production).

Nests

Providing nests for your ducks and geese to lay their eggs in helps keep the eggs clean and protects them from being cooked by sun, washed by rain, or frozen in cold weather. Eggs laid in nests are easier for you to find than eggs hidden in the grass, but the latter are more difficult for predators to find. As

Lighting Layers

If your waterfowl live naturally and forage for much of their food, you likely won't get the number of eggs quoted by sellers and listed in the Guide to Duck Breeds on page 73. To lay really well, mature hens must be fed carefully formulated rations and be housed under at least 14 hours of light year-round.

To keep your birds laying during the shorter days of late fall, winter, and early spring, you can use electric lights to augment natural daylight. Inside the shelter, use one 40-watt bulb (10-watt fluorescent) for every 150 square feet (46 sq m); outside the shelter, use one 100-watt bulb (25-watt fluorescent) for every 400 square feet (122 sq m). If your duck yard doesn't have electricity, you can provide lighting with 12-volt bulbs designed for recreational vehicles, powered by a battery connected to a solar recharger.

Use an inexpensive reflector from the hardware store to direct the light appropriately, and hang the bulbs about 6 feet (2 m) off the ground to reduce shadows. Adjust your lighting times as needed to maintain 14 hours or more of light per day. Even a minor reduction in light hours can throw off the production of laying ducks. To avoid risks associated with too-early laying, do not begin your lighting program until your ducks reach 20 weeks of age.

a rule of thumb, furnish one nest per three to five females.

The nest can be in the shape of a box, an A-frame, or a barrel on its side and braced to prevent rolling. A doghouse makes a good goose nest; for ducks, a larger doghouse with doors at opposite ends might be partitioned in half to create two nests. A good nest size is 12 inches square (30 sq cm) for a duck, 18 inches (46 sq cm) for a Muscovy, and 24 inches (61 sq cm) for a goose; the center of the top should also be 12, 18, or 24 inches (30, 46, or 61 cm) high, respectively. The precise size is not critical, provided the nest meets the following standards:

- Tall enough for a layer to enter and sit comfortably
- Wide enough for her to turn around, since waterfowl don't like to back out
- Small enough to offer a feeling of protective seclusion
- Separated physically or visually from the next nearest nest
- Situated in an area that is well protected from predators
- Open at only one end, so the layer can keep an eye on who might be coming
- Large enough to accommodate an abundance of soft nesting material. A thick layer of nesting material, such as shavings, dry leaves, or straw, will keep the eggs clean, reduce breakage, and prevent eggs from rolling out of the nest.

As laying season approaches, ducks and geese will start investigating available nesting sites. If they see an egg already in a nest, they will consider it a good safe place to lay and will be inclined to deposit their own eggs there. To encourage use of the nests, rather than hiding eggs in the bushes or clumps of grass, place an artificial egg in each nest. Hobby shop eggs and toy eggs available around Easter work well and, for ducks, so do golf balls. Sometimes fake eggs aren't enough to entice a duck or goose to lay in the place of your choosing. If you keep track of the laying habits of your bevy or gaggle, you'll know when you are not getting your proper quota and must initiate an egg hunt.

Layer Problems

Ducks and geese encounter few problems associated with laying eggs. Still, for the health and safety of your layers, it's important to be vigilant.

Predators. A duck or goose on the nest is not as mobile as a bird that moves freely on the ground or in water. Freedom from predators in the laying yard is therefore essential. Predators may be attracted not only to the layer but also to her eggs in the nest. The predator may peck holes in the eggs and eat them right there (as crows do) or may consider them a takeout meal (as a skunk would).

Egg binding. When an egg becomes stuck and cannot be laid, the hen is egg bound. It may happen when a duck or goose tries to lay an egg that is too big. The bird might, for example, be laying its first egg and not be quite ready, as might happen when young hens are raised under artificial light. Ducks or geese that tend to lay double-yolked eggs may also become bound. Double-yolkers occur when one yolk catches up with another, and both become encased in the same shell. Thin-shelled eggs resulting from a dietary inadequacy, such as low calcium intake, may cause egg binding, as may obesity.

Suspect egg binding if a duck or goose is listless, stands awkwardly with ruffled feathers, and has a distended abdomen. If you press gently against the abdomen of an egg-bound bird, you will be able to feel the egg's hard shell.

Sometimes you can lubricate your finger with a water-based lubricant or petroleum jelly, then work your finger around the stuck egg to help it pass. Standing the bird in a tub of warm (not hot) water may relax her muscles enough to allow the egg to pass. Breaking the egg is a drastic option that carries the risk of injuring the duck or goose with sharp shards of shell.

Prolapse of the oviduct. Also known as eversion or blowout, prolapse is protrusion of the oviduct through the vent. It occurs when a duck or goose strains to pass an egg, damaging the tissue so it does not recede back inside after the egg is laid. It is caused by the same factors

A nesting site should offer adequate space, privacy, and protection.

Parts of an Egg

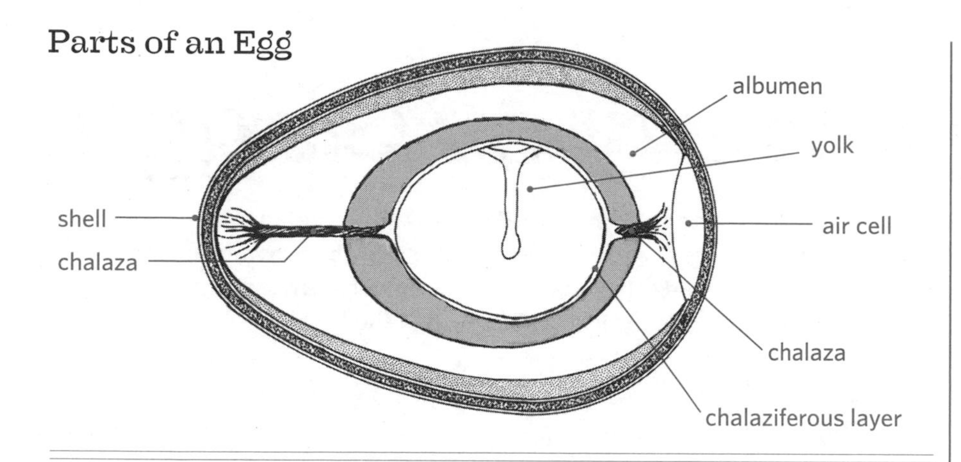

that cause egg binding or by weakened muscles due to prolonged egg production or excessive mating. Treatment may be successful if only a small portion of oviduct is everted and the treatment is begun right away. Gently wash the protruding part; coat it with a relaxant, such as a hemorrhoidal ointment; and push it back into place. Move the affected hen to a warm, dry, secluded pen and reduce the protein in her diet to discourage laying while she heals.

Eggs for Eating

Just before an egg is laid, it is coated with a moist film called the bloom. If you ever find an egg just after it's been laid, it will still be wet with bloom. The bloom dries quickly to form a natural protective coating.

Eggs left in the nest for a long time tend to get dirty through the layers' comings and goings. But washing eggs removes the bloom and creates a risk that contamination will penetrate the porous shell. Lightly soiled eggs can be brushed clean. A mildly soiled egg can be washed in water that is warmer than the egg is, but should be eaten soon because it will not keep as well as an unwashed egg. Extremely soiled eggs should be discarded.

To avoid wasting a lot of eggs through soiling, keep conditions tidy in your waterfowl yard. Gather eggs each morning and discard any with cracked shells or holes pecked in them.

Egg anatomy. If you break open an egg and examine its contents, you will find two stringy white spots on the sides of the yolk, one of which is usually more prominent than the other. These spots are called chalazae. As the egg was being formed, a layer of dense egg white (called the chalaziferous layer) was deposited around the yolk. At the two ends of the egg, the ends of the chalaziferous layer twisted together to form a cord that anchored the yolk to the shell. When you break open an egg, the chalazae break away from the shell and recoil against the yolk. Sometimes you may also see a red or brown spot, called a blood spot, on the yolk. A blood spot is caused by minor hemorrhaging in the oviduct. Although it is harmless, it is not appetizing. The tendency to lay eggs with blood spots is inherited. If you can identify the duck laying such eggs, do not hatch her eggs, as her offspring are likely to lay blood-spotted eggs.

Storing eggs. Refrigerated duck eggs can be kept as long as two months but are best used within two weeks, while they're still at their peak of freshness. Excess eggs to be used for baking can be stored in the freezer during the laying season for use during the off-season. Since whole eggs burst their shells when frozen, break the eggs into a bowl, add a teaspoon of honey or a half teaspoon of salt per cup (depending on your preference and the recipe you will use the eggs in), and scramble the eggs slightly, taking care not to whip in air bubbles. Pour the mixture into ice cube trays and freeze it solid, then remove the cubes from the trays and store them in well-sealed plastic bags. The frozen cubes take about 30 minutes to thaw and can be stored for up to one year at a temperature no higher than 0°F (–18°C).

Duck Eggs versus Chicken Eggs

Duck eggs are not much different in flavor from chicken eggs and are more popular in many countries. When fried or boiled, duck eggs have a firmer texture that might be described as slightly chewy. Duck eggs have proportionally larger yolks compared with chicken eggs and therefore lend extra richness when added to batter for baked goods. Some people have a sensitivity to undercooked chicken eggs that can cause cramping but have no such reaction to duck eggs; the opposite is also true.

Eggs for Hatching

Ducks are at their prime age for breeding from 1 to 3 years of age; geese take a little longer to mature and are at their prime from 2 to 5 years of age. Yearling geese lay eggs, but those eggs won't hatch as consistently during the first year as in subsequent years. A drake is fertile by the age of 6 months and peaks after the third season. A gander reaches his prime of fertility at 3 years of age and peaks at 6 years. A duck or goose will lay eggs regardless of whether or not a male is present, but the eggs will not be fertile and therefore will not hatch.

Housing a Bevy or Gaggle

As picturesque as it may be to have ducks and geese wandering around your estate, giving them their own confined area is a wise move. Otherwise, they will congregate on your doorstep, rat-a-tat for attention on your door, and leave their slippery calling cards on your porch. Geese will try to chase away your visitors and otherwise amuse themselves by generally destroying anything that catches their attention.

Space

The larger the area in which waterfowl can roam, the less trouble they are likely to get into and the cleaner they will remain. An area that is too small for its population will soon become muddy, mucky, smelly, and fly-ridden. The more area available for grass to grow, the better chance the grass has to survive against flat-footed waddling and the more widely duck and goose droppings will be spread, which will allow them to dissipate naturally instead of building up to create a problem.

So, although the minimum yard space requirement is about 10 square feet (3 sq m) per duck and twice that for geese, rather than considering minimums, provide as much roaming area as you can. Better yet, divide up the available yard space into paddocks so each patch of pasture can rest and rejuvenate while the birds are foraging elsewhere.

An ideal waterfowl yard has a slight slope and sandy soil for good drainage. The less ideal the area, the more space you should provide per bird. The yard should have both sunny and shady areas, letting the birds choose for themselves whether to rest in the warmth of the sun or away from its harsh rays. The yard should include a windbreak, which might consist of a dense hedge, the side of a building, or a portion of solid fence.

Minimum Living Space per Bird

(sq ft./sq cm)

	Age in weeks			
	1–3	**4–8**	**9–17**	**17+**
Duck	1.0/30	3.0/90	3.5/105	4.0/125
Goose	1.5/45	4.5/135	6.0/180	9.0/275

Fencing

Fencing is essential to keep waterfowl safe from predators and prevent them from roaming too far afield. The fence must be sound and well maintained, as waterfowl are much more adept at finding ways to get out than they are at finding their way back in. How closely spaced the fencing material needs to be depends on whether or not you wish to protect vegetation on the other side. Geese especially, but ducks, too, will graze through the fence as far as their long necks can reach.

A 3- or 4-foot (1–1.25 m) fence will confine nonflying breeds. A 5- to 6-foot (1.5–1.75 m) fence would be more appropriate for Muscovies and the lighter goose breeds. If you can keep geese grounded until they reach maturity, they will have a hard time getting enough elevation to clear a fence. If you'll be confining hatchlings, secure a 12-inch (30 cm) strip of tight-mesh wire along the bottom of the fence. Ducklings and goslings look a lot bigger in their downy coats than they really are, and they can slip through incredibly small spaces. The trouble is, after grazing on the wrong side of the fence, slipping back through with a full belly may not be quite as easy.

Perhaps the best fencing for waterfowl is narrow-mesh field fencing, stretched tight between posts spaced 8 feet apart (2.5 m), and augmented with electrified wire. If the electric wire is no more than 2 inches (5 cm) above the fence, any predator that attempts to climb over the field fence will get a shock when it reaches for the electric wire, causing it to drop off the fence.

Predator Control

As much as you like your ducks and geese, countless predators like them even more. Foxes, weasels, raccoons, and skunks have notorious appetites for waterfowl, their eggs, and their young. Dogs will chase and kill adult waterfowl for sport, and cats prey on downy hatchlings. A sturdy fence goes a long way toward keeping out most predators.

If you have a bunker mentality and the yard isn't prohibitively large, a concrete footer all along the fence line offers a permanent, though expensive, way to discourage burrowing animals. A less expensive option is to bury the bottom of the fence below ground. A distance of 4 inches (10 cm) underground is usually sufficient, unless you live in an outlying area where a predator might dig for hours undisturbed, in which case going down 18 inches (46 cm) might be more appropriate. Depending on your soil type, you might find it easier to cut and lift a 12-inch- (30 cm) wide strip of sod along the outside of the fence line, and clip or lash a 12-inch-wide length of wire mesh to the bottom of the fence, extending out horizontally along the ground away from the yard. Replace the sod on top, and the mesh will get matted into the grass roots to create a barrier that discourages digging. To prevent soil moisture from rusting the buried portion of a fence, use vinyl-coated mesh or brush plain metal mesh with tar or asphalt emulsion.

Keeping out flying predators is nearly impossible, unless your yard is small enough, or you are ambitious enough, to cover the entire yard with netting. A less expensive alternative, that may not be visually acceptable to any nearby neighbors, is to criss-cross the yard with single strands of wire placed high enough for you to walk underneath, and from the wires hang strips of old sheets, obsolete CDs or DVDs, or anything similar that flutters in a breeze and thereby discourages flying predators from landing. Still, you might find that crows or bluejays peck holes in your eggs and owls or hawks swoop down and make off with hatchlings. Aerial predation can be further minimized by providing a roofed shelter.

Shelter

Waterfowl aren't keen on staying indoors, but for their own protection, it's wise to provide them with a shelter and train them from a young age to go in each night. Once they get into the routine they are likely to go in by themselves and, if you are running late, call for you to shut the door. Since most predators roam during the night, a well-built shelter will keep your birds safe during the most critical hours. It will also provide a refuge from icy wind, scorching sun, and pelting rain.

The shelter need not be spacious, since the birds will be spending only their nights in it. About 3 square feet (1 sq m) of floor space per light breed duck, 5 square feet (1 sq m) for heavy ducks, and twice those amounts for light and heavy breed geese should be plenty. For only a pair or a trio, a large doghouse-style shelter with a low roof is inexpensive and easy to clean. For a larger bevy or gaggle that requires lots of floor space, make the roof high enough for you to easily enter for cleaning.

The best floor for the shelter is packed dirt. If your area has a serious predator problem, the expense of putting the shelter on a concrete foundation, if not an entire concrete floor, may be warranted. Small windows with hardware cloth screens will let in air and allow the birds to see out, but take care to avoid drafty conditions. A cold wind blowing through feathers would remove body heat trapped by the protective undercoat of down.

Shavings on the floor help control droppings and simplify shelter cleanout. Fresh shavings also keep eggs clean if they are laid inside the shelter. Do not provide water for your ducks and geese while they are inside. The combination of the boredom of confinement and the attraction of water to play in will result in a soggy, smelly mess. Since they have no water, they also should be given nothing to eat, but don't worry — they will get by just fine overnight and eagerly look forward to breakfast.

Because waterfowl don't sleep at night but take catnaps, much as they do throughout the day, they will tend to be restless while confined. Lights, including moonbeams, shining through a window of the shelter and creating shadows on the walls can stir them up and cause a lot of quacking or honking. Where lights shining in from outside

The best way to fence out predators is to stretch narrow-mesh field fencing between posts spaced 8 ft. (2.5 m) apart and add electrified offset wires along the bottom and top.

create shadow problems inside the shelter, hang some low-watt lightbulbs in the shelter, making sure to arrange them to avoid the birds themselves casting shadows as they move about. Protect the wiring in conduit, and hang the bulbs high enough that geese can't dismantle your electrical efforts and get electrocuted in the bargain.

Waterfowl don't like being indoors and should not be confined any longer than is necessary. Even in the severest climate, as long as they can find a place away from the cold wind, they prefer to be outdoors. You may think you are doing them a big favor by breaking open a bale of straw to rest on and protect their feet from cold, but don't be surprised if they hunker down in the snow right next to the straw. Their double coat of feathers and down keeps them comfortable in weather that might send you scurrying for home and hearth. To keep their feet warm, waterfowl stand on one foot with the other tucked up, then switch and stand on the other foot while warming the first.

Because they don't like to be indoors, you'll need to train your bevy or gaggle from the start to go inside at night. Have a plan for getting them from their favorite resting place to the shelter door, perhaps by enlisting help or arranging doors and gates to funnel them in the desired direction. If the first time out they lead you around the barn and through the garden, be prepared to take that same circuitous route every evening from then on.

Even if your birds are well trained and go in without a hitch night after night, the time will come when they will balk. If you once let them get away with not going in, they will try it night after night, trying your patience as well. But, as obstinate as they can be about not wanting to go in at night, should you come home late one evening to discover no ducks or geese in the yard, don't panic. Chances are good they've gone in on their own and are impatiently waiting for you to come and shut the door.

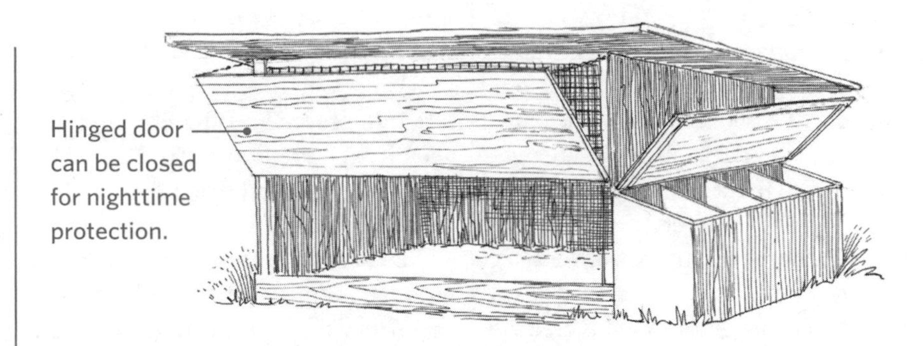
Hinged door can be closed for nighttime protection.

If you don't have an empty building to convert into a duck house, you can build a simple portable structure such as this one. The attached nests, accessible from the outside, make egg gathering easier.

A small shelter provides nighttime protection.

A Pond to Play In

Ducks and geese love to splash and play in water, and for the larger breeds water is essential for mating and egg fertility. A pond is also helpful for cleaning and conditioning plumage, and Sebastopol geese need one to maintain their good looks. Of all the various breeds, Muscovies are the only ones that seem not to mind if no pond is available for bathing and mating. For all the rest, and even for Muscovies, furnishing at least a small pond is a good idea for several reasons:

- It helps them keep themselves clean
- It improves egg fertility
- It lets them more comfortably endure hot and cold weather
- It helps them evade predators
- It lets them use up excess energy in play
- It gives you the pleasure of watching them float and frolic

Providing water is especially important if you intend to raise the large, deep-breasted breeds, which have a hard time mating without pond water to make them buoyant. And in evading predators, a goose will swim tantalizingly just offshore when tracked by a fox, and a duck will duck underwater when buzzed by a hawk.

The best reason to provide a pond is that waterfowl love water. Sure it's hard to tell whether a duck is happy or a goose is joyful, but who can doubt it after watching a duck or goose scoot across the water, disappear beneath the surface, pop up somewhere else, flap its wings, and start again? And even if we can't say for sure the bird is happy, it certainly makes us happy to watch. The pond needn't be expensive or elaborate. You can build a simple pond using a guide such as *Earth Ponds* by Tim Matson.

Waterfowl are happier and healthier when they have access to a pond.

Any waterfowl pond requires certain basic features to be safe for ducks and geese. It must have easy access in and out. The smallest swimmers may have serious problems if they can't get over the rim to get out. The pond must be easy to clean or large enough to cleanse itself of waterfowl droppings. For these two reasons, a children's wading pool is not entirely suitable, unless you can install a drain for ready cleaning and a hinged ramp that lets smaller birds climb in and out with ease, no matter what the water level.

The water should be free of any chemicals that can harm waterfowl. If your hatchlings get through the fence and make a beeline for your neighbor's chlorine-laden swimming pool, it could well end up being their last swim.

Fish and waterfowl may be compatible in a large pond or small lake. However, in a small backyard pond, fish won't appreciate waterfowl deposits or stirred-up mud, won't take kindly to frequent draining and cleaning, and certainly won't like being swallowed by some duck or goose. In a larger body of water, bass and other fish may eat hatchlings, and snapping turtles can grab a grown duck by a foot and pull it underwater.

Introducing Home

If your first waterfowl are babies, confine them indoors until they are old enough to go out on their own. Baby waterfowl with no parents don't have the sense to come in out of the pond or out of the rain and will get soaked through and take a chill. Until they are fully feathered (about 7 weeks old for Mallard-derived breeds, 8 weeks old for geese, and 14 weeks old for Muscovies), continue providing heat as required by the prevailing weather.

Grown or partially grown waterfowl can be turned loose in a safe yard as soon as you bring them home. If they are of a breed that tends to fly, clip one wing of each bird so it doesn't take a notion to explore the neighborhood. After your ducks and geese get well acquainted with their digs and learn they will always have plenty of good food to eat, fresh water to drink, and space to play, they will be content in their new home.

Preventing Frostbite

The knobs on the heads of African and Chinese geese, and the large caruncles on the faces of Muscovy drakes, are subject to frostbite, making winter shelter for these breeds essential in cold climates. Suspect frostbite if a bird seems depressed and its knob first turns pale and then patchy. Given time, a knob discolored by frostbite usually returns to normal. To prevent frostbite, confine Muscovies and knobbed geese indoors during icy, cold, blustery weather.

The feet of all waterfowl are susceptible to frostbite, which may occur when ducks and geese congregate around a feeder surrounded by ice or packed snow. The sign of frostbite is pale feet that later turn bright red. Keep the area around feeders clear of ice and snow, or spread sand or straw over the frozen ground. Providing open water for swimming also helps prevent frostbitten feet. Try to keep at least a portion of the pond free of ice, perhaps by installing a recirculating pump.

Feeding Ducks & Geese

The nutritional needs of waterfowl vary by species, age, purpose, and level of production. Ducks have slightly different needs from geese, growing birds have different needs from mature birds of the same species, and layers and meat birds have different needs from one another.

Under natural conditions, Mallard-derived ducks satisfy 90 percent of their nutritional needs by eating vegetable matter; the remaining 10 percent comes from animal matter, such as mosquitoes, flies, and tadpoles. The diet of a Muscovy leans more toward meat, whereas geese are entirely vegetarian.

If you have only a few ducks or geese, you have no desire to push them for egg or meat production, and you have plenty of space for them to forage, most waterfowl can meet nearly all their nutritional needs on their own. Supplementing their diet with grain or commercially formulated ration will help them through the winter months and will give fertility a boost early in the breeding season. It's a good idea to offer a little grain at least once a day, year-round, just to keep your waterfowl tame and used to your presence among them.

Commercial Feeds

Feed processors offer complete rations that are blended to meet the requirements of ducks and geese. Whether these feeds are available in your area depends largely on demand; if you are the rare person in your neighborhood who keeps waterfowl, you'll probably have a hard time finding commercial rations. In this case, look for a farm store that carries a national brand and is willing to order for you.

If you can't obtain duck or goose ration, you may have to make do with poultry feeds that are more widely available. Unless you expect your waterfowl to meet a high level of meat or egg production, most ducks and geese will do nicely on commercial chicken rations that have sufficient protein for their age. If you feed your waterfowl rations that were not originally formulated for ducks or geese, avoid medicated formulas, since the coccidiostats in these rations are of no use to waterfowl.

In general, you'll have a choice of six basic formulations:

Starter ration is formulated to get ducklings and goslings off to a healthy start during their first two weeks of life. The protein level of 18 to 20 percent is ideal for sturdy development.

Grower ration has a protein level of 15 to 18 percent and is designed to promote growth. Ducks and geese raised for meat should be fed a higher level of protein for rapid growth; those to be kept for breeders should get less protein for more evenly balanced growth. Birds to be butchered stay on this ration until the day of slaughter (usually at 8 to 12 weeks of age). Waterfowl to be raised to maturity are switched over to developer ration at about 9 weeks of age.

Developer ration is intended for growing birds that will be kept as layers or breeders. It has a protein level of 13 or 14 percent for more moderate growth. This ration is also suitable for mature waterfowl that are not breeding or laying.

Breeder-layer ration has a protein level of 16 to 20 percent and includes vitamins and minerals that layers and breeders need to remain fit. Breeders and layers should be switched from developer to breeder-layer ration two or three weeks before they are expected to begin laying. Make the change gradually by mixing greater quantities of the breeder-layer ration into the developer ration until the switch is complete.

High-protein concentrate is a commercial ration with a whopping protein content of 34 percent. Unlike the first four rations, which are complete feeds, this ration is intended to be combined with grains. By varying the ratio of concentrate to grains, as specified on the label, this ration can be used in place of any complete feed except starter.

Protein Requirements

Age in weeks	Stage	Protein content (%)
0-2	starter	18-20
3-8+	grower	15-18
9-20	developer	13-14
20+	breeder-layer	16-20
20+	maintenance	14

Mixed grains, sometimes called scratch grain or chicken scratch, usually consist of a mixture such as corn, milo, oats, and cracked wheat. Mixed grains should never be fed as the sole ration but can be used in combination with high-protein concentrate, fed in moderation to foraging waterfowl to keep them tame, and added to the ration in cold weather to provide an energy source for warmth. Whole grains are more nutritious than cracked grains, won't spoil as quickly, and, if dropped, will sprout and eventually be eaten — or can be deliberately sprouted to give your birds a welcome treat.

A range trough feeder.

Choosing the Right Form of Feed

Commercially prepared feeds come in three forms: mash (ground-up rations), pellets (compressed mash), and crumbles (crushed pellets).

Pellets are easiest for mature waterfowl to eat and result in the least amount of wasted feed (and, therefore, less wasted money). For birds that aren't yet old enough to handle pellets, crumbles are acceptable, but more feed will be wasted. Mash results in the greatest amount of wasted feed. This waste can be reduced by moistening the mash, but avoid spoilage by mixing only as much as the birds will eat before the next feeding and thoroughly clean out the trough before adding more feed.

Although not every ration is available in all three forms, you should have little difficulty obtaining the pellet form. The chief problem is getting a satisfactory pellet feed for the youngest birds, because they cannot eat pellets of the size adult birds eat. If you can't find small pellets, about 1/8 inch (0.5 cm) in diameter, you will have to settle for a crumble form for ducklings.

Storing Feed

To discourage raiding rodents, store feed in clean trash cans with tight-fitting lids. Secure the lid with a bungee cord if there's any chance of raccoons wandering by. Because commercial rations gradually deteriorate and because you have no idea how long the feed has been stored at the farm store, purchase no more than you will use in about two weeks. If you get home and find that the bottom of a bag has gotten wet and the feed inside is moldy, return it or toss it out. Moldy feed can be toxic to waterfowl.

How Much Feed Is Enough?

The easiest way to feed waterfowl is to provide the rations by free choice, meaning they have access to their feed anytime they wish to eat. Young waterfowl feeding on starter or grower ration should definitely be fed free choice. However, if you have older waterfowl that spend most of their time foraging, they don't need any more commercial ration than they can eat within 15 minutes twice a day, morning and evening.

A growing duckling will eat 1 or 2 ounces (28 or 56 g) of commercial ration per day, gradually increasing to about 8 ounces (226 g) per day as it reaches full size. A mature duck of a high producing layer breed will consume about 6 ounces (170 g) of feed per day, whereas a duck that is not a particularly prolific layer may eat 5 to 7 ounces (142 to 198 g) per day when not laying and 12 to 15 ounces (340 to 425 g) when laying. Geese of the heavier breeds may eat up to four times those amounts. Of course, these are only guidelines. The actual amounts your birds will eat depend on many factors, including the temperature, level of activity, age, and size. Birds naturally eat more when they are colder, more active, older, or of a larger breed. All waterfowl will eat less when they have access to high-quality forage.

Waterfowl Feeders

Feeders designed for other poultry can be used for ducks and geese. Hanging tube feeders can be adjusted for height as the ducklings or goslings grow, thereby minimizing waste. Rubber livestock feed pans with tapered sides are also suitable. Avoid any trough design that a duck or goose can easily tip by stepping on the rim and, of course, any style the birds can't dip their bills into. A feeder used for free-choice rations should be protected from rain, which is easily done by building a little roof over it or by putting it in the entry to the waterfowl enclosure. If you put it in the enclosure, remove it at night when the birds are shut in. Fill the feeder only with as much as your waterfowl will consume in one day. Clean out any leftover feed at the next filling.

If the feed gets wet, clean the feeder within a few hours. Freshly dampened feed will not hurt ducks and geese. Damp feed left standing for more than a few hours, especially in warm weather, may turn moldy and become harmful. In the course of eating, ducks and geese tend to get feed in their water and water in their feed. To keep both water and feed clean, space feeders and waterers 6 to 8 feet (2 to 2.5 m) apart for mature birds. For ducklings and goslings, place feeders and waterers on opposite sides of the brooder.

If your yard has no shade, you can create some with a beach umbrella — your geese will thank you for it.

Forage

In place of teeth, waterfowl have serrations called lamellae around the edges of their bills. Ducks use theirs to strain food out of water. Geese, like eternally teething toddlers, use theirs to chew on anything handy. Lamellae allow ducks to be foragers and geese to be grazers despite their lack of teeth. Given sufficient space with succulent vegetation, waterfowl can survive nicely on whatever they scrounge up. A grassy pasture is ideal; so is a lawn, provided it is not maintained with toxic chemical sprays. You would please your waterfowl to no end by planting your garden with extra leafy greens such as kale, cabbage, lettuce, alfalfa, or clover. Leftover fruits and vegetables make welcome additions to the waterfowl diet, provided they're fresh and not spoiled or moldy, and they are chopped or crushed into bite-sized pieces. Raw potato peels are not readily digestible and should not be offered.

Ducks make great gardeners that will scarf down emerging weeds, weed seeds, and pests of the bug, slug, and snail variety — provided they aren't let in until after the veggies have grown past the succulent stage and the ducks are fenced away from tender greens and ripening strawberries. Also, keep them out while watering the garden or after a rain, as the ducks will dig muddy holes by dabbling in the smallest puddles. In general, one pair of grown ducks per 500 square feet (152 sq m) of garden will do more good than harm.

Geese can get along almost exclusively on succulent greens and are often used as cheap labor for weeding crops. Although any breed can be used as weeders, Chinese are most often used, because they are energetic and active yet light enough not to cause soil compaction or otherwise inflict much damage. A pair of mature geese in the average backyard garden would likely do more harm than good, but a few goslings can work perfectly as springtime weeders. Goslings begin to graze as young as 1 week of age but must be enclosed at night for warmth until they feather out at about 6 to 8 weeks.

Geese are especially adept at controlling troublesome grasses, such as crabgrass and Bermuda grass. They have been used to weed many different cash crops, including strawberry fields, but must be removed before the berries ripen, because they like strawberries almost as much as we do. Geese can be used in a vineyard or orchard to clean up windfall fruit, which otherwise attracts insect pests, but should not be put in with young trees or vines, as they will strip away the flexible bark. Geese will keep an irrigation ditch flowing freely by cleaning out grass and water weeds. They are also good at controlling weed growth along fence lines.

Weeder geese must have an abundant source of drinking water and a shady area where they can get out of the sun. Shade might be provided by shrubs and trees, or by a man-made device, such as a large beach umbrella. Weeders will

An automatic waterer ensures your birds won't go thirsty.

concentrate their efforts near the source of water and shade, both of which may need to be moved daily to ensure the entire area gets weeded.

Allow goslings into the garden only after the veggies have grown beyond the young succulent stage, when they become less appetizing than sprouting weed shoots. Should the goslings get ahead of weed growth and not find enough to eat, they will start in on your garden. As succulent weeds become scarce, your geese may require supplemental feeding, but take care not to overfeed them, or they will tackle weeds with far less enthusiasm. To ensure the geese weed actively throughout the day, feed them only in the late afternoon or early evening.

Sprouting Treats

When forage is sparse, your ducks and geese will enjoy sprouts. Use only seeds that are free of harmful chemicals. Easy sprouters include:

- adzuki beans
- millet (unhulled)
- mung beans
- sunflower seeds (black oil or gray striped)
- wheat (hard winter)

Water

Having evolved in water environments, waterfowl cannot store water in their bodies and so, for their size, drink relatively large amounts of water. They may become ill if they are deprived of drinking water for more than a few hours. Laying breeds will stop laying and go into a molt when deprived of water.

Unless your waterfowl yard has a free-flowing source of potable water, you will need to provide plenty of clean drinking water. The two most important features of a proper water trough are its ability to be easily cleaned and the ability of ducks and geese to dip in their entire heads. Waterfowl keep their bills free of caked mud and feed by squirting water through their nostrils. The water trough must be easy to clean to prevent an accumulation of sediment from all that face washing.

> Waterfowl cannot store water in their bodies so they drink relatively large amounts of water.

Using an automatic watering device is an ideal way to ensure that your gaggle or bevy has a continuous flow of fresh drinking water. Hog waterers work well for geese, and chicken waterers are fine for ducks.

In an area where temperatures dip below freezing, providing unfrozen drinking water can be a challenge. Water lines can be wrapped with heating tape, or a livestock heating device can be used in the trough, although don't be surprised if your geese work determinedly to dismantle any device they get their bills on. A recirculating pump on a pond will help maintain an area of open water. If nothing else, place a rubber livestock feed pan beneath a faucet left open to a steady drip.

Waterfowl like to bill water out of a trough, causing muddy puddles to form. The area around a waterer must therefore be able to drain quickly. Placing the waterer on a platform of wooden slats or hardware cloth and set on a bed of deep gravel will considerably improve drainage.

Grit

The digestive system of a bird starts with its crop, where swallowed food is stored. From there it moves into a tough organ called the gizzard, where food particles are rubbed against pebbles and sand the bird has eaten, thus grinding up food matter so it can be digested. Eventually the grit, too, gets ground up in the process and must be continually renewed.

Ducks and geese that spend time foraging pick up natural grit in the course of their daily activities. To make sure they get enough for proper digestion, insoluble grit should be available at all times. Granite grit may be purchased at a farm store. If you prefer, you can substitute clean coarse sand or fine gravel from a clean riverbed.

Mating

The lighter breeds of both duck and goose have no trouble breeding on land, but breeding on water helps the male get a better grip, minimizing the wearing away of feathers from the hens' backs. The heavy breeds need a lake or pond (a shallow stream will not do) to get sufficient buoyancy for good fertility. The mating ratio, or number of males per female, is also an important factor for ensuring good fertility.

Ducks

Mallard-derived breeds tend to bond for the season. As soon as the ducks start nesting, the drakes band together in a bachelor group, and next year's pairings may or may not be the same as this year's. If too few drakes are available to go around, some of the drakes will bond with two ducks. Even if you have an equal number of ducks and drakes, one drake may take on two ducks and leave another drake with none.

During the breeding season, a duck will try to keep other ducks away from her drake, and the drake will run off other drakes. But while all this posturing is going on, actual mating occurs indiscriminately among and between the various pairs, and any time one drake mounts a duck, the others will join in. So, despite the apparent preferences for mates, one drake of a light, active breed can cover eight to ten ducks, one midweight drake can handle six to eight ducks, and a heavy breed drake can cover up to five ducks.

Ducks also can be kept in pairs, although maintaining fewer drakes not only reduces the feed bill but also offers hens some relief from high libido males. If the hen starts losing feathers on her back and head, give the duck a rest by separating her from the drake and letting him in only briefly every week or so for breeding.

If your yard is home to more than one breed, the ducks will not necessarily bond within their own breed and will almost certainly interbreed. To avoid having mixed-breed ducklings, the various breeds must be separated from one another at least three weeks before eggs are collected for hatching, or nesting is allowed to begin, and kept apart until nesting starts.

This small brooder house with a wire-covered run is ideal for your broody mama and her babies. When allowed to range, hen-mothered ducklings learn to forage for their food at a young age.

Muscovies

Muscovies do not form similar allegiances; rather, the two sexes live separate lives throughout the year. The drakes seem to have insatiable sex drives, though, and won't discriminate between Muscovy ducks and others. Any offspring from the mating between a Muscovy and another breed will be a mule that is normally incapable of reproduction.

Muscovy drakes use their long, sharp claws and powerful wings in violent attacks on one another. This aggression can be minimized by supplying each Muscovy drake with at least five ducks and by providing plenty of living space for a bevy that includes more than one drake. If you have no need or desire for ducklings, peace and calm will reign in a yard housing only sweet-tempered female Muscovies.

Geese

Geese generally mate in pairs or trios, although a gander of a light breed may take on as many as six females, and one of a heavy breed can handle up to four. Once a goose and gander have bonded, they may chase away other females expressing an interest in joining them but may readily bond with daughters they raise themselves. Bonded geese will remain tightly bonded as long as they can see or hear each other. If one is removed or dies, the other will mourn. How deep the mourning and how long it continues vary by how long they have been together. The surviving older goose or gander may mope around, lose interest in eating, and eventually die. A surviving younger goose or gander may one day stop mourning and take on a new mate.

During the brooding period, the mother duck or goose leaves the nest only occasionally.

Nesting

The easiest way to get ducklings and goslings is to leave the hatching to your waterfowl. When a duck or goose becomes reclusive, builds a nest to hide her eggs in, spends increasingly more time on her nest, and lines the nest with feathers pulled from her breast, she's getting ready to hatch her eggs. Gathering eggs in a nest and sitting on them until they hatch is called setting or brooding, and the resulting ducklings or goslings are the brood.

A duck or goose is likelier to get broody and stay broody if she can find a secluded place to make her nest. The best way to ensure that is to furnish your waterfowl yard with suitable nesting sites. A marauder safe nest has a closable front that can be latched at night. A nest with a latchable front must have air vents. Where temperatures get high during the spring and early summer, place nests in a shady area, such as beneath a tree or shrub, or pile brush on top. Since waterfowl like to brood in seclusion, cover the opening of each nest with brush to improve the hens' sense of security.

If your waterfowl shelter is large enough, you might partition off a section for nests, although waterfowl do not like to nest in close proximity to one another, but prefer to have a choice of nesting spots scattered around the yard. For successful hatching, provide one nest per female. If nests are in short supply, two females may lay in the same nest and may even set side by side to hatch the eggs, but the resulting hatchlings will probably get trampled in a squabble over whose they are.

A nest against the ground is better than one with a man-made bottom because soil helps retain the moisture necessary for a successful hatch. Such a nest, however, may not be safe from predators that can tunnel through the soil to get at the eggs or the female. Hardware cloth fastened to the nest bottom, then topped with a layer of soil, ensures both good humidity and protection from diggers. If you use buckets or barrels as nests, block both sides to prevent rolling and tamp clean soil into the bottom to form a level floor. Supply each nest with nesting material in the form of dry leaves or straw, and each female will shape her own nest.

A duck or goose will usually continue to lay in the same nest, provided it remains undisturbed. If one day she enters to find all the eggs gone, she is likely to look for a different place to lay. If you wish to hatch more ducklings or goslings than your waterfowl can hatch naturally, you'll need to collect some of their eggs to hatch in an incubator. Mark a couple of eggs and leave them in the nest to keep the duck or goose coming back to lay more. Later you can let the duck or goose accumulate a nestful of eggs to hatch, or you might help her along by adding eggs from another bird to her nest. Be sure to remove the ones you marked, because they will be old and will rot rather than hatch in the heat under her breast.

Setting

When a goose has 6 to 12 eggs in the nest and a duck has 12 to 18, she will probably get the urge to set. If a duck or goose accumulates many more eggs than that number (as can easily happen when more than one female lays in the same nest), she may not be able to cover them all with her body; as a result, few or none will hatch. A duck or goose stops laying soon after she starts setting. During the brooding period, she will stay on the nest most of the time, leaving only occasionally to eat, drink, or swim. While she is off the nest, which may be as long as an hour, she will cover the eggs with feathers and nesting material to keep them hidden and warm.

When a duck or goose comes back from her swim and hunkers back down on her nest, moisture from her feathers will get on her eggs to keep them from drying out. As she settles back into the nest, she reorganizes the eggs by rolling them with her bill or paddling them

Artificial Incubation

To develop into ducklings or goslings, waterfowl eggs must be subjected to a specific amount of heat and humidity for a specific length of time. Most mother ducks or geese know what to do. But even when they do everything right, you may one day wish to hatch eggs in an incubator for one reason or another:

- Your chosen breed may not be reliable setters or good mothers
- You may wish to keep the hens laying and not setting
- You may want more hatchlings than your hens are able to hatch
- Local wildlife or household pets may disturb the nests
- Your yard may not be conducive to encouraging broodiness

An incubator may be either a still-air or a forced-air model, the main differences being that the former is generally smaller and has only a heat element, whereas the latter can be much larger since it has a fan to circulate heat throughout. A still-air incubator can be used to hatch duck eggs, but only the largest models will successfully hatch goose eggs. A forced-air incubator can be used for both duck and goose eggs, and will probably have a handy device that automatically turns them.

An incubator's capacity is measured in terms of how many eggs it will hold. Some incubators specify duck or goose eggs; others specify only chicken eggs. To estimate the equivalent number of waterfowl eggs, figure 75 percent capacity for duck eggs, 60 percent for Muscovy eggs, and 40 to 30 percent for goose eggs. For automatic turning, the egg-holding trays must be correctly sized, as specified by the manufacturer, according to whether you are hatching duck or goose eggs. To get good results hatching eggs in an incubator, carefully observe the manufacturer's instructions.

with her feet. Sometimes during this process, an egg will roll out of the nest. A good broody will roll the egg back in with her bill, but a lazy mother will let it go. A duck or goose that is sloppy with her eggs may thus lose them one by one until none is left to hatch. To avoid this problem, add a sill at the front of the nest that is low enough for the female to step over but high enough to keep eggs from rolling out. A solid wall at the back of the nest will give the duck or goose a strong sense of security, since she'll then need to watch for predators approaching from only one direction.

A drake will stand guard while his duck lays an egg, but after she starts nesting, he loses interest and wanders off on his own. This attitude is probably a protective ploy to draw attention away from the nesting female. A gander, on the other hand, becomes fiercely protective, standing guard and warding off all intruders throughout the brooding period. This apparent mean streak is reserved for outsiders; when the goslings arrive, the gander shares equal responsibility for protecting them from harm. So strong is a gander's parental instinct that, even without a mate, he may adopt an orphaned brood of goslings or ducklings.

If all goes well, the hatchlings will appear in approximately 28 days if they are of a Mallard-derived breed, 30 if they are geese, or 35 days if they are Muscovies. The hatchlings will remain hidden under their mother for the first day or two before venturing out to explore the big world. Even after leaving the nest, they will stay close to their mother and periodically crowd around her for warmth. On cold or rainy days, she may open her wings like an umbrella to provide protection. In early spring when the weather is still cold, or in rainy weather, confining the family to a shelter will help ensure the hatchlings' survival.

Mothering Instinct

Ducks

Breed	Mothering
Ancona	good to fair
Appleyard	variable
Aylesbury	variable
Buff	good to fair
Campbell	fair to poor
Cayuga	good to fair
Magpie	good to fair
Muscovy	variable
Pekin	variable
Rouen	variable
Runner	fair to poor
Saxony	variable
Swedish	good to fair
Welsh Harlequin	fair to poor

Geese

Breed	Mothering
African	good to fair
American Buff	good
Chinese	poor
Embden	good to fair
Pilgrim	best
Pomeranian	good
Roman	best to good
Sebastopol	poor
Shetland	best
Toulouse	good to fair
Toulouse, dewlap	fair to poor

Duck & Goose Health

Your ducks and geese are likely to remain healthy if you feed them a balanced diet, provide a continuing supply of clean, fresh water for drinking and bathing, and furnish adequate facilities for comfort and sanitation. Diseases become a threat mainly where waterfowl are kept in crowded, unsanitary conditions.

Contagious Diseases

If your bevy or gaggle exhibits several symptoms and sudden deaths occur, you can be pretty sure a disease is involved. Since the signs of various diseases can be confusingly alike, seek a diagnosis from your state's poultry pathology laboratory, located through your county's Extension office. The lab will probably want to examine several sick and recently dead birds to determine the cause of the illness.

By discovering the cause, you can learn about suitable treatments and vaccines to prevent further outbreaks and whether or not survivors will be carriers that continue to spread the disease.

Noncontagious Conditions

Health problems in a gaggle or bevy are more likely to involve a noncontagious condition than a serious disease. Following are some of the most common noncontagious conditions.

Signs of Trouble

- Weak and listless behavior
- Ruffled feathers
- Crusty or sticky eyes
- Unusual body movements
- Loss of appetite
- Increased thirst
- Change in the color or consistency of droppings
- Inexplicable deaths

Crop Impaction

Geese that graze on fibrous plants, including grass that has gotten tall and coarse, and on degrading feed sacks made of woven polypropylene can suffer from crop impaction. Geese are natural grazers, but tough, fibrous vegetation is difficult for them to digest and instead wads up in the crop. An affected goose loses interest in grazing, steadily loses weight, and may die unless the wad is surgically removed. You can check for impaction by feeling the crop. It should feel squishy. If, instead, you find a tight hard lump, the crop may be impacted, blocking food from passing through to the digestive tract. Prevention involves mowing the grazing area to keep it in the vegetative stage and providing other feed at times of year when vegetation stops growing.

Drowning

As unbelievable as it may seem, ducks and geese can drown. Among domestic waterfowl, drowning most often occurs when they get into water they can't get out of. Ducklings or goslings may jump or fall into a kiddy wading pool, for instance, and by the time they've finished splashing around, the water level is too low for them to climb back out. Even older ducks can drown if, for example, they've gotten into a swimming pool or livestock water tank and can't climb up the side to get out. Water that's contaminated with chemicals (including swimming pool chlorine), oil, sludge, muck, or mud can also lead to drowning.

Hardware Disease

Waterfowl don't discriminate between tasty treats, such as bugs and slugs, and nasty things, such as shards of glass or bits of wire. Swallowing small, sharp objects may merely cause irritation or may poke a hole through some part of the digestive tract. The end result may be depression or death. Prevention involves meticulously patrolling the waterfowl yard to remove any small, sharp objects lying around and to check that such things aren't tossed there in the first place.

Lameness

The most common affliction of waterfowl is lameness, because they have structurally weak legs. Their legs are made for flying and swimming, not walking. Lameness may occur if you grab a bird by the leg or it pulls free a leg that's been caught in a fence. Lameness may also result from a glass sliver, a thorn, or a stick lodged in the footpad.

The feet of waterfowl kept on dry, hard-packed ground can develop abscesses that harden into calluses on the bottom of the pad. This condition, known as bumblefoot, may involve one or both feet and most often affects the heavier breeds. Treatment involves washing the affected foot, cleaning it with a bactericide, pressing any pus out of the abscess, and removing the hard core if one is present. Isolate the bird in a quiet, secluded place with clean litter or fresh grass and clean swimming water. To prevent this problem, keep feed and watering areas clean; cover hard surfaces with clean litter; and provide several paddocks, rotating the flock to periodically rest the vegetation in each paddock.

Penis Paralysis

Phallus prostration, or penis paralysis, is a sad result of the domestication of waterfowl. Under natural conditions, ducks and geese breed at maturity and have a short breeding season. Domesticated waterfowl do little to sustain themselves. They start breeding at a younger age, the males are often expected to breed more females than they would under natural conditions, and they breed for a longer period. As a result, sometimes a drake or gander overexerts his male organ to the point that he can't pack it in anymore. Eventually it becomes dirty and scabby, dries out, and perhaps gets infected, and the drake or gander may lose weight or die. This condition may be a genetically inherited weakness of the muscles, possibly aggravated by some dietary deficiency. Usually, by the time the condition is noticed, nothing can be done but to put the drake or gander out of his misery.

Sticky Eyes

Crusty, sticky eyes may be the sign of a low-grade infection due to the duck or goose not being able to fully submerge its head in water and thereby flush out its eyes and nostrils. If swimming water is not available, make sure drinking water containers are deep enough that your birds can get in their entire heads. Isolate an infected bird, clean its eyes with an eyewash recommended by your vet, and provide a nutritiously balanced diet to reduce susceptibility to infection.

Poisoning

The most common materials — salt used to deice winter paths, insecticides sprayed around your (or your neighbor's) yard or garden, rat poison, and bait set out for slugs, snails, or other garden pests — can poison ducks and geese. Perhaps the most common cause of poisoning is botulism from spoiled food scraps or contaminated water. If you feed kitchen scraps to your waterfowl, first sort out anything that is spoiled or rotting. Any water your waterfowl have access to must be free of excessive droppings, an accumulation of decaying leaves, and dead animals.

Signs of botulism poisoning appear within 8 to 48 hours of consuming the toxin. In mild cases, the only sign may be a brief bout with weak legs before the birds return to normal. In severe cases, the birds may appear sleepy and unable to hold their heads erect, due to paralysis of the neck muscles. Their wings and legs may become paralyzed, too, causing them to lie on their sides.

If the poisoning is not so severe the birds cannot drink, replace all sources of water with a solution of dissolved Epsom salts in the proportion of 1 pound (500 g) per 15 gallons (57 L) of water. By the time signs are first noticed, however, chances are the affected waterfowl are beyond help. The most prudent course of action is to avoid conditions that could lead to botulism poisoning.

Managing Ducks & Geese

The basic daily activities of waterfowl are eating, napping, swimming, and mating. If they have plenty to eat and a safe place to rest, they tend to be fairly content. After eating their fill, many breeds are happy to laze around. The active breeds tend to entertain themselves by investigating things on the ground (sometimes taking them apart) or by flying up to get a better view from overhead.

Restricting Flight

Ducks and the lighter goose breeds will fly over a low fence when startled or attracted to something on the other side of the fence. If you can keep your young birds from flying, they probably won't attempt to fly over the fence after they mature. Muscovies are unique in liking to fly up to perch on a fence post or tree branch, and they don't always come back down on their own side of the fence. You can stop waterfowl from flying in two ways, one of which is permanent, the other temporary.

The permanent method is to clip a wing tip. The tip of the wing is called the pinion, and removing it at the first joint is called pinioning. Pinioning throws the bird's two wings out of balance, making it unable to fly. When hatchlings are pinioned, they seem to barely notice, and the wing heals rapidly. Pinioning gets more difficult as birds get older. Pinioning mature waterfowl should be done by a veterinarian or other waterfowl professional. Pinioning is one of the recommended Fish and Wildlife means of marking captive-bred Muscovies.

The temporary method of restricting flight is to clip the flight feathers of one wing, which serves the same purpose as pinioning. If you clip the feathers of young waterfowl to keep

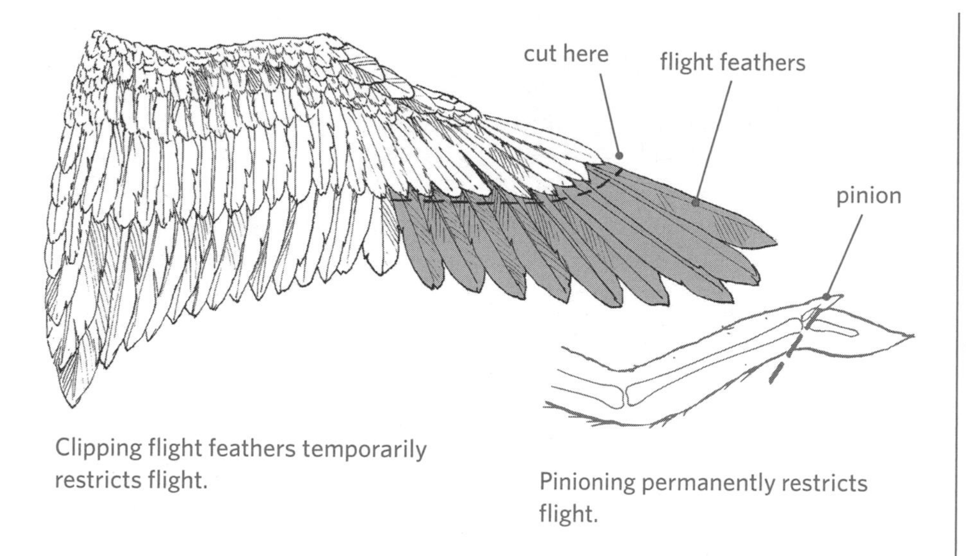

Clipping flight feathers temporarily restricts flight.

Pinioning permanently restricts flight.

them grounded while they grow, you probably will not have to do it again once they mature. Except, of course, for Muscovies.

Molting

Over time, a bird's feathers get worn or broken. Its entire set of plumage is therefore periodically renewed in a process called molting. Young birds go through two consecutive molts. After maturity, they experience one complete molt each year, generally in late summer or early fall after breeding season but before cold weather sets in. They may also go into a molt at other times, as a result of some change in their environment or diet.

During the molt, wild Mallard drakes take on the camouflage pattern of the females, making it easier for them to hide when they drop their wing feathers and can't fly. As soon as the wing feathers have been renewed, the males go through a second, partial molt to resume their normal color pattern. All drakes of Mallard-derived breeds molt in a similar manner, although the color change may not be as obvious.

Your first clue that molting has started may be piles of feathers in the yard, perhaps leaving the impression a predator has gotten one of your birds. You need only make a head count to be sure they are all there. The molt will last six to eight weeks, during which ducks that are laying will decrease in production, or stop laying altogether, because they can't efficiently produce both feathers and eggs.

Aggression

Mallard-derived ducks and female Muscovies tend to be more reserved than geese and male Muscovies, both of which can get pretty aggressive. This behavior has a connection to species characteristics: Geese and Muscovies are large and clumsy, can't easily hide, and therefore resort to aggression to ward off enemies. The best way to keep them from getting mean in the first place is to take time to greet them whenever you meet. Notice how they gabble and posture while greeting one another; it's the polite thing to do. Another way to minimize aggression is to ensure that dogs, children, and all others are taught to move quietly among the birds and never, ever tease them.

Geese get naturally aggressive during breeding season, when a gander will fiercely defend his mate while she is on the nest, and later will defend his offspring with the same fierceness. When a gander is thinking about attacking, he will stretch his neck full length, throw his head back, and peer at you with one eye. This posture indicates he isn't too sure about the wisdom of attacking. If you move toward him waving your arms and making a loud noise, he's likely to back off.

The gander to beware of is the one that makes a beeline toward you with his neck stretched forward, head down, making hissing sounds. This gander is ready to bite. The worst thing you can do is run away. The best thing you can do is show him who's boss, but that doesn't mean giving him a swift kick with your boot or otherwise engaging in a tussle. Fighting with a gander will only make him meaner.

When a gander comes at you with the intent to bite, clap your hands and stamp your feet. If he keeps coming, swing your arms to make yourself look as big as possible, and run toward him. If that doesn't work, extend your arm with a pointed finger (creating the appearance of a long goose neck with a beak at the end), and move menacingly toward him. Tell him firmly, "Don't you dare!" If he gets close enough to bite, smack him on the bill. If the gander is young, only one or two smacks on the bill should be enough to let him know which of you is the bigger gander.

If the smack doesn't work, step to one side as he rushes at you and grab him by the neck. Grasp the back of his neck with one hand and lift him off the ground, placing your other hand beneath his body. This action causes a gander to suffer a great indignity he is not eager to repeat.

Although you can keep your geese from getting mean enough to attack, you cannot stop their innate need to display aggressive behavior. It's part of the triumph ceremony, the purpose of which is to strengthen family bonds. A well-trained gander will pretend to attack you, running at you full tilt but turning back at the last moment. He'll then return to his mate with his wings outstretched (the gander's equivalent of thumping himself on the chest), honking triumphantly, and the two will enthusiastically discuss the gander's brave deed. If they have goslings, the young will circle around and join the celebration.

Two ganders or drakes may fight if the yard is too small for the number of birds, if there aren't enough mates to go around, if nesting sites are too few or too close together, or if a new bird is introduced into the group. Once the problem is resolved — a larger yard is provided or some birds are removed, additional nesting sites are provided, and the newcomer finally makes friends — peace and calm usually return.

Avoiding fights between, and with, Muscovy drakes is especially important, as their powerful wings and sharp claws make a formidable combination. When a Muscovy is excited or angry, its head feathers rise to create a sort of crest. The male, being larger to start with, raises a higher crest than the female. At the same time, the Muscovy may alternately jerk in and bob out his head; the rapidity of jerking and bobbing indicates the degree of anger or fear. Crest raising and head bobbing are also signs of sexual arousal, and slow head bobbing is often used as a form of social greeting. Muscovies wag their tails to indicate some unpleasant encounter has been successfully resolved, and also for mysterious and unfathomable reasons as they go about their daily business.

Males versus Females

A good way to avoid aggression in your waterfowl population is to ensure that you have an appropriate proportion of males and females. Being able to distinguish the males from the females is therefore essential. Some color breeds make it easy, because the male has a different color pattern from the female (except briefly during the molt, when males may look like females).

Even when the male and female are of identical color, all duck breeds except the Muscovy give you two other clues. The fully feathered male has curly feathers, aptly called drake feathers, on top of its tail. And, whereas the male speaks in a barely audible, hoarse whisper, the female quacks raucously.

Muscovy females can also quack loudly, but they usually do so only when startled or frightened. Otherwise they make a pleasant whimpering sound, while the male hisses somewhat like a goose. Male and female Muscovies of all varieties are nearly identical in color but are easy to distinguish from one another because the male is nearly twice the size of the female.

Distinguishing a goose from a gander is not easy unless they are autosexing, which applies only to Pilgrims and Shetlands. Otherwise you need to recognize subtle differences in voice and body posture. The gander's usual sound is a shrill alarm, while the female's is a repeated comfort sound at a much lower pitch. Both geese and ganders honk loudly when disturbed and hiss when angered. In stance, the gander walks more erect and in appearance has a slightly thicker neck and coarser head. The goose carries herself closer to the ground, and her neck is usually curved. When they walk together — and mated pairs don't often stay apart for long — the gander walks ahead and the goose follows. In the African and Chinese breeds, the ganders have somewhat larger knobs on their heads than the females do.

The only way to be 100 percent certain whether you have a goose or a gander is to vent-sex the bird, meaning you must examine its sexual parts. Catch the bird and turn it on its back, taking great care not to get whacked by a wing, sliced by the claw of a paddling foot, or bitten on the back of your arm. Having someone help you hold the bird can be a great asset. Sit down, or hunker on the ground, with the bird's back over your knee, its head tucked under your arm, and its vent (the orifice under the tail through which the bird eliminates and lays eggs) facing away from you.

With one hand on each side of the tail, use your fingers to push the tail downward while using your thumbs to press and pull apart the sides of the vent. If you are lucky, the bird will reveal its sex organ. If you are unlucky, it will first show its indignity by squirting a foul green mess in your direction. If the bird is tense, it will reveal nothing until you can get it to relax its vent muscles by inserting a finger and letting the muscles hold it tightly until they finally grow tired and relax. If you work things right, a penis will unwind from the vent of a gander, while a goose will reveal frilly folds of pink skin. To avoid getting hurt when you're ready to release the

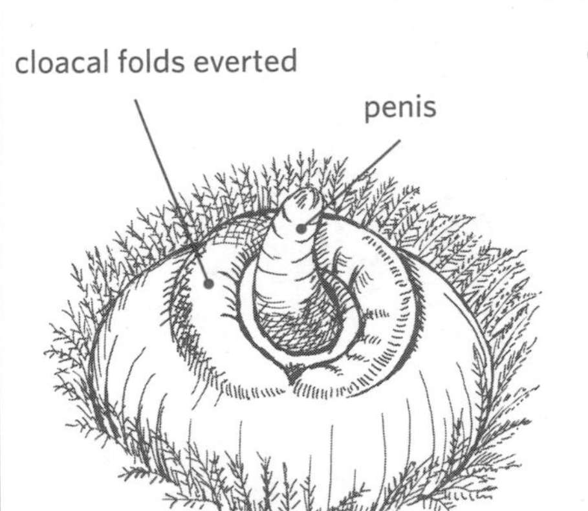

Male genital organ

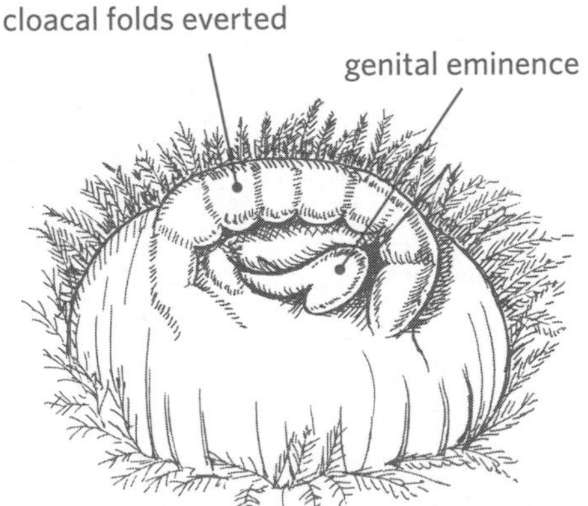

Female genital organ

bird, turn it over and set it with its feet on the ground, holding your hand on its back for a moment before letting go. The bird will then stand up and move away from you.

Catch and Carry

Unless you have ducks that are tame pets, catching waterfowl can be a challenge. A poultry net (something like a butterfly net) comes in handy for the purpose, although if you have a sizable pond, the crafty creatures will swim out to the middle and wait for you to leave. If you have more than a few ducks or geese, you have to take care not to cause a stampede in which one could get seriously hurt, and you do not want to get hurt yourself if your waterfowl find themselves trapped in a corner and launch en masse into the air to get around you. Move slowly and try not to frighten them into flight. If one crouches — a sign it is about to spring into the air — you may be able to catch it just as it leaves the ground, especially if you have a net.

Herding your bevy or gaggle into a small area gives them less space in which to give you the runaround. The small area should be a place they're already familiar with; wary waterfowl do not herd easily into unknown territory. Muscovies are especially difficult to herd and do not readily go into enclosures, but they tend to run away less readily than other breeds, making them easier to catch.

Ducks and geese don't move as agilely after dark, and they are easily confused by the beam of a flashlight. A duck will try to get away from the beam, letting you use the light to herd ducks from a distance into an enclosure. A goose tends to move toward the beam, giving you a chance to grab it when it gets near enough.

The safest way (for the bird, that is) to grab a duck or goose is by the neck. The neck is much stronger than a leg, which can easily snap if you pull in one direction while the bird tries to move in another direction. Once you have a bird by the neck, lift it with one arm carrying the weight of its body and the other arm over the wings so they can't flap. If you are carrying a Muscovy, watch out for those claws. A small person who has trouble carrying a large bird under one arm would be wise to carry a Muscovy with one hand on each wing joint, where the wing joins the body, keeping the bird's back toward you and its claws away from you. A goose can be held in a similar manner, or by hooking both wings under one arm while using the other arm to carry some of the weight.

When handling ducks and geese, take great care not to frighten or excite them. (You'll be able to tell they're extremely frightened if they wobble instead of waddle or move around by using their wings as crutches.) Whenever you work around waterfowl, remain calm and move with deliberation. Handle your birds gently, and reassure them before releasing them. If you spend enough time with your waterfowl, catching one won't be a stressful ordeal.

A safe way to carry a goose is with one arm underneath the goose for support and the other arm above the goose, securing its wings.

Transporting

The ideal way to transport a duck or a goose is in a pet carrier, such as one designed for a cat or dog. Make sure the carrier is of adequate size for the bird. A sturdy cardboard box also works, provided it can be sealed so the bird can't break out and has air holes large enough so the bird can breathe but can't get its head through.

For short trips, you can transport a duck or goose in a feed sack. For a goose, cut off the tip of one corner of the sack, put the goose inside with its head through the hole, and tie off the sack's opening. For a duck, drop the duck into the sack and tie the hole with enough slack to admit air but not enough for the duck to escape. If you aren't going far and the temperature is not high, you can carry two or three ducks in one sack. Otherwise, put only one bird in each.

Any of these carriers can be loaded into the back of a pickup truck. If the bed has no camper cover, place the containers close to the cab, where the draft is reduced. If you're traveling in a car, avoid using the trunk, where exhaust fumes could kill your birds. Put the carrier on the back seat or on the floor.

When transporting a pair of mated geese, let them ride side by side, where they can see each other. If you transport them within earshot but not sight of each other, by the time you reach your destination you will be deaf thanks to their lovesick calling. Geese are more content when they can see where they're going and can peer out the window at passing scenery.

CHAPTER 4

Rabbits

Rabbits are easy to raise, make no noise, and reproduce year-round to yield an abundance of lean, flavorful, healthful meat. Rabbits can be kept just about anywhere, since most regulating agencies don't classify them as livestock. Rabbits grow fast and require so little space they can be kept in a carport or on a back porch.

So why doesn't everyone have rabbits? Well, it's called the Easter Bunny Syndrome — rabbits are fuzzy, friendly, and so awfully cute, how could you possibly think about eating one? But here's a good way around that issue: Let yourself get attached to your breeders while raising their many anonymous look-alike kits for meat.

Rabbit Families

Rabbits belong to the lagomorph group of mammals. Lagomorphs are subdivided into two families, pikas, and rabbits and hares. Pikas are small short-eared animals that live in mountainous areas. Hares are closely related to rabbits but are usually larger, have longer legs and ears, and run with long, high leaps.

The rabbit group is further divided into 25 species. Wild rabbits native to North America include the eastern cottontail, the desert cottontail, and the marsh rabbit. The tame rabbits raised in North America as pets and for fur and meat are descended from wild European rabbits.

From fossils, scientists have determined the rabbit has been around for about 30 to 40 million years and has changed little during that time. When humans arrived on the scene, wild rabbits became part of their diet. Exactly when people decided to raise rabbits rather than hunt them in the wild is unknown. We do know that early sailors took rabbits along on voyages because they were easy to feed and care for and could be used as meat. These early sailors introduced rabbits to some of the new places they explored.

Early European settlers probably brought rabbits with them to the New World, but because wild rabbits and other game were plentiful here, settlers didn't need to raise many rabbits. However, that trend changed in the early 1900s with the Belgian hare boom. The Belgian hare is a breed of rabbit, not a hare, as the name might imply. In the beginning of the twentieth century, a legion of aggressive promoters led people to believe raising Belgian hares could be a profitable pursuit. Some of the newly converted breeders were successful, but many others who invested in these rabbits lost money. Although Belgian hares didn't do all their promoters had promised, they did increase interest in rabbits in North America.

Since then, other breeds have been imported to North America and several American breeds have been developed. Forty-five breeds are currently recognized in North America by the American Rabbit Breeders Association (ARBA), of which some are more suitable than others for meat production.

Rabbit Family Tree

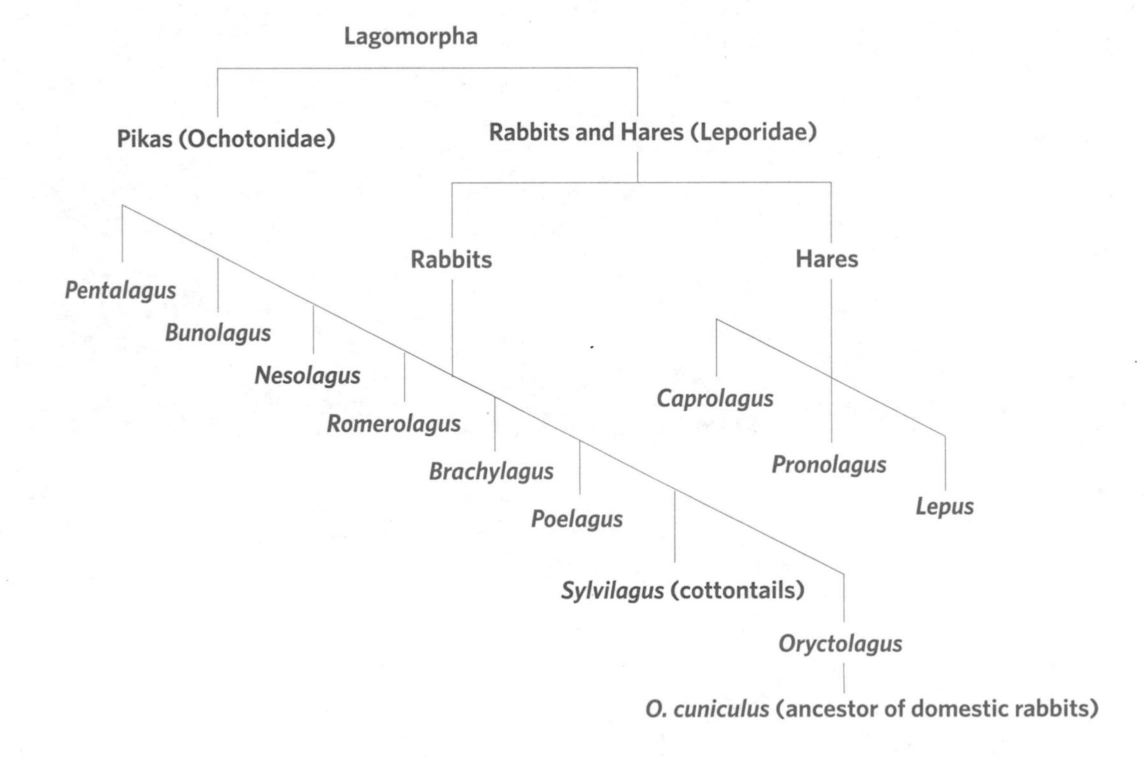

Choosing a Breed

Your choice of breed will affect many things, including the size of the cage you will need and when your young rabbits will be ready for their first breeding. A good way to learn about the different breeds is to become a member of ARBA (for contact information, see page 336). With your ARBA membership, you'll receive a guidebook that has pictures and a written description of every breed. The association also publishes a *Standard of Perfection*, which gives detailed, up-to-date descriptions of each breed.

Besides reading about the different breeds, find out where you can see rabbits in your area. If an agricultural fair is held in your community, check out the rabbit section, where you can meet local breeders. Rabbit shows held under the auspices of ARBA usually include a great variety of breeds. These shows are called sanctioned shows and usually attract exhibitors from a fairly large area. If you become a member of ARBA, the magazine that comes with your membership will list upcoming sanctioned shows in your area. Your membership also provides a yearbook that lists all ARBA clubs and members. If you contact the club nearest you, you can obtain a calendar of its scheduled events, including shows where you can see the different breeds and meet knowledgeable breeders who will be able to help you get a good start.

When you visit a rabbit show, just look and learn the first few times you go — don't buy yet. It won't be easy, because you are sure to see several rabbits you'd like to take home. Try to put off a purchase until you find a breed that appeals to you and will fill your needs. Although any breed can be raised for meat, some breeds are better suited to the purpose than others, so keep your goals in mind.

Parts of a rabbit

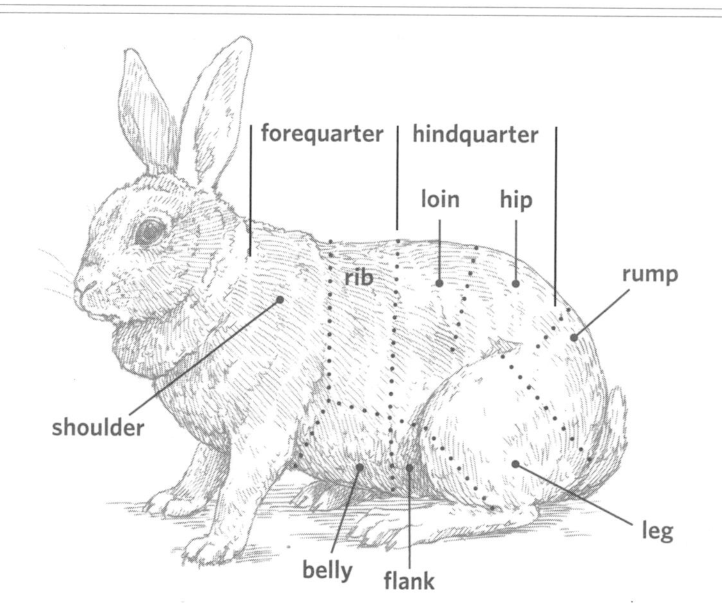

Quick Guide to Rabbit Breeds

Breed	Mature weight of buck in lbs. (kg)	Mature weight of doe in lbs. (kg)	Color(s)	Average kits/litter
American Chinchilla	9 (4)–11 (5)	10 (4.5)–12 (5.5)	blue and black	7–10
Californian	8 (3.6)–10 (4.5)	8½ (3.9)–10½ (4.75)	white with black feet, tail, ears, and nose	6–8
Champagne D'Argent	9 (4)–11 (5)	9½ (4.3)–12 (5.5)	silver	8–10
Florida White	4 (1.75)–6 (2.75)	4 (1.75)–6 (2.75)	white	5–6
New Zealand	9 (4)–11 (5)	10 (4.5)–12 (5.5)	white, red, or black	8–10
Satin	8½ (3.9)–10½ (4.75)	9 (4)–11 (5)	several	5–15

American Chinchilla

Californian

Champagne D'Argent

Florida White

New Zealand

Satin

Meat Breeds

For meat production, the large breeds, weighing from 9 to 11 pounds (4 to 5 kg) at maturity, generally convert feed to meat at the most efficient rate. They also yield an ideal fryer weighing 4 pounds (2 kg) or more at 8 weeks of age, the preferred age for harvest, and they dress out with less waste than other breeds — provided you start with quality breeders and not sale barn bunnies or freebies. Not all large breeds are ideal for meat production, and not all meat breeds grow large. The following popular breeds are worth looking into.

American Chinchilla. Developed in the United States as a dual-purpose fur-bearing and meat rabbit, the American Chinchilla is now a rare breed. It is a hardy rabbit and the does have good mothering instincts.

Californian. Developed in Southern California as a fine-boned, plump breed with full hips and a meaty saddle, ribs, and shoulders, the Californian rabbit is the second most popular meat rabbit.

Champagne D'Argent. Originating in the French province of Champagne, the Champagne D'Argent is among the oldest breeds. *Argent* is the French word for silver; the name of this breed means "silver rabbit from Champagne." Babies are born completely black and gradually turn silver as they mature. The breed is known for having medium-fine bones, resulting in a greater proportion of plump, solid meat.

Florida White. Developed in Florida, the Florida White looks like a New Zealand but weighs about half as much, making it a good choice for a small family. This compact but meaty breed has a meat-to-bone ratio that's at least as good as other meat breeds. The kits, though not as large, mature more quickly than those of other meat breeds.

New Zealand. This breed, despite its name, was developed in the United States and is considered the top meat-producing rabbit, known for its full, well-muscled body. Although New Zealands come in three colors, the white variety is generally considered to be more muscular than the red or black.

Satin. Developed in the United States, the Satin rabbit is known for its long silky fur that comes in many colors — including black, blue, chocolate, copper, red, and white — making it a dual-purpose breed suitable for both fur and meat production. The does are easy breeders with excellent mothering instincts, and the kits have a good growth rate with one of the best meat-to-bone ratios.

Crossbred. Having a buck and doe of two different breeds results in vigorous offspring that grow more quickly than either parent. Always select a smaller buck and a larger doe. A popular combination is crossing a Californian buck with a New Zealand doe. Never use a larger buck with a smaller doe, as the doe may have trouble giving birth to the larger offspring.

Recognizing a Good Animal

After deciding which breed you want, you must learn how to tell if a particular rabbit will make good stock. For the most part, that means making sure the rabbit is healthy and exhibits the proper characteristics of its breed. Pay special attention to the meat qualities of the rabbits you review for potential purchase.

Evaluating Health

Good health is the most important quality to consider when selecting your stock. Examine all of the following traits:

- Eyes should be bright, with no discharge and no spots or cloudiness.
- Ears should look clean inside; a brown, crusty appearance could indicate ear mites.
- Nose should be clean and dry; a discharge could indicate a cold.
- Front feet should be clean; a crusty matting on the inside of the front paws indicates the rabbit has been wiping a runny nose and may have a cold.
- Hind feet should be well furred on the bottoms; bare or sore- looking spots may indicate the beginning of sore hocks.
- Teeth should line up correctly at the front, with the top two teeth slightly overlapping the bottom ones.
- General condition of the rabbit's fur should be clean; the body should feel smooth and firm, not bony.
- Rear end, at the base of the rabbit's tail, should be clean, with no manure sticking to the fur.

Breed Characteristics

Each rabbit breed has distinctive features that make it different from all other breeds. These features include colors, markings, body type (the proper size and shape for its breed), fur type, and weight. The more you know about the unique characteristics of your chosen breed, the better job you will do of picking out a good individual. Detailed descriptions of breeds are found in the *Standard of Perfection*, as well as materials from breed clubs.

Most meat-producing breeds are 15 to 17 inches (38 to 43 cm) long and share the following characteristics:

- Slightly tapered from front to back
- Wide, deep body
- Plump loin
- Well-rounded hips and rump
- Hard flesh
- Little waste when dressed

Rabbit Prices

How much you can expect to pay for a rabbit varies by breed, age, availability, and popularity; it can range from $10 to more than $100 per animal. Because so many factors affect the price, it is difficult to give an average. However, if you are buying a purebred rabbit of one of the common breeds, you should be able to find a good-quality animal for $20 to $50. Make sure the purchase price includes the pedigree.

A meat-producing rabbit is well rounded at the rear and along the top (left), and slightly tapered front to back (right).

Making the Purchase

Before leaping into rabbit raising, make a good evaluation of your needs and preferences. Having the wrong type or number of rabbits can put a damper on your enthusiasm. Once you know what you're looking for, it's time to find stock. If you've done your homework, you'll probably already know where to buy stock.

How Many Rabbits?

If you are new to raising rabbits, you might start with a trio: one buck and two does. Three rabbits is a reasonable number for a beginner to care for, and a trio can become the start of a breeding program. Since a doe has eight teats, by having two does and breeding them at the same time, if one has more than eight kits and the other has fewer, you can balance out the litter size.

If you purchase a trio, buy animals that are not too closely related; you can determine their bloodlines by looking at their pedigrees. It's acceptable for the rabbits in your trio to have some relatives in common, but they should also have some differences in their family trees. They should not be brother and sisters; three rabbits from the same litter are too closely related to be used in a breeding program.

How Long Does a Rabbit Live?

Few rabbits live out their full, natural lives. Many get taken by predators, including raccoons and dogs. Rabbits raised for meat have a short life of 3 months or less. Rabbits kept as breeders are most productive between 6 months and 3 years of age, after which they may become rabbit stew. When properly fed and protected from harm, a rabbit may live an average of 6 to 8 years

Two does and one buck make an ideal start.

Choosing the Right Age Range

Rabbits have a better start in life if they remain with their mother until they are 8 weeks of age. Rabbits that have not nursed for a full eight weeks may not thrive, so before you move rabbits to a new home, be sure they have been weaned from their mother and are eating well. Weaned rabbits will quickly learn to recognize you and anticipate your visits, but older rabbits will be ready for breeding sooner. Each age range has advantages and disadvantages.

Two to three months. As a beginner, consider buying rabbits that are 2 to 3 months of age. A baby rabbit is cute, fun, and easy to handle, and you'll have the pleasure of watching it grow to adulthood. On the other hand, you'll have a wait of six months or more before your first weaned kits will be ready for the freezer.

Four to five months. Baby rabbits are adorable, but you can't always be sure exactly how they will look when they grow up. An advantage of purchasing a slightly older rabbit is that you can get a better idea of what it will look like when it matures. Rabbits that are 4 to 5 months old are still young, but their appearance will not change much further. On the other hand, older rabbits that have not been handled much will take a little more patience to learn to trust you.

Six months or older. With a 6-month-old rabbit, you'll have a pretty accurate picture of what type of adult it will be. An older rabbit will also be ready for breeding sooner. However, you'll miss the fun of watching it grow up. If it has not been handled much it may be mistrustful, and if it has been mistreated it may be downright unfriendly.

Two years or older. If you plan to breed your rabbits for meat production, don't buy animals that are more than 2 years of age. You will have only one good year before you will be looking for replacement breeders.

The best way to determine a rabbit's age is by checking its pedigree paper against its ear number. Although anything else is pretty much a guess, signs of a young rabbit include:

- A narrow lip cleft
- Smooth, sharp toenails
- Soft ears that bend easily

Making Your Selection

The breeder's knowledge and experience can be very helpful to you if he or she knows what you're looking for. You should know the answers to the following questions before you visit the breeder:

- Do you want a buck or a doe?
- What age rabbit are you looking for?
- About how much can you spend?

The breeder may have several rabbits that fit your criteria. Ask the breeder to remove the prospects from their cages so you can get a closer look. Placing the rabbits on a table will give you a chance to see them next to one another and to compare them in terms of size, markings, body type, and so on. Ask which rabbit the breeder thinks is the best and why. Use the information you've learned as well. Many of the conditions you'll want to know about require handling the rabbit. If you are not an experienced handler, ask the breeder to help you check the rabbit's teeth, sex, and toenails.

If you're interested in a young rabbit, ask to see its mother and father so you can get an idea what the rabbit should look like at maturity. Ask to see written records for the parents that indicate how well they have performed. These include the parents' pedigrees and records that provide information such as how many kits the mother has raised, how many kits the buck has sired, and perhaps information on how well the parents have placed at rabbit shows — which tells you how closely they conform to the ideal traits for their breed.

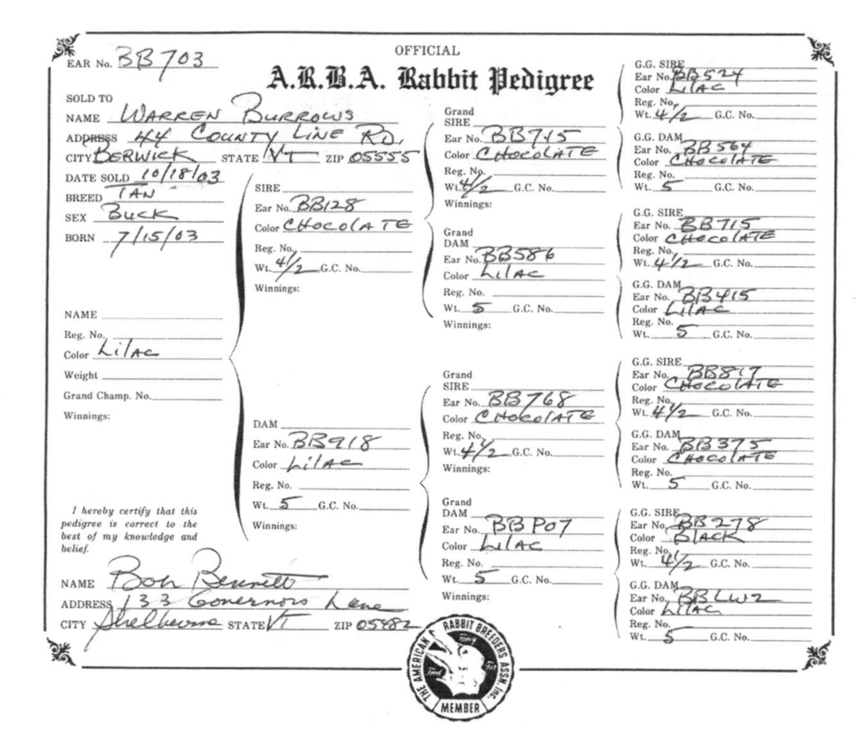
OFFICIAL
A.R.B.A. Rabbit Pedigree

EAR No. BB703
SOLD TO
NAME WARREN BURROWS
ADDRESS 44 COUNTY LINE RD.
CITY BERWICK STATE VT ZIP 05555
DATE SOLD 10/18/03
BREED TAN
SEX BUCK
BORN 7/15/03

NAME
Reg. No.
Color LILAC
Weight
Grand Champ. No.
Winnings:

SIRE
Ear No. BB128
Color CHOCOLATE
Reg. No.
Wt. 4 1/2 G.C. No.
Winnings:

DAM
Ear No. BB918
Color LILAC
Reg. No.
Wt. 5 G.C. No.
Winnings:

Grand SIRE
Ear No. BB715
Color CHOCOLATE
Reg. No.
Wt. 4 1/2 G.C. No.
Winnings:

Grand DAM
Ear No. BB586
Color LILAC
Reg. No.
Wt. 5 G.C. No.
Winnings:

Grand SIRE
Ear No. BB768
Color CHOCOLATE
Reg. No.
Wt. 4 1/2 G.C. No.
Winnings:

Grand DAM
Ear No. BB P07
Color LILAC
Reg. No.
Wt. 5 G.C. No.
Winnings:

G.G. SIRE Ear No. BB524 Color LILAC Reg. No. Wt. 4 1/2 G.C. No.
G.G. DAM Ear No. BB564 Color CHOCOLATE Reg. No. Wt. 5 G.C. No.
G.G. SIRE Ear No. BB715 Color CHOCOLATE Reg. No. Wt. 4 1/2 G.C. No.
G.G. DAM Ear No. BB415 Color LILAC Reg. No. Wt. 5 G.C. No.
G.G. SIRE Ear No. BB817 Color CHOCOLATE Reg. No. Wt. 4 1/2 G.C. No.
G.G. DAM Ear No. BB375 Color CHOCOLATE Reg. No. Wt. 5 G.C. No.
G.G. SIRE Ear No. BB278 Color BLACK Reg. No. Wt. 4 1/2 G.C. No.
G.G. DAM Ear No. BBLW2 Color LILAC Reg. No. Wt. 5 G.C. No.

I hereby certify that this pedigree is correct to the best of my knowledge and belief.
NAME Bob Bennett
ADDRESS 133 Governors Lane
CITY Shelburne STATE VT ZIP 05482

A typical pedigree paper should be included in the animal's price of purchase.

Don't make a final decision until you are sure the animal you are interested in is healthy; has no eliminations or disqualifications; meets the size, color, and other characteristics required of its breed; and meets your own criteria, including price. If you have examined the animal carefully and it has passed your inspection, the parent stock looks good, the rabbit meets your criteria, and you and the breeder agree on price, you have found your rabbit.

Final Purchase Details

Once you've found your rabbit, here are some details to attend to before you take it home.

Pedigree papers. If you are purchasing a purebred rabbit, the price should include the pedigree paper. An organized breeder will have a pedigree paper ready. (See pages 134–35 for more about pedigree and registration.)

Feed. Ask the breeder what kind of food the rabbit is currently receiving, how much it receives, and how it's fed. This information will help you smooth the rabbit's transition to its new home. If you plan to use a different feed from that used by the breeder, make the change gradually. Ask the breeder to supply you with about 1 pound (0.5 kg) of the rabbit's current feed. Give this feed to the rabbit for the first day or two. Then begin to mix the breeder's feed with yours, gradually making a complete changeover to the new feed.

Guarantee. Ask about the breeder's policy concerning problems that may arise with your new rabbit. Most breeders will guarantee the health of their rabbits for at least two weeks. After two weeks, it becomes difficult to know whether a problem started at the breeder's rabbitry or was acquired after the rabbit was sold.

Handling Your Rabbit

Repeated gentle handling will make your rabbit easier to handle, but you must observe certain points of rabbit etiquette. Being picked up can be scary for a rabbit. When you lift a rabbit, you take away its most effective method of defense — the ability to run away from danger. With practice, your skill will increase.

If your rabbit is frightened, it will try to run away. Its toenails help it grip surfaces, enabling it to run faster. When you lift the rabbit, you become the only surface it has to grip. This desire to grip often results in scratches for the handler and a panic situation for both the handler and the rabbit. Many new owners become discouraged when their rabbit scratches. Remember, your rabbit is scared — it's not mad at you or scratching you on purpose. Make things easier on yourself by wearing a long-sleeved shirt when you handle a rabbit.

Rabbits have powerful hind leg and back muscles. If not properly supported, a rabbit can kick out and break its own back, causing paralysis.

Some breeds handle more easily and respond better than others. Most rabbits enjoy being stroked, especially when they are young. Remember always to talk calmly and quietly. Loud noises easily startle rabbits.

The time you spend learning proper handling skills will pay off in many ways. Rabbitry chores are easier if your animals cooperate during cage cleaning, grooming, and breeding.

Getting Acquainted

A new rabbit has many adjustments to make. Give it a few days to get used to you and its new surroundings before you handle it a lot.

Get to know your rabbit in a setting where both of you are comfortable. A good place to get acquainted is at a picnic table covered with a rug, towel, or other covering that will give the rabbit secure, not slippery, footing. Your rabbit will be able to move safely around on the table, and you can safely pet it without having to lift it. Rabbits will usually not jump off a table; nonetheless, never leave the rabbit unattended when it is out of its cage.

Once your rabbit seems comfortable on the table, start to practice picking it up. The table offers a handy surface on which to set the rabbit safely back down.

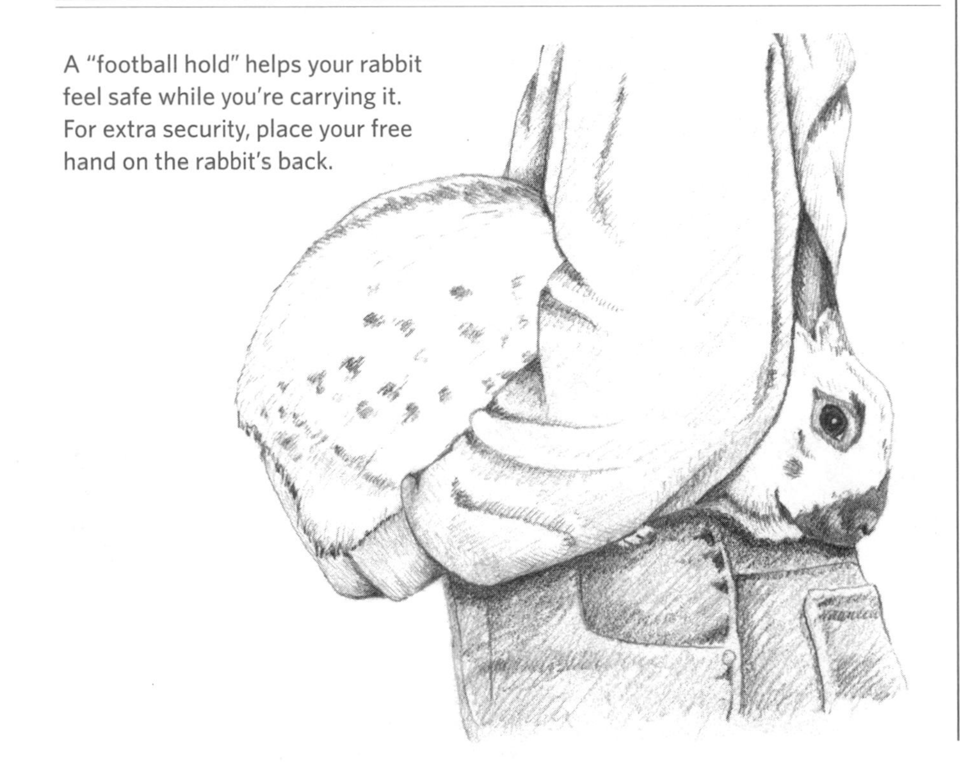

A "football hold" helps your rabbit feel safe while you're carrying it. For extra security, place your free hand on the rabbit's back.

Lifting Your Rabbit

The best way to lift a rabbit is to place one hand under it, just behind its front legs. Place your other hand under the animal's rump. Lift with the hand that is by the front legs and support the animal's weight with your other hand. Place the animal next to your body, with its head directed toward the corner formed by your elbow. Your lifting arm and your body now support the rabbit; it's like tucking a football against you for a long run. Place the animal gently back on the table and repeat this lift.

Practice handling skills often, but for short periods of time — about 10 to 15 minutes daily is all it takes. Have a short practice session each day until you and your rabbit are comfortable with each other.

Once you feel at ease, begin to move around while holding your rabbit. You may need to steady and secure your animal with your free hand until it gets used to being carried. At first, just walk around the table; if your rabbit becomes frightened, you have a safe place nearby to set it down promptly.

If you need extra control, you can pick up your rabbit with one hand by grasping the loose skin at the nape of its neck, providing support under its rump with your other hand. This method can damage the fur and flesh over the rabbit's back, however, and is especially harmful to the more delicate coats of Satin rabbits. Once you have lifted the rabbit, move it close to your body where it will feel more secure. After it's tucked in football-style, you can use a one- or two-hand carry. The rabbit's behavior will help you decide whether to use one or both hands.

Even the most mild-mannered rabbit can have a bad day, so be prepared to handle any situation in a calm, controlled manner. For instance, if an overactive bunny struggles and gets out of control, drop to one knee as you work to quiet the animal. Lowering yourself lessens the distance for the rabbit to fall and provides a position from which you can easily set the animal on the ground, if necessary. After a short rest on the ground, carefully and securely lift the rabbit again.

Turning Your Rabbit Over

After you master lifting and carrying your rabbit, learn how to turn it over. In the future, as part of your health maintenance program, you'll need to get your rabbit on its back in order to examine its sex, teeth, and toenails, so it's worth your time to help your rabbit become accustomed to being held on its back.

Turning a rabbit over puts the animal into an unnatural position. First, its feet are off the ground, so it cannot run away, and second, its underside is now exposed. Your rabbit has good reasons to resist this type of handling — a wild rabbit in this position is probably about

How to Hold a Rabbit

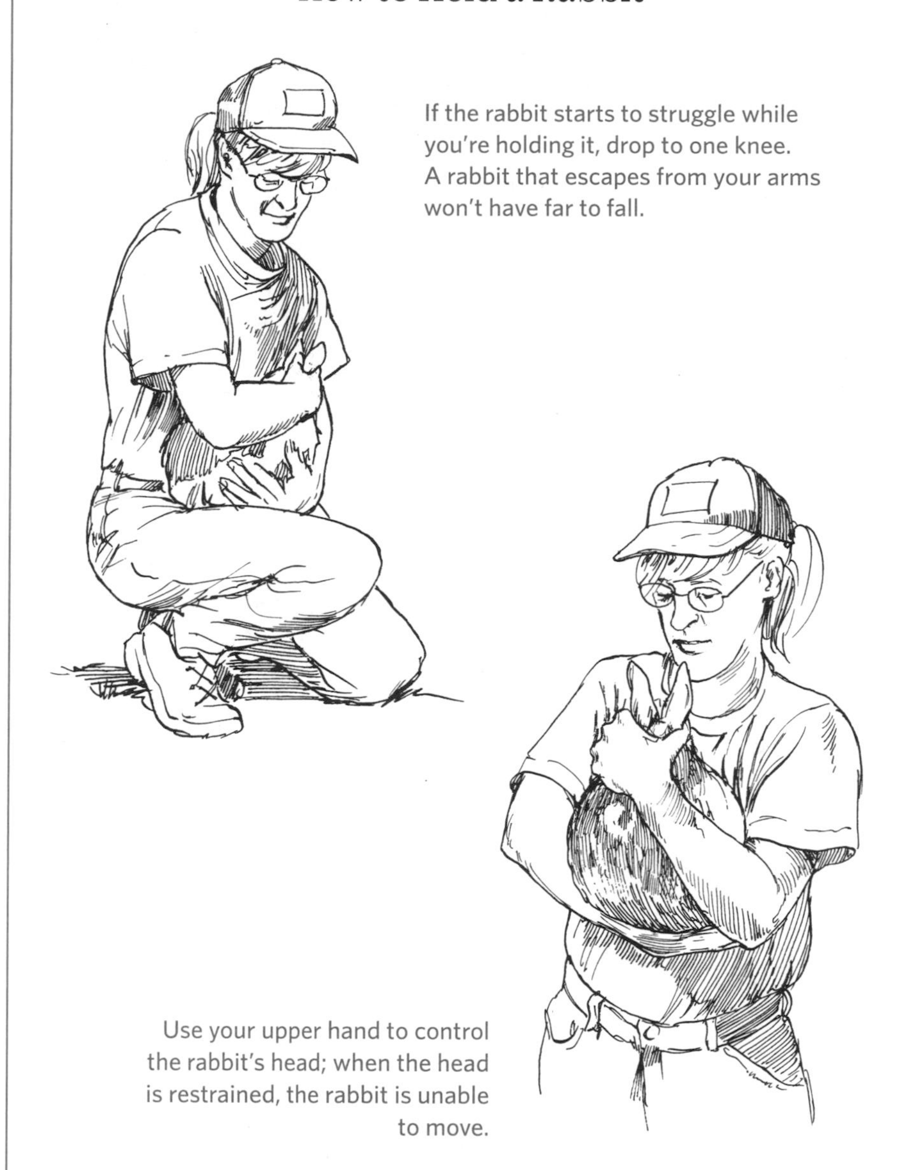

If the rabbit starts to struggle while you're holding it, drop to one knee. A rabbit that escapes from your arms won't have far to fall.

Use your upper hand to control the rabbit's head; when the head is restrained, the rabbit is unable to move.

An easy trick to getting the rabbit on its back is to cradle it against your chest and then slowly lean forward over a table, until the rabbit's back is on the table's surface.

to be eaten by a predator — so be especially careful and patient. Again, a rug-covered picnic table is a good place to practice.

To turn the rabbit over, use one hand to control the head and the other hand to control and support the hindquarters. Place the hand that holds the rabbit's head so you are holding its ears down against its back while you reach around the base of the head. If you prefer, place your index finger between the base of the rabbit's ears and then wrap your other fingers around toward its jaw. With your other hand, cradle the rump. With your hands in place, lift with the hand that is on the head and at the same time roll the animal's hindquarters toward you. Try to do this movement in a smooth, unhurried manner.

If your rabbit has cooperated, you will now be able to let the table support its hindquarters so the hand that was holding the rump is free to check the rabbit's teeth and toenails. If your rabbit fights against this procedure, keep trying, but do so in a place where it hasn't far to fall, and try to support it securely.

Part of the trick to successfully turning your rabbit over is being able to grasp its head so it cannot wiggle away. A way to gain some extra control is to grasp the lower portion of the loin instead of cradling the rump.

Another approach is to pick up your rabbit and allow it to rest against the front of you instead of tucking it against your side. The animal will face upward with its feet against you. When it is in this upright position, place one hand on its head, as suggested above, and keep your other hand supporting its hindquarters. Now bend forward slowly at the waist and lower the animal to the table, where it should arrive in the proper turned-over position.

If you need to turn your rabbit over for a closer look and no table is handy, let the animal rest on your forearm. Having the rabbit in this position gives you better control over the hindquarters, because they are tucked between your elbow and your body. This maneuver is easier if you sit in a chair. You can use your legs instead of the table to support your rabbit. You may find that when you are seated you can hold your rabbit more securely for such procedures as trimming toenails.

Meat Rabbits

Rabbit meat tastes a bit like chicken; some people can't tell the difference. Rabbit, however, is all white meat that is leaner than chicken, beef, pork, and even turkey — making it perfect for anyone on a fat- and cholesterol-reducing diet. Rabbits are harvested at one of three stages, depending on their age.

A fryer is a young rabbit under the age of 12 weeks, by which time it will have reached 4 to 5 pounds (2 to 2.5 kg) live weight (about 3 pounds [1.5 kg] for a 10-week-old Florida White). The meat is tender, fine grained, and bright pearly pink, and can be cooked in the same ways as a chicken fryer. Since older bucks tend to taste, well, bucky, they are best harvested at the fryer stage.

A roaster is older and larger than a fryer, generally grown by people who want more meat than a fryer provides. Its live weight is usually between 5½ and 8 pounds (2.5 to 3.5 kg), typically close to 6 (2.75 kg). The meat is firm, coarse grained, slightly darker in color, and less tender than a fryer, and the fat is of a creamier color and stronger flavor than that of a fryer. Cooking with liquids or oil helps keep a roaster moist and tender.

A stewer is generally older than 6 months, weighs more than 8 pounds (3.5 kg) live weight, and has more fat and tougher meat than a roaster. The best cooking methods for a stewer are braising and stewing. Of course, a younger rabbit makes great stew, but moist cooking is required to tenderize the meat of an older rabbit.

Rabbit Giblets

Rabbit giblets include the heart, kidneys, and a liver that is large in relation to the animal's size. Even if you don't ordinarily care for liver, you might enjoy rabbit liver. Compared to other kinds of liver, especially liver sold at the supermarket, rabbit liver has nearly no odor, is far more tender, and has a sweet, clean flavor. The kidneys, too, are tender and somewhat sweet.

Feed Conversion

The average feed conversion ratio for a rabbit is about 4, meaning it gains 1 pound (0.5 kg) of weight for every 4 pounds (2 kg) of feed it eats — assuming, of course, the rabbits eat everything they are fed and don't scratch it out of the hopper onto the ground. Assuming you start with a pair of 4-month-old breeders, by the time you have your first litter of fryers they will have eaten

Rabbit Meat Facts

- A rabbit dresses to between 50 and 55 percent of live weight, not counting giblets.
- Only 7 to 8 percent of a rabbit is bone.
- Rabbit meat has little marbling; fat surrounds the muscle and can be easily peeled away.
- Rabbit meat is higher in protein than any other meat (nearly 21 percent compared to the next highest, turkey and chicken, at 20 percent).
- A doe that produces five litters a year, averaging eight kits per litter, yields 120 pounds (54 kg) of meat or more each year.
- Since rabbits are raised in off-the-ground hutches, their meat is cleaner than that of animals living on the ground.

Parts of a Butchered Rabbit

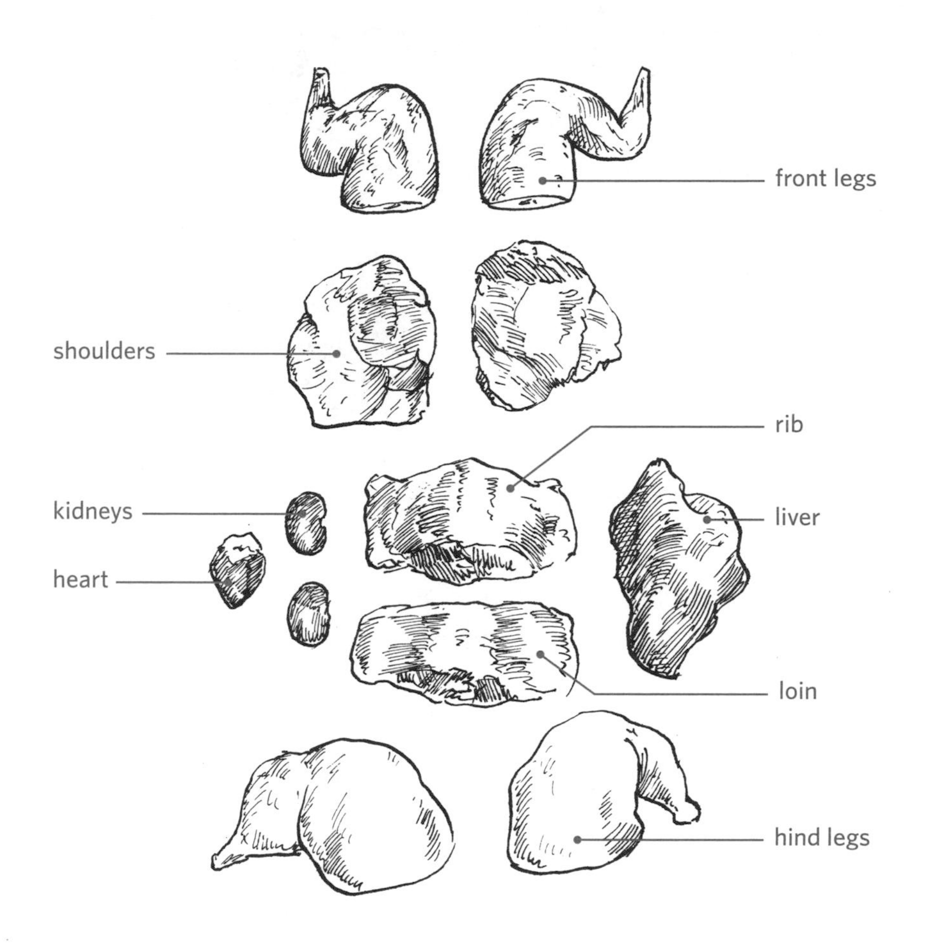

a total of about 130 pounds (59 kg) of rabbit pellets.

Can you save money by raising your own meat rabbits? That depends on the price of feed, how efficiently your rabbits grow, and the weight at which they are harvested. Using averages (and for simplicity's sake, ignoring the cost of housing, water, electricity, and so forth) do the math: By the time a rabbit fryer reaches the live weight of 4 pounds it will have eaten at least 16 pounds of pellets. At 50 percent of live weight, a 4-pound rabbit dresses out to approximately 2 pounds. Your break-even cost can be calculated by comparing the purchase price of a rabbit with the cost of rabbit pellets. At current rates, a whole fryer sells for approximately $6 per pound, or about $12 for a 2 pounder. The current rate for one brand of all-natural rabbit pellets is about 25 cents per pound, or about $4 for 16 pounds, which is one-third the value of the meat. Even if you pay 50 to 100 percent more to purchase certified organic feed, it's still a good deal.

Let's see what happens when we factor in the cost of feeding the breeder buck and doe. By the time the pair is 2 years old they should have produced seven litters averaging eight kits per litter, or 48 fryers weighing about 4 pounds each live or 2 pounds dressed, for a total of 96 pounds of rabbit. At a purchase price of $6.00 per pound, that comes to a total market value of $576.00. By that time, you would have fed the 48 fryers plus two breeders approximately 840 pounds of rabbit pellets at a cost of $210.00 dollars. Your total cost of producing 96 pounds of rabbit meat is a little less than one-third its market value, and your cost per pound is about $2.20.

Finding a Processor

Rabbit processors are few and far between, and most of those that do exist process for commercial rabbitries that produce meat for sale. Butchering your own rabbits is easy and takes little more than 15 or 20 minutes from start to finish. If you don't have someone nearby who can show you how to kill, skin, and dress a rabbit, you can easily learn from an illustrated book such as *Storey's Guide to Raising Rabbits* by Bob Bennett.

Rabbit Housing

Rabbit housing, called a hutch, can be the largest single expense in rabbit keeping. You can buy one or build your own from largely recycled materials, but be sure it meets the needs of your animals. Proper housing will contribute greatly to the health and happiness of your rabbits.

An all-wire hutch is the first choice of an experienced rabbit raiser. Wire hutches come in two basic styles. The first type is intended to be suspended so the manure falls to the ground. The second type includes a pull-out tray the manure falls into. These cages, too, can be hung, but they can also be stacked on top of one another. Stacking saves space but can make it a little harder to reach each cage.

Protection from Weather and Predators

With proper care, rabbits have fewer health problems when raised outdoors. However, if you keep your rabbits indoors, you must have excellent ventilation. Outdoor housing must be adjustable to accommodate changes in weather. Where you live determines what weather conditions you will have to deal with. Rabbits are most comfortable at temperatures of 50 to 69°F (10 to 20°C). In hot climates, keep the hutch in the shade and make sure it has plenty of air circulation around it.

Hutch Size

Your rabbit will spend most of its time in its hutch and will need space to move around, as well as space for feeding equipment, and a doe needs room for her nest. A hutch 30 × 36 × 18 inches high (76 × 91 × 45 cm) is suitable for most meat breeds, although a 24-by-36-by-18-inch (61 × 91 × 45 cm) hutch is adequate for the smaller Florida White.

Wrap plastic sheeting around three sides of the cage to protect the rabbit from cold winds and precipitation.

Rabbits do well in cold weather and can survive temperatures well below freezing. However, they need to be protected from wind, rain, and snow. If your cages are outside, you'll need to add protection as the temperature drops. Most outdoor cages can be enclosed by stapling plastic sheeting around three sides. If the weather gets cold in your region, cover the front with a flap of plastic as well. Do not enclose the whole cage so tightly that it is difficult to feed and care for your animal or that the ventilation is poor. Outdoor cages should also have a secure roof to keep out rain and snow. Try to take advantage of the winter sun. If your outside cages are movable, place them in a location that receives a lot of direct sun. Make sure the cage is not exposed to the wind.

A sturdy, secure hutch will contribute to the safety of your rabbits. Most cages are built on legs or hung above the ground, which protects them from dogs, cats, rodents, raccoons, and opossums. Keeping your cages in a fenced area or indoors is even better but may be impossible when you're just getting started.

Ease of Cleaning

Clean cages are important to the health of your rabbits. If your cages are easy to clean, you will be able to do a better job of caring for your rabbits. Cages should have wire floors that allow droppings to fall through to a tray or to the ground. Choose all-wire cages or wood-framed cages that do not have areas where manure can pile up. Plan your cages so you can easily reach into them and clean all parts.

Wire Choices

As you look at cages, you will discover many types of wire. The most common wire used for the sides and tops of hutches is 14-gauge wire woven in a 1 inch by 2 inch (2.5 × 5 cm) mesh. Smaller wire mesh also works well but will probably be more expensive. Sometimes you may have a choice of 14- or 16-gauge wire. The 14-gauge wire is heavier and stronger, and should be used for the large breeds.

You may also need to choose between wire that is welded before being galvanized and wire that is welded after being galvanized. Wire that is galvanized after welding is stronger and smoother for the rabbit's feet.

The floor of the cage should be made of 14-gauge welded wire woven in

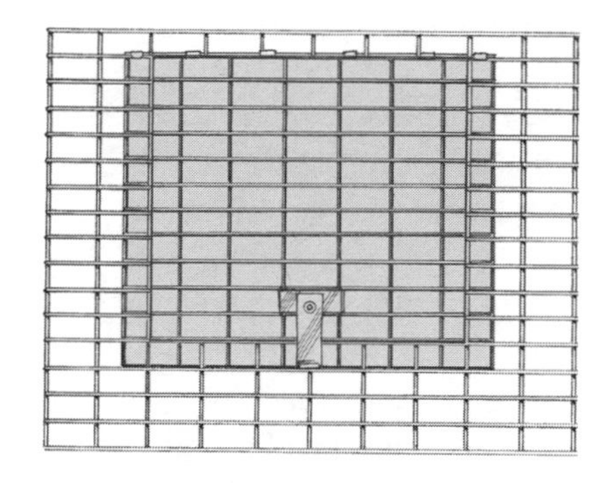

The door should overlap the cage on three sides. If it opens out, attach it to the cage on one side; if it opens in, attach it at the top.

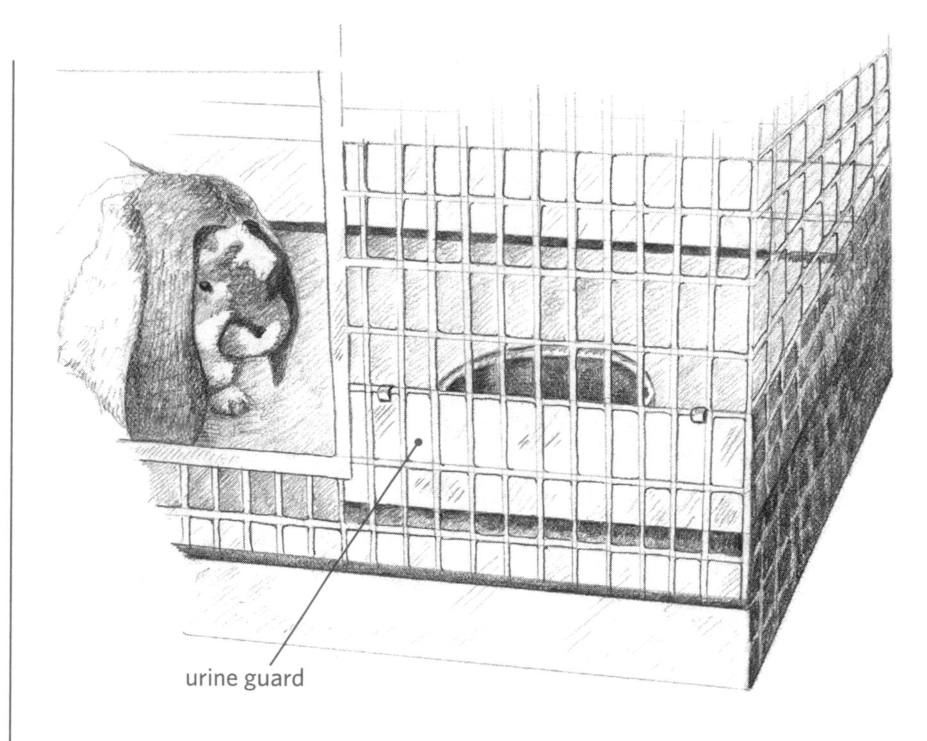

A urine guard helps keep the area around the cage clean.

a ½ inch by 1 inch (1.25 × 2.5 cm) mesh. This smaller mesh gives more support to the rabbit's feet but still allows manure to pass through. Never use larger wire for the floor of the hutch. Rabbits can get their feet caught in the openings and break or dislocate their rear legs.

Door Designs

The door opening is cut into the front of the cage. The door itself should be larger than the opening so it overlaps the opening on all sides. Doors should be placed toward one side of the front of the cage, so there is enough space left on the other side for the feeder and waterer.

A door that is attached at the top and swings into the cage when opened will fall back to a closed position even if you forget to latch it, and the rabbit cannot push it open. Cages with doors that open toward the outside make it easier for you to reach into the cage, but if you forget to latch the door, your rabbit may push it open and tumble out of the cage.

Check the door and door opening for sharp edges. You don't want a cage that will scratch you or your rabbit. Some cage doors are lined with metal or plastic to protect you from sharp edges.

Feeders

Most rabbit cages are displayed with an attached feeder. However, the list price of the cage usually does not include the feeder. If the feeder doesn't come with the cage, you'll need to buy one. (See page 122 for information about feeding equipment.)

Baby-Saver Wire

Some manufacturers offer cages made from baby-saver wire, a specially designed wire mesh that can be used for cage sides. The upper portion of baby-saver wire has 1 inch by 2 inch (2.5 × 5 cm) mesh, and the bottom 4 inches (10 cm) has ½ inch by 1 inch (1.25 × 2.5 cm) mesh. Having this smaller mesh next to the hutch floor prevents babies from slipping through the wire, keeping them safely inside the hutch. Baby-saver wire is more expensive than 1 inch by 2 inch (2.5 × 5 cm) wire mesh, but makes a wise investment for a cage housing a producing doe.

Urine Guards

A urine guard is a common extra feature on rabbit cages. Four-inch (10 cm)-high metal strips attached around the inside base of the cage help direct the rabbit's urine so it falls directly below the cage floor. A urine

Managing Manure

Besides keeping urine from splashing where it is unwanted, urine guards help keep rabbit manure confined directly beneath the hutch for handy removal to the garden or compost bin. Rabbit manure is higher in nitrogen and phosphorus than most livestock manure and contains more potash than most. It can be applied directly to garden plants without danger of burning the plants or causing illness to humans who eat the resulting vegetables.

guard also helps prevent baby bunnies from becoming stuck in or falling through the sides of the cage. Urine guards are not, however, a substitute for baby-saver wire unless they extend all the way around the bottom of the cage. Cages with urine guards are more expensive than those without them.

Buying a Cage

For a small rabbitry, buying cages is usually less expensive than building them. The price suppliers charge for cages is based on the cost of the cage materials. By purchasing materials in large quantities, suppliers get a discount and can charge less per cage. When you purchase materials to build just two or three cages, your cost per cage will likely be as much as or more than that of a ready-built cage.

Purchase the best possible cages you can afford; your rabbits will, after all, be spending most of their time there. You will need one cage for each rabbit, plus you may want to have one or two spares in case you don't get around to butchering all the kits when they need to be removed from the doe's hutch. Rabbit cages are available from several sources:

Mail-order rabbitry supply companies. Rabbitry supply companies specialize in cages and equipment for rabbit breeders. They generally offer the best quality, selection, and price. Most rabbitry supply companies have online catalogs or will send you a free catalog if you request one. A list of several suppliers is included in the appendix.

Farm supply stores. The same store where you buy your feed may also sell cages. Prices will probably be fairly reasonable and the quality acceptable.

Local rabbitry suppliers. Many small local suppliers also sell cages. These folks are often rabbit enthusiasts who build and sell cages and other supplies as a part-time business. However, it may be a challenge to find one of these suppliers near you. Talk with other rabbit raisers to find a supplier that serves your area.

Pet shops. Pet shops often carry rabbit cages, but their products may be more expensive than from other sources, and unless they specialize in rabbits, their supply may be limited. Ask a salesperson to assist you in selecting a cage, and if the selection seems limited, ask to browse through the stock catalogs and special-order a cage.

Other rabbit raisers. Many rabbit keepers begin with used cages. This option is usually a less expensive way to get started, but before you use secondhand equipment do the following:

- Check carefully for holes, rusted wire, chewed wooden areas, and sharp edges. Repair any problems before you introduce your rabbits to their new home.
- Carefully clean and disinfect the cage before you use it. Start by scraping off old manure and hair. Then wash the entire cage with a solution of bleach and water (mix 1 part household chlorine bleach with 5 parts water). Set the cage in a sunny spot and allow it to dry thoroughly before you put a rabbit into it.

Purchase the best possible cages you can afford; your rabbits will, after all, be spending most of their time there.

Rabbit Carriers

If you need to transport a rabbit from one location to another, you'll need a rabbit carrier. Like cages, the best rabbit carriers are made of wire mesh. Purchase a good rabbit carrier if you plan to transport your rabbits often or for long distances. A carrier designed for transporting meat-sized rabbits is 12 inches (30 cm) high, has a top that opens for safe and easy removal of the rabbit, and has a wire floor with a tray below to collect droppings.

For infrequent use or short trips, you can use a standard pet carrier designed for transporting cats or small dogs. Whatever type of carrier you use, it should not be so roomy the rabbit bounces around in transit, risking bruises or other injuries.

Building a Hutch Frame

If your all-wire cage will be used outdoors, you will need to adapt it to protect your rabbit from various weather conditions. The most common way to protect a hutch is to build a wooden framework into which you set the cage. A wooden frame has features that make it adaptable to a variety of situations:

- The sides can be left open for good air circulation.
- Removable, solid wooden sides or plastic sheeting can be attached for protection against cold weather.
- The cage can be removed from the frame for easy cleaning.
- The cage can be removed from the frame to move an expectant doe to a warmer location for a winter kindling.

The following plan for a single-hutch frame can be expanded to accommodate a multiple-unit hutch.

Materials

- Four 10-foot lengths of 2 × 4 lumber
- One 8-foot length of 2 × 4 lumber
- One 4-by-5-foot sheet of exterior-grade plywood (¾" thickness)
- 12d common galvanized nails
- 6d common galvanized nails
- 8 L-brackets
- Wood screws

Tools

- Saw
- Tape measure
- Hammer
- Screwdriver
- Pencil

Cut the Legs

1. Cut two 60-inch lengths from one of the 10-foot 2 × 4s to form two front legs.

2. Cut two 56-inch lengths from a second 10-foot 2 × 4 to form two (shorter) back legs.

Cut the Support Pieces

3. From each of the two remaining 10-foot 2 × 4s, cut:

- One 35¼-inch piece (for the side of the cage-support frame)
- One 38-inch piece (for the front and back of the cage-support frame)
- One 44⅛-inch piece (for the front and back roof support)

4. Cut the 8-foot 2 × 4 in half. These pieces will be the side roof supports. You will later recut them to length.

Preassembled versus Do-It-Yourself

Suppliers often give you a choice of buying the cage assembled or unassembled. Assembled cages are more expensive and, if you order them from a catalog, cost more to ship. If you choose to save money and buy an unassembled cage, you'll need a tool called J-clip pliers. If you have one or just a few cages to assemble, you may be able to borrow J-clip pliers from a local rabbit raiser. If you are going to assemble lots of cages, purchase one of these tools.

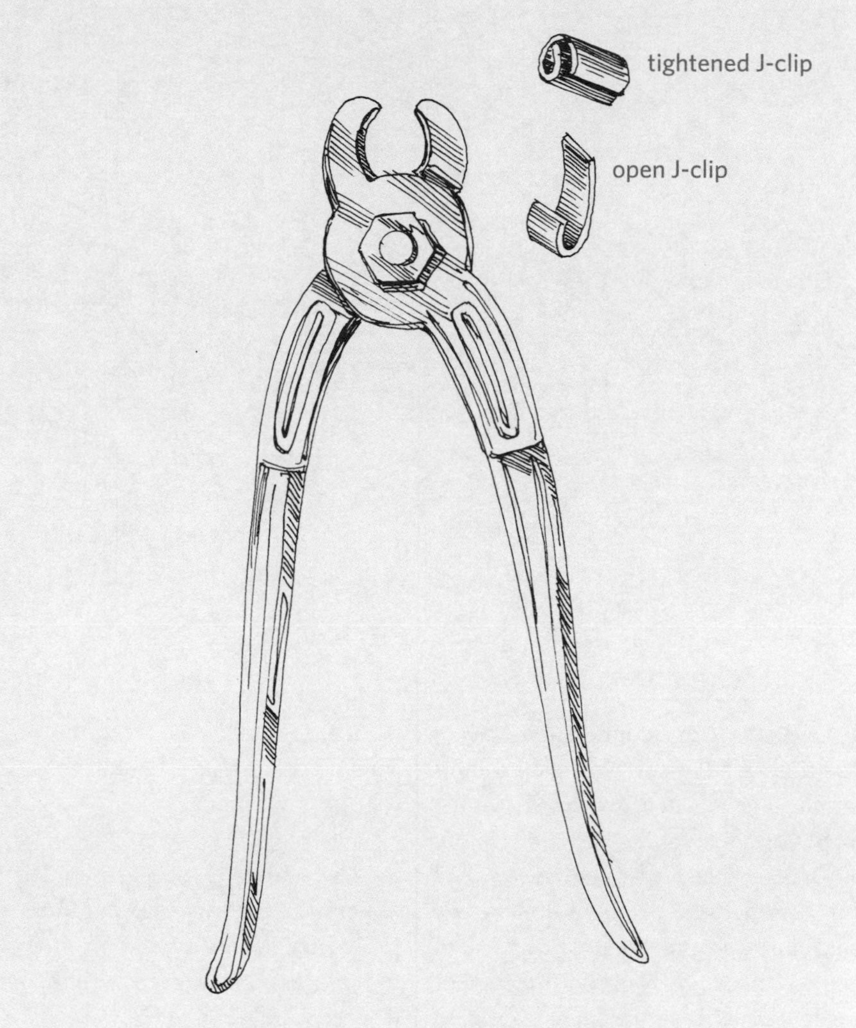

J-clip pliers are used to assemble a cage.

Frame for Single Hutch

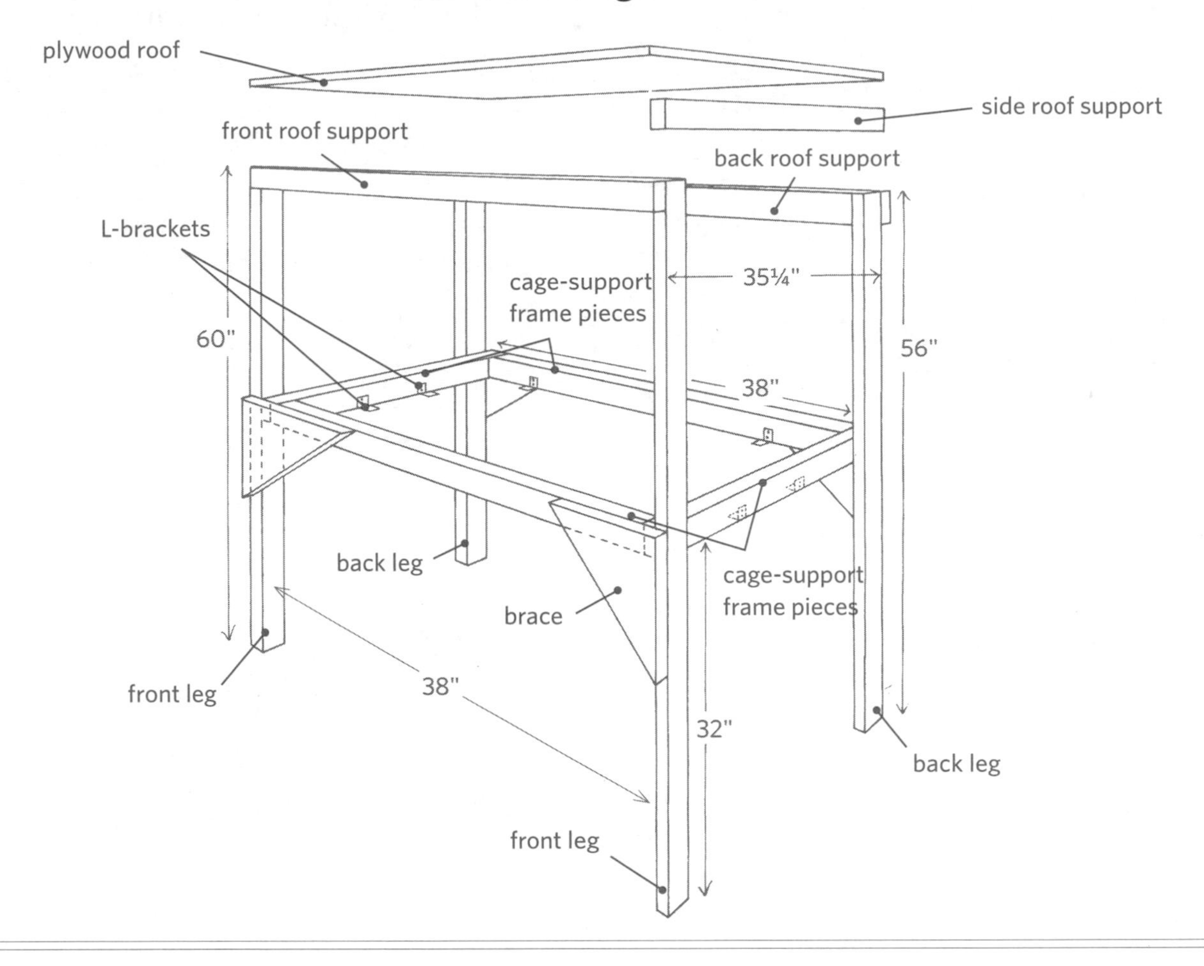

Put It All Together

5. Lay the four cage-support frame pieces on edge on a flat surface. They should form a large rectangle, with the 38-inch front and back pieces between the 35 ¼-inch side pieces. Drive two 12d nails through the outside face of a side piece into the end of the back piece to make the corner. Nail the three other corners together in the same fashion to complete the cage-support frame.

6. Screw the L-brackets to the inside face of the cage-support frame, as shown in the illustration above. The bottom of each bracket should be flush with the bottom edge of the frame.

7. Mark one end of each of the legs with a pencil to indicate the bottom and to avoid confusion during assembly. Measure up from the bottom of each leg 32 inches, and draw a straight line across the inside face of each leg at this point. This mark represents the height of the cage-support frame.

8. Use 12d nails to attach the legs to the ends of the cage-support frame (as shown above, making sure that the bottom of the frame is even with the lines marked on the legs.

9. Use 12d nails to attach the front and back roof supports to the legs. The top edge of the front roof support should be slightly higher than the top of the front legs to prevent the front legs from being in the way when it comes time to put on the plywood roof.

10. Take one of the side roof supports and hold it in place against the legs on the right side of the hutch. With a pencil, mark the angled cut you'll need to make at the side roof support's ends so it fits snugly between the tall front legs and the short back legs. Cut the piece to size, and attach it with 12d nails to the front and back roof supports. For added strength, nail it to the tops of the legs as well. Repeat this procedure for the left side.

11. Use scraps of plywood to cut four isosceles triangles, with the equal sides being 12 inches long. Nail these braces to the hutch with 6d nails, as shown.

12. Use 6d nails to attach the 4-by-4-foot plywood roof. The roof should overhang all four sides, to help keep rain or snow out of the hutch. A large overhang along the front will give extra protection to the attached feeders.

Feeding Your Rabbit

Providing a healthy, balanced diet is key to successful rabbit raising. In the past, rabbit owners mixed different kinds of feeds to get the proper nutritional balance for their animals. Today, stores carry feed that is already balanced. However, a good feeding program requires much more than simply filling up the feed dish. Several types of ration are available to suit specific needs.

Commercial Rabbit Feed

A wild rabbit picks and chooses what it eats to get a balanced diet. Your caged rabbits depend on you to satisfy their nutritional needs. Feeding commercial rabbit pellets is the best and easiest way to provide proper nutrition. A great deal of research goes into producing a balanced ration, and feed companies are well qualified to formulate rations. Commercial feeds vary in quality and price, and different feeds are available in different locations. Talk with other breeders and your feed store clerk to learn about brands of feed that are available in your area.

Rabbit Feed Guidelines

Type	Optimal feed content
Mature rabbits and developing young	12–15 percent protein 2–3.5 percent fat 20–27 percent fiber
Pregnant does and does with litters	16–18 percent protein 3–5.5 percent fat 15–20 percent fiber

Caution

Never feed greens to young rabbits. Greens can give the youngsters diarrhea and can kill them.

Rabbit feed is available in several types, each intended to meet the needs of a rabbit at a different stage in its life. The ingredient that usually varies among these types of feed is the protein content. Mature animals usually need feed that provides 12 to 15 percent protein; active breeding does usually need 16 to 18 percent protein.

The label on a feed package will tell you not only how much protein but also how much fiber and fat the feed contains. For optimal health, follow the feed guidelines in the chart at left.

Salt

Salt is important to a balanced feed. You may see advertisements for inexpensive salt spools that can be hung in your rabbit's cage. Using such spools is not necessary, and they will cause your rabbit's cage to rust. Commercial rabbit pellets include enough salt to meet a rabbit's needs.

Hay

Rabbits enjoy good-quality hay, and it adds extra fiber to their diet. If you feed hay, be sure it smells good, is not dusty, and is not moldy. And remember that feeding hay will add more time to your feeding and cleaning chores.

If you choose to feed hay on a regular basis, place a hay manger in each cage. Hay mangers are easy to construct from scraps of 1-by-2-inch (2.5 × 5 cm) wire mesh. If your cages are inside and are all-wire construction, you can just place a handful of hay on the top; the rabbits will reach up and pull it through the wire. Do not place hay on the floor of the cage because it will soon become soiled. The droppings that collect on it will promote disease and parasites.

A covered metal garbage can keeps pellets clean and dry and keeps mice out; a scoop is handy for feeding.

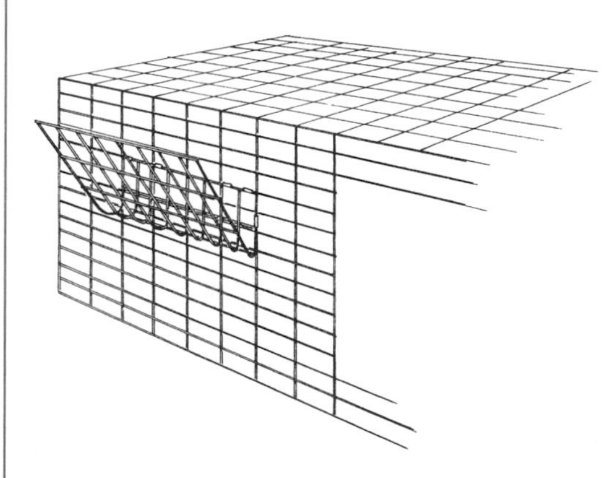

To make a simple hay manger, bend a scrap of wire mesh and attach it to the sides of your cage with J-clips. Make sure no sharp wire edges protrude into the cage.

Special Feeds and Treats?

Experienced rabbit raisers often give their animals special feeds, such as oats, sweet feeds, corn, and sunflower seeds. However, any additions to the diet can upset a rabbit's digestion. It is best to keep to a simple diet of pellets and water, perhaps supplementing with some hay. Avoid additional feeds until you have several years' experience in successful rabbit raising.

Many new rabbit raisers like to pamper their animals by giving them fresh greens and vegetables as treats. But the fact is that many of these treats can be harmful to rabbits. Because young rabbits have especially sensitive digestive systems, it's best not to give them any treats. Absolutely do not give young rabbits *any* greens, including lettuce — greens can cause diarrhea. As a rabbit matures, its digestive system becomes hardier; still, treats should never become a large part of your rabbit's diet.

What might you give as an occasional treat to adult rabbits? They'll enjoy the following:

- Alfalfa
- Apples
- Beets
- Carrots
- Lettuce
- Pumpkin seeds
- Soybeans
- Sunflower seeds
- Turnips

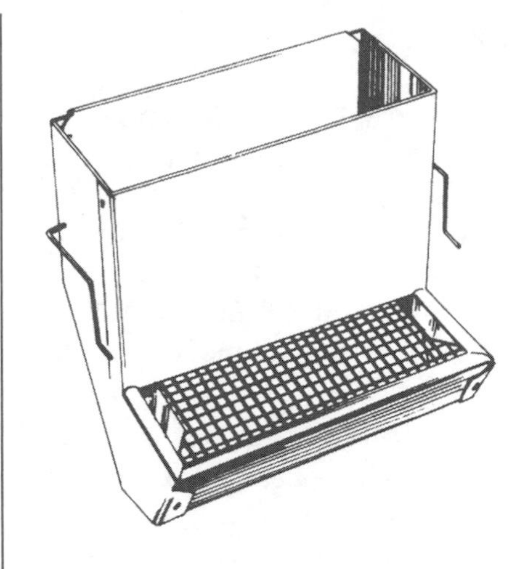

This self-feeder clips to the cage wire and can be filled from outside the cage.

To keep costs low, make a simple feeder from a 12-oz. (340 g) tuna fish can. Tack the can to a 10 in. (25 cm) square of ½ in. (1.25 cm) plywood to make your feeder hard to tip.

Feeding Equipment

Pellets are usually fed from crocks (heavy plastic or pottery dishes) that sit on the cage floor or from self-feeders that attach to the side of the cage. If you choose to use a crock, select one that is heavy enough that your rabbit cannot tip it over and waste its feed. To make feeding easier, place the crock close to the door of the cage. Never place the crock in the area that your rabbit has chosen as its toilet area. Because rabbits can soil the food in crocks, feed only as much as will be eaten between each feeding time.

Self-feeders attach to the wall of the cage and are filled from outside the cage. If you use a self-feeder, choose the type that has a screened, not solid, bottom. The screen allows fines (small, dusty pieces that break off from the pellets) to pass through. Otherwise, fines tend to build up because rabbits will not eat dust from feed. An advantage of self-feeders is that young rabbits are less likely to climb in the feed.

Watering Equipment

Your rabbits cannot thrive unless you provide them with a constant supply of clean, fresh water. Each day a rabbit will drink about 1 quart (1 L) of water, more in hot weather. As with feed, providing water can be done in several ways.

The ideal type of waterer is a plastic water bottle that hangs upside down on the outside of the cage and has a metal tube that delivers water to the rabbit. This type of waterer keeps water cleaner

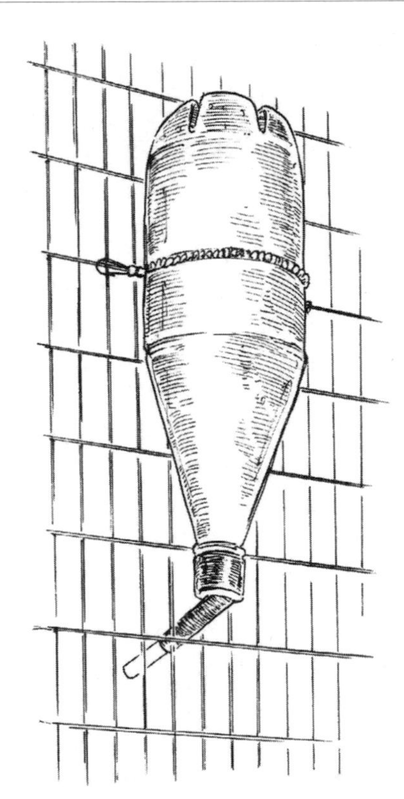

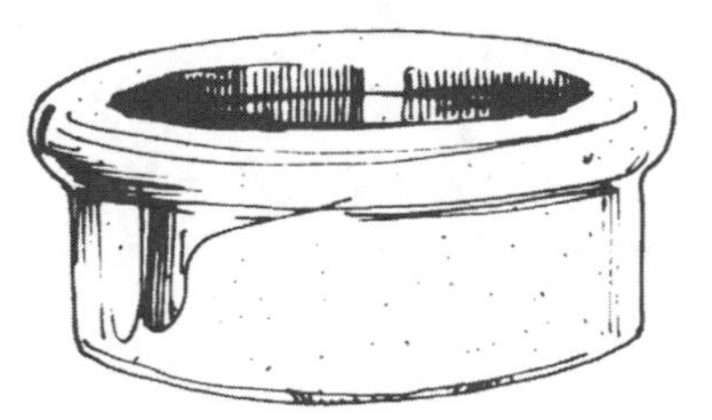

A water bottle (top) is cleaner than a crock (bottom) but will freeze and crack in cold weather.

than a crock does and cannot be tipped over, but water bottles freeze easily and therefore cannot be used outdoors in cold weather.

If you live in an area where freezing temperatures are common and you keep your rabbits outside, you might try heavy plastic crocks. You can remove frozen water from these crocks without breaking them or waiting for them to defrost. To save time during winter feeding, hit the plastic crocks against a solid object to dislodge the frozen water, then refill them.

When to Feed

The feeding habits of domestic rabbits are similar to those of wild rabbits. Wild rabbits are nocturnal — you don't see them much during the daytime, but they often appear in the early morning or late afternoon to have their big meal of the day.

If our tame rabbits could pick a mealtime, they would probably agree with their wild relatives and choose late afternoon. Using this information, most rabbit breeders feed the largest portion of their rabbits' daily ration in the late afternoon or early evening. Although different breeders develop schedules that work for them, two feedings a day — one in the morning and one in the late afternoon — seem to be most common. If you choose to feed twice a day, serve a lighter feeding in the morning and a larger portion in the afternoon.

Whatever feeding schedule you choose, it's important to follow it every day. Your rabbits will settle into the schedule that you set, and variations in that schedule may upset them. And even if you feed your rabbits only once a day, check their water twice a day to ensure they never run out. Remember, your rabbits need clean, fresh water at all times.

Although different breeders develop schedules that work for them, two feedings a day seem to be most common.

Feeding Chart

	Amount of pellets needed each day	
	Florida White	**Other meat breeds**
Buck	3–6 oz. (85–170 g)	4–9 oz. (113–255 g)
Doe	6 oz. (170 g)	9 oz. (255 g)
Doe, bred 1–15 days	6 oz. (170 g)	9 oz. (255 g)
Doe, bred 16–30 days	7–8 oz. (198–226 g)	10–11 oz. (283–312 g)
Doe, plus litter (6–8 young), 1 week old	10 oz. (283 g)	12 oz. (340 g)
Doe, plus litter (6–8 young), 1 month old	18 oz. (510 g)	24 oz. (680 g)
Doe, plus litter (6–8 young), 6–8 weeks old	28 oz. (794 g)	36 oz. (1 kg)
Young rabbit (weaned), 2 months old	3–6 oz. (85–170 g)	6–9 oz. (170–255 g)

How Much to Feed

Knowing how much to feed is probably the most complicated part of feeding rabbits. No set amount is right for all animals, but the chart above gives some average amounts.

To use the guidelines in the chart, you'll need to measure how much you feed each animal. You can make your own measuring container from an empty can. You'll need a scale that measures weight in ounces or grams. To adjust for the weight of the can, place the can on the scale and set the weight dial back to zero. If this is not possible with your scale, remember to figure the weight of the can into your measurements. Add pellets to the empty can until the scale reads the number of ounces or grams you want to feed, then mark a line at that level. Repeat this procedure as many times as necessary to ensure that you give each of your rabbits the correct amount of feed at each feeding time.

Feed guidelines are exactly that: guidelines only. Some rabbits need a little more feed, and some need less. Adjust each rabbit's ration according to its physical condition. Some reasons a rabbit's feeding may need to be adjusted are these:

- A young, growing rabbit needs more feed than a mature rabbit of the same size
- Rabbits need more feed in cold weather than in hot
- A doe that is producing milk for her litter needs more feed than a doe without a litter

Form the habit of checking each rabbit weekly by running your hand down its back to feel the condition of the backbone. The bumps of the individual bones that make up the backbone should feel rounded. If the bumps feel sharp or pointed, your rabbit could use

A homemade feed measure is an inexpensive way to determine how much to feed each rabbit.

an increase in feed. (However, extreme thinness may indicate a health problem.) If you don't feel the individual bumps, most likely too much fat is covering them and the rabbit should receive less feed.

Stroking your rabbit takes only a few seconds, and if you check regularly, you can adjust the ration before your rabbit becomes too fat or too thin. If you need to adjust the amount you feed, make the change gradually. Unless the rabbit is very thin or very fat, increase or decrease by about 1 ounce (28 g) each day. You should see and feel a difference in one to two weeks.

Overfeeding is a more common problem than underfeeding. If you feed your rabbits too much, caring for them costs more than necessary. More important, overweight rabbits are less productive than animals of the proper weight. Fat rabbits often do not breed at all, and an overweight doe that does become pregnant often develops health problems.

Feeding Time Is Checkup Time

Feeding time is for more than just feeding. Any time spent with your rabbits gives you an opportunity to observe many important details:

- Has a youngster fallen out of a nest box?
- Is an animal sick or not eating well?
- Does a cage need a simple repair before it becomes a major repair?
- Does a rabbit need extra water because of extreme heat or freezing conditions?

A lot can happen to your rabbits in 24 hours. The expenditure of time to feed, water, and observe your rabbits at least twice a day is a worthwhile investment.

To determine whether you need to increase or decrease the ration, stroke each rabbit from the base of its head along the backbone, feeling the backbone beneath the fur and muscles.

Breeding Rabbits

Producing your own rabbits for meat is an exciting part of keeping rabbits. Before you begin your breeding program, make sure your rabbits are mature enough and in good health to ensure the best odds of success.

When to Breed

There are two main signs that your doe is ready for breeding: She rubs her chin on her food dish or on the edge of her cage, and her genital opening looks reddish. Rubbing is how rabbits leave their scent to mark territory and to let other rabbits know they're around. If the doe's genital opening is pale or light pink, she is probably not ready to mate.

The larger the breed, the more time it takes for the rabbits to reach maturity. Medium- to large-size meat breeds — those that mature at 6 to 11 pounds (3 to 5 kg) — are generally ready to breed at 6 months of age. Producing young is strenuous for the rabbit, so take the time to check each animal's health before you consider breeding it. Breeders should be in good physical condition, not too fat and not too thin, and free from diseases, especially coccidiosis.

Putting the Doe and the Buck Together

Bucks are almost always willing to breed; does sometimes resist. When you are ready to put the male and the female together for breeding, always take the doe to the buck's cage. Rabbits, especially does, are territorial. If the buck is put into the doe's cage, the doe may respond by defending her territory from the buck rather than mating with him. By taking the doe to the buck's cage, you avoid arousing her territorial instinct, and she is more likely to mate.

When a doe and a buck are placed together, the buck is usually ready to mate and chases after the doe. If the doe is ready to mate with the buck, she will not run away and will raise her hindquarters after the buck starts the mating motions. To hold his position on the doe, the buck grabs a mouthful of her fur. Sometimes he pulls out some of the doe's fur, but he usually does not hurt her. You will be able to tell when mating has occurred because the buck will make a noise (usually a groan or a squeal) and then fall off to the side of the doe.

Hutch Cards

Hutch cards, as their name suggests, are attached to the rabbits' cages. They are kept so you can write down important information along the way (sample shown at right). Since the cards can be easily soiled or lost, copy the information into a permanent notebook on a regular basis — perhaps once a month.

Hutch cards help you keep track of how productive your breeders are. For a doe, a hutch card records such information as which buck she is bred to, when her litter is due, and how many young she kindles. For a buck, the card lists which does he is bred to, when they kindle, and how many young he has fathered.

Hutch cards are often available free from feed companies. You may be able to pick them up at your feed store, or you may have to write to the feed company and request a supply. You can also make your own.

Hutch Card

Doe's name: ____________________

Doe's number: ____________________

Buck's name: ____________________

Buck's number: ____________________

Date litter is due: ____________________

Number born: ____________________

Number weaned: ____________________

Comments: ____________________

Be There for Mating

Do not leave your rabbits unattended during mating. By watching, you can be sure mating has taken place and you can separate the animals if problems arise.

Sometimes the doe does not seem to want to mate with the buck. She will not raise her hindquarters and will run around the cage to try to get away from him. She may change her mind and be ready to mate after a few minutes, so leave them together for a short while. However, if the doe does not show interest within 10 minutes, take her back to her own cage and try again in another day or two.

Occasionally a doe that is not ready to mate will fight with the buck. If the buck and the doe fight, separate them and try again another day. If the doe is upset, she may bite. Be prepared: Have a pair of heavy gloves ready in case you need to remove the doe quickly.

After Mating

Once mating has taken place, return the doe to her own hutch. Next, write down when the doe will kindle (give birth). You can expect the litter to be born in 28 to 35 days; the average length of gestation is 31 days. You might want to keep this information on a hutch card attached to the doe's cage.

Feeding Pregnant Does

Fatness causes problems, such as pregnancy toxemia, in pregnant does. Resist the urge to give extra feed to the doe during the early stages of pregnancy. Most breeders keep the doe on her normal ration for the first half of the pregnancy (15 days). During the second half of the pregnancy, increase the doe's feed gradually. (See the chart on page 123 for guidelines.) Just before a doe kindles, she will eat less, and you can decrease her feed during the last two days.

The Nest Box

Your doe will need a nest box in which to have her litter. A nest box gives the kits a cozy place to snuggle and confines them so they won't stray and become chilled. Nest boxes come in many sizes and types. The larger the breed, the larger the box required. Nest boxes can be made of wood, wire, or metal.

Wooden nest box. Homemade nest boxes are commonly made of wood. Use ¾-inch (2 cm) plywood. Many does love to chew on the edges of the nest box, and plywood holds up well. Use wood glue and nails to assemble the nest box. Drill a few holes in the bottom to permit drainage, which helps keep the box from becoming damp. A damp nest box can contribute to diseases in young rabbits.

Nest Box Sizes

- Florida Whites: 18 inches long, 10 inches wide, 8 inches high (46 × 25 × 20 cm)
- Other meat breeds: 20 inches long, 12 inches wide, 10 inches high (51 × 30 × 25 cm)

Some breeders in cold climates place a cover over part of the nest box to help retain heat. A disadvantage of a partial cover is that it traps moisture.

Wooden nest boxes need to be thoroughly disinfected between litters. Use a bleach-water solution (1 part household chlorine bleach to 5 parts water) to sanitize the box and let it dry thoroughly in the sunshine. Store nest boxes where they won't be contaminated by other animals, such as dogs, cats, or rodents.

Wire nest box. You can build a nest box from ½ inch by 1 inch (1.25 × 2.5 cm) wire mesh. Line the box with cardboard in cool weather to prevent drafts. Discard cardboard liners after each litter to help keep the nest box clean.

Metal nest box. An advantage of metal nest boxes is that they are easy to clean. During cold weather, many breeders use cardboard liners as insulation.

Nesting Materials

A doe needs clean, dry, soft materials in the nesting box so that she can make a nice nest. Hay and straw are the most common nesting materials. Place about 2 inches (5 cm) of wood shavings in the bottom of the nest box and add lots of clean hay on top. Use less bedding in the summer. In winter, insulate the bottom of the box with layers of cardboard, followed

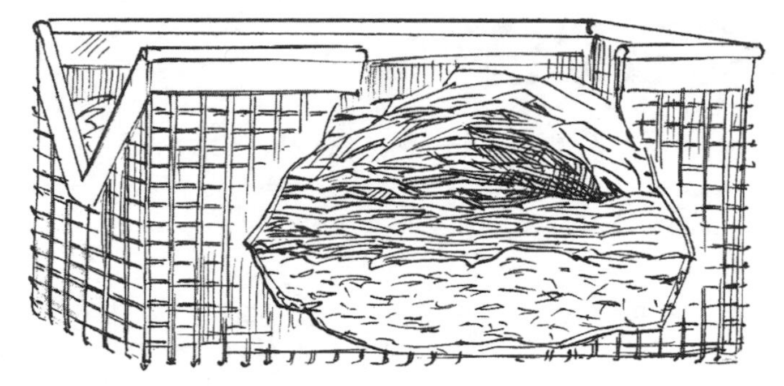

In cold weather, use several inches or centimeters of shavings and fill the box completely with straw.

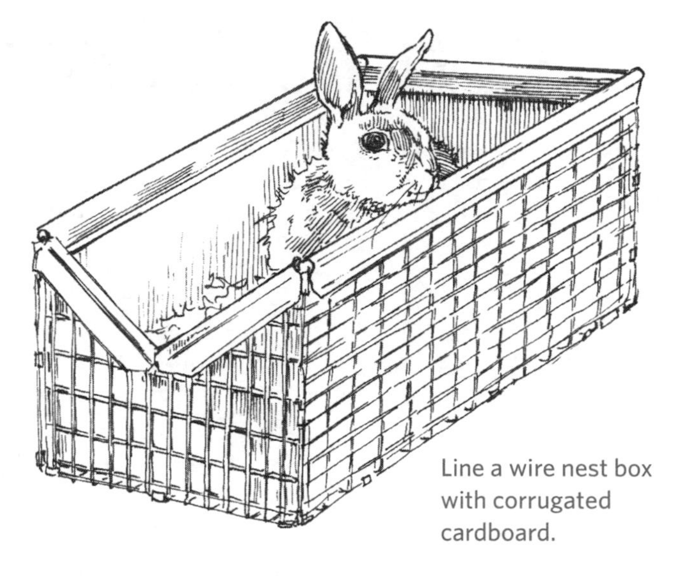

Line a wire nest box with corrugated cardboard.

In warm weather, keep nesting material to a minimum. Too much may cause overheating of the litter. Use 1 in. (2.5 cm) of wood shavings, covered with a few handfuls of straw.

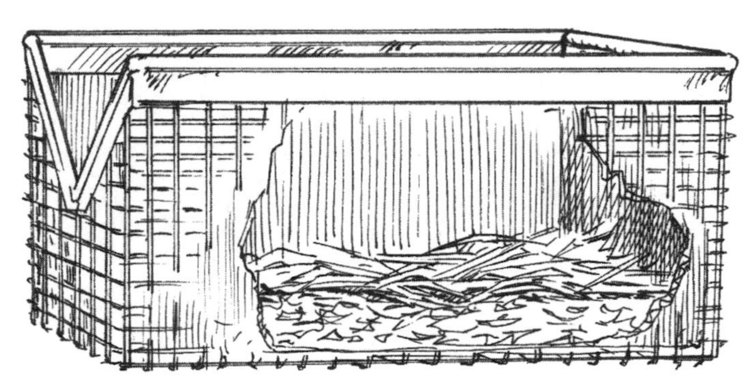

by wood shavings, with hay packed tightly on top.

Putting the Nest Box in Place

Place the prepared nest box in the doe's cage on the 27th or 28th day after mating. If you put the box in too early, the doe may soil the nest with droppings or urine. Any later may be too late for a doe that kindles early. Do not place the nest box in the part of the cage the doe has chosen as her toilet area.

Cold-Weather Kindling

Although healthy adult rabbits don't suffer when it's cold, newborn rabbits are vulnerable to cold temperatures. If you're expecting a litter during cold weather, be sure the doe has a well-bedded nest box. You may want to move the doe into a warmer location, such as a cellar or a garage. After the litter is about 10 days old, the cage, with the doe and the nest box of babies, can be moved back outside. Some breeders keep the box of newborn rabbits in their warm homes. At feeding times, they take the nest box out to the doe or bring her inside to nurse her litter.

When the doe starts to carry straw around the hutch, it's almost kindling time.

Birth & Aftercare

As the big day approaches, the doe will begin to prepare a nest. Some does jump into the nest box as soon as it is placed in the cage and rearrange the bedding to suit themselves. You may see the doe carrying a mouthful of hay around the cage as she decides where to place it. This indicates that kindling time is near.

The doe also uses fur that she pulls from her own chest and belly to pad the nest. Pulling the fur not only provides soft nest material but also exposes her nipples, making nursing easier for her babies after they're born.

Some does, like some people, leave everything to the last minute. They show little interest in the nest box when you first put it in. You may feed your doe one evening and see an untouched nest box only to come back the following morning and see that it's all over: The nest is made, fur is pulled, and babies have already arrived.

The doe will eat less the day or two before kindling. This change in appetite is normal. She also may act more nervous than usual.

Checking on the New Litter

Some does do a neat job of kindling, and you may not even notice the litter has arrived. If you look closely, you will see that the fur in the nest appears all fluffed up. On closer observation, the fur seems to move.

Be sensitive to the new mother's anxieties. Be sure to keep the doe's area quiet. If she appears nervous when you approach to check on the litter, distract her first. A nervous doe may jump into the nest to protect her babies, unintentionally stepping on and injuring them. Take a handful of fresh hay, a slice of apple, or another of the doe's favorite treats if she needs a distraction. Once she is occupied, check things out.

Gently place your hand into the fur nest. You should feel warmth and movement. If the doe remains calm, part the fur and take a peek. You should see several babies cuddled together for warmth. Newborn rabbits have no fur, and their eyes will be closed. Take a quick count. Counting may be a challenge, as the bunnies will be piled on top of one another.

Check the new litter for babies that did not survive. If you find any dead babies, remove them, as well as any soiled bedding.

Fostering

Medium- and large-size does usually have eight nipples from which their young will nurse. If a doe has more babies than nipples, some of the babies won't get as much to eat as others. To prepare for this situation, many rabbitry owners plan the breeding so more than one doe kindles at a time. If one doe has an unusually large litter, some of the babies can be moved into the nest box of a doe that has had a small litter. Most does will accept additional babies if they are about the same size as their own and smell like their own.

Some people go to great lengths to trick the doe into thinking these additional babies are her own. One method is to put a strong-smelling substance on the doe's nose. The idea is that a little dab of perfume or vanilla makes everything smell the same to the doe, and therefore she will not notice the extra babies. But you can be successful in fostering babies without using special scents. Here's how: Rabbits nurse their young only at night. If you need to move babies to another nest box, move them in the morning. The babies cuddle in with their new littermates, and by feeding time, they will all smell the same to the doe.

Feeding the Doe

Keeping the new mother in good health is the most important thing that you can do for the babies. Give the doe a limited amount of feed for the first few days after kindling. She will continue to eat and drink, but she will eat a little less than her normal amount. After a few days, her appetite will increase. She will then require an increased amount of feed so her body can meet its own nutritional needs and produce milk for the growing litter. The doe's need for water will also increase.

Caring for the New Litter

Check the litter every two or three days to make sure the bunnies are healthy and remain in the nest with their siblings. It's normal for the young to react to a human hand invading their nest, so don't be surprised if the babies jerk away from your touch. Just be careful that they don't injure themselves. Do not let individuals squiggle away from the warmth of the group.

Occasionally a baby will fall or crawl out of the nest box. Mother rabbits do not carry their babies like cats carry their kittens. If you find a young bunny out of the nest box, place it back with its littermates. In addition to needing the warmth of the group, babies that stray from the nest are likely to miss mealtime. The mother will nurse only one group of babies. Absent young won't get fed and will soon die.

In cold weather, kits that leave the nest too early are at particular risk for dying of exposure. Rabbits are born without fur, and if they leave the warmth of the nest, they will soon become chilled. On cold days, begin your morning chores with a quick check for straying babies. If you find a baby outside the nest and it still feels warm to the touch, place it back with its littermates. If the baby feels cold to the touch, it may need some extra help regaining its body temperature (see the box on page 130).

Baby rabbits begin to grow fur within a few days of birth and by 2 weeks will be completely furred.

Nest Box Maintenance

Most does will enter the nest box only at feeding time. Once in a while, however, a doe spends more time than necessary in the box. (She is more likely to spend excessive time in a nest box that is too big.) As a result, the nest box becomes dirty and damp — a perfect setup for disease. If you see droppings in the nest box or the bedding materials feel damp, clean the nest box.

1. Remove the nest box from the doe's cage.

2. Remove and save any clean and dry fur the doe has pulled. Place this fur in a container — a cardboard box or a clean bucket with some clean bedding in the bottom — and gently place the litter in this temporary home.

3. Remove the soiled bedding and replace it with clean bedding.

4. Use your hand to form a hole in the clean bedding. Line this area with some of the saved fur, place the babies in it, and cover them with the remaining fur.

5. Return the clean nest box and litter to its location in the doe's hutch. Your bunnies now have a clean and healthy nest to grow up in.

Ten-day-old babies in the nest. Newborns should be handled rarely, if at all, so the doe will not be upset by the intrusion in her nest and so you avoid passing your scent to the young.

Caring for Orphans

Sometimes a doe dies after giving birth. In that case, you may wish to try to feed and care for the babies until they are mature enough to take care of themselves. To feed the young, make up the following mixture and store it in the refrigerator.

- 1 pint (473 mL) skim milk
- 2 egg yolks
- 2 tablespoons (30 mL) Karo syrup
- 1 tablespoon (15 mL) bonemeal (available in garden supply centers)

Use an eyedropper or drinking straw to feed this mixture to the babies twice a day. Feed them until they stop taking the milk (usually about ¼ ounce [5 to 7 mL]).

In addition to keeping the kits warm and fed, make sure they urinate and defecate regularly. Stimulate elimination by stroking their genitals with a cotton ball after you feed them. Follow these procedures until the young are 14 days old.

Eye-Opening Time

The babies should open their eyes at 10 days of age. It's important to watch for this event. Sometimes a baby cannot open its eyes even though it is old enough to do so. This situation usually indicates an eye infection. An infection may affect one or both eyes. Eye infections may occur in one or several littermates. Tending to this condition as soon as possible is your best means of avoiding blindness.

A baby that cannot open its eyes usually has dry, crusty material sealing the lids together. Use a soft cloth, cotton ball, or facial tissue soaked in warm water to wash and soften the crusted eyelids. Then use your fingers to gently separate the eyelids. Gently wash away any remaining crusty material. If the edges of the eyelid look reddish, place a drop of over-the-counter eyedrops for humans (Systane is a good choice) in the eye.

Check the babies for several days and repeat this procedure as necessary. Once the eye is cleaned and opened, most rabbits recover quickly.

If you see pus when you open an eye as directed above, the bunny has a more serious condition. After gently washing away the crusty material and pus, use medicated eyedrops containing the antibiotic neomycin; these eyedrops are available from your veterinarian. You will need to clean the eye and apply medication for several days, after which the infection should clear up.

If a 10-day-old kit cannot open its eyes, it is usually because they are sealed with crusty material.

Some rabbits have a normal-looking eye after recovering from an infection of this type. However, some may develop a cloudy area on part or all of the eye. Cloudiness usually means the rabbit will be partially or totally blind in that eye

Handling the Kits

As the kits approach 3 weeks of age, they will begin to come out of the nest box. At this age, they also can find their way back into the box, so you don't have to worry about constantly watching for strays. The bunnies will begin to eat pellets and drink water. You'll need to start providing more food and water, even though the babies are still getting milk from their mother. Baby rabbits continue to nurse until they are about 8 weeks old.

Bunnies are at their cutest from 3 to 8 weeks of age. This age is an excellent time at which to begin to handle them. Be especially careful when you pick them up for the first time. Young rabbits are easily frightened, but they adapt quickly. Try to spend a few minutes each day handling the babies. Rabbits that are handled properly as youngsters will be more easily handled throughout their lives. Of course, you don't want to get attached to rabbits you plan to eat, but you also don't want them to be difficult to handle when the day comes.

Methods of Warming a Chilled Bunny

Share your body heat. Tuck the baby inside your shirt so it can rest against you and gain warmth. It's a wonderful experience to feel a cold, almost lifeless bunny slowly regain consciousness.

If you have a heating pad, turn it on, cover it with a towel, place the baby rabbit on it, and place the pad and rabbit in a small box. Make sure the box provides protection if you are taking the rabbit into an area where there are pets, such as dogs or cats.

Sexing the Litter

It takes some practice to sex young rabbits accurately. Try to sex the litter at around 3 weeks of age, when you are beginning to handle the youngsters on a regular basis. Part of the young rabbits' handling experience should include getting used to being turned over for examination, which is necessary for sexing.

The small sexual openings on young rabbits make it difficult to be absolutely

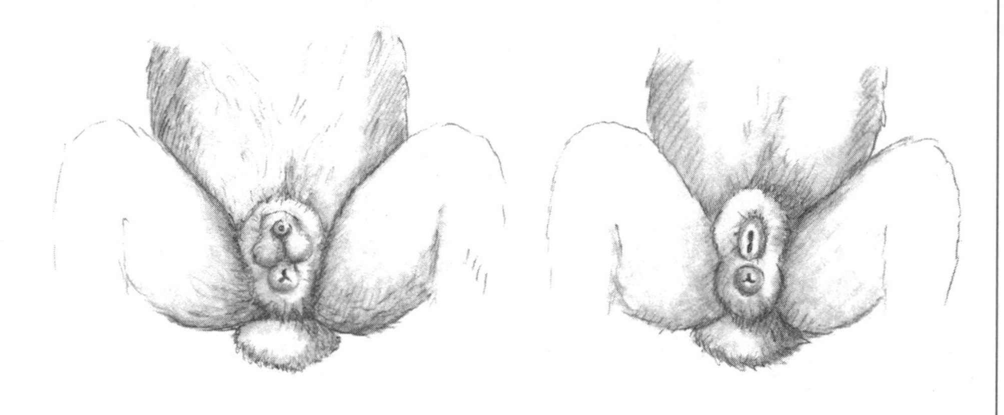

When the litter has reached 3 weeks of age, examine each young rabbit to determine its sex.

sure which sex an animal is. To help you gain confidence in sexing your litters, start to check them at an early age and reexamine them weekly. Use a black marking pen to write a *B* in an ear of those you believe are bucks and a *D* in an ear of those you believe are does. Each time you reexamine the litter, check to see if you agree with what you thought when the rabbits were a week younger. You may find that you will change the letter in the ears of several animals. If you continue to check through weaning, you will find that sexing young rabbits becomes easier as the rabbits get older. Even experienced rabbit breeders can make mistakes in sexing young rabbits, so don't become discouraged.

Hutch Maintenance

As the litter grows, keeping the hutch clean will become more of a challenge. The crocks used to feed and water the litter will become soiled frequently, as young bunnies climb into them and leave their droppings behind. Feeding smaller amounts more often will save more feed than feeding a large amount once a day. Giving several small feedings will also give you more opportunities to clean out the feed dishes. The cleaner the litter's environment, the healthier it will be.

When the babies are out and about at 3 weeks of age or so, you can remove their nest box, unless the weather is very cold. Once the babies start eating pellets, they also produce droppings that dirty the box. If you choose to leave the nest box in the cage longer, be sure to clean it and replace the bedding whenever it becomes soiled.

Weaning and Separating the Litter

As the litter reaches 8 weeks of age, the young rabbits are ready to be weaned, or removed from their mother. By now, young rabbits are used to eating pellets and drinking water, and the doe's body needs a rest before she is ready to raise another litter.

In addition, by 8 weeks of age, the youngsters begin to exhibit adult behaviors. They express instincts about territory and breeding. Young bucks will begin to chase after their sisters. Rabbits can mate and produce litters before they are fully grown. Because having a litter at a young age is stressful for a doe, and because it's not desirable for close relatives to breed, part of the weaning process includes separating the litter by sex.

Many breeders feel that the weaning process is easier on the mother if not all the young rabbits are removed at once. Since most litters are about half bucks and half does, a good plan is to remove the young bucks first. A few days later, the young does may be removed. This two-part weaning process allows time for the doe's body to adapt gradually to the need for reduced milk production.

As rabbits mature, they do best if they have their own private space. Rabbits that are housed together often fight to decide who will be the boss of the hutch. Although rabbit fighting seldom leads to life-threatening injuries, torn ears, scratches, and bite injuries are common. Young rabbits being raised for meat can be kept more than one to a cage, but the cage should provide adequate space, and each cage should house just one sex. Rabbits can be kept more than one to a cage until they are up to 4 months of age. Be alert to any fighting and separate those involved.

The Doe's Refuge

You may wish to allow the nest box to remain a little longer than three weeks, more for the doe's sake than for the litter's. As the babies explore their hutch, they pester their mother for extra attention. A nest box with a partial top provides a place for the doe to get away from the demands of the litter. Once the babies are big enough to climb on top of the box, the doe's hideaway no longer serves its purpose, and it's time to remove the nest box.

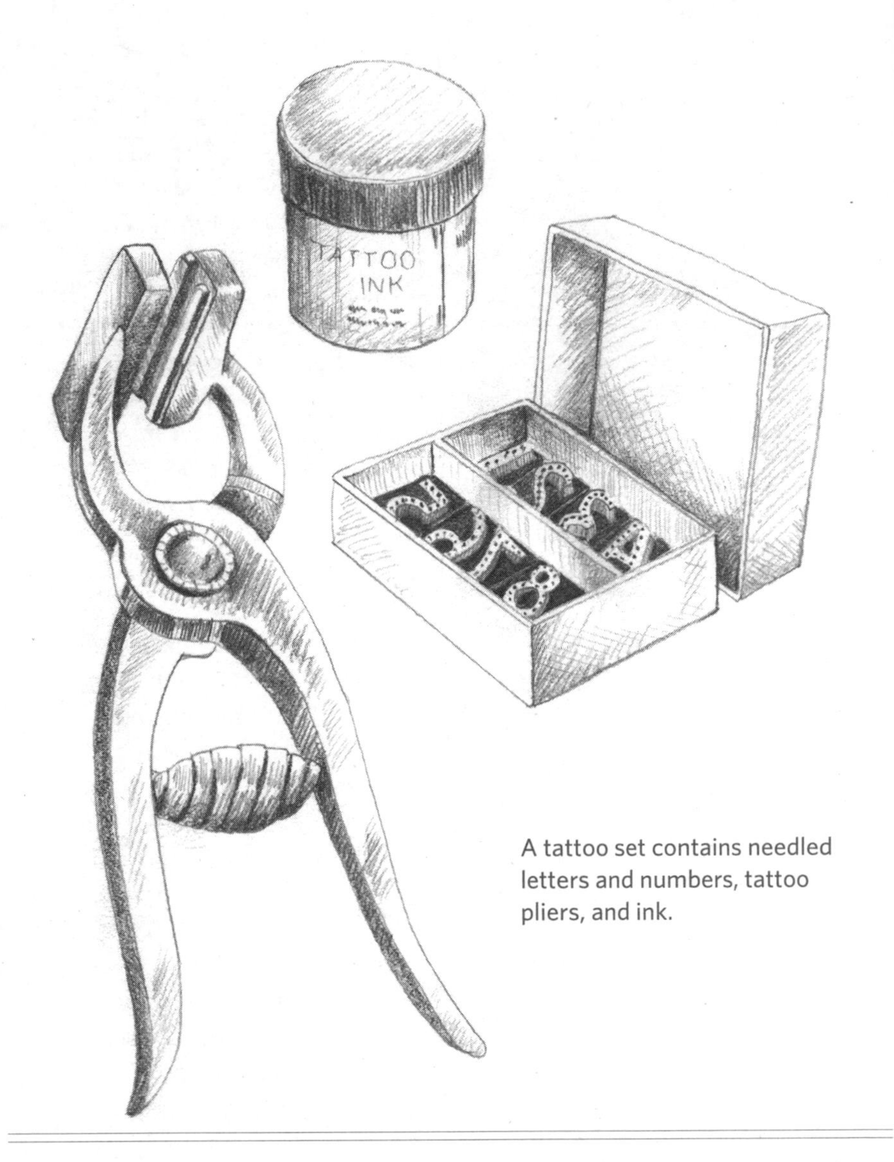

A tattoo set contains needled letters and numbers, tattoo pliers, and ink.

Tattooing

Tattooing is generally done at weaning. Of course, you do not have to tattoo rabbits destined to become meat, but you might want to tattoo any you plan to keep as future breeders to help you keep track of who is who. And if you want to register any offspring kept as future breeders, they must be tattooed.

If you have a small rabbitry, or won't keep many of your breeders' offspring, you may not want to spend money on tattooing supplies. In that case, ask other breeders if they will tattoo your animals for a small fee.

The tattoo may be a number, a word, or a combination of letters and numbers. Some breeders use short names, such as Joe or Pat, so the tattoo names the animal as well as identifies it. Others develop systems that provide more information about each animal. For example, in one system, a letter stands for a certain buck, another letter stands for a certain doe. Thus, if we had an animal with the tattoo TB3, we would know that it was a buck whose father was named Thumper and whose mother was named Bambi. The possibilities are endless. Develop a tattoo system that works best for you. Most tattoo sets are limited to five numbers or letters per tattoo, so be sure your system takes this limitation into account.

The letters and numbers in a tattoo set are made up of a series of small needles. When the tattoo pliers are closed on the rabbit's ear, they make a series of small punctures. Tattoo ink is then rubbed into the area. The tattoo ink that goes into the punctures becomes permanent, while the ink on the ear's surface soon wears away. Tattooing is a bit painful for the rabbit, but only for a short time.

To learn how to tattoo your own animals, observe others when they tattoo. Choose rabbits that are not show quality for your first tattoos. Tattooing has no one right way. Here is one successful technique:

1. Practice placing the numbers and letters in the tattoo pliers. Use a piece of paper to check what you have in the pliers. It's very easy to get a 21 when you really wanted a 12. By checking first, you'll avoid an incorrect tattoo.

2. Restrain the rabbit. The rabbit will pull away when the pliers are clamped, and it could be injured. If you have someone to help you, one person can restrain the animal and the other can do the tattooing. The holder can wrap the rabbit in a towel so only its head and ears stick out. The towel will keep the rabbit from moving and from scratching the holder. If you have to tattoo alone, you might place the rabbit in a narrow show carrier. The carrier keeps the rabbit confined and allows you to use both hands for the tattooing. One hand secures the left ear and the other holds the tattoo pliers.

3. Use a little rubbing alcohol on a cotton ball to wipe out the ear before tattooing. Cleansing the ear surface allows the ink to adhere better.

4. Look at the inside of the rabbit's left ear. You'll be able to see some of the

Selecting Future Breeders

If you're going to keep any of the bunnies for future breeding, move them to separate hutches and attach a card that identifies the rabbit housed there. The hutch card should include when the animal was born, what sex it is, and who the parents are. This information will come in handy when the rabbit is ready to be bred. In deciding which animals in the litter are best suited to be breeders, apply the same qualifications that you would use in evaluating an animal to purchase.

blood vessels that are just below the surface of the skin. Find a space where no large vessels cross, and plan to place the tattoo there. Even with careful planning, you may hit a blood vessel. Hitting a vessel will cause some bleeding that can be easily stopped by applying some gentle pressure to the area. Bleeding can wash the tattoo ink out of the punctures, so avoiding blood vessels usually makes for a more successful tattoo.

5. Clamp the pliers down on the rabbit's ear in the spot you've chosen. This part is the hardest, because no one wants to hurt an animal. Take comfort in the fact that the pain is brief. Release your grip on the pliers immediately after you close them, and the worst is over. Just how hard to close the pliers to ensure a good tattoo is difficult to say. The thin ears of young rabbits need only the pressure used to pretest the tattoo on paper. Older animals have thicker ears that require a slightly firmer grip.

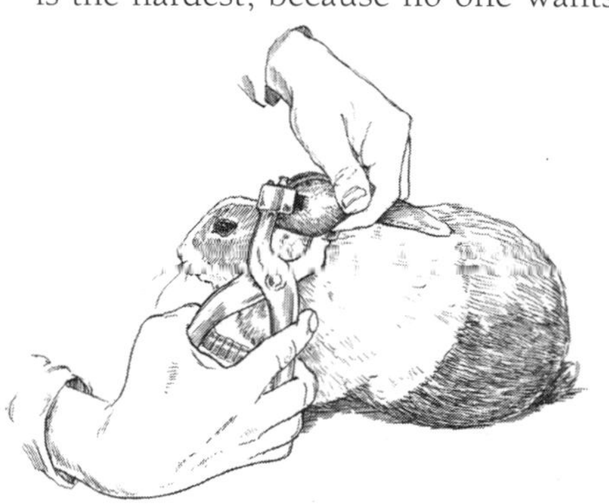

Tattooing, Step 5.

6. Give the rabbit a few seconds to calm down, and then look inside the ear. Your tattoo should appear as a series of dots.

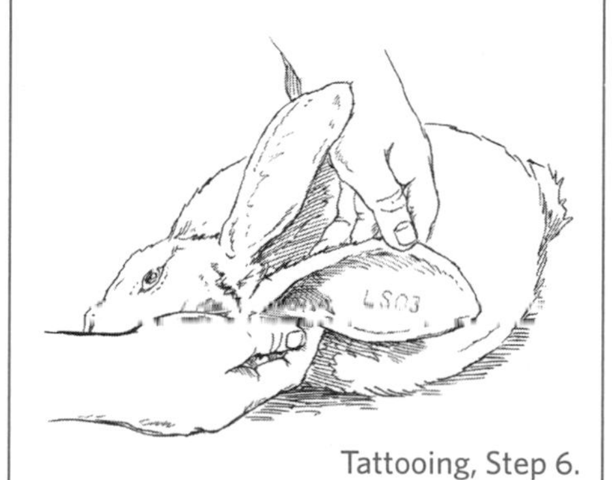

Tattooing, Step 6.

7. Now apply the tattoo ink. An old toothbrush works well for this purpose; the bristles help work the ink into the small punctures. Most rabbits do not seem to mind this part of the procedure.

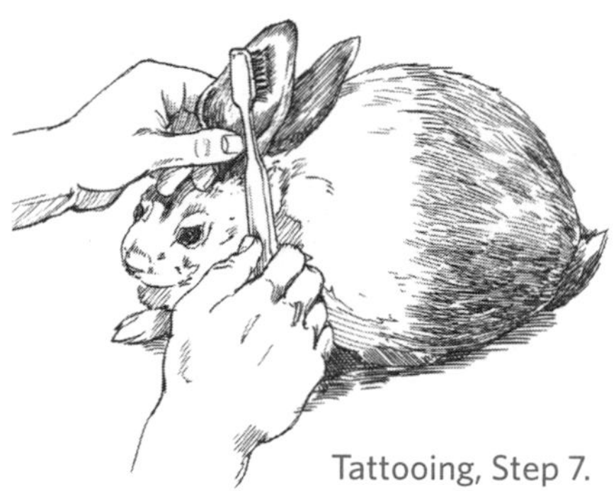

Tattooing, Step 7.

8. After the ink is applied, some breeders apply a thin layer of petroleum jelly. The jelly forms a seal that helps keep the ink in the punctures. The excess ink will wear off normally, or the mother may clean it off while grooming her baby. If the ink has not worn away by the time your rabbit is entered in a show, you will need to remove it to make the tattoo readable. Use a tissue and a little petroleum jelly to wipe the excess ink away.

Keeping Records

If you own one or two rabbits, remembering all the important details about them is easy. But as time goes by and breeders are replaced, it will be difficult to recall everything about each animal. Keeping accurate written information about your rabbits is therefore important for the following reasons.

- You'll have information to help you decide which doe gets bred to which buck.
- Your rabbitry will be more productive because your records will show which animals are your top producers; you can use this information to help you decide which animals to keep in your herd.
- You'll know how much your rabbits cost you and how much meat they produce for you.
- You'll have the information you need to make a written pedigree for each animal.

Written information includes hutch cards (described on page 125), pedigree papers, and registration papers. Of course, you need not write a pedigree for each meat animal, but you should certainly write a pedigree for any stock you keep for future breeding.

Rabbit Registration

Certificate of Entry in the Stud Book of

AMERICAN RABBIT BREEDERS ASSOCIATION, INCORPORATED — No. 430 K

NAME Marquette — EAR NO. BB755 — BREED Tan — SEX Buck — COLOR Black

BORN 7-1-09 — WEIGHT 4½ — WINNINGS Best of Breed ARBA Conv 2009 — DATE REG. 6-11-2010

NAME OF BREEDER Robert Bennett — ADDRESS Wyckoff, N. J.

PURCHASED FROM — ADDRESS

REGISTRAR'S NAME Charles Lyons — ADDRESS Rutland, Vt.

DESCRIPTION: TYPE Good — BONE Fine — COLOR Very Good Marking — BODY Good — FUR Nice Fly-Back

HEAD Well Shaped — EARS 4" — EYES Clear — CONDITION Very Good — BALANCE Good

REG. REMARKS

OWNER Robert Bennett — STREET 624 Lawlins Road — TOWN Wyckoff — STATE N. J.

DATE FILED 6-29-10

SIRE Fee — REG. NO. — COLOR Blue — WT. 4½ — GRAND CHAMPION NO. Ear # BB764

GR. SIRE Seminar — REG. NO. 1574 G — COLOR bBlack — WT. 4 — GRAND CHAMPION NO. K 756

G. G. SIRE Covors Image — REG. NO. 3988 E — COLOR Black — WT. 4½ — GRAND CHAMPION NO. K-757

G. G. DAM Best Bett — REG. NO. 3975 E — COLOR Black — WT. 5 — GRAND CHAMPION NO.

GR. DAM Lady Bell II — REG. NO. 3989 E — COLOR Black — WT. 4½ — GRAND CHAMPION NO.

G. G. SIRE Emporor — REG. NO. — COLOR Black — WT. 4 3/4 — GRAND CHAMPION NO.

G. G. DAM Belle — REG. NO. — COLOR Black — WT. 4½ — GRAND CHAMPION NO.

This Certificate denotes QUALITY for our Registrars are instructed to register no animal disqualified in any section.

DAM Test — REG. NO. — COLOR Black — WT. 4½ — GRAND CHAMPION NO. Ear # CE3

GR. SIRE Jaguar — REG. NO. — COLOR Blk. — WT. — GRAND CHAMPION NO.

G. G. SIRE In England — REG. NO. — COLOR — WT. — GRAND CHAMPION NO.

G. G. DAM Cover Boy — REG. NO. 3994E — COLOR Black — WT. 4 — GRAND CHAMPION NO.

GR. DAM Coquette — REG. NO. — COLOR Blk — WT. 4½ — GRAND CHAMPION NO.

G. G. SIRE Admission — REG. NO. — COLOR Blue — WT. 5 — GRAND CHAMPION NO.

G. G. DAM — REG. NO. — COLOR — WT. — GRAND CHAMPION NO.

Ed Pufer Jr — *Secretary*

Should any animal have been admitted to entry or transfer through misrepresentation or fraud the Executive Board, on discovery of the same, shall declare the entry or transfer void, together with any entries that may have been made of descendants of such animals where based upon said entry. The American Rabbit Breeders Association (Incorporated) is not responsible for any erroneous statement upon which this certificate is based.

The registration certificate of the winner of best of breed at a national ARBA convention show.

Pedigree Papers

A rabbit's pedigree paper is a written record of its family tree — who its parents, grandparents, and great-grandparents were. Usually, pedigrees also include such information as the color, tattoo number, and weight of each animal. You can use the information on a pedigree to help you make decisions that will affect your breeding program. Are you trying to increase the weight of your rabbits? If you breed your doe to a buck whose pedigree shows relatives with high weights, you will have a better chance of producing heavier rabbits than if you breed to a buck whose relatives are on the light side.

Filling out a pedigree can be confusing at first. You have to take information from the buck's pedigree and combine it with information from the doe's pedigree. Pedigree forms use the term sire for the father or buck, and dam for the mother or doe. If you should someday decide to sell one or more of your breeders, those with pedigrees usually bring a higher price.

Sources of Blank Pedigrees

- The American Rabbit Breeders Association (ARBA) sells pedigrees in books of 50.
- Most breed associations have pedigrees that feature an illustration of their breed.

Registration Papers

Many people mistakenly think pedigreed animals are the same as registered animals. A registered rabbit must have a pedigree, but it has to meet additional requirements as well:

- The rabbit must be examined by a registrar who is licensed by ARBA. The registrar goes to the rabbitry and checks to see that the rabbit has no disqualifications for its breed. The registrar will need a copy of the rabbit's pedigree.
- The rabbit's owner must be a member of ARBA.
- ARBA will charge a registration fee.
- The rabbit must have a permanent tattoo number in its left ear.

When you raise meat rabbits for yourself and your family, you may not want to go through the trouble and expense of registering any offspring you keep as future breeders. But you may want to consider registration if you are thinking of someday showing or selling rabbits.

Rabbit Health

The following four major factors contribute to your rabbits' health: housing, environment, observation, and nutrition. A problem in one area can lead to problems in one or more of the others.

Housing. Hutches should offer protection from extremes in temperature, from wind, and from rain and snow. Hutches must be cleaned often to prevent disease-causing germs from flourishing.

Environment. Prevent situations that stress or upset your rabbits, such as exposure to loud noises, encounters with wild or frightening animals, and summertime heat.

Observation. One of the most important things you can do each day for your rabbits is to observe them. Through observation you'll learn what is normal for each of your rabbits, and you'll be better prepared to identify abnormal behavior. Changes in the way a rabbit acts or looks often indicate that it is not feeling well. The sooner a problem is spotted, the better the chances are of solving it before it becomes serious.

Nutrition. Feed the proper amounts of the right foods on a regular schedule. Be familiar with the amount of food each rabbit normally eats. A change in eating habits is often one of the first signs that a rabbit is not feeling well. And don't forget your rabbit needs fresh water at all times.

A Rabbit First-Aid Kit

If you start with healthy rabbits and take good care of them, they should have few health problems. If a problem arises, early treatment will increase your rabbit's chances for a full recovery. Having the necessary things on hand will help you start treatment as soon as you notice a problem. Assembling a rabbit first-aid kit is easy; you may already have some of the supplies in your home. Every couple of months, go through the kit and check the condition of your supplies. Discard and replace expired medications.

For items you have to purchase, try to find other rabbit owners who would also like to make a first-aid kit for their rabbits. Most medicines come in fairly large amounts, so it helps if you can share the supplies and the cost with several others.

A Health Checklist

Occasionally an animal becomes sick. If you suspect a rabbit is not well, the following checklist will help you pinpoint what might be wrong.

The rabbit is eating less than its normal amount of feed.
☐ Yes ☐ No

The rabbit's eyes look dull, not bright.
☐ Yes ☐ No

Discharge is coming from its eyes or nose.
☐ Yes ☐ No

The rabbit's droppings are softer than usual.
☐ Yes ☐ No

The rabbit's coat looks rough.
☐ Yes ☐ No

The rabbit is dirty and has manure stuck to its underside.
☐ Yes ☐ No

The rabbit is having trouble breathing or is breathing rapidly.
☐ Yes ☐ No

If you checked "yes" for one or more of the above items, the rabbit may be ill. Begin a more thorough evaluation, or consult with your veterinarian to identify the problem.

Preparing a Rabbit First-Aid Kit

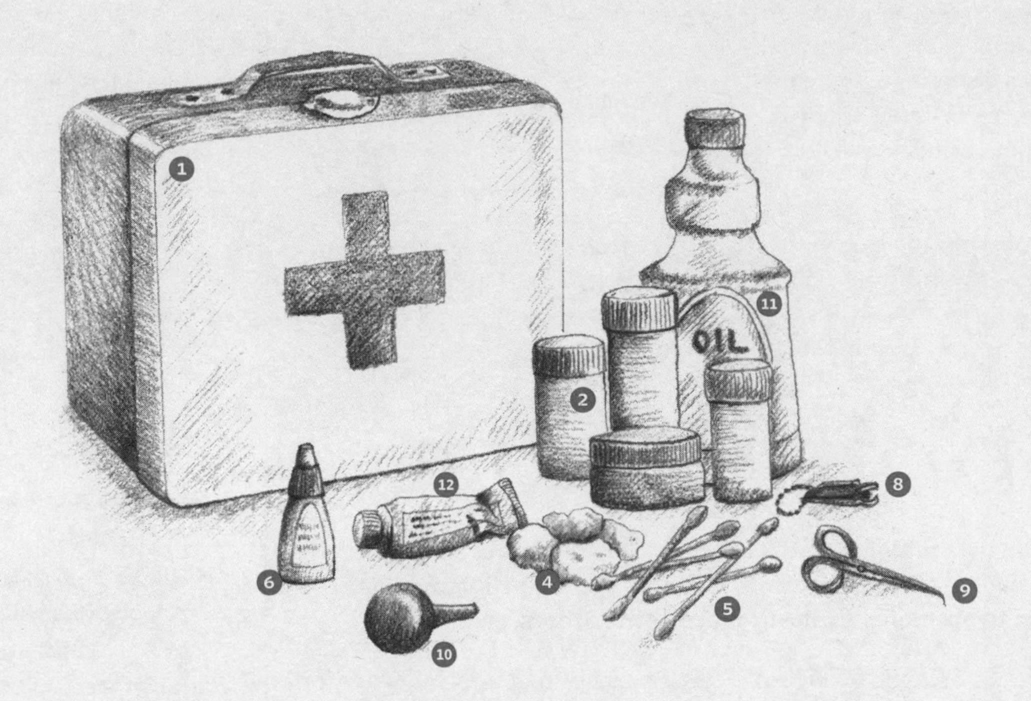

Be prepared for emergencies by keeping a fully stocked first-aid kit on hand. Your rabbit first-aid kit should contain the items listed below.

1. A container to hold your first-aid materials; a school lunch box is about the right size
2. Several small bottles or jars; empty containers from prescription drugs work well
3. Labels for marking each container with the following information:
   - Name of the medicine
   - What the medicine is used for
   - How the medicine is given
   - How much medicine is given
   - How often the medicine is given
4. Cotton balls to clean wounds or apply medicines; store cotton balls in a plastic bag
5. Cotton swabs to clean wounds or remove ear mite crust; store swabs in a plastic bag
6. Saline solution wound wash
7. Disposable rubber gloves
8. Nail clippers for keeping nails trimmed. Clippers may also be used to trim teeth. Use ordinary nail clippers intended for humans or, if you have a large breed, use dog nail clippers.
9. Small pair of sharp scissors to trim hair from around wounds
10. Plastic eyedropper to give liquid medicines or put oil into ear to treat ear mites. Wash well between uses. Store in a small plastic bag.
11. Miticide or cooking oil to treat ear mites
12. The following products, purchased at a local farm store or drugstore:
    - Antibiotic cream for wounds
    - Bag Balm or Preparation H for sore hocks
    - Eyedrops for nest-box eye infections
13. The phone number of your veterinarian, if she is knowledgeable about rabbits, or the phone number of an experienced rabbit owner in your area.

Colds and Snuffles

When a rabbit has a cold, it exhibits some of the same symptoms as humans do, such as sneezing and a runny nose. Unfortunately, 90 percent of the rabbits that appear to have colds actually have a much more serious disease called snuffles, which is caused by *Pasteurella multocida*. Snuffles can be treated, but it's difficult to cure. You're better off putting your time into preventing this illness, and the best prevention is to keep your rabbitry clean and well ventilated.

Sneezing is usually the first sign of a cold or snuffles, but many things other than disease can make a rabbit sneeze. If you hear sneezing, be sure to look for other signs of illness. A white discharge from its nose or matted, crusty fur on the front paws (caused by a rabbit wiping a runny nose) indicates that the rabbit probably has a cold or snuffles.

Rabbits and humans don't communicate colds to one another, but rabbits communicate colds to other rabbits. If you suspect that a rabbit has a cold or snuffles, isolate it from your other rabbits and thoroughly disinfect its cage before moving another rabbit in.

If you want to try to treat the sick rabbit with antibiotics, contact your veterinarian for advice. Many rabbitry owners, however, would recommend that you simply have the rabbit put down. The risk of contagion is high, and snuffles can infect an entire rabbitry in no time at all.

Ear Canker

Ear canker is caused by tiny mites that burrow inside the rabbit's ear. A rabbit with ear mites will shake its head and scratch at its ears a lot, and the inside of its ears will look dark and crusty.

Although a miticide-containing oil is available from the vet, mites can easily be treated with common household oils. Mites breathe through pores in their skin. If they come into contact with oil, it blocks these breathing pores and they soon die.

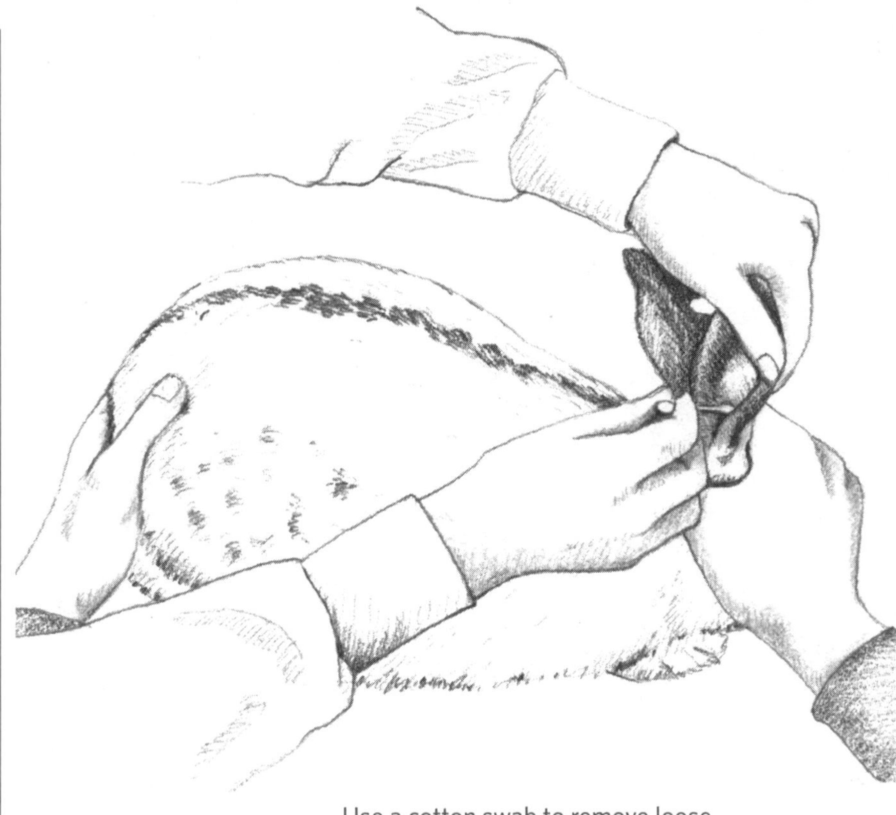

Use a cotton swab to remove loose, crusty material from the ear.

1. Use an eyedropper to put several drops of miticide-containing oil or olive, cooking, or mineral oil into the rabbit's ear. Gently massage the base of the ear to help spread the oil around.

2. Use cotton swabs to remove some of the loose, crusty material.

3. Repeat this treatment daily for 3 days, wait 10 days, and repeat the treatment for 3 more days. Wait another 10 days and repeat the treatment again for 3 days.

Ivermectin, either injected under the skin or swabbed into affected ears, is another effective treatment for ear mites. Consult your veterinarian about the proper formulation and dose. Be aware that there will be a withholding period if you are treating rabbits raised for meat.

If you see mites on one rabbit, check all your rabbits. Mites are more likely to appear in rabbits that live in dirty cages. To prevent ear canker, keep your cages clean.

Long Toenails

Properly trimmed toenails are important because they decrease the chances your rabbits will be injured. Long toenails can get caught in the cage wire and cause broken toes and missing toenails. The time spent trimming toenails will also decrease your likelihood of being scratched while handling your rabbits.

Trimming toenails is not hard to do, but many rabbit owners put off this task because they are afraid of hurting their rabbit. Learning to trim toenails is easier if you have a helper — one person

Coccidiosis

A common disease that causes diarrhea is coccidiosis. The disease can kill rabbits if it goes untreated. Coccidiosis often occurs in young litters of rabbits. The protozoans that cause this disease are present in droppings and in soiled feed and bedding. A cage that was a clean home for one doe gets soiled much more easily when she has to share that space with eight growing youngsters. But older rabbits living alone may also contract coccidiosis. No matter what their age, rabbits kept in clean living conditions are much less likely to suffer from this disease.

Treatment of coccidiosis has two parts: thorough cleansing of the hutch and medical therapy for the animals.

One of the best preventives against coccidiosis is to keep your rabbit's cage clean.

Symptoms

- Soft droppings (diarrhea)
- Rough-looking fur
- Animal not growing as well as you think it should
- A potbelly, although the rabbit is rough-feeling over the backbone

Housing Treatment

- Remove all soiled bedding and food.
- Scrape all built-up manure from the cage at least every three days.
- Clean cage with a bleach-water solution (1 part household chlorine bleach to 5 parts water). Let dry thoroughly.
- If the bunnies are old enough (5 to 6 weeks), separate them into smaller groups, for which you'll need additional cages.

Medical Treatment

Sulfaquinoxaline is used to treat coccidiosis. It should be available in most farm supply stores, probably as a treatment for coccidiosis in calves, pigs, or chickens. If the product label doesn't give directions for treating rabbits, use the dosage suggested for poultry or mink.

Sulfaquinoxaline is mixed in water when given to animals. Mix up a batch of medicated water, following the instructions on the product label. Provide the medicated water for seven days, or as recommended on the label. The treatment plan is generally as follows:

- Give sulfa-treated water for seven days
- Give plain water for seven days
- Give sulfa-treated water again for seven days
- Repeat this sequence two to four times a year

Repeated treatment is necessary because during the first few days of sulfa treatment, the original organisms that made the rabbit sick are killed, and the rabbit will seem to be much better. The organisms that cause coccidiosis produce eggs, however, and a second treatment period is needed to kill organisms from the eggs that later hatch.

If the animal is being raised for meat, it must not be given medication within a certain number of days before it is sold. Follow the instructions on the label.

Other Causes of Diarrhea

Coccidiosis is not the only trigger for diarrhea in rabbits. A rabbit may also have diarrhea because it lacks fiber in its diet, has experienced a sudden change of feed, is suffering from a disease of the digestive system, or has a bacterial infection. Administering tetracycline (by mixing it in with the rabbit's water) may cure some cases of bacterial diarrhea. Consult your veterinarian about this treatment and others.

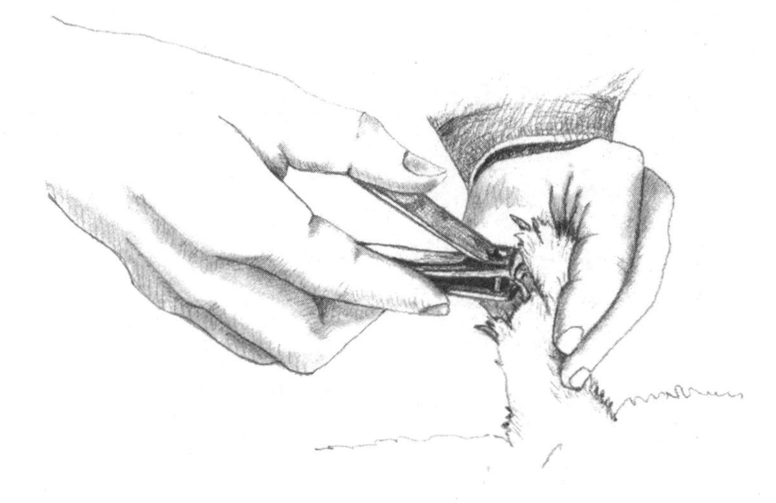
Use toenail clippers to trim your rabbit's nails.

can hold the rabbit while the other does the trimming. For most breeds, human fingernail or toenail clippers work well; larger breeds sometimes need dog toenail clippers.

The person holding the rabbit should sit with the rabbit supported on his or her legs and turned over for trimming. The person doing the trimming can now use both hands — one to push the fur back to see the nail, the other to do the trimming.

If possible choose a white rabbit for your first trimming job. When you look at a white toenail, you will be able to see a pinkish line extending from the toe. This line is the blood vessel that brings nourishment to the growing nail. Toenails that need trimming have a clear white section after the pink. This white area can be trimmed without hurting the rabbit; trimming the pink area will hurt the rabbit and cause bleeding. Leave about 1/4 inch (6 mm) of nail beyond the end of the blood vessel.

On colored toenails, seeing where the blood supply ends is more difficult. If you are not sure, trim only a little at a time. If you cut too closely and the rabbit begins to bleed, use a clean cloth to apply direct pressure and stop the bleeding.

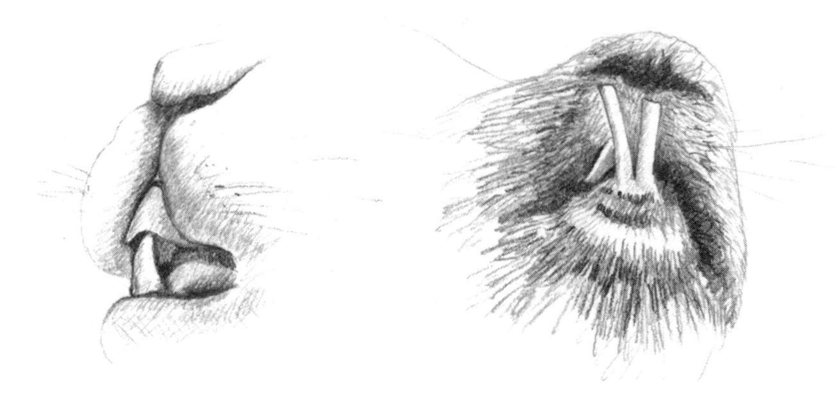
Properly aligned teeth (left) are much healthier for a rabbit than teeth that are overgrown because they are not properly aligned (right).

Malocclusion

A rabbit's teeth grow continuously, about 1/2 to 3/4 inch (13 to 19 mm) per month. In a normal rabbit, the top front teeth slightly overlap the two bottom teeth. Normal chewing keeps a rabbit's teeth at a proper length. If the teeth do not meet properly, this normal wearing down doesn't occur and the teeth can grow overly long. This condition is called malocclusion, buckteeth, or wolf teeth. The top teeth may curl around and grow toward the back of the rabbit's mouth, like a ram's horns. The bottom teeth may grow out in front of, instead of in back of, the top teeth. These bottom teeth may grow so long they stick out of the rabbit's mouth and may even grow as far as into its nostrils. The rabbit may have a hard time closing its mouth. When it tries to chew, these long teeth can stick into its mouth and cause it pain; they can also make it difficult for the rabbit to pick up its food.

Signs of Malocclusion

Watch for these two warning signs of malocclusion in your breeders:

- A rabbit is losing weight but does not seem sick
- A rabbit drops its food or seems to have difficulty chewing when it eats

If you suspect malocclusion, remove the rabbit from its cage to check its teeth carefully. Good handling skills will really pay off, because you must turn your rabbit over to examine its teeth. To help your rabbit comfortably eat its food, you will have to trim these overgrown teeth. It's easiest to have two people for this job — one to hold the rabbit and another to trim.

1. Use fingernail clippers to trim the teeth as close to normal length as possible. Be careful not to cut shorter than normal, or you may cause bleeding and discomfort. You may have to trim each tooth more than once to get it down close to normal size. You may need a larger tool for some large breeds. Your rabbit may not like having you stick the clippers in its mouth, but the actual clipping does not hurt it, because the teeth have no nerves.

2. Check the rabbit every two or three days to see how its teeth are coming along. Trimming should enable your rabbit to begin to eat normally again.

Some rabbits develop malocclusion from an injury, as when a rabbit chews on its wire cage. Whatever the cause, once the teeth are out of position, they rarely return to the proper position. In most cases, the teeth must be clipped every two to three weeks for the rest of the rabbit's life. Most cases of malocclusion are inherited. To avoid perpetuating this trait, animals with malocclusion should never be used as breeders.

Sore Hocks

The hock is the joint in a rabbit's hind leg between the upper and lower leg bones. Because the rabbit carries most of its weight on its hind feet, the area between the hock and the foot suffers a lot of wear and tear. Although the rabbit has fur on this part of the leg, it is not always thick enough to protect it. Sometimes the fur wears away and the unprotected skin develops sores. A sore hock can cause much discomfort. Causes of sore hocks include the following:

- Poor breeding of rabbits with thinly furred footpads
- Small feet compared to body size
- Large breed carrying a heavy weight, putting stress on the hock area
- Nervousness, resulting in frequent foot stamping
- Solid hutch floor with wet (from urine) bedding

Suspect sore hocks if your rabbit shows signs of being uncomfortable when it moves. Your rabbit will put its foot down and then quickly reposition it, as would a person trying to walk barefooted over sharp stones.

If you suspect a rabbit has sore hocks, remove the animal from its cage and examine its hock area. Are there areas where the fur has been worn off? Are these areas bleeding or infected? Also examine the front feet for sores.

If you spot sores, clean the hock area and apply a medicated cream, such as Bag Balm or Preparation H.

You can help your rabbit recover more quickly by checking its cage for things that may have caused the sore hock, such as sharp spots on the cage floor or wire that has been used upside down (on a ½ by 1 inch [1.25 × 2.5 cm] wire floor, the ½-inch [1.25 cm] wire should face upward to create a smoother surface). Place a board or piece of carpeting in the cage to keep your rabbit's feet off the wire. Keep the cage clean to reduce risk of infection.

Sore hocks take a long time to heal and, if untreated, can become infected. The sooner you start treatment, the better your chance of success will be.

Sunstroke and Heatstroke

Hot weather can be hard on rabbits, and high temperatures can kill them. The fur coats that keep rabbits cozy in the winter sometimes provide too much warmth during the summer months. To keep your rabbits comfortable during a heat wave:

- Place outside hutches in shady locations.
- Provide good ventilation. Remove plastic sheeting or the hutch cover. If your hutches are inside, open windows or provide a fan to increase air circulation.
- Provide lots of cool, fresh water.

During a hot spell, check your animals often during the day. It's normal for a rabbit to stretch out in the coolest part of its hutch. During hot weather, supply a wet towel for the rabbit to lie on during the hottest part of the day to help keep it cool. If your weather is particularly humid, a wet towel can't evaporate enough to cool a rabbit, in which case put a rabbit cooler in each hutch (described on page 141).

If a rabbit is breathing heavily and its muzzle is dripping wet, it may have

Rabbit Coolers

Be prepared for extreme heat — use empty plastic soda bottles to make rabbit coolers. Fill the bottles two-thirds to three-quarters full with water, and keep them in your freezer. In periods of extreme heat, lay a frozen bottle in each cage. If you are traveling with rabbits in hot weather, place a cooler in each rabbit carrier. The rabbits will cool down by stretching out alongside their coolers.

heatstroke or sunstroke. You must act immediately to save the rabbit's life.

1. Move the rabbit to a cool location.

2. Wipe the inside and outside of the rabbit's ears with ice cubes. Because the blood vessels in a rabbit's ears are closer to the surface than those in other parts of its body, ice applied to the ears will cool it faster than applying ice to any other spot. Wrap the ice in a cloth so you can hold it more easily.

In an advanced case of sunstroke, the rabbit may be almost lifeless. In this situation, every second is important. To cool the rabbit as quickly as possible, gently place it in a container of water. Use room-temperature water, not cold water, so you don't give the rabbit a sudden shock. Do not let the rabbit's head go under the water; just wet it up to its neck. Once it is wet, place it in a quiet, shaded area to recover.

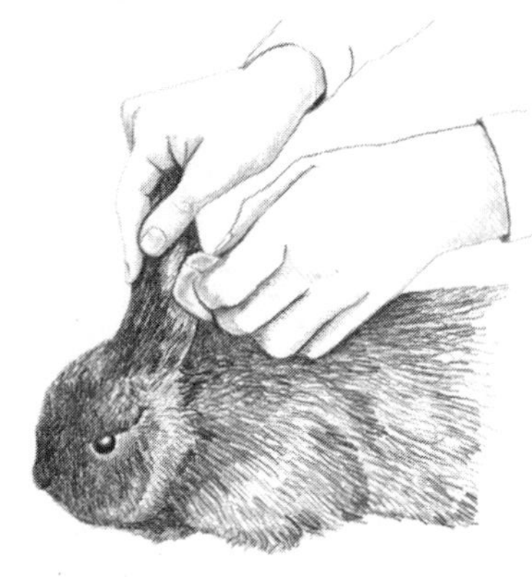

Wipe the inside and outside of your rabbit's ears with an ice cube to counteract heatstroke.

CHAPTER 5

Honey Bees

Our relationship with honey bees began thousands of years ago, judging by a Spanish rock painting circa 6000 BC showing a man climbing a tree and removing a wild nest. Beekeepers eventually began hiving bees in hollow tree trunks (gums) and straw baskets (skeps), destroying the nest to collect the honey. The insects were encouraged to reproduce by swarming, the only way to populate new nests provided by the beekeeper.

The honey bee is not native to the New World, so aboriginal Americans have no history with this insect, although Mayan and other native civilizations kept other types of bee. Bee skeps first voyaged to the Americas on the ships of early Spanish explorers and English settlers.

Why keep bees? The benefits of keeping honey bees include harvesting products of the colony (honey and pollen) and employing the insects in the service of pollination. Honey has been claimed to have numerous health properties, and its sweet taste is lovely in tea or coffee, and in a variety of desserts. A cosmopolitan pollinator, the honey bee is known for servicing a number of crops, particularly fruits, nuts, and vegetables. Contracting hives to a grower for pollination can be a profitable business, but it comes at the sacrifice of producing a honey crop. Keeping bees is also a relatively environmentally benign activity, because no land needs to be owned nor is preparation (such as plowing or fertilizing) required, as is the case for many crops.

And then there is the appeal of the craft itself. Go to any gathering of beekeepers and listen to them talk (even about their challenges) with enthusiasm and pride. Go into a beeyard on a pleasant day and sit there, immersed in the calm serenity of the scene. Watch the bees coming and going, sometimes even indulging in what is called "playtime." If beekeeping is truly in the blood, it is impossible to resist the allure of the craft. For want of a better term, some call the passion that arises in some novices "bee fever."

Bee Selection

Two popular subspecies of the Western honey bee are Carniolan bees (*Apis mellifera carnica*) and Italian bees (*Apis mellifera ligustica*). The two are considered gentle and productive, but they have a few differences that cause various beekeepers to choose one over the other in specific geographic areas.

Carniolan Honey Bees

The Carniolan bee originated in Carniola, which is now part of Slovenia. It is the second most popular honey bee in North America. Carnis are dark brown to black (sometimes described as dusky brown-gray) with lighter brown stripes and are sometimes called gray bees. They fly at cooler temperatures than Italians, making them ideal for northern climes and areas where the early spring nectar flow is strong. They also fly in wetter weather, making them more productive where Italians remain in the hive to avoid spring showers.

Carniolans respond to environment conditions by adjusting brood rearing to the nectar flow. During times of dearth, and as winter approaches, they stop rearing brood; during times of plenty, brood rearing gears up. This system has the advantage that a great number of workers during times of heavy nectar flow produce more honey, while fewer bees in the hive during long winters and times of dearth use up less stores. On the other hand, during summer dearths, the weaker colony is more likely to be robbed, and during times of plenty the rapid population buildup easily leads to swarming. Carniolans therefore require more careful management than Italians.

Who's Who in the Hive

In a honey bee colony there are three castes: a fertile female (queen), sterile females (workers), and males (drones). Division of labor occurs within castes, which often have specific duties depending on age. As the word *caste* suggests, these three individuals in a colony cannot change much, and each will not survive without the contribution of the rest.

Italian Honey Bees

The Italian originated in the continental part of Italy and is the most popular honey bee among commercial beekeepers in North America. It is readily identifiable by yellow and brown bands on the abdomen. Italians are adaptable to most climates, from cool to subtropical, but are less able than Carniolans to cope with hard winters and cool, wet springs. They also have a greater tendency to drift.

Italians aggressively raise brood and continue doing so in times of dearth and as winter approaches. Their high population going into winter requires lots of stores to keep all those bees warm and fed. Consequently, the hive tends to be weakened as spring approaches, making Italians less ready than Carniolans to take advantage of a strong early nectar flow. However, once brood rearing ramps up it continues throughout the summer. During times of poor nectar flow, instead of foraging over longer distances, they are more likely to rob honey from weaker hives. Italians do best in climates where favorable weather throughout the summer is conducive to a continuous flow of nectar.

Beekeeper Talk

brood. Eggs, larvae, and pupae, collectively, that develop into baby bees.

brood chamber. The main body of a hive, where the brood are raised.

colony. A family of bees living within a single hive.

dearth. A scarcity of nectar.

drift. To mistakenly return to another hive after foraging.

flow. An abundance of nectar.

hive. The place where a bee colony lives.

hive body. The brood chamber.

nectar. Sugary water produced by plants, used by bees to make honey.

nuc. A nucleus colony of bees, purchased with their drawn comb and brood.

pollen. Yellow protein powder produced by plants, fed to brood.

rob. To take honey from another hive.

stores. Honey or nectar used by bees as food.

super. A box placed on top of the brood chamber for the collection of honey.

swarm. To leave the hive in large numbers, led by a queen, to start a new colony.

Quick Guide to Honey Bees

Factor	Carniolan	Italian
Climate	does not thrive in hot weather	thrives in hot weather
	able to withstand harsh winters	less suitable for cold northern latitudes
	suitable for cool maritime regions	less suitable for cool maritime regions
Foraging activity	good where early flow is strong	not as active during early spring flow
	forages in cool, wet weather	does not forage in cool, wet weather
	starts foraging earlier in morning	overall forages more aggressively
	forages later in evening	travels shorter distances
	adapts to summer nectar dearths	suffers during summer dearths
Brood rearing	varies with pollen availability	strong brood rearing disposition
	increases rapidly in spring	increases more slowly in spring
	slows during dearths	continues during dearths
	stops in autumn	continues in autumn
Overwintering	requires fewer stores	requires more stores
Robbing behavior	less prone to robbing; easily robbed	tends to rob other hives
Swarming behavior	swarms readily when crowded	moderate tendency to swarm
Drifting behavior	low rate of drifting	high rate of drifting
Resistance	resistant to diseases	somewhat resistant to diseases
	resistant to insect pests	less resistant to insect pests
Temperament	gentler	gentle

When a Bee Stings

A bee won't normally sting unless it has been squeezed, stepped on, or swatted at, or it feels its colony is being threatened. A drone has no stinging apparatus. A queen has a stinger, but rarely will she use it on a human, and if she does, the stinger has such small barbs that she can pull the stinger back out and go on her way.

A bee sting is most likely to come from a worker bee, which has a barbed stinger that cannot pull out of a human's tough skin. When a worker stings and then flies away, the stinger gets ripped out of its body, rupturing the abdomen, and the bee will die. It is easy to see why these insects are not eager to sting, but they will do so if threatened.

The most important thing you can do when you are stung is to remove the stinger immediately, preferably by scraping it off.

Getting Started

Learning to be a beekeeper takes time, and like many activities, you get back proportionally what you put into it. The largest time investment will be in the learning phase: reading and attending bee schools, workshops, and beekeeper meetings. Plan on a steep learning curve the first year or so, and expect continuing challenges after that. Part of the fun of keeping bees is that you never stop learning.

It's best to begin with two bee colonies, rather than just one. As a novice, you may lose a colony during the first season; don't worry, even an expert will occasionally lose a colony. Having a colony in reserve will provide you with a cushion against a discouraging disaster that might bring your beekeeping activities to a premature end. The second unit would then provide an invaluable resource — bees and brood to keep your beekeeping operation going in case of emergency. Having two colonies is also useful for comparative purposes; the shortcomings of one colony are easier to detect when you have another to compare it to.

Visiting the Bees

A bee colony should never be completely ignored. Chores not done at the proper time cannot usually be done effectively later, if at all. Plan on inspecting your bees an average of every two weeks during the active season, perhaps more often as the new season is getting under way and less often in the inactive part of the year. Each visit can be quite brief, depending on the season and your reason for being at the hive. Some visits may last a minute or two; others involving a specific task might take 20 to 30 minutes per hive at the longest. Most inspections that involve opening the hive are a substantial disruption to colony life, so they should not be done without an important purpose. Have a goal in mind before disturbing your bees.

Visit your bees every two weeks or so during spring and summer, and always have a specific purpose for your inspection.

Inside the Colony

In the wild or in the hive, a honey bee nest is made up of parallel combs filled with hexagonally shaped cells. The hexagon is the structural form that provides the most strength using the least amount of material, a proven mathematical axiom. The comb is constructed solely of beeswax, produced by the bee's own body, and serves as both the pantry for the colony's food and the cradle for the brood.

The central combs usually have the young bees and/or brood in the middle, surrounded by a band of pollen and then honey. As combs are built out from the center of the nest, less brood and more pollen and honey is often found on each. The side combs generally are full of honey and are excellent insulators, maintaining the heat necessary to keep the brood developing.

Seven Basic Tips for Getting Started

The following suggestions will minimize problems as you begin keeping bees.

1. **Start with new equipment** of standard (Langstroth) design and dimensions. Used and homemade equipment can come with problems you aren't experienced enough to recognize or handle.

2. **Do not experiment** during your first year or two. Learn and use tried-and-true methods. Master them. Later you will have a better basis for comparison if you choose to experiment in future years.

3. **Before buying a beginner's outfit,** know how each piece of equipment is used and be sure you need it.

4. **Get Carniolan or Italian bees.** These honey bees are the two most commonly available in the United States. Bees acquired from a competent producer should be gentle and easy to handle. In future years you can experiment with other races or strains to form comparisons.

5. **Begin with a package of bees or a nucleus colony (nuc)** rather than an established, fully populated one. Establishing a nuc or package will help you gain confidence.

6. **Start early in the season, but not too early.** All beekeeping is local, so seek guidance from local beekeepers and bee inspectors about timing in your specific area. Advice about managing honey bees from those in other geographic areas, even if they are successful, is fraught with risk.

7. **Recognize that your colonies will not produce a surplus of honey the first year,** especially those developed from package bees. The first year is a learning time for the beekeeper and a building time for the new honey bee colonies.

A Bee's Life

Like many other insects, the honey bee goes through complete metamorphosis. Each caste undergoes a developmental process where the body radically changes in both form and function. Beginning with an egg, which hatches into a small worm (larva), the individual then moves into a resting phase called the pupa. During the pupal process, tissues are rearranged into the final stage, the adult. Thus, the individual bee has specific lifestyles: eating and developing and contributing to the colony's maintenance and growth.

Development takes place in the cells of the comb. When an egg hatches, the workers first provision it with a nutritious jelly in a process called mass feeding. As the larvae develop, the adults provide food as needed, a process called progressive feeding. At the end of the larval period, the bees cap each brood cell with wax and each larva then reorganizes its tissues through a process called pupation. Once this change or metamorphosis has occurred, the fully formed adult chews away the cap of its cell and emerges to join the other adults in the colony.

The Queen

The queen is the reproductive center of the colony. She is the only female capable of laying fertile eggs, without which the colony could not grow in population. She is also the longest-lived individual, sometimes surviving several years, and the largest in size, with a long, tapering body. Because her job is to lay fertile eggs, the queen has about 160 ovarioles, or ovary tubes (workers have four at most) and a single sperm storage device, the spermatheca.

Although the queen lacks some characteristics of the workers, including wax glands, pollen baskets, and a scent gland, she is still able to sting. This ability helps her control rivals but is apparently not used in general defense. Queens also have notched jaws or mandibles, which differentiates them from workers that have smooth ones.

Egg production by the queen ebbs and flows with colony development and is based on many factors, including prevailing weather conditions, the state of the local vegetational complex, and the makeup of the population within the hive. In the active season, some queens may lay up to 2,000 eggs a day if a large population is deemed needed,

Meet the Honey Bee Family

Within the insect class, honey bees belong to the order Hymenoptera, which includes ants, bees, and wasps. The large super family of bees (Apoidea) includes not only honey bees but also bumblebees, carpenter bees, orchid bees, and many others. Honey bees are in the genus *Apis* and include the European or Western honey bee, *Apis mellifera*, kept for honey production and to ensure the pollination of various crops around the world.

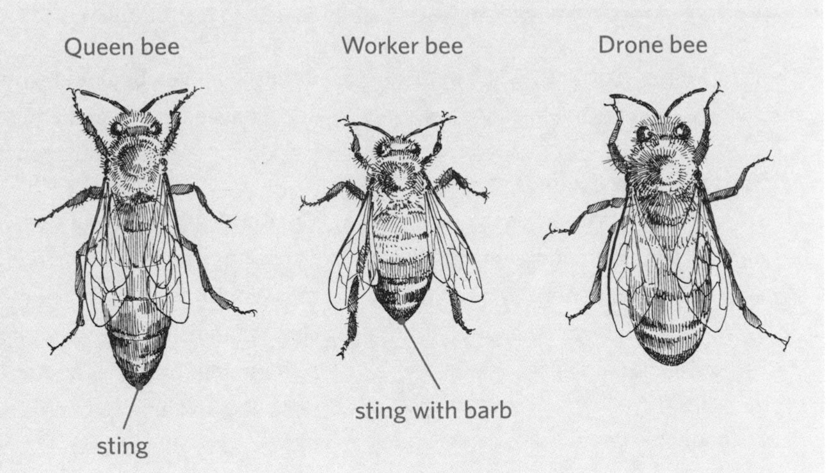

but when the conditions change, egg laying does as well.

The queen doesn't actually rule a colony, but she certainly exerts control. Her most important task is to use a suite of chemicals or pheromones called queen substance to prevent workers' ovaries from developing to the point where they produce eggs. Workers do, rarely, produce eggs but they cannot be fertilized, and they result exclusively in drones that cannot sustain a colony of laying workers. Because no workers develop to replace workers that die, a colony raising a worker-produced brood rapidly loses population.

Another way the queen exerts control is by determining the sex of her offspring, producing either drones from unfertilized eggs or workers from fertilized eggs. So she is responsible for the eventual population makeup of the colony.

Each colony usually has only one queen, but sometimes you might find two or more. The usual explanation is that both a mother and a daughter who is about to replace her are present for a short period on the same comb. Occasionally a number of virgin queens may appear in a swarm, especially among Africanized honey bees (see Glossary).

The Worker

The worker honey bee is a fully formed female, but her ovaries are small and kept from developing by contact with the queen's pheromones. The worker population is responsible for all the activities for which honey bee colonies are known, including honey production, pollen collection, temperature regulation, and defense. Workers number in the tens of thousands, making this caste the most populous in the colony.

A worker begins life as a fertilized egg laid by the queen and hatches into a larva in 3 days. The larva is fed for about a week, and then pupates in 12 more days, emerging as an adult after a total of 21 days. The importance of understanding this 3-week development cycle is that it lets you monitor a colony's population either for honey production or pollination. It takes at least 6 weeks to build up a colony from a maintenance population level to one that can be counted on to be productive.

The Drone

The drone is the male bee, a remarkable organism in his own right. First of all, he is haploid, the result of parthenogenesis or virgin birth. That means he has half the number of chromosomes as a female bee. He is the genetic mirror image of his mother and he has no father. His only known job is to mate with virgin queens, providing them with a great many copies of himself in the bargain.

During the active season a colony has at most only a few thousand drones. At the first sign of stressful environmental conditions, the colony disposes of its drones by banishing them from the hive.

Development Rates

Bee	Egg	Larva	Pupa	Total	Life span
Queen	3 days	5½ days	7 days	15½	up to several years
Worker	3 days	6 days	12 days	21	weeks to months
Drone	3 days	6½ days	14½ days	24	40–50 days

The Varroa Mite

When you think of the members of a honey bee colony, three individuals come to mind: queen, worker, and drone. In the last two decades, however, a fourth has intruded itself into the hive, the Asian mite known as *Varroa destructor.*

The varroa mite is not a run-of-the-mill parasite. It is an exotic organism completely new to its host, and thus the Western honey bee has little natural defense against the mite's depredations. Well over 90 percent of infested colonies are at risk of dying. The mite is locked into the honey bee life cycle in ways not fully understood. It is totally dependent on the honey bee and cannot live isolated from its insect host for long.

Flying bees carry mites, an activity called phoresis. Mites in a colony that is weakened by starvation or predation may mount a worker bee and travel with her to a healthy hive. Robber bees may be infested as well. Finally, drones, which are preferred by mites, are universally accepted into colonies and can enter a large number during their lifetimes, thus becoming a primary distributor of mites among hives.

Varroa mites on a pupa.

The reproductive phase of the varroa mite shows how tightly its existence is woven into the fabric of honey bee colonies. Not only are certain types of cells more attractive to female mites, but certain specific ages of the larvae are as well. Mites prefer drone brood because it takes longer to develop, giving the parasite more time to produce additional mites per brood cycle than is possible in worker brood.

Unlike many other diseases and pests the varroa mite is not a passing phase. It is a permanent resident in all colonies in affected regions. It cannot be eradicated from the nest. This futile goal of many beekeepers has resulted in the use of increasingly more toxic treatments, which are now considered counterproductive to honey bee health.

Bees versus Wasps

Although bees and wasps may look similar, they are not. Bees are vegetarians, consuming only sweet plant juice, called nectar (carbohydrate), and pollen (protein). Wasps also imbibe nectar, but they take their protein from other organisms; they are carnivorous and often prey on other insects (including pests). Many people in the general public don't realize that bees and wasps are both beneficial insects that should be protected.

A Worker Bee's Jobs

During its life a worker goes through a set series of predetermined tasks depending on its age and colony need. A typical sequence:

House bee. Tasks during the early part of a bee's life include cleaning brood cells, attending queens, feeding brood, capping cells, packing and processing pollen, secreting wax, regulating temperature, and receiving and processing nectar into honey.

Nurse. These bees feed the larvae a nutritious jelly called brood food.

Comb builder. Glands on the bee's belly secrete wax flakes that she molds into comb.

Honey maker. Workers receive relatively raw nectar from foraging bees and place it in the comb. They fan their wings in unison to evaporate excess liquid and they also add enzymes, which chemically transform the nectar. The result is honey.

Forager. Two to three weeks after emergence, the house bee begins to fly from the colony preparing to become a forager. Field bees are usually guided by scouts to the best nectar sources nearby. Once they begin foraging on a certain crop, field bees tend to be faithful to this plant, visiting it exclusively. One well-known characteristic of honey bees is their so-called dancing. Their regular, repetitive movements communicate many things, from the location of a food source to a call for a grooming session from a nearby sister bee.

Housing Bees

The movable-frame beehive is the foundation of modern beekeeping. Most managed beehives are of the type patented by the Rev. L. L. Langstroth in 1851. The dimensions he used, a box accommodating 10 frames, 9⁹⁄₁₆ inches (243 mm) high (often called full-depth), have become standard in the United States.

Other systems are also available and make sense in some situations. These alternatives include hives based on eight frames, which allow for easier lifting for the operator, and the shallower (6⅝ inch/168 mm) Dadant square box, for people wishing to interchange equipment extensively. Another system that is rapidly gaining adherents is the top bar hive, a radically different design employed by beekeepers who prefer a small-scale approach.

Hive Elements

The standard box or hive body is constructed of wood, although other materials such as plastic and even concrete have been used in some areas with success. Initially stick to tried-and-true standard wooden equipment; experiment with other configurations only as you gain confidence and experience. Most hive bodies are made of pine or cypress, both of which will provide years of service.

Painting the exposed surfaces will lengthen the box's life. The majority of hives are painted white, which allows hives in full sun to stay cool; however, white hives are much more visible in the landscape. In cold climates, black or other colors may increase heat absorption from the sun. Some paint stores and recycling centers give away off-color or leftover paint.

Supers. Supers are boxes that in construction are identical to those used for the hive body, and are stacked on top of the hive body for the collection and harvesting of honey. The word *super* refers to the superior position of the additional boxes on top of the hive, the natural place where a strong colony stores honey.

Boxes are available in several depths. Commercial beekeepers with lifting equipment, and hobbyists living in temperate climates, often use two full-depth, 9⁹⁄₁₆-inch-high (243 mm) boxes to make up the basic hive. Full-depth boxes allow bees to cluster into a ball to keep warm in winter, and provide room for storing the honey needed for winter food. During nectar flow, supers are placed on top of the basic hive. Since a full-depth super full of honey can

The Top Bar Hive

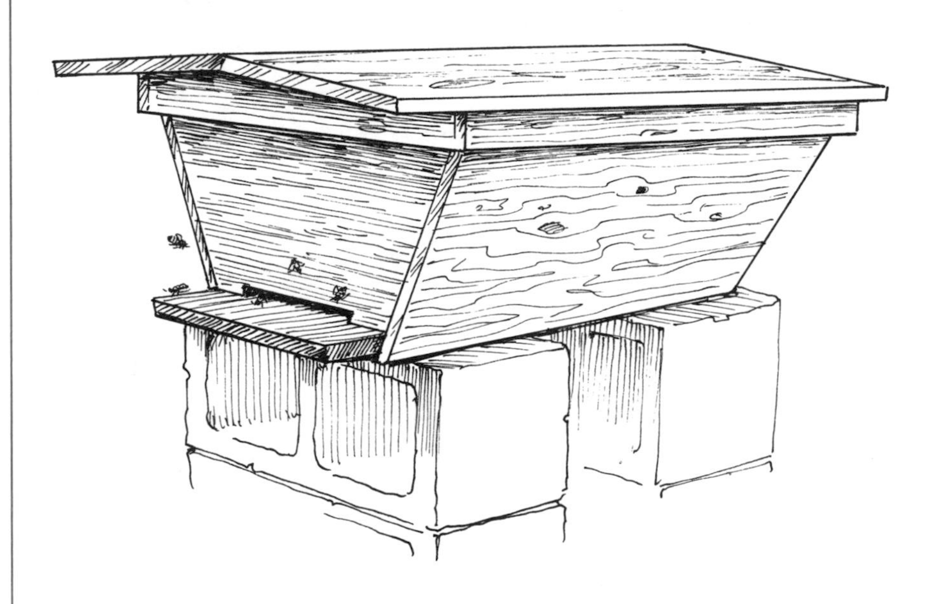

The top bar hive is inexpensive, flexible, and offers the following advantages:

- The brood is generally placed toward the hive's front entrance and the honey is located in the rear. Since you don't have to dismantle the whole hive to examine the brood or take off honey, working the hive is less stressful on the bees and they are less likely to become defensive.
- The top bars butt against each other, doubling as a cover and thereby reducing material requirements and conserving weight. An outer cover of tin or cardboard is necessary, however, to protect the colony from moisture.
- The hive can be mounted on a stand at waist level. Working the hive is therefore more comfortable because you don't have to continually bend over.

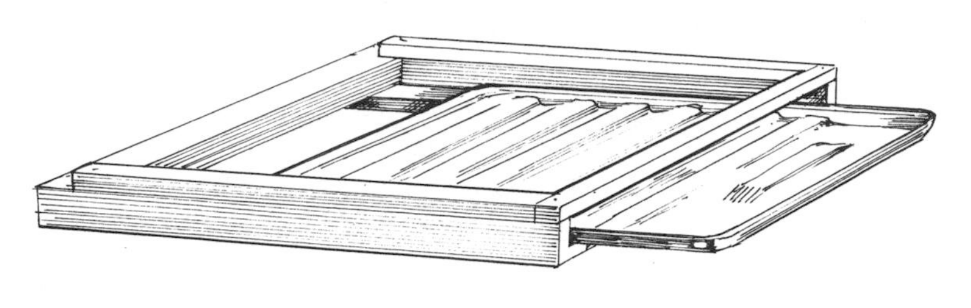
sliding bottom board

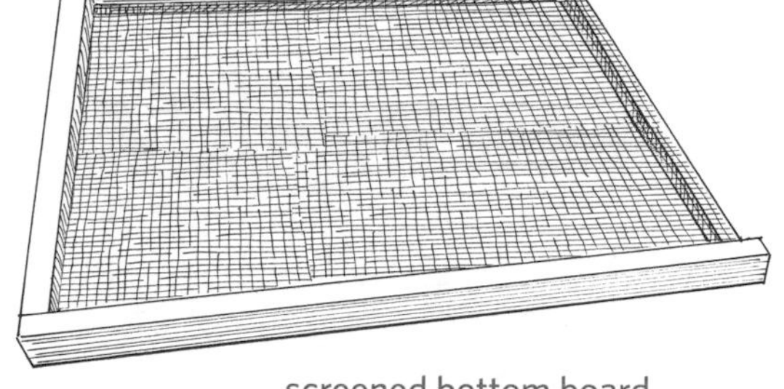
screened bottom board

pollen trap

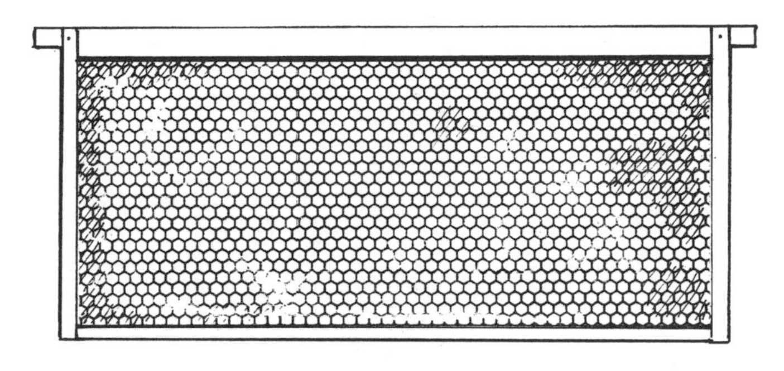
A frame is fitted with a sheet of beeswax foundation on which bees draw out the walls of the cells.

weigh 100 pounds (45 kg), shallower boxes can be used as supers.

In warmer climates, where bees don't need to store up much honey, some beekeepers use a single full-depth brood chamber, while others prefer several medium-depth 6⅝-inch (168 mm) chambers. Medium-depth boxes are lighter in weight and, when used as both brood chambers and supers, let you standardize with boxes of all the same size.

Frames and foundation. Ten combs fit in the standard hive body (brood chamber), each surrounded by a frame. The frame is designed to have a critical gap, or bee space, of about ⅜ inch (10 mm) on all sides between it and the interior wall of the hive. The bees do not build into or glue up this space.

Frames can be either wooden or plastic. The latter is preferred by beekeepers who are concerned about beeswax contamination due to mite treatments. Each frame has a top bar, a bottom bar, and two side bars assembled into a rectangle that holds the comb.

The bottom board, the floor of the hive, has a number of functions. Bottom boards available in supply houses are usually reversible, providing the option of a small entrance to conserve colony heat during winter and a larger one for the rest of the year.

A sliding bottom board allows you to monitor varroa mite fall. A screened bottom board controls the mite population by letting mites fall through to the ground.

You can modify the hive floor to fit other accessories such as feeders, pollen traps (for scraping pollen off bees' legs as they enter), or dead bee traps (to measure colony mortality).

Frames are fitted with sheets of beeswax foundation, the template from which bees build their nest. Foundation is commercially produced out of beeswax or plastic, and embossed with a pattern of hexagonal cells. The bees then draw out the walls of the cells based on the template (foundation). You can purchase reinforced foundation that has embedded vertical wires with hooks.

Queen excluder. The queen excluder is a series of parallel bars spaced so worker bees can squeeze between them, but the larger queen cannot. The excluder is placed above the brood chambers to prevent the queen from entering the honey supers and laying eggs there. Excluding the queen keeps honey supers from being contaminated with brood, since most beekeepers prefer to work with solid combs of honey.

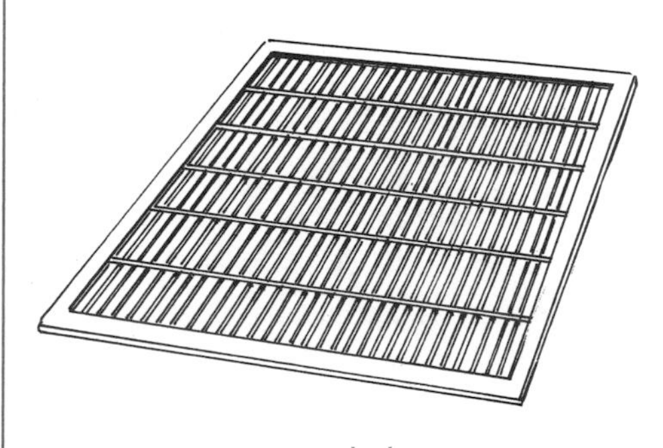
queen excluder

Langstroth Hive Bodies and Supers

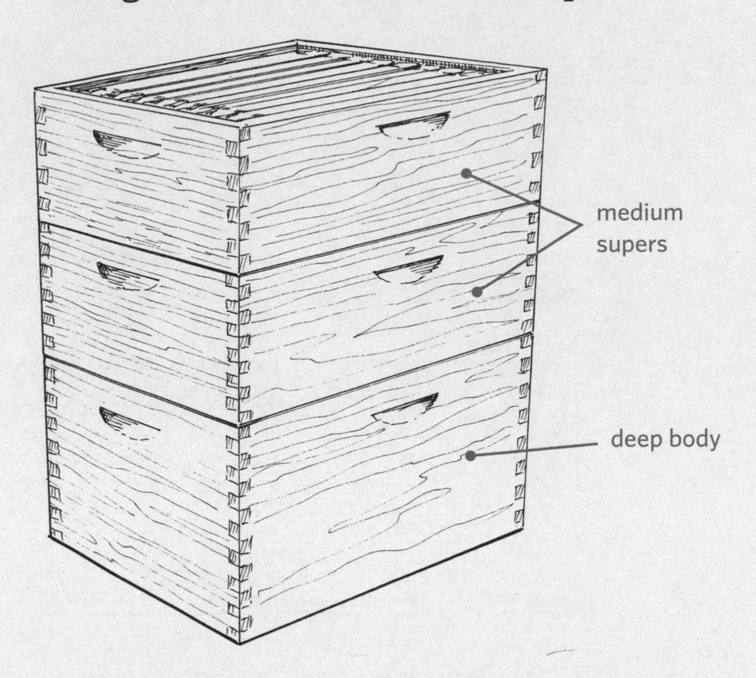

The Langstroth system consists of four-sided boxes with standard interior dimensions to accommodate standard-size frames of honey comb. A deep (9⁹⁄₁₆ inches [243 mm]) box might be used as a brood chamber, where the queen lays her eggs and the bees care for the developing larvae, but when used for honey stores, a deep box full of honey weighs 100 pounds (45 kg), making it difficult to handle. Instead, medium boxes (6⅝ inches [168 mm]) are more typically used as supers for honey collection and harvesting.

During early summer the hive consists solely of brood chambers. Commercial beekeepers tend to prefer one or two deep boxes for brood, as shown on the left. During nectar flow, a queen excluder and one or more supers are placed on top for the collection and harvesting of honey. Hobbyists who prefer to have brood chambers and supers of the same size will standardize on all mediums because they are lighter and easier to handle; three medium boxes, as shown on the right, provide a similar amount of brood space as two deep boxes.

Covers. Most supply houses sell telescoping outer hive covers and ventilated inner hive covers. Both are standard and useful for the novice. Many large-scale beekeepers opt for what is called a migratory cover, which is often nothing more than a piece of plywood used without an inner cover. These covers are cheaper than the telescoping variety, but not as secure. An advantage, however, is that they allow hives to be stacked together tightly when moved by truck.

The inner cover usually has a rim and an oblong hole in the center to accommodate a bee escape device. The rim is notched, offering the option of increased ventilation.

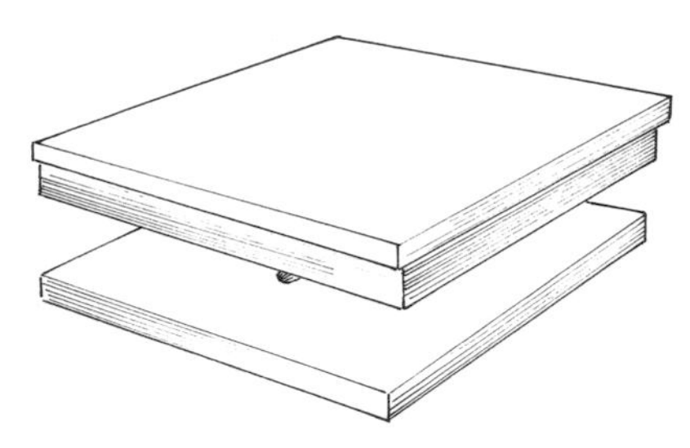

Telescoping outer cover (top) slides down around a ventilated inner cover (bottom), which provides a barrier between the top cover and the bees. The inner cover makes the top cover easier to remove.

Hive stand. In most areas a hive stand is essential. It keeps a wooden hive off the ground, protecting it from dampness, dry rot, and termites and other insects. Either concrete blocks or pressure-treated wooden rails might be used to construct a stand. In areas where ants are a problem, the hive stand can be a rack perched on narrow legs that are inserted into cans of oil or water, which keep these insects at bay.

Feeders. You'll need a feeder to furnish food for a new colony while it's getting established and to feed an established colony when stores run low. The Boardman feeder, an inverted glass jar that

Types of Feeders

A Boardman feeder helps a new colony get established and can be used to feed any colony that is running low on stores.

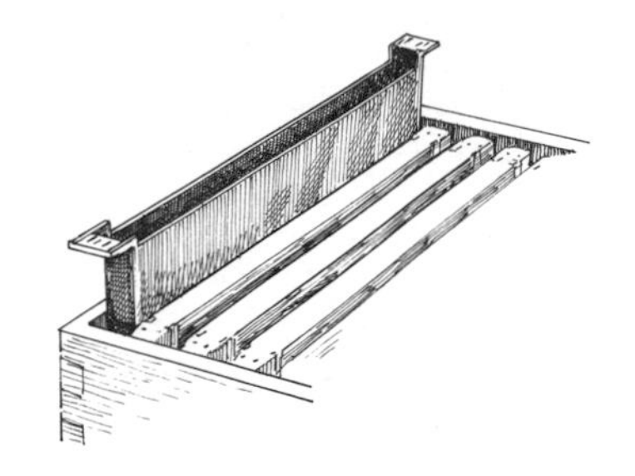

A division board feeder replaces one of the 10 honey comb frames.

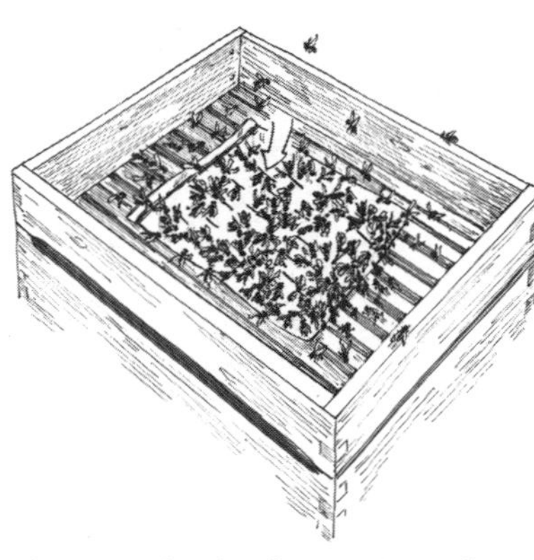

An inexpensive feeder consists of a zipper bag filled with syrup and placed on top of the frames, and punctured with holes or small slits so the syrup can seep out for the bees to collect.

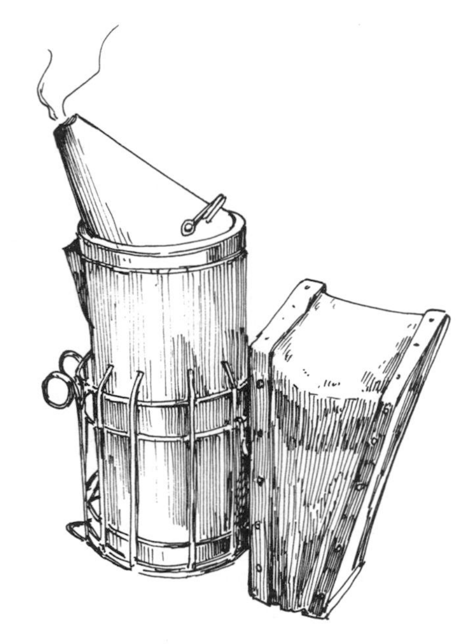

A good smoker and chemical-free fuel are indispensable for controlling bees while you work the hive.

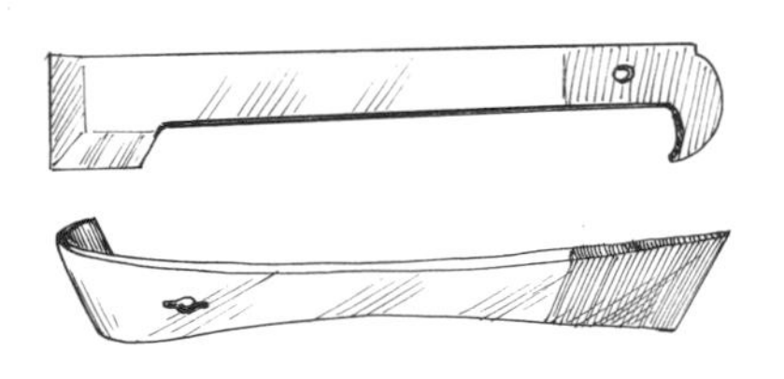

Two common hive tools.

fits in the entrance of the hive, is a good choice for the beginner. A major advantage of the Boardman is that the syrup level can be monitored and the syrup supply replenished without dismantling the hive. This feeder has several disadvantages, however. It doesn't hold much food, and if used with a weak colony, it may incite robbing behavior.

Other types of feeding arrangements include modified frames (division board feeders) and top feeders, some of which are quite elaborate. An inexpensive alternative is to use a small plastic bag filled with syrup, punctured with small holes, and placed over the frames.

Smoker and fuel. The smoker is a most important and essential beekeeping tool, although you may find some beekeepers who claim not to need it. Smoke suppresses defensive behavior, and you should have a smoker lit any time you are in the beeyard.

Smokers usually come in two sizes, 4 by 7 inches (10 × 18 cm) and 4 by 10 inches (10 × 25 cm). Stainless steel models last longer than those made from regular sheet metal. Some smokers have a shield to prevent direct contact of the hot barrel with the beekeeper's clothing.

You'll need fuel to burn in the smoker. Practically anything that burns is good smoker fuel, from pine needles to dried dung. Avoid any material that has been chemically treated, such as coated burlap or pressure-treated wood. Chop seasoned ½ to ¾ inch (1 to 2 cm) limbs into 3 to 4 inch (8 to 10 cm) lengths. Seasoned chips or chunks from logs or stumps also work well.

Hive tool. Sometimes called the universal tool, the traditional steel hive tool is hard to beat for the variety of jobs it can do. Its main jobs are to remove frames from a hive and to scrape debris off hive parts. Several shapes have been devised, and you can paint yours in bright colors so you can find it easily when you leave it behind in the field. In a pinch, you could use a screwdriver or something similar to manipulate a hive, but you have to take great care to avoid damaging your equipment. The best approach is to buy two tools, one for each of your new hives.

Beekeeper Garb

A wide range of protective clothing is available to beekeepers. A good idea is to wear more when you begin a hive manipulation, because it's usually easier to take some off than to put more on. Depending on your reaction to stings, you may decide to be fully protected or not. Only one item is essential, however: the veil. Although you can recover from the effects of stings no matter where they might occur on your body, a sting to the eyeball will cause severe damage, perhaps resulting in blindness. Never manipulate bees without wearing a veil.

A complete beekeeper outfit covers you from head to toe. The crucial part is the veil, which you should always wear when working a hive.

Veil and hat. A variety of headgear is sold to novice beekeepers to support that critical accessory, the veil. Wear a veil at all times to protect your face and head from stings where they are most dangerous. Veils are designed to be worn with helmets and hats, or in some cases without a hat (the Alexander model). The square folding type veil is more durable over time because you can easily pack it away when you are not using it.

Gloves and coveralls. Gloves give the novice confidence. They are often abandoned by experienced beekeepers but can become necessary in an emergency. Several kinds of gloves are available, including rubber ones, but the traditional canvas or leather glove with a long cuff designed for beekeeping is best. Coveralls range from all cotton to a space-age model made from breathable foam.

Where to Set Up the Hive

When thinking of potential sites, consider the bees' preferences as well as your own. Bee studies done in temperate locations, where nests are usually found in hollow trees, reveal certain natural preferences. The insects choose an elevated location, about 10 feet (3 m) off the ground, with a southerly exposure and not too much sun or shade. But in setting up your hives, you need to consider accessibility and the potential concerns of neighbors while applying a measure of common sense.

Tips on Hive Location

Out of sight is out of mind. The closer your hives are to your residence the better. As a beginner, keep your colonies nearby so you will be encouraged to visit them often. Casual visits are important even if you don't open the hives. You can learn a lot simply by observing the entrance.

Don't invite vandalism. Isolated beehives can attract the attention of vagrants, bored kids, and thieves. Colonies have been stolen in their entirety, pelted with rocks, tipped over, targeted with gunshot, and tossed into rivers. Keep yours where you can see them easily, but camouflage your hives in shrubbery and/or locate them behind fences and buildings.

Provide some room to work. Manipulating a colony takes space, so don't put yours too close together. Always allow enough space between colonies so each can be worked without disturbing

In Times of Drought

Many areas where honey bees are located experience dry times during the course of the year. When intermittent creeks cease to flow and tree leaves show signs of moisture stress, bees start collecting water from leaking faucets, bird baths, pet dishes, and swimming pools. Drinking from a chlorinated swimming pool won't harm the bees, but this practice can disturb humans in the vicinity. When honey bees become more noticeable to the general public, complaints start rolling in.

Don't let your bees become trained to a swimming pool or other unsuitable watering place. Once they establish a water foraging pattern, it's almost impossible to change. Prevention is the only cure for this problem. At times when your bees are likely to forage in nearby urban areas during a dry spell, you must furnish them with a reliable source of clean water.

others nearby (you'll have to decide for yourself how much space is enough, based on your mobility and needs).

Stay on the level. A hive must be on level ground. Honey is heavy and moving it can be difficult. Experienced beekeepers use aids, such as a wheelbarrow, a hand truck on inclined planes, and a mechanical lift to transport filled supers.

Neighborly Concerns

Neighbors are a constant concern for a beekeeper, and it's important to keep them on your side, especially if you are in an urban or suburban area. The following procedures may help you avoid potential problems caused by beekeeping activity.

Look up local regulations dealing with keeping bees or any animals in your neighborhood or area. If the restrictions are intolerable, consider writing up a specific ordinance for your situation. Your local and state beekeeping associations will help in this regard.

Place hives away from foot traffic and building entrances. Orient hive entrances so they face away from likely human foot traffic.

Erect a 6-foot (2 m) barricade between the hive and the lot line. Use anything bees will not pass through, such as dense shrubs and solid fencing. Potential problems occur any time bees are flying close to the ground and across the property line of a neighbor.

Provide a watering source. If no pond or stream is nearby, set out a tub of water near your hives. Add wood floats to prevent the bees from drowning. Change the water periodically to avoid stagnation and mosquito breeding. To prevent stagnation, a better watering device is one that trickles water down a wooden board or slowly drips onto an absorbent material such as a cloth, keeping the surface damp.

Minimize robbing by other honey bees. In times of dearth, bees may seek out weaker colonies and rob their honey. Robbing bees are usually quite defensive and are more likely to sting passersby. Manipulate your hives only during nectar flows, if possible. To prevent robbing, keep extraneous honey or sugar water from being exposed by avoiding sloppy feeding techniques, exposing combs during manipulation, and taking off honey inappropriately. Use entrance reducers to minimize the likelihood of stronger colonies robbing weaker ones.

Prevent swarming. When a colony gets so big that the hive starts feeling crowded, a large group of bees may swarm away, looking for another place to live. Although swarming bees are the most gentle, a large, hanging ball of bees often alerts neighbors to beekeeping activities and may cause undue alarm.

Keep no more than three or four beehives on a lot less than half an acre. If you desire more colonies, find a nearby farmer who will host bee colonies in exchange for some honey.

Work your bees when neighbors are not in their yards. Bringing attention to your beekeeping activities only invites trouble.

Bribe the neighbors. Give each neighbor a jar or two of honey every year and you are less likely to get complaints about your bees.

Rooftop Hives

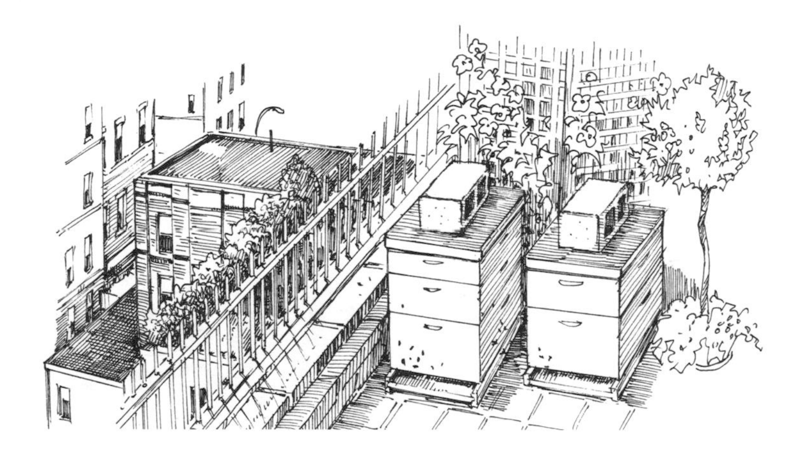

One of the best locations for honey bees is on a roof. Many apiaries thrive on city rooftops from Paris to Chicago. The benefits to a rooftop location include the fact that your hives avoid interference by humans and other mammals — such as bears. And rooftop hives are easy to monitor. Urban bees forage in parks, gardens, street trees, and balcony flower boxes: any nectar source within approximately 1 mile (1.5 km) of the colony. City bees can be more productive than country bees.

The challenges of keeping bees on a roof include the difficulty of lifting honey-filled supers by hand, since you can't use a truck or other big mechanical device. In addition, a roof can be sweltering in summer and/or extremely cold in winter, when you'd rather not be up there yourself. Fortunately, bees are good at maintaining their own temperature. Be sure to supply barrels of water during heat spells, and make sure the hive covers are firmly held in place by cinder blocks or other weights in case of high winds.

Installing New Bees

Outside of beekeeping circles, finding a good source of honey bees can be difficult. You can locate reliable sources by subscribing to specialized beekeeping publications and joining local and national beekeeper associations. Bees can be obtained in a package, as a nucleus hive, as an established colony, from a swarm, or from a wild or feral colony, but as a beginner you should consider only the first two options.

Package Bees

The best way for a novice to begin is with package bees. In the United States most packages originate in the South and the West so they can be ready for the needs of more northerly beekeepers in early spring.

A package is easier for a beginner to handle than a nuc because the bees are not yet a true, organized colony but just a bunch of insects in a box. They lack a home to defend and have limited organization and little sense of purpose. They are, therefore, among the least defensive of honey bees.

Bees in a package have been stressed in many ways. They have been separated from their original queen, endured the rigors of a trip through the mail or other delivery method, and perhaps suffered other indignities such as being left out in the rain or hot sunshine. Since you can't know what your package bees have been through, transfer them into a hive without delay.

Watching your package develop into a fully functioning honey bee colony can be extremely satisfying. But be aware that only rarely will a colony established from package bees produce a honey crop the first season.

Installing Package Bees

Any brand-new hive that receives a package is considered a fragile work in progress. The bees have been treated roughly and will have a tough time as they transition from a purposeless mass of insects in a screened box to a productive colony in a beehive. They need a lot of tender loving care as they get to know their new queen and establish a feeding regimen. Below are some tips and guidelines.

Storing and Feeding a Package

Install a package as soon as possible after it arrives. Any delay will result in fewer live bees in the colony. Although packages are designed to be stored for a limited time period — a maximum of a week might be possible — every day of delay is critical.

If a package needs to be stored, feed the bees by dripping sugar syrup (1 part sugar mixed with 1 part water, by volume) through the screen. Add just enough syrup to dampen the bees. Turning the package on its side (so the sugar syrup goes through the screen)

What Is a Package?

A package is a screened cage packed with honey bees. The unit is really an artificial swarm, taken from a large colony and literally shaken through a funnel into a screened box. A caged queen comes included in the package, along with a can of sugar syrup to sustain the bees on their journey. Blocking the entrance of the queen cage is a candy plug, a sugary mixture which the worker bees will ultimately eat to release her into the hive. The package typically contains about 11,000 bees (3,500 bees to 1 pound/0.5 kg).

Package bees can no longer be expected to be free of diseases and pests. At least some parasitic mites are probably found in any package, and some packages have more than others. Nevertheless, the unit is still not as potentially problematic as a nucleus or an established colony, both of which come with comb that may carry mites. You should not have to worry about managing varroa mites until your colony develops a sizable population.

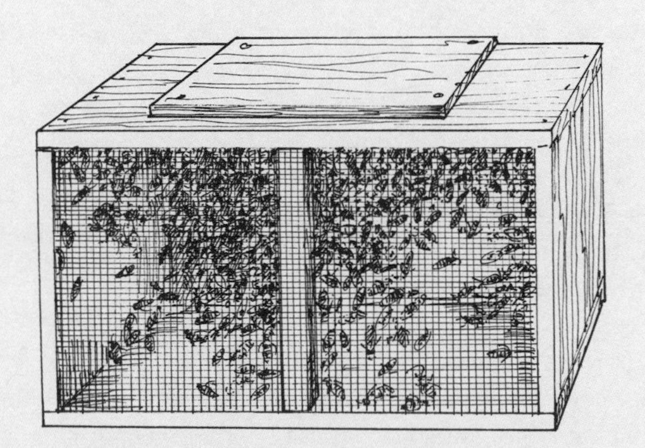

Feed Properly

Do not feed package bees by brushing syrup on the screen with a paint brush, as some people recommend. The danger of damaging honey bees' delicate mouth parts and limbs with the brush is too great. Instead, tip the package and drip enough syrup on the screen to dampen the bees, shown below.

makes the feeding process easier. Some package designs have an enclosed bottom to hold syrup temporarily.

Getting Ready

Before installing your package bees, assemble all your equipment and beekeeper clothing.

1. Make sure everything is laid out and ready for the installation. Tools needed include a hive tool, a pair of pliers for removing the food can of sugar syrup, and a sharp knife or thin nail for puncturing the candy plug in the queen cage. You also need your veil and gloves, a smoker, and the feeder.

2. Fill the feeder with sugar syrup and have it ready.

3. Remove four frames from the center of the hive, and lean them against the hive body.

4. Plug the entrance to the hive with straw or leaves so the bees can't leave easily. Do not seal off the entrance entirely.

Smoker Basics

The smoker is a primary tool for controlling bees while you work. Here's how to properly operate a smoker.

Lighting the Smoker

1. Have your fuel handy.
2. Crumple up one-half to three-quarters of a sheet of newspaper so it will fit into the smoker, but don't wad it so tightly that it won't burn easily.
3. Light one end of the paper, put it into the smoker, and start puffing the bellows. At the same time, use your hive tool to push the paper down into the smoker and get it burning well.
4. Add some small wood chips on top of the paper. When they are burning nicely, add bigger chips until you have filled the smoker. The smoke produced should be cool to the touch. Guard against hot smoke or flames. To cool the smoke, pack some green material like grass clippings on top of the burning mass in the canister.
5. Close the lid.

If the fire goes out, shake everything out of the smoker and start again. Don't get frustrated — you may need three or four tries to get it right.

After half an hour, about half the fuel will be burned and you should fill the smoker again. When you have finished for the day, plug the snout of the smoker with a wad of paper. The fire will go out and preserve the half-burned fuel to make the next lighting easier.

Using the Smoker

Using a smoker requires experience and patience and is an art form itself.

1. Put a hook on the bellows of the smoker so you can hang it from your belt while you move around, or hang it from the end of the opened hive so it will be handy while you're working with the bees.
2. As you approach, give the hive a couple puffs of smoke in the entrance. If you are standing between two closely spaced hives, give both a little smoke in the entrances.
3. Use your hive tool to pry up the cover. Waft smoke over the frames and gently direct it into the colony.
4. Replace the cover and wait at least a minute before removing the cover again.

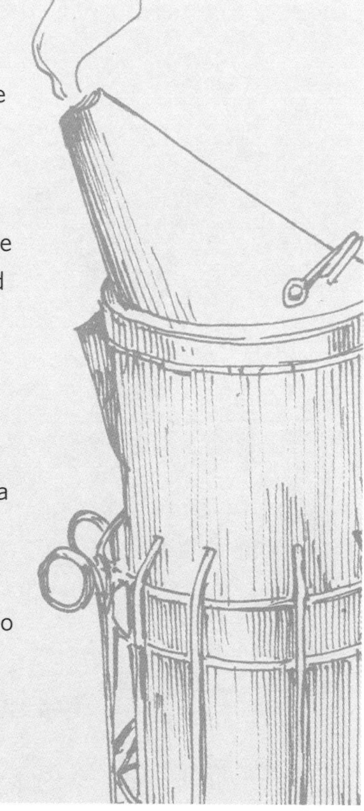

Installation

If possible choose a calm, sunny day to install the package. The best time to set up a colony is in the morning (9 a.m. to 11 a.m.) before bees become too active. Read the following steps several times before proceeding so you have the sequence clear in your mind.

1. Put on your veil and light your smoker in the unlikely case it is needed. Since package bees, like swarms, are not usually defensive, coveralls and gloves are not necessary.

2. Use the hive tool to pry the lid gently off the package in one or two places, as you might open a paint can. This method usually does not allow bees to escape, although some may. Don't worry about escaped bees; they will rejoin the colony in due time.

3. Remove both the sugar syrup can and the queen cage, and then invert the package lid over the empty spaces thus created to keep the majority of the bees inside. Ignore any bees that take flight; concentrate on the job at hand.

Installation, Step 3.

4. Examine the queen and make sure she's alive. A colony established with a dead queen will lose many workers and they may abandon the hive. If she is dead, call the producer and arrange to have another queen sent immediately. The package can usually be stored for a few days as noted (see page 156). Some queens are caged with worker attendants who feed and groom her.

5. Remove the cork or rotate the metal disk to expose the candy in the end of the queen cage. Make a hole in the candy with a small nail, being careful not to injure the queen inside.

6. Install the queen cage by wedging it, candy side down, between frames near the inside wall of the hive body. Avoid entirely blocking the screen on the cage so the bees can feed the queen through it. Properly orienting the queen cage can be challenging, so try not to obsess too much about it. Indirect introduction of the queen allows the bees to become organized while freeing her gradually, reducing the risk she will be lost or killed in the initial confusion.

7. In one smooth motion, invert the package and dump the bees into the space in the hive body created when you removed the frames (see illustration at top right). Don't worry about getting them all in; only a critical mass (with the queen) is needed to attract the outliers.

8. Replace the frames in the hive gently so as not to damage the bees.

9. Install the feeder.

Installation, Step 7.

10. Put the empty package near the entrance. Bees attached to it or flying around outside the hive will rejoin the colony on their own.

After They Are Installed

The next day, make sure the hive entrance is somewhat open for the bees to enter and exit; do not remove the plug entirely, however, so the small population can effectively guard the entrance. Leave the unit alone for at least a week to allow the colony to release the queen by eating through the candy plug. When a week has passed, briefly open the hive to check only on the queen's status by briefly opening the hive (see First Inspection on page 159).

Continue to fill the feeder as it becomes emptied. Package bees must be continuously fed until they stop taking syrup, which can take as long as eight weeks. Carefully observe the syrup levels in the feeder to determine when you can stop replenishing it. Err on the side of giving too much rather than too little.

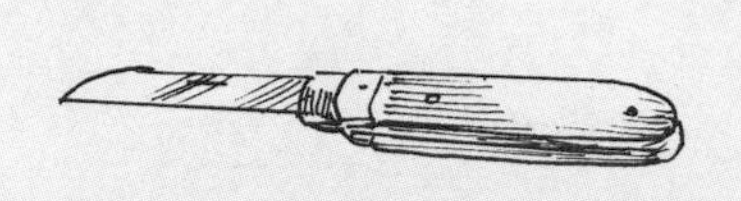
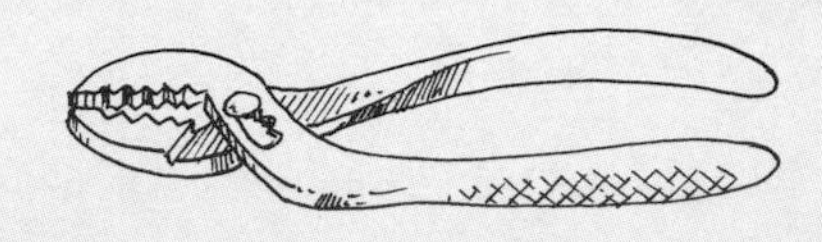
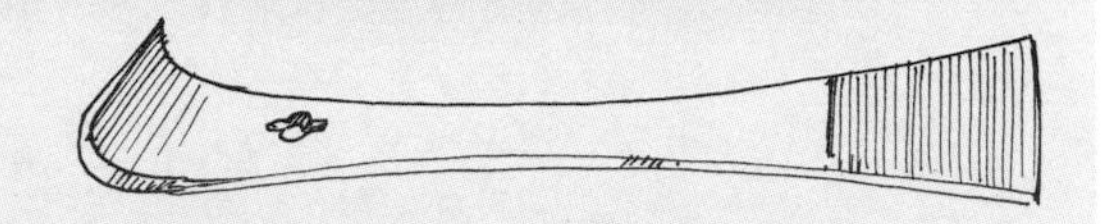

Tools needed to install package bees include (left to right) a sharp knife, a pair of pliers, and a hive tool.

Managing Package Bees

Because it is a fragile unit, an installed package requires more attention than an established colony. The bees' progress must be carefully monitored so you can lend assistance as needed.

The population is small, about 11,000 bees to begin with, and workers live only four to five weeks. Three weeks are needed for a functioning adult to develop from an egg and maybe two more before it begins foraging, so initially the colony's population will rapidly decline.

If established on beeswax foundation, the comb must be drawn (built out from the embossed base). The colony will recruit certain bees for this task, and they won't be available for other duties. The new queen should begin to lay eggs as soon as possible to begin a suitable brood nest. Nectar and pollen may be scarce, so your most important concern is with the food supply. The mantra should be feed, feed, feed!

First Inspection

Your first inspection of the bees should occur a week after installation and be brief. You have one objective: to see if the queen has been released and is functioning.

1. On a sunny day, light the smoker and don protective gear.

2. Observe the activity around the hive before attempting to remove the cover. Are bees actively entering and exiting? Do they have pollen pellets on their legs? Does the activity look normal and relaxed, or do the bees appear agitated or excited? Take all this in, but remember the primary objective of your visit.

3. Check your smoker to ensure the smoke is cool, not hot. The smoke gets hot when too little fuel is in the smoker. Add a little fresh green grass to the smoker to cool the smoke if necessary.

4. Puff several times, sending smoke into the entrance; remove the cover and again puff several times into the top of the hive.

5. Replace the cover, wait at least two minutes, and remove it again. Locate the queen cage. It should be empty. Take it out and put it aside for the time being.

6. If the bees have built comb around the cage, attaching it to the foundation, scrape it off with your hive tool, as it will interfere with subsequent construction.

7. Now search for the queen. Remove an outside frame and prop it outside against the hive to provide room to work the rest.

8. Slowly take out each frame, one at a time, examine it, and replace it.

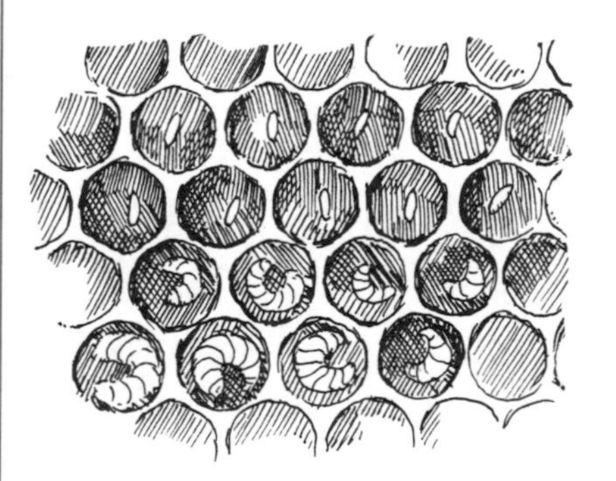

First Inspection, Step 8.

9. When you have finished your inspection, and either found the queen or determined that she is not present or functional, replace the combs and close up the hive.

You don't have to actually see the queen. It's enough to see evidence that she is on the job, such as drawn cells and eggs or young larvae in jelly. When you are satisfied the colony is functioning properly, replace the combs and close up the hive.

If you do not see the queen, nor any eggs or larvae, the queen may still be functioning, but you would be wise to begin planning to get a replacement. Check the colony again in a day or two. If you still find no sign of brood, assume the colony is queenless and acquire another queen as soon as possible.

Ongoing Monitoring

If you find signs of a functioning queen (in the form of eggs), begin to monitor the colony's progress on a weekly basis. Always make these visits brief. You just want to linger long enough to determine the population's characteristics, such as number of frames of food and brood, absence of disease, brood-to-adult ratio, and food supply.

You must continue feeding for at least three weeks, stopping only when the bees no longer consume the syrup. As time goes on, the colony should begin to fill with bees. At that point, the colony is established and can be managed like any other colony in a beekeeper's operation.

Nucleus Colony

A nucleus colony, or nuc, is a small but complete colony containing brood, adults, and a laying queen. It may comprise three to five frames and be delivered in a throwaway or returnable box. Many beekeepers have success with nucs, and they are widely used to increase colony numbers.

As a functioning colony, a nucleus is much less fragile than a package. Because it is small, the residents are not generally defensive, but the beekeeper must be on guard because the unit is much better organized than any package colony. It also comes complete with combs, which has advantages and disadvantages: The frames should contain

A nucleus colony (nuc) with an entrance that can be completely sealed during transit

stored food, which is an edge over the use of package bees that require continuous feeding, but the combs also increase the risk that a disease may be introduced to the colony.

Some suppliers will install a nuc if you bring a standard hive body to the seller's establishment. If the nuc is delivered in a reusable container, then the frames must be relocated to the center of a waiting brood chamber. To do so, lightly smoke the bees and transfer all the nuc frames into the center of the hive body. A small, functioning colony should now be in the middle, flanked by five to seven frames of foundation or drawn comb. Install a feeder for the colony and put on the cover. Even though the frames contain stored food, it is still important to monitor food reserves and provide food when necessary.

Because it already has brood, the nucleus colony will probably contain a population of varroa mites. If the seller has done a good job managing them, the mite population should be small, but don't count on it. Begin to monitor the mite population immediately to determine its potential to adversely affect the nuc's development.

Swarm

Many beekeepers have begun their careers by catching swarms. However, this method is not recommended for a beginner. The window for collecting and installing a swarm successfully is small, and although swarming bees do not usually sting — because they are disoriented and not protecting an established nest — they can be defensive in rare instances. Of particular concern is a hungry or dry swarm that has consumed its food supply, and you have no way to know in advance if this is the case. As a beginner, stick with an artificial swarm in the form of packaged bees.

Established Colony

Purchasing established colonies is risky for the beginner. A fully functioning unit can be defensive, and an inexperienced beekeeper may simply not be up to the task of manipulating it. One big risk is the potential for acquiring a devastating varroa mite population, which will affect not only that specific hive but others nearby as well. Another disadvantage is that the history of the unit is not known. It might have been neglected and be substandard; a novice will not have the experience necessary to detect deficiencies.

The Colony's Year

- As the active season begins, the increased pollen and nectar resources from the field stimulate the queen to begin laying large numbers of eggs. This balancing act continuously ensures the colony has the requisite number of workers to warm and care for the developing brood.
- As the population grows, the hive becomes constricted for space and drone eggs appear. This overcrowding eventually causes the colony to begin swarming preparations.
- Queen cells are constructed and the colony begins to rear queen replacements.
- The old queen is readied for flight and the scouts begin to seek out new nesting sites.
- The swarm, headed by the old queen, moves on a warm day and establishes itself in a temporary bivouac before it decides to move to a new location.
- A virgin queen is allowed to emerge in the parent colony and fly in order to mate so the colony can again develop a stable population.
- Both parent and child colony begin preparations for the inactive period — winter in temperate regions. Through this season, the colony maintains a steady population level, ready to begin anew next year.

Working a Colony

Before manipulating a colony keep a few things in mind. The better the weather, the more pleasant the experience will be. Honey bees begin to cluster (form dense groups in the shape of a ball) when the temperature drops to about 57°F (14°C). You shouldn't manipulate a colony when a cluster has formed.

Working a colony with a cluster disorganizes the bees, and they may not have enough time to reorganize before the temperature drops again, which can lead to life-threatening complications. If a colony must be manipulated during cold weather, do it early in the day.

Bright, sunny, warm days are much better than cloudy days for inspecting colonies. Bees can be highly defensive in the rain and on hot, muggy days, especially when thunderclouds are present. During these periods, worker bees may stop foraging, and that means too many bees are in a colony with nothing to do except deal with the beekeeper. Also static electricity from such unstable weather conditions may make bees irritable.

Tips for Working a Colony

- Choose a bright, warm, calm day, ideally during a nectar flow when field bees are absent from the hive.
- Dress appropriately (see Dress for Success on page 162).
- Work slowly, calmly, smoothly.
- Use a cool smoke, and don't over-smoke the hive.

Bee Savvy

The best time to work bees is during a nectar flow, when the maximum number of worker bees is out foraging. You can tell a nectar flow is in progress by examining the frames for fresh nectar: It is so liquid it can be shaken out of the cells. You may also notice increased activity in nearby fields and bees on open blooms, but these observations are less reliable than looking at the combs and seeing fresh product.

Of course, it is not always possible to work bees during nectar flow. And if the flow should suddenly cease due to inclement weather, close up the hive and leave as soon as possible.

Honey bee colonies are sensitive to vibrations. Do not bang on the hive or jar the bees. They might overlook a single instance, but doing it again is a mistake. Other repetitive vibrations coming from lawn mowers or cars and trucks also can set off a colony's defensive behavior.

Bees have good and bad days. If you have trouble each time a colony is visited, step back and analyze what might be the matter. What is the weather like? Is the colony being bothered by animal nuisances? Enlist another beekeeper to come, watch, and give pointers on technique. In general, treating a colony like a good friend brings dividends: The bees will often respond in like manner.

Late-Season Management

The late season really begins in late summer: July and August, to be exact. At this time the beekeeper begins to anticipate and prepare for the coming winter. In the north (United States Department of Agriculture hardiness zone 6) the target date is August 1. A month or so later is appropriate for beekeepers in zone 7. This season is critical. Late honey may flow in various regions, complicating things in unexpected ways.

Your objective at this time of year is to ensure that a viable population of honey bees goes into winter with a good chance of surviving. Young bees are important, but even more significant are good healthy populations of winter bees. These overwintering insects are adapted to storing nutrients for a long

Avoid Nighttime Inspections

Some of the worst experiences are not with flying bees but with crawling bees at night. Instead of flying in the dark, they may crawl up from the ground or otherwise get under your clothes or veil.

If you must manipulate a colony at night, use a red light — bees are not attracted to it as they are to white light. Some beekeepers manipulate the bees in the dark via red light if the daily temperature is extremely high. They have the advantage of working at a cooler temperature, but they must still contend with crawling bees.

Dress for Success

Photographs often show a beekeeper in the field, fully clad in suit, gloves, helmet, and veil. That degree of protection is sometimes appropriate, but not always, especially when it's hot. Some beekeepers wear little or no protection, but that is not recommended either. The best advice is to wear what makes you feel comfortable and at ease around the hive. What you wear can vary from occasion to occasion; experience will be the key to dressing for success when managing your hive. Here are a few tips:

- Wear light-colored, smooth fabrics
- Avoid red or black, which can trigger defensive responses
- Don't wear a sweater, flannel shirt, knit socks, or other fluffy fabrics in which a bee's delicate hooked feet can get permanently caught
- Tie or tuck in long hair
- Tuck pant cuffs into high boots or socks
- Remove rings, watch, and other ornaments — their movement might glitter in the sun, provoking defensive behavior
- Don't wear odors that might attract the insects, like aftershave, hair spray, or deodorant
- Other products may also be offensive, including gasoline and petroleum products that give off fumes

period of time. Summer bees cannot overwinter as easily, since they lack specific structures known as fat bodies.

Winter bees are made by the queen. Thus, the beekeeper must take pains to ensure she is up to the job. Brood rearing naturally slows down at this time, making it hard for you to detect a failing queen. On the other hand, honey bees are good at detecting a failing queen and you may see signs of supersedure cells (those destined to produce queens that will replace the current queen) being constructed.

A supersedure cell cup on the face of the comb indicates the colony is about to replace the queen.

If you have any doubt about the queen's condition, give serious thought to requeening (either by purchasing queens or letting the colony requeen itself). Some beekeepers requeen in late summer or early fall on a regular annual basis, for a number of reasons: New queens lay at a higher rate than older ones, so the resultant population is more numerous; late summer requeening allows for multiple chances for queen acceptance; and a first-year queen is much less apt to swarm the following spring. Further, a substandard population in the late season can hinder a colony's preparations for winter. By combining weak colonies into a stronger unit, the colony has a better chance to survive the coming harsh conditions. The larger colonies can always be split in the spring.

Employing good varroa mite management techniques during the late season is particularly important. Parasitized honey bees are not good candidates for winter survival. The mite population is usually large at this time, fueled by all the brood the colony produced since beginning the active season. Many mites are protected in the brood cells and not susceptible to chemical exposure. Having a break in the brood cycle at this time reduces the brood population available to be parasitized, and makes the mites vulnerable to chemical control. You can create a break in the cycle by requeening, or by dividing the strong colonies, treating and requeening the splits, then letting them overwinter outside in mild climates or in a shed, garage, or cellar in a harsher climate.

Winter Management

Winter management begins in the late fall. Once a cluster forms (at around 57°F [14°C]), little can be done to manipulate hives with any degree of success. During this period beekeepers turn to repairing equipment and getting the extraction area shipshape for the coming active season.

In areas where mice are present, install mouse guards to prevent the rodents from entering hives when the bees are clustering above the bottom board. In areas with fierce winds, add a windbreak, such as a bale of hay or the odd piece of plywood.

Whether or not to wrap colonies with tar paper, straw, or other materials, as is done in some regions, is a subject of much debate. Honey bees do not attempt to warm the entire space of their hive; they only warm their cluster. The cluster produces a great deal of warm, humid air that must be exhausted, so the hive should never be closed entirely. If this air is not removed, moisture may condense on the inner surfaces of the hive, which is highly harmful to bees. Provide an upper entrance during winter. Also, sometimes the outer cover is slightly propped up by small wedges to increase ventilation.

You can protect a colony from severe cold by using a top insulation board. No airspace lies between it and the

Winter Preparations

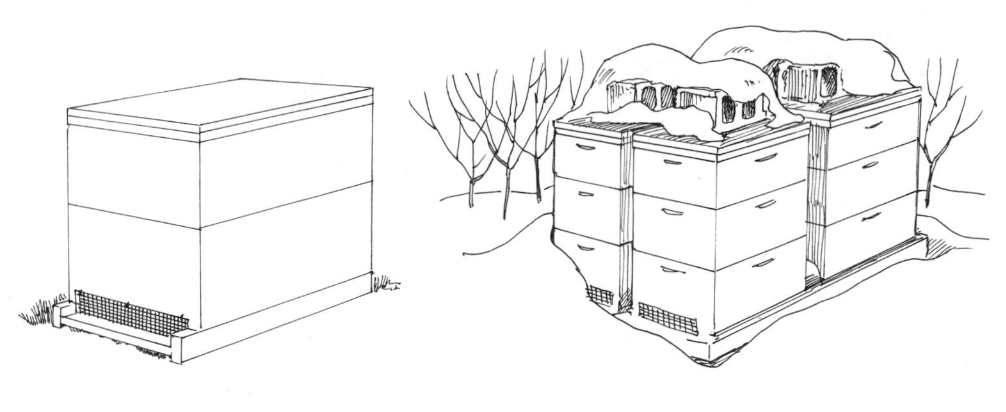

Install a mouse guard, consisting of ½ inch (13 mm) hardware cloth, to prevent mice from entering the hive.

Use cinder blocks to anchor hive covers against blowing wind, especially if covers have been propped up to accommodate an insulation board.

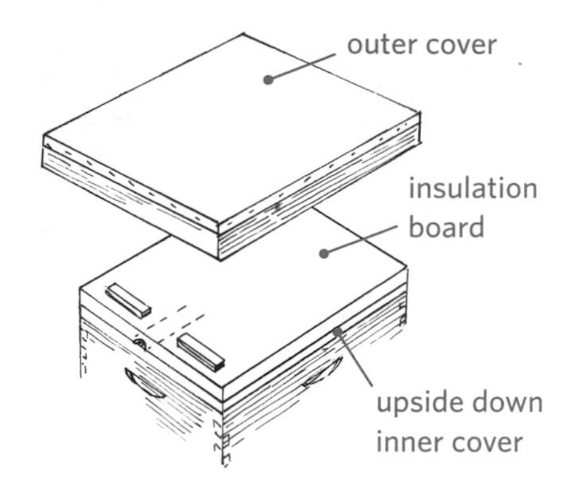

Install a top insulation board to protect the colony from severe cold.

inner cover, which has been turned upside down. Some beekeepers simply put an empty super filled with a fiberglass-type material on top of a hive. This material must be protected by a screen or the bees will attempt to remove it. Do not fill supers with hay, straw, or leaves, as these materials can trap a lot of moisture.

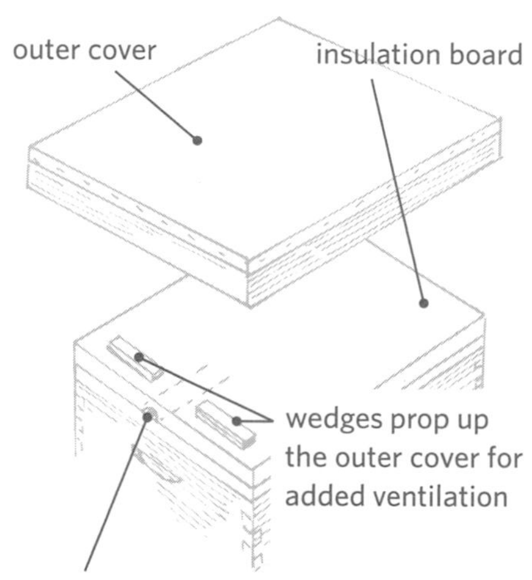

A groove in the insulation board improves ventilation and gives bees access to the slot (upper entrance) in the center of the inner cover.

Requeening

Beekeepers have a wonderful management tool: the ability to change a single individual, the queen, and thereby influence the behavior and direction of an entire honey bee colony. Thus, one of the universal answers to any problem is to requeen. More often than not, this action solves not one but a whole host of problems.

All queens are not equal, however. A queen's egg-laying potential varies based on the conditions under which she was reared, her age, her lineage, her history, and perhaps most significant, the kind and number of sperm she carries. Of these, the only factor a beekeeper can judge readily is her age; all other characteristics aren't known until her resulting colony can be observed.

One of the debates in beekeeping circles regards whether beekeepers should purchase queens or let the bees rear their own when making splits or replacing the current queen. This issue has no clear answer, although it makes sense that a commercial producer would raise a queen in a more controlled environment than is possible in a colony left to its own devices. The question then is whether producers should be held to any standard. Beekeepers would be better served generally by asking specific questions about rearing conditions, but many don't know what to ask — and they get few definitive answers if they do.

Monitoring Mites

You can monitor the degree of varroa mite infestation in a hive by counting the mites that fall off bees. For this purpose you'll need a sticky board, available from a beekeeping supplier. The sticky board is covered with adhesive so the mites will stick to it. You can make your own sticky board using heavy white freezer paper from the grocery store. Cover the surface facing the bees with a thin layer of petroleum jelly (Vaseline). Place a sheet of ⅛-inch hardware cloth on top of the sticky board so your bees won't stick to it.

Place the sticky board on the hive's bottom board. Mites that fall off the bees will fall through the hardware cloth and stick to the sticky board. Leave the sticky board on the bottom board for a day or two, then remove the sticky board and count the mites. A magnifying glass will help you identify the mites among other hive debris. Divide the total number of mites by the number of days they were collected.

If you collect more than 50 mites per day, it's time to think about treating your colony against varroa mites. For information on current treatment methods contact your local beekeeper club or county Extension agent.

Managing Nutrition

Honey bees are vegetarians and consume only plant juice (nectar) for their energy needs and pollen for development. Colonies cannot survive without honey or nectar stores. Honey bees also need 10 essential amino acids for growth and development; pollen provides these in varying amounts. Certain pollens from fruit trees, for example, are much more nutritious than the pollen of other species such as pines.

Pollen also contains enzymes important to bee health and development, sterols for hormonal production, minerals, and vitamins. If workers do not receive enough pollen, their brood food (hypophryngeal) gland is affected, resulting in reduced longevity. Queen bees that do not receive enough pollen can have early supersedure, and virgins can fail to adequately mate. The lack of pollen may also be responsible for a reduction in the number of drones available for mating and/or a decrease in their potential sperm production.

Managing honey bee nutrition is one of the beekeeper's most important jobs. Supplementing colony nutrition is, like beekeeping itself, more art than science. The food generally has to be prepared from ingredients purchased elsewhere. It then has to be administered to the colony at the correct time and under optimal conditions to be sure it is eaten.

Feeding Carbohydrates

Monitoring carbohydrates in a colony is relatively easy. It is generally done by estimating the number of frames of nectar or honey. Another way is to heft a colony — get an estimate of its weight. The best food is combs of honey or nectar. Beekeepers routinely shift these among colonies to ensure all have adequate resources.

The traditional supplemental bee feed is cane sugar crystals mixed with water and administered as needed. The usual proportion is 1 part sugar mixed with 1 part water, but the strength should vary depending on conditions. In spring, the solution should be weaker and more nectarlike to stimulate greater population growth. In the inactive time of the year, particularly during cold weather when bees have difficulty evaporating excess water, the solution should be stronger, with more sugar than water.

Mixing Sugar Syrup

The strength of sugar syrup should be varied, depending on conditions.

1:2

This light syrup is made using 1 part sugar to 2 parts water (for example, 1 cup sugar to 2 cups water). It is used in late winter and early spring to stimulate the queen to lay eggs.

1:1

This medium weight syrup is made using 1 part sugar to 1 part water (for example, 1 cup sugar to 1 cup water). It is used as artificial nectar to feed brood larvae in spring and summer or to get the bees to draw comb.

2:1

This heavy syrup is made using 2 parts sugar to 1 part water (for example, 2 cups sugar to 1 cup water). It is used in fall or early winter as a honey substitute to feed your bees.

Colony Collapse Disorder

First identified in 2006, Colony Collapse Disorder (CCD) was responsible for the death of many hives, especially those managed by large-scale commercial pollinators. Symptoms of CCD include the complete absence of adult bees in colonies, with little or no build-up of dead bees in or around the colonies; the presence of capped brood in colonies; and the presence of food stores, both honey and bee pollen, which are not immediately robbed by other bees and may also be ignored by wax moth and small hive beetle.

No one single cause for CCD has been identified; most see it as the possible result of a number of things, perhaps acting together. CCD appears to be less of a concern to many stay-at-home, small-scale beekeepers. If you suffer large-scale bee loss and can find little reason for it, carefully document the symptoms to be able to clearly communicate them to others. Include in your written statement details about management decisions and environmental conditions.

Harvesting Honey

The optimal time for harvesting will depend in part on the nectar flows in your area and in part on your honey preferences. Depending on its source, some honey is stronger tasting than other honey. Follow this basic rule of thumb: Do not take the honey until you are sure the colony does not need it for the winter.

If you have more or less continuous nectar flows throughout the season, you may choose to wait and take off the honey at the end, or you may take it at intervals through the season. If you are able to separate varietal flows (from different flowers), the latter may be desirable. If you have large crops, you may wish to spread the work over more than one extraction and you may wish to minimize the number of supers you own by reusing them during the season.

If you have a single early flow, you will normally take off honey once, at the end of your season.

If you have both an early and a late flow, you can take one of two approaches. You can take off honey after each flow, or wait and take it only after the final flow. The former approach will give you two separate crops, probably of different color and taste, while the latter will give you a blend. The former will also be more laborious while the latter will allow you to own fewer supers.

How Much Honey to Take Off

Simply stated, take off only honey that is surplus to the needs of the colony. To establish what these needs are, ask yourself these questions.

- Is the main nectar flow finished?
- Are minor flows yet to come?
- How long before winter begins?
- How much will the colony consume between now and then?
- How much will they need to make it through the winter?
- Though honey may be in the supers, do the hive bodies also contain the requisite amount?

Your goal is to come to the end of the season with as much honey or stores in the hive bodies as the bees require for overwintering in your area. Requirements vary throughout the country. The practices of beekeepers in your area, modified by your experience as time goes by, are your guide.

Removing the Bees

Removing the bees so you can take the honey crop may be done in one of two principal ways: by brushing or using escapes.

Brushing is easy, inexpensive, and relatively quick. If you are harvesting from many supers, however, brushing becomes tedious, and it's difficult to keep the bees from returning to the frames even as they are being brushed. Brushing also can provoke robbing.

Before you begin, place an empty hive box on a solid bottom board a little distance from the colony, where you can put the brushed combs one by one after the bees are removed. Cover it with a damp towel. If you don't have an extra super, improvise with a cardboard box or other container, but be sure bees can't get access to the just-brushed combs.

Several kinds of bee brushes are available. Most beekeepers prefer those with plastic bristles, but they are harder on bees than softer ones made of natural bristles.

The bee escape is an elegant piece of equipment that acts as a one-way valve. Bees can exit through a set of

Is It Honey Yet?

A key factor in deciding when to remove honey is assessing its moisture content. Generally, honey must be at or below 18.6 percent water before it is ripe enough to harvest. When bees determine the moisture content is correct, they cap it over with comb wax. Thus, a rule of thumb is to remove only honey from frames that are two-thirds capped.

As in other aspects of beekeeping, err on the side of harvesting too little rather than too much. Honey will absorb moisture from the air during processing, so be conservative here — process honey only after it has been capped and limit its exposure to warm, moisture-laden air during extraction and bottling.

flexible springs but cannot return. The inner hive cover has an oblong hole that accommodates commercially available (sometimes called Porter) bee escapes.

Alternatively, you can construct escape boards. Be sure the hive has no cracks or other ways robbers might enter above the bee escape. Use duct tape to seal any areas of concern. With the escape board in place overnight, the vast majority of bees should leave the super, and you can then remove it as a unit, fairly free of bees.

Before using this method, check the combs in the super you are harvesting to make sure it is free of brood. If brood is present above the escape board, removing the bees is more difficult because they are reluctant to leave the developing offspring.

Brushing Bees

Time is of the essence to prevent robbing of the honey by bees from other colonies. Move the brushed combs inside a hive body or box as soon as possible.

1. Remove the frame and hit the bottom of the frame against the ground to remove as many bees as possible.
2. Quickly, gently, but firmly brush off the remaining bees with a sweeping motion. You can't get all the bees off; don't waste time trying. Just remove the majority.
3. When the frame is mostly free of bees, transfer it to the empty box or container.

Bee Escapes

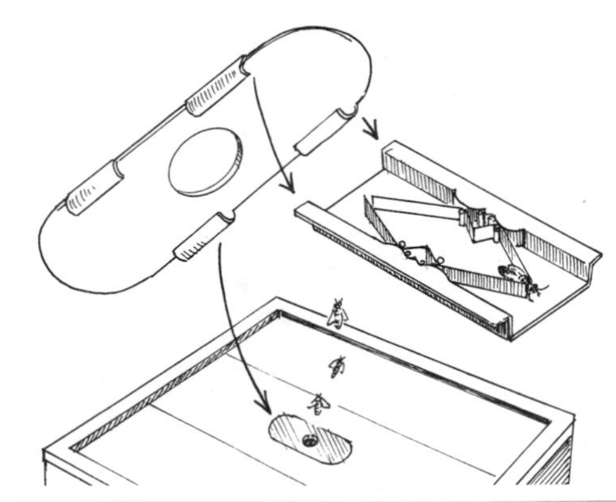

The Porter bee escape consists of a plastic oblong board that fits into the slot in the center of the inner hive cover, and a set of springs that have just enough tension for the bees to pass through but are too close together for a bee to return.

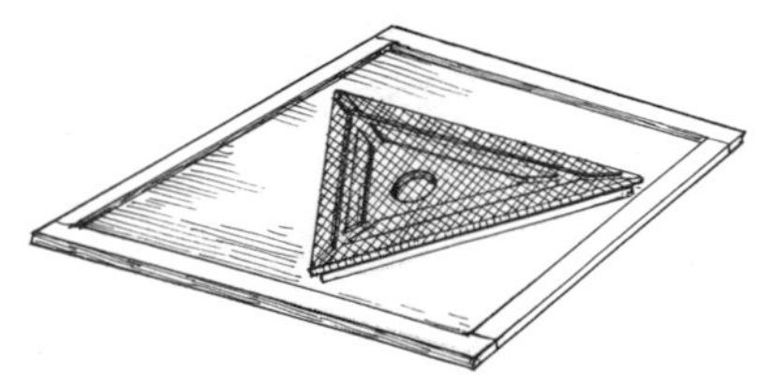

This bee escape board uses a double triangle design that lets bees easily leave the super but not so easily return.

Removing Honey from the Comb

The easiest ways to remove honey from the comb are to crush the comb and strain the honey, or use an extractor. For a small amount of honey, manually crushing the comb is less expensive than acquiring an extractor; if you have a large amount of honey to harvest, crushing the comb is not practical.

Straining Honey

The simplest way to crush honeycomb is to put it in a nylon stocking and then squeeze it, forcing the honey to separate from the comb. Squeezing and forcing takes a long time and can waste a lot of honey. Another disadvantage of this method is that the comb is destroyed and cannot be reused.

Honey is hygroscopic, which means it absorbs water from the air. The longer the honey is exposed to air during straining, the more moisture the honey will absorb. Warm honey flows better and can be processed more quickly, but warming the honey can result in the absorption of more moisture. Running a dehumidifier will mitigate this situation to some degree.

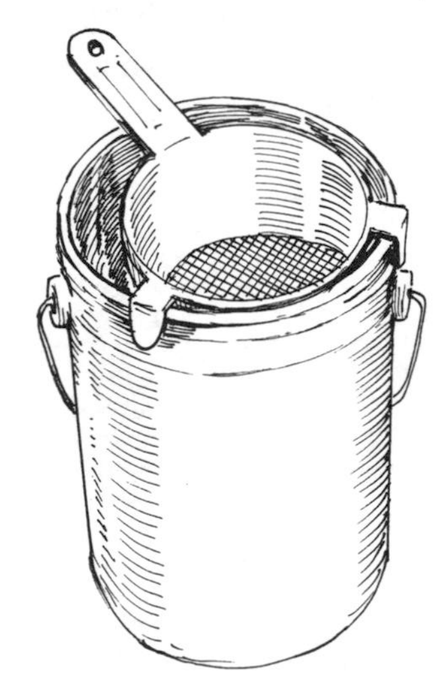

Use a strainer to separate honey from crushed comb.

Using an Extractor

Extractors are of two kinds: radial and centrifugal. To use an extractor, you must first remove the cappings from the frames. Both styles of extractor spin the uncapped frames at a high speed, slinging the honey out of the comb onto the extractor walls; from there it drips to the bottom of the holding tank. Many models are available, holding from two to sixty or more frames. Some extractors spin the supers themselves, frames and all.

An extractor is expensive, but you don't necessarily need to purchase one. Some associations lend extractors or even extract for their members. In some areas, private custom extractors are available to do the job for a fee.

Honey Extractors

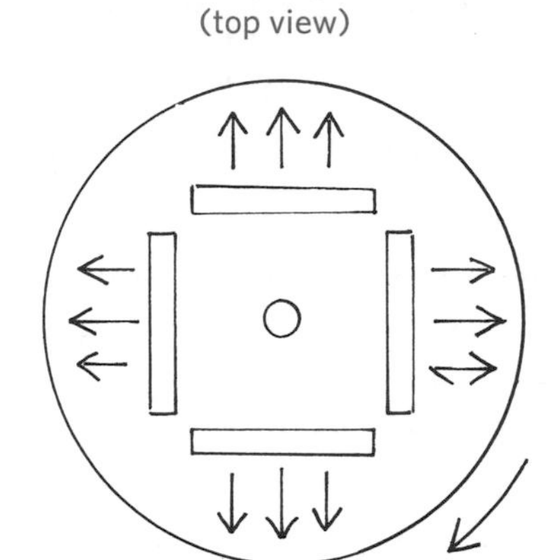

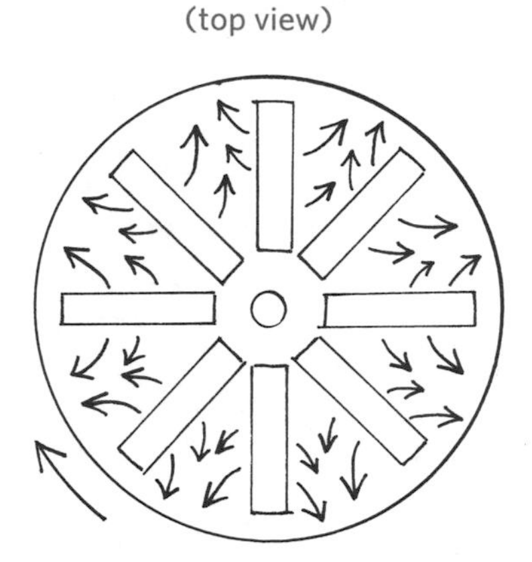

An extractor, whether centrifugal or radial, literally slings honey out of the comb.

Storing the Crop

Honey can be stored for a long period. Some discovered sealed in ceramic vases in sunken Greek and Roman ships was still edible after thousands of years. Honey straight from the hive is, like wine, a living organism filled with enzymes, proteins, and sugars, all capable of changing the honey's properties over time. Honey that is highly processed by heating and filtering, on the other hand, is killed and converted into nothing more than dead syrup.

Avoid Excess Moisture

Too much moisture will destroy honey. Once the moisture level rises above 19 percent, natural yeasts present in honey begin to ferment, producing alcohol. Honey allowed to stand for a long period might crystallize — a process that separates out the solids and thereby increases the liquid portion, promoting fermentation.

Many consumers believe crystallization ruins honey, but it is the fermentation, not the crystals themselves, that leads to spoilage. If a distinct alcoholic odor is not discernible, then fermentation has not begun. You can easily reliquefy crystallized honey by immersing the container in warm water.

Control Temperature

Temperature can affect honey in several ways. Most commercial honey has been flash-heated to 160°F (71°C). At home it can be brought to 120°F (49°C) for easier liquefaction and filtration, but should not remain at this temperature for long. Warm temperatures denature enzymes in the honey and also convert the fructose into hydroxymethylfurfural. At high levels, this organic compound, derived from dehydrated sugars, is toxic to bees.

The optimal crystallization temperature is 57°F (14°C). For best results, store honey above or below that temperature. It can even be frozen with no ill effects.

Stored in bulk, honey is slower to granulate than honey kept in smaller jars. Thus, when storing honey for extended periods, use larger containers. As a small-scale beekeeper, you can use 60-pound (27 kg) or 5-gallon (4 L) food grade buckets with tight-fitting lids and bottle the honey in smaller quantities over time as you consume it.

Warning: No Honey for Infants!

Raw honey is a potential source of *Clostridium botulinum* spores, which are widely found in the soil and in dust. By 1 year of age, most children develop resistance to this normally benign strain of *botulinum*. But in an infant's immature gastrointestinal tract, bacteria from these spores may grow and multiply, making a toxin the baby's undeveloped immune system is unequipped to handle. The result is infant botulism.

Although infant botulism is rarely lethal, it is unpleasant and easy to avoid by not sweetening baby food with raw honey, and not attempting to soothe a crying baby by coating a pacifier with honey. At greatest risk are infants under the age of 6 months, although to be on the safe side, children up to the age of 12 months should not be given raw honey.

Comb and Chunk Honey

Many people like to eat honey in the comb. Producing quality section comb honey is not easy. It requires large, established populations of honey bees and strong honey flows. The insects do not like to work in the close quarters demanded of them in frames designed to hold comb honey sections. When they are forced to do so, the result is a lot of incomplete sections of questionable quality. Different kinds of technologies have been developed to ensure better comb honey with less labor. The production of comb honey is not generally recommended for beginners.

Chunk honey, however, is easier to make using cut comb, and reduces the amount of extraction needed. Simply cut a thin piece of comb from a frame, place it in a jar, and cover it with strained honey.

Can you save money on honey by keeping bees? Aside from your initial purchase of equipment and bees, annual expenses include the cost of requeening and any health care treatments required. With good management, including not removing all your bees' stores and feeding them syrup — and assuming you live in a good nectar flow area — the cost of producing your own honey is negligible compared to the price of store-bought honey, and you will have the pleasure of enjoying a superior product.

Beeswax

Beeswax is unique and made only by honey bees. It is literally the foundation for the bee nest, becoming both the cradle for future bee generations and the colony's food warehouse. The elegant form of the six-sided cell uses the least amount of material for maximum strength and has been copied for many human applications. A recent spectacular example is the James Webb Space Telescope, partial successor to the aging Hubble Space Telescope.

Beeswax is an ingredient in many cosmetics, creams, and salves, and a component of wood and metal polishes.

Most of the world's beeswax is recycled into foundation and given back to the bees to manufacture comb. It is also an ingredient in many cosmetics, creams, and salves, and a component of wood and metal polishes. It is a good dry lubricant used by carpenters and others in woodworking applications. Of course, it makes a superior candle that is used for many purposes including religious services.

During the year, you will likely accumulate wax from scrapings of the hive and frames. Wax cappings from the honey crop and the wax rendered from combs that are being renovated also provide good sources of wax, which should be collected and saved. Considering that bees produce at most only 1 pound (0.5 kg) of wax for each 7 pounds (3 kg) of honey, it would be a shame to let it go to waste. The Internet is full of useful information on how to melt the wax into blocks and how to use the resulting beeswax.

Beeswax is easy to melt and can be readily damaged by overheating. The best way to make the finest wax blocks is to use a solar wax melter, which is little more than an insulated box covered with glass and set in direct sunlight.

Honeydew

Honey bees produce a variant of honey called honeydew. It is a sweet material collected and processed just like honey, although its source is not flowers but other insects.

In certain parts of the world, forests support large populations of plant-sucking insects that plug into the circulatory system of a plant by drilling through the bark. They have to suck a lot of juice out of a plant's vascular system to get the appropriate nutrition, and in the process, they excrete surplus liquid in the form of a sweet juice, which honey bees then collect and modify in a manner similar to making honey.

The resultant honeydew generally has a markedly different flavor than honey. Major areas of honeydew production include pine forests of Germany, western Turkey and nearby Greece, and the beech forests of New Zealand. Other regions, such as the Panhandle of Florida, also have honeydew flows that support the honey bee population but are not sufficient to be harvested by beekeepers.

Pollen

Pollen can be collected from a hive and later fed back to the bees or used as a health food. Specialized pollen traps at the hive entrance force worker bees to clamber through small openings. In the process, any pollen balls located on the last pair of legs are literally scraped off.

Because traps are hard on a colony, do not place them on a hive that has a low population. Even on a populous

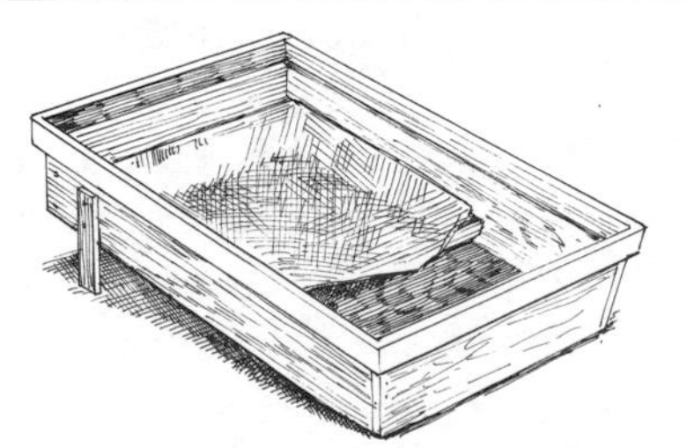

Beeswax is easy to melt in a solar wax melter, consisting of a box with a glass cover. Old combs, cappings, and other hive scrapings placed in an upper tray are melted, accumulate at the bottom of the box, and, when cooled, are removed as a solid cake.

colony, remove them periodically to give the bees a chance to replenish their protein reserves.

Pollen is fragile and quickly loses its nutritional value when exposed to the air. For use in bee feed, mix the fresh pollen with sugar and freeze it until it is needed.

Dried pollen is often touted as a perfect food for humans, but little research has been done to back up this claim. One reason is that all pollen is not the same and, just like honey, has radically different characteristics based on its plant source. Bees try to consume a diet of mixed pollen rather than pollen from a single plant source. A standard dose of pollen for various ailments and applications therefore does not make sense.

Pollen is a magnet for environmental pollutants, whether particulates from smokestacks or pesticides. Since you have no way to know the source of collected pollen, you cannot easily ensure its purity for either bee or human use.

To Your Health

Beeswax, pollen, honeydew, and honey: All are generally considered to be health supplements or natural foods. Little direct scientific evidence supports many of the health claims for these products, but interest in their properties is growing. The American Apitherapy Association provides information on the use of honey bee products in a branch of human medicine called apitherapy. In some countries where modern medicine is not available or too expensive, apitherapy is the basis for what has been called green medicine. Cuba, for example, has a significant body of physicians engaged in this endeavor.

This alternative therapy employs bee products in a variety of ways to treat human health problems ranging from simple wounds to chronic conditions such as multiple sclerosis. Honey has been a remedy and treatment for human ailments for centuries. It has been used to treat sore throats, coughs, and other conditions, including hangovers. Some people find that eating honey produced by local bees reduces pollen allergies.

Honey has been used as a topical dressing for burns, wounds, and other injuries. It contains an enzyme known as glucose oxidase that, as a by-product, forms hydrogen peroxide, which kills many kinds of bacteria. Further, its ability to attract water leaves little moisture available for infection-causing bacteria to thrive on — honey literally dehydrates the bacteria to death. Some patients suffering from a pernicious form of bacterium, methicillin-resistant *Staphylococcus aureus*, have had remarkable results when treated with honey.

CHAPTER 6

Goats

Goats are easy to handle and transport, and they produce delicious milk, healthful low-fat meat, and garden fertilizer. If you raise dairy goats, each doe will give you about 90 quarts (85 L) of delicious fresh milk every month for 10 months of the year. Each bred doe will produce one kid or more annually; some does give birth to twins or even triplets year after year. From each meat wether (castrated buck), you will get 25 to 40 pounds (11 to 18 kg) of tasty, lean meat. Every day, each goat will drop a little more than 1 pound (0.5 kg) of manure, which makes good fertilizer for the garden.

Goats do not require elaborate housing. All they need is a shelter that is well ventilated but not drafty and provides protection from sun, wind, rain, and snow. Each goat requires at least 15 square feet (4.5 sq m) of space under shelter and 200 square feet (61 sq m) outdoors. A miniature goat needs at least 10 square feet (3 sq m) under shelter and 130 square feet (40 sq m) outdoors. You'll also need a sturdy fence — don't underestimate the ability of your goats to escape over, under, or through an inadequate fence.

Goats are opportunistic eaters, meaning they both graze pasture and browse woodland. When allowed to harvest at least some of their own food by grazing or browsing, goats remain healthier and cost less to maintain in hay and commercial goat ration. Each year the average dairy goat eats about 1,500 pounds (680 kg) of hay and 400 pounds (181 kg) of goat ration. Meat goats do well on hay and browse, with little or no commercial ration unless they're growing youngsters or nursing mothers.

Goats are social animals that like the company of other goats, so you'll need at least two. If you're getting goats for milk, you will be breeding them every year and, if you can't resist keeping some of the kids, your herd may grow larger than you initially expect.

Choosing the Right Goat

More than 200 breeds of goat may be found worldwide. Each breed has characteristics that are useful to humans in different ways. Some are more efficient than others at turning feed into milk or meat. Some breeds are small and produce less milk or meat than larger breeds but are easier to keep in small spaces.

Dairy Goats

A dairy goat, sometimes called a milk goat, is one that produces more milk than it needs to nurse its kids to weaning age. In the United States, there are six main dairy breeds: Alpine, LaMancha, Nubian, Oberhasli, Saanen, and Toggenburg. Two additional breeds are Sable, which is a Saanen of a different color, and Nigerian Dwarf, a miniature dairy breed.

All these different choices give you lots to think about. Pick a breed that looks good to you in terms of coat color and ear style — long floppy ears, pointed ears, or no ears. Look also at the goat's size in terms of your own size (smaller people tend to prefer smaller breeds) and the amount of space you have for keeping goats. And consider the milk output — it makes little sense to spend money feeding does that produce more milk than you can use.

Meet the Goat Family

Scientifically goats belong to the suborder Ruminantia — that is, they are ruminants, like cows, deer, elk, caribou, moose, giraffe, and antelope. Ruminants are hoofed animals with four-part stomachs. Within the suborder Ruminantia, goats belong to the family Bovidae, which includes cattle, buffalo, and sheep. Of the six species of goat, five remain wild and only one, *Capra hircus*, is domesticated.

Remember, you'll want at least two does so they can keep each other company, but they needn't be the same breed. If milk from two does is more than you need, you can slightly reduce their output by milking once a day instead of twice. In estimating milk output, consider that you can breed both does at the same time, meaning you'll have lots of milk at some times of year and none at other times, or stagger their breeding to even out the milk flow year-round.

Your climate is yet another consideration. Alpines, Oberhaslis, Saanens, and Toggenburgs all originated in the Swiss Alps and are therefore referred to as the Swiss breeds or European breeds. These goats are closely related and are similar in shape. They all have upright ears and straight or slightly dished faces. They may or may not have wattles consisting of two long flaps of hair-covered skin dangling beneath their chins. These breeds thrive in cool climates.

LaManchas, Nigerian Dwarfs, and Nubians, on the other hand, originated in warmer climates and are therefore grouped together as tropical or desert breeds. The Nigerian Dwarf and the Nubian originated in Africa, and the LaMancha comes from the West Coast of the United States. In general, these breeds are better suited to warmer climates than the shaggier Swiss breeds.

Assessing a Doe

If you buy a young female, or doeling, you can't tell for sure how much milk she will give when she matures, but you can get a good idea by looking at her

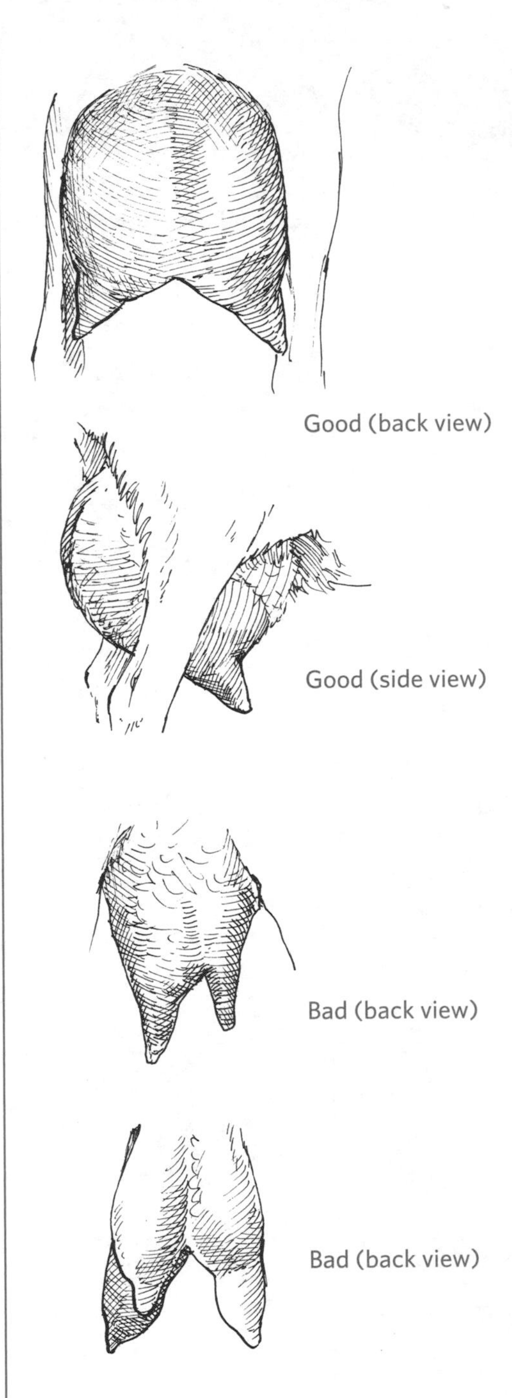

A proper udder is round and well attached and has two teats of equal size. A poor udder has asymetrical teats and poor capacity.

Parts of a Goat

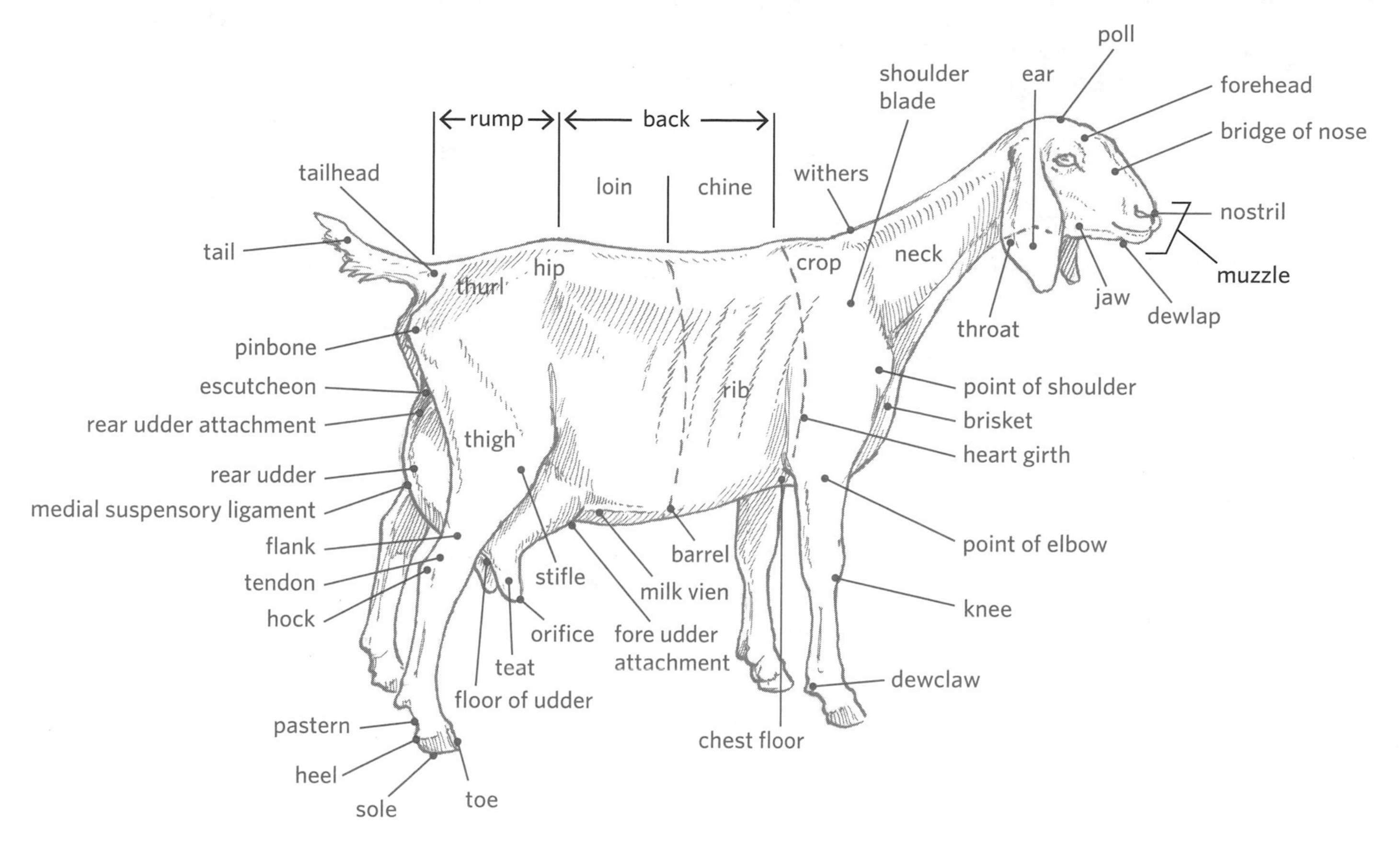

Quick Guide to Dairy Goat Breeds

Breed	Size	Milk/year in lbs./kg	Butterfat (%)	Coat color
Alpine	medium large	2,000/907	3.5	several acceptable
LaMancha	medium large	1,700/771	3.9	any color
Nigerian Dwarf	small	750/340	5.0	any color
Nubian	large	1,600/726	4.0	any color
Oberhasli	medium	1,600/726	3.6	brown and black
Saanen	medium large	1,900/862	3.5	white, cream
Sable	medium large	1,900/862	3.5	any color except white
Toggenburg	medium	1,950/885	3.2	brown and white

Goats and Tin Cans

Being opportunistic eaters does not mean goats eat things like tin cans. A goat learns about new things by tasting them with its lips. Young goats like to carry things around in their mouths, as puppies do. If you see a goat with an empty can, it could be playing with it or eating the label, which after all is only paper made from wood, and goats do like to browse on woody plants. Although the goat may look cute carrying a can, it's a bad idea to leave such things where a goat can find them; the goat may cut its lips or tongue on the sharp rim.

Alpine

LaMancha

Nigerian Dwarf

Nubian

Oberhasli

Saanen

dam's milk records. A doe's dairy character also gives you a fair idea if she will be a good milker. Characteristics of does that prove to be good milkers include:

- A soft, wide, round udder
- Teats that are the same size, hang evenly, and are high enough not to drag on the ground or get tangled in the doe's legs when she walks
- A well-rounded rib cage, indicating that the doe has plenty of room for feed to fuel milk production
- A strong jaw that closes properly, so the doe has no trouble eating
- Strong, sturdy legs
- Soft skin with a smooth coat

Goats born without horns are called polled. Most dairy goats are born with horn buds that will eventually grow into horns. Kids with horn buds are usually disbudded, because mature dairy goats without horns are easier to manage and are less likely to injure their herd mates or their human handlers.

Dairy Breeds

Alpine. Originating in Switzerland, the Alpine averages the highest annual milk output of any breed. This goat has a long neck and a two-tone coat that comes in several color patterns, each of which has a specific name. Generally the front end is a different color from the back. A mature doe weighs at least 135 pounds (61 kg) and stands 30 inches (76 cm) high at the withers.

LaMancha. Developed in California, the LaMancha is the calmest and friendliest of the dairy breeds and comes in just about any color a goat can be. LaManchas are easy to recognize because they have only small ears (elf ears) or no visible ears at all (gopher ears). A mature doe weighs 130 pounds (59 kg) or more and is 28 inches (71 cm) or more at the withers.

Nigerian Dwarf. Originating in West Africa, the Nigerian Dwarf is a miniature dairy goat that produces less milk and meat than a full-size breed, but it is ideal for a small family with a small backyard, provided the doe comes from milking stock and not pet stock. The Dwarf has smaller teats than a full-size goat, making it difficult for someone with large hands to milk, and because it

Sable

Toggenburg

Pints and Pounds

The amount of fat and other milk solids in a doe's milk varies with the doe's breed, age, and diet. Early in lactation, milk has more fat and solids than it contains toward the end of the lactation period (which makes sense, because at the beginning the doe has kids to nurse), and the milk of does fed commercial rations is generally higher in solids and fat than milk from the same does that are grassfed. So weighing milk tells you more about a doe's output than measuring it by volume.

But when you produce milk for your family, you need to know how much each doe is milking volume-wise. For that, remember the old saying designed to help people remember how much water weighs: "A pint's a pound the world around." Although milk weighs slightly more than water, because it contains more solids, and because the exact solids content varies by goat and season, you can get a good approximation of volume by remembering that 1 pint of milk weighs approximately 1 pound. Since 1 quart has 2 pints, 1 quart weighs 2 pounds; 1 gallon has 4 quarts and weighs 8 pounds. A doe that averages 1,600 pounds of milk in a year therefore produces about 200 gallons of milk (1,600 divided by 8).

is lower to the ground, milking requires a smaller pail that fits underneath the goat. Nigerian Dwarfs come in all colors. Mature does weigh 30 to 50 pounds (14 to 23 kg) and stand about 22 inches (56 cm) high.

Nubian. Developed in England, the Nubian is popular among makers of cheese and ice cream because its milk is so rich. This goat comes in many colors, most typically bay or black, and is the most energetic, active, and talkative of the dairy breeds. You can tell a Nubian from any other dairy goat by its rounded face (called a Roman nose) and long floppy ears. A mature doe weighs 135 pounds (61 kg) or more and is 30 inches (76 cm) or more at the withers.

Oberhasli. Coming from Switzerland, where the breed is quite numerous, Oberhaslis are relatively rare in North America. This breed is known for its gentle disposition. Its coat color is reddish brown with black markings, although some does are solid black. A mature doe weighs at least 120 pounds (54 kg) and is 28 to 32 inches (71 to 81 cm) high.

Saanen. From Switzerland, the Saanen is the most popular breed worldwide, often referred to as the Holstein among dairy goats. It is a big goat with an extremely mild temperament and a white or cream-colored coat. A mature doe weighs 135 pounds (61 kg) or more and stands 30 inches (76 cm) at the withers.

Sable. Developed in the United States, the Sable is simply a Saanen that is any color, or color pattern, except solid white or cream. Colors typically range from tan through black. Like the Saanen, the Sable doe weighs 135 pounds (61 kg) or more and stands 30 inches (76 cm) at the withers.

Toggenburg. Originating in Switzerland, the Toggenburg is a gentle, friendly goat. It has white ears, white face stripes, and white stockings setting off a brown coat that can range in shade from soft brown to deep chocolate. A mature doe weighs 120 pounds (54 kg) or more and is 25 inches (63 cm) at the withers.

Goats for Meat

In many countries more goats are kept for meat than for any other purpose, and many people prefer goat meat to any other. Since slightly more than half of all goat kids are male and only a few mature bucks are needed for breeding, most young dairy goat bucks are raised for meat. Surplus goats of any breed can be used for meat, although a breed developed specifically for meat puts on more muscle and grows more rapidly than nonmeat breeds.

The main meat breed in the United States is the Boer, known for rapid growth, large size, high-quality meat, and uniformity of size, meat quality, and color. Other meat breeds include Kiko, Savannah, Spanish, and Tennessee Fainting Goat (Myotonic).

Raising meat goats is big business, and buying a young animal to raise for slaughter may likely cost more than its value as hamburger. Economically speaking, it makes more sense to keep a couple of milking does and raise their male offspring for meat. Since you have to breed your milking does annually anyway, you might consider breeding them to a Boer or other meat buck to produce fast-growing, meaty crossbred kids.

So why not milk a Boer doe, or a female of some other meat breed? For one thing, they don't produce as much milk as a dairy breed. For another, they don't milk as persistently. Does developed for meat production lactate only long enough to wean their kids, while dairy goats produce milk for nearly a year, and sometimes longer. Finally, a lot of Boer and Boer crosses have faulty teat structure that makes them difficult to milk.

Before meat goats became big business in the United States, the Nubian was considered a dual-purpose goat raised for both milk and meat. Breeding a Nubian doe to a Boer buck to produce meat kids makes so much sense that some professional meat goat producers do it. Another popular breed for crossing is the LaMancha. One advantage to this plan is that you may have an easier time finding a Boer buck at stud than a buck of your chosen dairy breed.

Another possibility is to opt for the dual-purpose Kinder goat (pronounced KIN-der), which is a cross between a Nubian doe and a Pygmy buck; the Pygmy is a miniature goat with a body that's chunky like a meat goat. The resulting kids are smallish, muscular goats with high feed conversion efficiency and a rapid rate of growth. The does typically give birth to three or more kids at a time and annually produce an average of 1,500 pounds (680 kg) of milk that is rich (3.5 percent average butterfat) and high is solids, making it superior for cheesemaking. The Kinder has longish ears that stick out to the side, and its coat can be any color. The Kinder's size is midway between the Pygmy and the Nubian; does weigh 115 pounds (52 kg) or more and are 26 inches (66 cm) at the withers.

Boer

Kinder

In many countries more goats are kept for meat than for any other purpose, and many people prefer goat meat to any other.

Making the Purchase

Even if the milk from one doe is plenty for your needs, the doe will need a companion, which can be another doe or a wether (castrated male). Since most dairies have too many young bucks, wethers are often available at a good price, and one can be raised to butcher for meat by the time your doe gives birth and has the companionship of one or more kids.

If you want milk year-round, you must have a second doe. A doe must be bred before she will produce milk, but during the two months just before she gives birth, she'll need a rest and won't be milked. By breeding one doe at the beginning of the breeding season and the other at the end, the first one will be back in milk by the time the second one is dry.

Getting a buck (intact male) as your first goat is not a good idea, and keeping a buck to service just a couple of does is not an economical proposition. A buck must be housed separately so he won't fight with other goats or breed does that are too young. During breeding season, a buck may become aggressive and hard to handle, and will develop a strong odor that gets on your skin and clothing when he rubs against you. Unless you have a lot of does to breed, keeping a buck is an unnecessary expense. You'd be better off finding

A healthy goat has a clean coat and bright eyes (left); a sick goat is hunched over with her tail down (right).

a buck owner nearby who is willing to breed your does.

Once you have selected the goat you wish to purchase, make sure the animal is healthy and sound. A healthy goat has a clean coat and bright, alert eyes. It should be as curious about you as you are about it. A good goat has a strong, wide back, straight legs, sound feet, and a wide, deep chest. Avoid a goat with a swayback, a narrow chest, a potbelly, bad feet, lame legs, or a defective mouth.

Good milkers need not be registered, although you may want registered does if you think you might someday wish to compete at shows or sell your does' offspring, since registered animals go for more than those that have no papers. A registered goat has official papers issued by an organization that keeps track of production records, show records, and pedigrees for that breed. Insist on receiving the registration papers when you pay for the goat; lots of buyers who are promised the papers will be sent later never get them.

Ask the seller to make a list of medications or vaccinations the goat has had and the date of each, and seek the seller's recommendations regarding future vaccinations. If you live in the same general area, ask about the seller's veterinarian — knowledgeable goat veterinarians are hard to find.

Ask what the goat has been eating and obtain enough of the same feed to last at least one week. If you plan to change the feed, make the change gradually by mixing the two together to ensure that your new goat remains healthy.

Before you take the goat home, ask the seller to trim hooves, remove horn buds, vaccinate, and perform any other necessary procedures that haven't already been done. Even the most routine procedures cause some degree of stress, which is considerably reduced when the animal undergoes them in familiar surroundings.

How Long Do Goats Live?

The normal life span of a goat is 10 to 12 years, but some goats live as long as 30 years. The productive life of a dairy goat, during which you can expect a reasonable amount of milk, is about 7 years.

Handling Goats

Working with goats can be frustrating or rewarding: frustrating if you try to work against the goats' nature, rewarding if you put their nature to work for you. By understanding why goats act the way that they do, you will better know how to treat them, and the experience will be more rewarding.

Goats are like cats in that they are curious and independent and do pretty much as they please, whether or not their behavior pleases you. If you know what pleases them, you can get them to do what pleases you and let them think it was their idea.

Goats are social animals. A goat should have at least one companion, which can be another goat or some other type of animal. No goat should be housed alone.

As soon as you put two or more goats together, one of them takes over. You can easily tell which goat is the herd boss — it's the one in the lead. The herd boss is usually the oldest doe, called the herd queen. The other goats won't move until the herd queen leads the way. If anything happens to the boss, the herd falls into confusion until a new leader emerges. When you visit a herd, if you don't give your attention first to the queen, she'll display jealous misbehavior.

Goats protect themselves by butting enemies with their hard heads. They also butt heads with each other in play and to determine the pecking order. Baby goats play by pushing each other with their heads, and they will try to push against your leg or hand. Don't let them, because as a young goat grows up, pushing turns into butting. If you teach your goats while they are young not to push or butt you, they will be easier to handle as they mature.

Goats will butt heads to determine the pecking order in the herd.

Goats and Other Animals

Goats get along well with other animals. They are often kept in the same pasture with cows or sheep. Goats do especially well with cows because they eat some plants that cows won't eat, whereas cows eat inferior hay the goats turn up their noses at. Keeping cows and goats together is an economical way to manage available feed.

Goats are sometimes kept with sheep because they are generally calm, while sheep are easily frightened. Sheep tend to remain calmer with goats around. However, goats and sheep are subject to some of the same parasites.

Goats get along well with cats and dogs, too. Cats are often kept as mousers in dairy barns, an arrangement that works out well. Treat your barn cats to a daily saucer of warm fresh goat milk and they'll be happy to stick around. Certain breeds of guardian dogs are used to protect goats. Donkeys are also sometimes housed with goats to fend off predators, especially coyotes and dogs, which a donkey will kick at or chase away.

Housing chickens with goats may be picturesque but is not a great idea. The chickens will nest in the hay and roost over the manger. Hay will be wasted because goats won't eat it once it's been soiled with chicken poop.

Goats and Stress

Any unusual, painful, or unpleasant experience causes goats stress. Such experiences include being chased by dogs, teased by insensitive people, or handled roughly. Many ordinary events in a goat's life are inherently stressful, including being weaned, castrated, disbudded, transported, isolated, or artificially bred. How a goat reacts to stress depends somewhat on its genetic background. Some breeds, especially Nubians, are more excitable than others. Reaction to stress also depends on individual temperament, past experiences, and familiarity with surroundings.

Developing a routine for managing your goats helps reduce stress. Goats like to be fed by the same person at the same time every day. If you are late, your goats will misbehave. The same goes for milking. If you fail to milk on time or send someone unfamiliar to do the milking, your goats may act up.

Oddly enough, although regular routine reduces stress, an overly rigid routine can cause worse stress. The trick is to make change part of your routine. Instead of always taking care of your goats by yourself, occasionally ask a friend or family member to come along and help. Then, if one of them takes over while you are away, your goats won't be upset by the presence of a stranger. Similarly, if you don't feed or milk at the exact same time every day, your goats won't be upset if you arrive early or late once in a while.

Because goats are naturally curious, not all new situations are stressful. Forcing a goat to confront a new situation, however, always causes stress. When a goat balks, give it a little time to check things out, and it will probably soon proceed on its own.

Preconditioning goats to new procedures goes a long way toward reducing stress. If you regularly take a doeling to the milk stand for a brushing and an udder massage, she will be comfortable with the idea of getting on the milk stand long before she starts giving milk. Handle a kid's feet frequently while it's young, and it won't act up when it needs its first hoof trim. Run the electric clippers when you handle a young goat — bringing the clippers gradually nearer until they eventually touch the goat's body — and when it's time to clip its coat, the goat won't be frightened by the noise and vibration.

Tips for Reducing Stress on Goats

- Ask a friend to help with your goats, so they get used to strangers.
- Give your goat time to check out a new situation.
- Handle a kid's feet frequently.
- Familiarize your goat with the sound of clippers.
- Brush your goat and massage its udder while it is on the milking stand.
- Talk or sing to your goats while performing chores.
- Start with the herd queen when performing a task on the whole herd.

Outfit your goat with a plastic collar that will easily break away if it snags anything.

Another stress reduction measure is to reassure each animal by repeating its name. Talk or sing to your goats while you milk, feed, groom, shear, medicate, or perform other chores. After a goat has had an unpleasant experience, talk calmly or sing quietly until the animal has calmed down.

Training your goats, especially the herd queen, to be cooperative and well mannered reduces stress for the entire herd. To minimize squabbles among the herd, start with the queen whenever you perform any procedure, such as feeding, grooming, milking, hoof trimming, and shearing.

Well-Behaved Goats

Goats that aren't handled become shy, and you will have a hard time getting them to come for milking, hoof trimming, or weighing. Handling goats to keep them friendly takes little time. Whenever you enter your goat house, greet each animal by name, starting with your herd queen. Scratch each

A well-behaved goat will learn to follow you when you talk gently, use its name, and put your hand on its collar.

goat's ears and face. Your goats will crowd around, happy to see you. If you always handle them in the same order, they will learn to come to you each in her proper turn.

As soon as your goats are big enough, give each one a collar. A plastic chain makes a good collar. It is sturdy enough to lead a goat by but will break if the goat gets hung up somewhere and pulls away, preventing the goat from being choked by its own collar.

A well-behaved goat will learn to follow you when you talk gently, use its name, and put your hand on its collar. A stubborn goat will plant all fours on the ground and refuse to budge. If the goat balks, grab one ear and pull firmly. A goat doesn't like to have its ears pulled and will usually come just to make you stop pulling. Don't pull by the collar, or you may choke the goat.

A frightened goat may rear up on it hind legs. If a goat rears, let go and move out of the way to avoid being hurt. Talk gently until the goat calms down, then try again.

Transporting Your Goats

A baby goat can easily be transported in a pet carrier. Many goats ride in the back seat of the family car. On a long trip, though, you'll have to stop once in a while and take the goat for a walk, as you would a dog.

For long distances, the back of a pickup truck works better for both the goat and the human passengers in the cab. The goat must not be able to jump out and must be protected from wind. The pickup bed should be covered with a camper shell or a sturdy stock rack wrapped in a tarp. Add a little bedding to keep the goat from slipping during curves or sudden stops.

Housing Goats

A goat's housing needs are simple: fresh air; a place to get out of hot sun, blowing wind, and cold rain and snow; a clean place to sleep; and safety from predators. You may think this makes them an easy-care species, but they do have their challenges and require good fencing and much patience.

Goats are incredibly curious creatures. They constantly check things out. If the fence develops a hole, in no time they will find the hole and wiggle through. If you accidentally leave a gate open, before you know it they'll be down the road checking out the neighborhood. Goats investigate not only with their eyes but also with their lips, which they use to test new foods and objects, including gate latches. If a latch moves, they'll keep working it until it falls open, and out they go.

Goats also chew on things. If you use a rope to tie a gate shut, a goat will chew through the rope, open the gate, and go exploring. Electrical connections are especially dangerous for a goat to chew. Since a goat will stand on its hind legs and stretch to investigate anything that looks interesting, make sure all electrical wiring and fixtures are well out of reach.

Goats are famous for their ability to jump. Kids love to leap against a wall and push off with all fours. If the wall has a glass window in it, the glass could shatter and the kid could be seriously cut. Goats also love to climb. Make sure your goats can't climb onto the roof of their house. Their sharp hooves could cause the roof to leak, and even though goats are surefooted animals, one could fall off the roof and break a leg.

Every time you visit your goats, check around for things that could harm them. A nail sticking out of the wall can rip open a goat's lip. A loose piece of wire can get wrapped around a goat's neck or leg. A rake or pitchfork lying on the ground can pierce a goat's foot.

The Goat House

Goats require a dry, clean shelter, which need not be fancy. Any sturdy structure will do as long as it provides shade and protection from rain, snow, and wind. Goats will keep one another warm down to temperatures as low as 0°F (–18°C), provided they can get out of wet and drafty weather. To find out whether your goat house is too drafty, go out on a cold or windy day and squat down to goat height. If you feel an uncomfortable draft, your goats feel it, too. Seal the gaps where the wind comes through. On the other hand, take care not to make your goat house too tight, since good ventilation promotes good herd health.

Goats suffer more in hot weather than in cold weather. Swiss breeds, because they originated in cool climates, suffer more in warm climates than desert breeds do. Using electric animal clippers to give them a trim will keep the long-haired goats cooler.

When temperatures get above 80°F (26°C), make sure your goats have shade and cool water. Stir up a breeze indoors by opening the doors and windows. To keep your goats from escaping through the wrong door, create a screen door of sorts by using a stock panel or other sturdy open-wire structure that lets in the breeze but keeps the goats from going through.

In warm climates, give your goat house a south-facing wall that can be removed in summer to increase air movement. In climates that remain mild, the house will need only three walls. Some large dairy herds are sheltered by only a roof and one or more half walls mounted with hayracks. Allow at least 15 square feet (4.5 sq m) of housing per goat, 10 square feet (3 sq m) per miniature. If you're building from scratch, plan now for future herd expansion.

In addition to the main area, you will need at least one smaller stall big enough to hold one goat. An extra stall will come in handy for housing a sick or injured goat, a pregnant goat that is about to kid, or kids you wish to wean.

You will need space to store feed, supplies, and dairy equipment away from the goats' living area. Separate the storage area from the goat area with a wall at least 4 feet high (121 cm); the half wall lets you watch your goats and your goats can watch you while you work in the storage area.

Allow one opening per goat for hay so all your goats can eat at the same time. Because they are herd animals, goats tend to all eat and sleep at the same time. However, they won't all drink at once; you'll therefore need fewer additional openings for water. One opening per six to eight goats should be adequate, depending on how rapidly they empty the bucket and how often you refill it.

Make the feed and water openings just big enough for a goat to push its head through but not big enough for its whole body to get through. An opening big enough for a doe's head is big enough for a kid to pop through.

Bedding

Packed dirt makes a good floor for a goat house because it lets urine and other moisture drain away. A plain dirt floor should be covered with a thick layer of bedding. Wood shavings (but not cedar, which is toxic) make excellent bedding. Straw also works fine, if you clean it out often. Otherwise it gets packed down so tight that you'll hurt your back removing it, unless you have a tractor with a front loader and tines. Waste hay that isn't moldy is the most common bedding for goats. Hay is the primary food of goats, but they tend to eat only the best parts and leave the rest. Using waste hay as bedding saves you money by making use of leftovers. The goats will help you out by pulling hay into their living area and spreading it around to make themselves cozy.

Goats that spend their days on wet or filthy bedding may develop a bacterial infection in their udders or hooves. Keep the bedding clean and dry by

A goat house should be sturdy and provide protection from the elements.

Goats should be able to reach their heads into, but not enter, the feeding area.

Layout for a Small Goat House

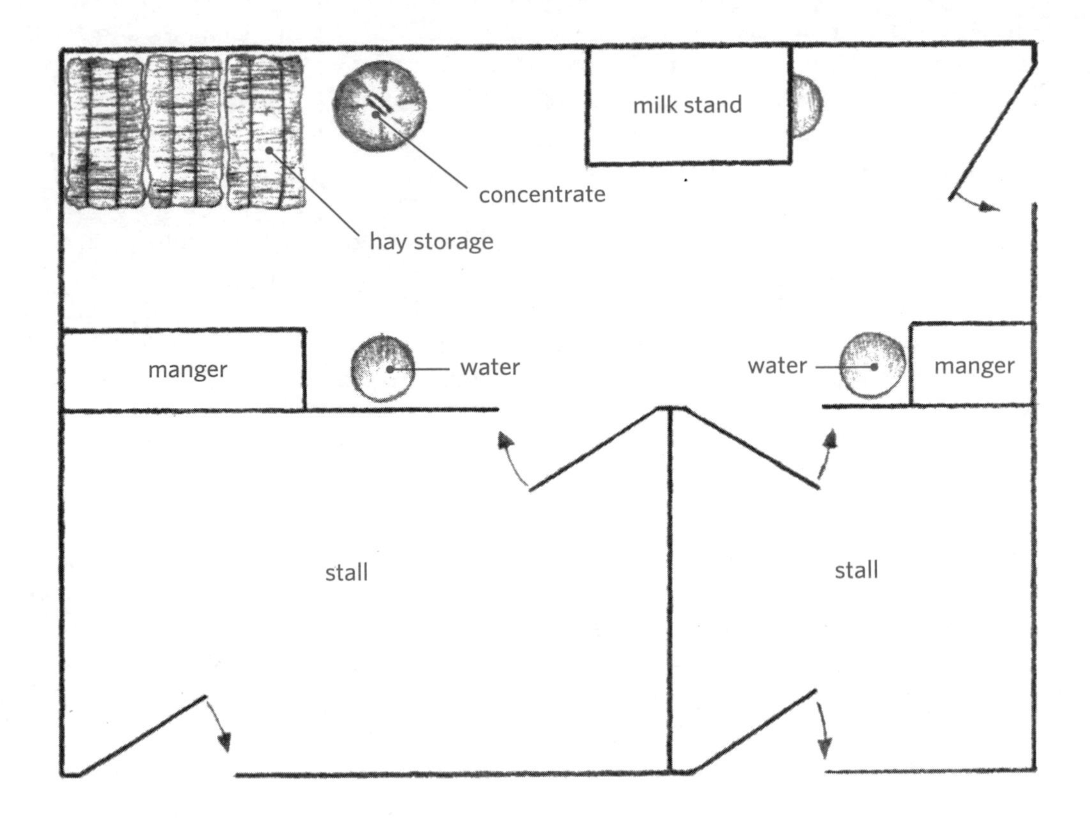

periodically removing and replacing it. Between cleanings, spread a fresh layer on top as often as necessary to provide your goats a clean place to sleep. Some people replace the bedding weekly; others replaced it only in the spring and fall. The more often you clean out the bedding, the easier the job.

Each day, each goat produces about 1¼ pounds (0.5 kg) of manure and 1¼ pounds (0.5 kg) of urine. Some of this manure and urine gets mixed into the bedding. Used bedding from a goat house makes good fertilizer for flower beds, vegetable gardens, or fruit and nut trees. To fertilize crops and gardens, spread the used bedding in the fall and work it into the soil so it is well rotted by spring planting time. Before mulching trees and shrubs, compost the bedding to avoid damaging plants with fresh manure.

The Goat Yard

Goats need space outside to wander around and get fresh air and sunshine. Allow at least 200 square feet (61 sq m) of outdoor space per regular-size goat, 130 square feet (40 sq m) per miniature. A yard this size would be considered a dry lot; any vegetation it might contain to start with won't last long after your goats arrive. If you want your goats to forage naturally, instead of feeding them everything they eat, they'll need additional space (see Grazing and Browsing on page 185).

A yard on the south side of the house gets more sun and stays drier than a yard on the north side, an important factor in avoiding damp ground that can lead to bacterial infections of the hoof. The yard should slope away from the house for good drainage. If your land is level or drains poorly, erect a wooden platform or a concrete pad in the yard to give your goats a dry place to stand and lie. A concrete pad also gives goats a place to scrape their feet, which helps wear down excess hoof growth.

Goats of all ages love to jump and play. An outcropping of rocks makes an ideal play area. A popular and inexpensive toy is an empty cable spool with a board securely nailed over the holes in both ends so that the goats can't slip during play and break a leg.

If you provide your goats something to climb on, take a good look to see what they can climb onto from there. Make sure the climbing object is far enough from their house that they can't jump onto the roof, and far enough from the fence that they can't jump over and escape.

A Goat Fence

Goat owners love to say that a fence that won't hold water won't hold a goat. Of course, that's a slight exaggeration — but only slight. Most goat troubles occur because of inadequate fencing. Goats are curious, agile, and persistent. If there's a way to escape, they'll find it. They can flatten their bodies and crawl under a fence or spring off the ground and sail over it. If they can't get under or over a fence, they'll lean on it until they crush it down.

Goats seem to believe the grass is greener on the other side of the fence, and they'll stretch their necks to eat that grass and whatever else is growing out there. They will nibble on trees and shrubs growing within 2½ feet (76 cm) of the fence. In doing so, they push against the fence until they bend it out of shape.

A goat loves to scratch its back by leaning sideways against a fence and walking along in one direction, then turn and go the other way to scratch the other side. Sooner or later, all that pushing and rubbing will knock down a flimsy fence. A properly built fence not only keeps goats where they belong, but it also protects them from predators, such as stray dogs, coyotes, wolves, and bears.

A goat-tight fence can be made of wood or wire. Your choice may be narrowed down by what's legally allowed in your area. Electric fence wire works best and can be used in conjunction with a nonelectric fence or to build an all-electric goat-tight fence. One strand of electrified wire placed 12 inches (30 cm) off the ground on the inside of a fence will keep goats from pushing against the fence. Another strand placed about nose high will keep goats from leaning on the fence or jumping over. A field fence made of woven wire needs the protection of these electrified strands to survive typical goat activity.

An empty cable spool is a popular toy.

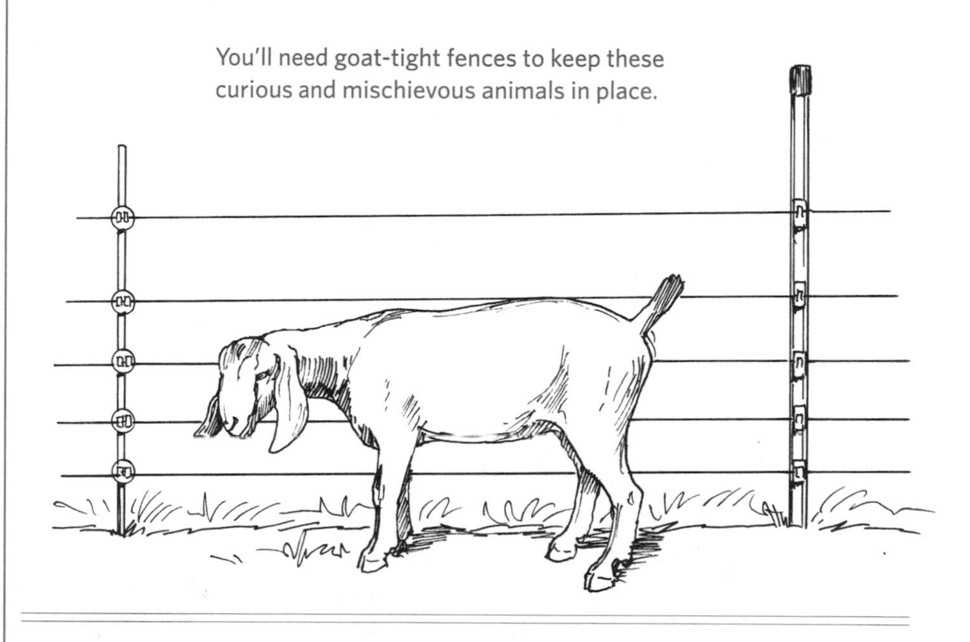

You'll need goat-tight fences to keep these curious and mischievous animals in place.

Electrified wire is the way to go if you wish to goat-proof an existing nonelectric fence. If you are building a new fence, it's cheaper to make it all electric from the start. Use high-tension smooth wire and a high-energy, low-impedance energizer. Goats will stay away from it because they don't want to get shocked; however, the fence must always be on. Your goats will test the fence constantly, and the moment the juice goes out, so will they.

An electric fence need not be as high as a nonelectric fence; up to 40 inches (1 m) will suffice. String the bottom wire 5 inches (13 cm) from the ground. String the second wire 5 inches (13 cm) from the first, the next wire 6 inches (15 cm) up, the next wire 7 inches (15 cm) up, the next wire 8 inches (20 cm) up, and the top wire 9 inches (23 cm) up. Connect every other wire to the energizer, and the alternating wires to the ground. For safety reasons, you must get all the connections right; if you aren't sure how to build an electric fence, consult an expert or a good book such as *Fences for Pasture and Garden* by Gail Damerow.

A nonelectric fence can be 4 feet high (1 m) for calm breeds but should be 5 feet high (1.5 m) for active breeds such as miniature goats and Nubians. Place corner post and gatepost bracing on the outside of the fence; if you brace on the inside, your goats will use the braces to climb up and out. Nonelectric options include:

- Wooden fencing, consisting of wide-board rails spaced close enough together that a goat can't slip between them or constructed stockade-style (but not picket-style, which can trap the hooves of a goat standing on its hind legs to peer over the fence).
- Stock fencing, made from 16-foot-(5 m) long welded-rod panels attached to steel or wooden posts.
- Chain link, which is about as goat-tight as you can get, but much more expensive than other types of fence and not allowed in some municipalities.

Where local regulations prohibit any type of fence that looks too agricultural, you may need to combine two types of fence. For instance, you might put up a rail fence that looks nice on the public side and fortify it with a woven-wire field fence on the goat side.

The Goat-Proof Gate

Goats are expert gate-crashers, so take special care in designing your gate. Make the gate as high as the rest of the fence. Use a latch that goats have trouble opening. A latch that requires two different motions, such as lifting and pulling, is more difficult for a goat to open than a latch that flips up.

No matter what kind of latch you use, secure it with a bolt snap, available from any hardware or farm store. Attach the snap to the gate with a short chain so you can't drop it, put it into your pocket, or otherwise misplace it.

Install the latch partway down the gate, on the side away from the goats. They won't be able to reach over the gate to work the latch, but be sure you can reach the latch from the inside — if a tall person installs the latch, a shorter person may have trouble reaching it.

Hinge the gate to open into the yard, toward the goats. Even if the goats manage to open the latch, they'll be pushing against the gate and keep it shut. Make the gate strong enough to support the weight of a goat standing on its hind legs to peer over the gate at the world outside the goat yard.

A goat-proof latch secured with a snap on a chain

Feeding Goats

Goats, like humans, need a balanced diet to remain active and healthy and to produce kids, milk, and meat. Like cows and sheep, goats are ruminants. Even among ruminants, goats eat the widest range of plants. Their ability to use plants other animals can't digest makes them popular as livestock worldwide.

The Goat's Digestive System

A goat's complex digestive system fills up one-third of its body and includes a stomach with four chambers — the rumen, the reticulum, the omasum, and the abomasum. These chambers are sometimes called the goat's four stomachs, although only one of the chambers — the abomasum — works like a human stomach.

Everything a goat eats goes into the first and largest digestive chamber, the rumen, which in a mature goat holds 5 gallons (19 L) of liquid and plant matter. You can see the rumen bulge out on a goat's left side. Fermentation in the rumen breaks down plant matter into digestible form. Fermentation also produces heat, which in cold weather helps keep a goat warm.

Periodically the goat burps up a soft mass of partially digested food for further chewing. This soft mass is called a cud. A relaxed goat chews cud for hours. Because the cud comes from the rumen, cud chewing is called ruminating.

The reticulum is next to the rumen, separated only by a partial wall. The reticulum acts as a pump, moving food back to the mouth for more chewing or passing it along to the omasum for further digestion.

The omasum removes moisture from the digesting food. This chamber is pleated, like draperies. The pleats increase the omasum's surface area to speed up the absorption of moisture from digesting matter that passes through.

Of the four chambers, the last one — the abomasum — is considered the true stomach and functions the most like the stomach of a human. The abomasum is the second largest part of the goat's digestive system. In the abomasum, protein breaks down into simple substances that help a goat grow and remain healthy.

Humans and other animals with simpler digestive systems need dietary fiber to stimulate digestion, even though their stomachs cannot digest the fiber itself. A goat's digestive system breaks down fiber into nutrients the animal needs for survival. Fiber is a goat's main food, and the goat consumes it in the form of grass, hay, twigs, bark, leaves, cornstalks, grain, and various other plant parts.

Grazing and Browsing

Whereas some animals are grazers, reaching down to munch on grass and other low-growing plants, and others are browsers, reaching up to snack on leaves and bark, goats are opportunistic feeders — they eat whatever is available. Given a choice, they wander from one food source to another, taking a nibble of each.

Because they eat such a varied diet, goat herds can live in diverse circumstances. Some goats roam entirely in wooded areas, others are kept on pastures, and still others are confined to barns and have all their food brought to them. Each arrangement has advantages and disadvantages.

Letting your goats browse or graze reduces feeding time, labor, and expense. Since feed is about 70 percent of the cost of keeping goats, letting them forage adds up to big savings. It's also more natural for the goats and is less likely to result in digestive problems compared to feeding grain and other bagged rations.

Allowing dairy breeds to browse a densely forested area, however, is not a great idea for two reasons. First, brambles and low branches can scratch udders. Second, some plants give milk an unpleasant flavor. Dairy goats are more often allowed to graze on improved pasture (pasture that is maintained by the owner), where their udders are safe from harm and weed control eliminates wild onion, garlic, mint, and other plants that give milk an unpleasant flavor.

Grassfed goats generally produce more milk, and some cheese makers find that cheese made from the milk has a cleaner, more natural flavor. But pasturing requires having a fair amount

A goat's digestive system fills up one-third of its body.

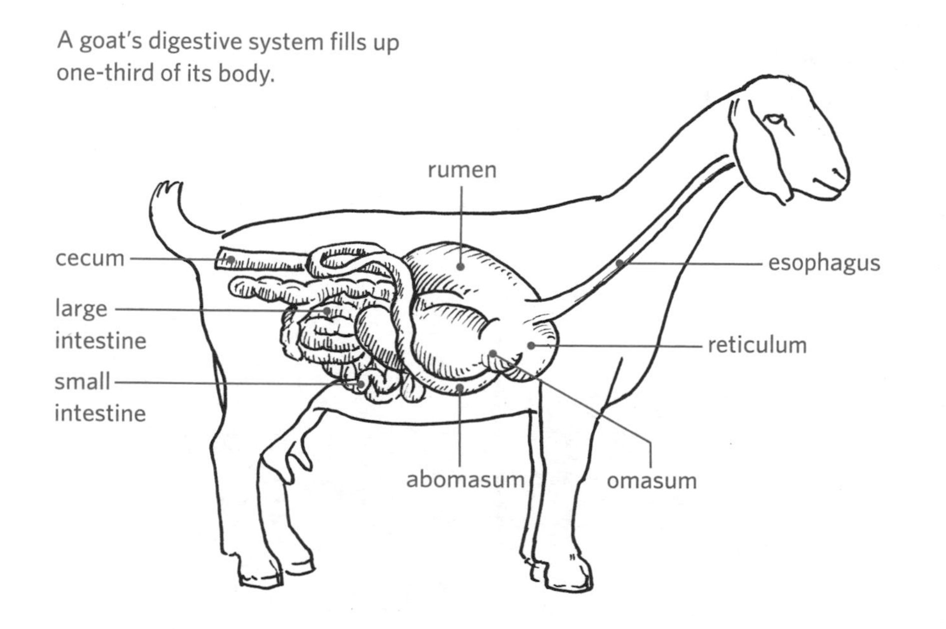

Grazing Systems

Systems for grazing dairy goats fall into three categories:

Continuous grazing. The goats graze the same pasture year-round, with the result that they trample or foul some of the pasture and unevenly graze the remainder, and they are continuously exposed to an unhealthful buildup of parasitic worms.

Simple rotation. The same amount of land is divided into four paddocks and the goats are moved from one to the other weekly, giving the unused paddocks a rest for regrowth and parasite control. If grazed unevenly, unused paddocks may need to be mowed to minimize parasites by exposing them to sunlight and to encourage fresh pasture growth.

Intensive rotation. The same amount of land is divided into smaller paddocks, and the goats are moved from one to the other based on pasture graze-down and regrowth. This system accommodates a greater number of goats, produces a larger amount of forage (because each paddock has a longer resting period during which to regrow), and results in the greatest degree of parasite control.

Given the option, goats will browse on shrubs and trees, much like deer.

of land, which for the health of the goats as well as the pasture, should be divided up by fences into no fewer than four small fields called paddocks. The goats are moved from one paddock to another as they graze down the previous paddock. How much of their diet they obtain by foraging depends on what's growing in the pasture, the seasonal growth rate as regulated by the climate, and the number of goats competing for it. During times of lush plant growth — generally when the weather is warm but not hot and rainfall is adequate — plant regrowth may keep up with the goats. At other times of year plant growth slows or stops, and the goats must rely more on hay than on pasture.

Because of the seasonal variation in plant growth, the milk from grassfed goats varies in water content, which is highest when growth is lush and lowest as the season progresses. For some cheese makers, the higher percentage of water is problematic because the same amount of milk produces less cheese. In commercial dairies, the goats are often confined to loafing sheds, where all their feed is brought to them. The confinement system also works for goats living on a small lot with insufficient land for browsing or grazing.

Feeding Hay

Whether goats browse, graze, or are kept in confinement, they need hay. Hay is nothing more than pasture plants that have been cut, dried, and stored loose or compressed into square or round bales.

On average, a goat eats 3 percent of its body weight in hay each day, which adds up to about 4 pounds (2 kg) for a large goat and 2 pounds (1 kg) for a miniature. If a square bale weighs 40 pounds (18 kg), two mature full-size goats will consume approximately 73 bales per year. Since a goat won't eat more hay than it needs, feed hay free choice — that is, always keep the manger full, so the goat can eat hay whenever it wants to.

Fresh green pasture plants contain a high percentage of water. If pasture is a goat's sole source of feed, the animal will have a hard time satisfying its hunger. Hay takes the edge off, making the goat less likely to scarf down poisonous plants or overeat proper pasture plants. In addition, fermentation in a rumen full of nothing but fresh pasture produces excess gas that cannot escape fast enough, causing the rumen to bloat dangerously. A goat that eats plenty of hay before going out to pasture will not graze frantically to curb its hunger and is much less likely to bloat. A goat that can come and go as it pleases, munching on free-choice hay and wandering out for a mouthful of pasture, has little chance of bloating.

Hay quality varies, as does a goat's nutritional needs. A growing goat, a pregnant doe, and a lactating doe all have higher nutritional needs than a mature wether or an open dry doe (one that is neither pregnant nor giving milk).

Legume hays, such as alfalfa, clover, soybean, vetch, and lespedeza, provide excellent nutrition for kids, pregnant does, and lactating does. Grass hays, such as timothy, red top, Sudan, bromegrass, and fescue are less nutritious. A good all-purpose hay is a 50-50 grass-legume mix.

Look for early-cut hay that is fine-stemmed, green, and leafy. Buy hay sold for goats or horses; hay sold for cows is often too stemmy. Goats, especially milking does, can be pretty persnickety about eating hay with coarse stems; they'll nibble down the tender parts and leave the rest for you to dispose of. Wethers will eat stemmy hay when forage is sparse, but if they have a choice they, too, will waste most of it. Stemmy hay doesn't even make good bedding.

Avoid coarse, unpalatable hay with seed heads. It was harvested past its prime.

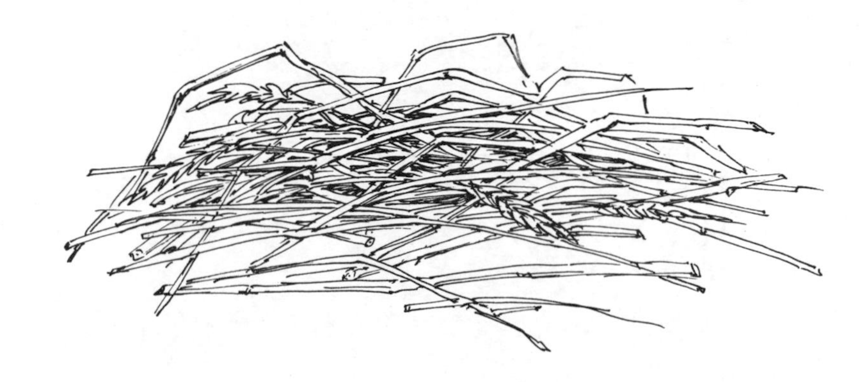

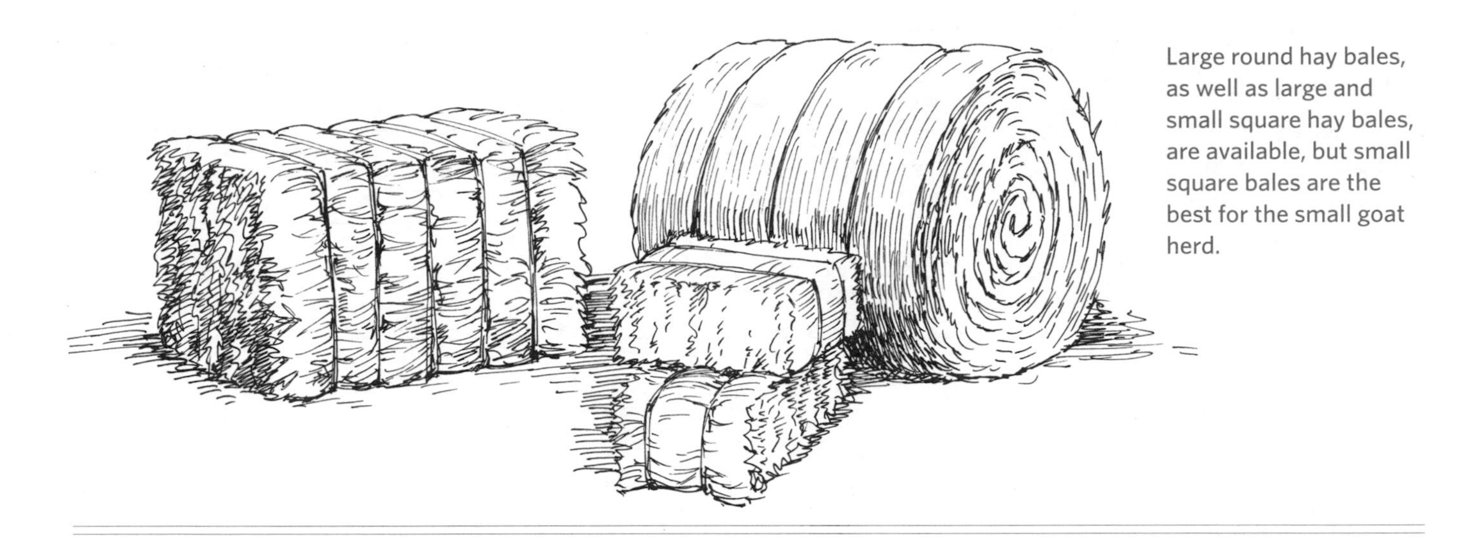

Large round hay bales, as well as large and small square hay bales, are available, but small square bales are the best for the small goat herd.

Hay sellers advertise through the classified section in the newspaper, especially from late spring through early fall. Various county agriculture offices keep lists of hay growers. The clerk at your farm store may know someone who sells hay, or the farm store may stock it.

Most goat owners prefer to handle small square bales, even though many hay growers prefer making large round or square bales that are easier to load by tractor for transport. Small square bales that are easy to move by hand are ideal for the small goat herd.

You can buy hay by the load delivered to your door. To save money, you can sometimes buy hay in the field, right after it has been baled. The grower will expect you to pick it up and probably load it yourself onto a truck you supply. If you do not have room to store enough hay to last a year, find a grower who will store if for you and let you pick it up as you need it. Expect to pay more for stored bales than for hay purchased in the field.

Keep your hay under cover and off the ground on pallets. Properly stored hay retains its nutrients for a long time, but one good rainstorm can ruin baled hay. Never feed your goats moldy or musty-smelling hay.

Unless the hay is exceptionally good or your goats are exceptionally hungry, about one-third of the hay will go to waste (so figure this into your calculation at the time of purchase). Remove leftover hay from the manger every morning and use it as bedding or toss it onto the compost pile. Don't put the hay directly onto your garden soil, or you risk introducing unwanted weed seeds.

Concentrate

A young growing goat and a mature doe that is producing kids or milk — especially a high-producing doe — needs more nutrients than even the most nutritious hay can provide. Such a goat requires a ration that contains grains and other nutrient-rich feeds combined into a dietary supplement variously called goat feed, goat chow, goat ration, or concentrate (because it is a concentrated source of nutrients).

Not all goats in a single herd require the same amount of concentrate at the same time. One doe may be dry while another is lactating. One may be still maturing while another is about to give birth. Even two does of the same size and age, both dry or both lactating, may require different amounts of concentrate to maintain the same body weight or to produce the same amount of milk.

Feeding guidelines are therefore nothing more than estimates. Always use your own best judgment. If you raise miniature goats, feed them approximately one-half the amount required by large goats. Keep written records to remember who gets what. Adjust concentrate levels according to the following four factors.

The quality of the roughage. Goats that eat fresh browse, green pasture, or good hay need less concentrate

Dietary Changes

Any time you change a goat's diet, you run the risk of disrupting the rumen's digestion activity. To keep your goat from getting sick, make all dietary changes gradually. Whenever you increase or decrease the amount of concentrate a goat gets, do it gradually over several days. If you change from one kind of hay or concentrate to another, mix the new feed with the old in increasing quantities until the switch is complete. If the goats have been kept off pasture for the winter, let them graze for only a short time each day while their rumens adjust to the fresh greens.

than goats that get little or no browse or pasture and poor-quality hay.

Each goat's physical condition. A well-conditioned goat is fleshed out but not too fat. A dairy goat is too fat when you can't feel her ribs. A meat goat is too fat when you can grab a handful of flesh behind the elbow. If a goat is too thin, feed more grain. If a goat is too fat, feed less grain.

The goat's age. Let kids nibble on concentrate as soon as they are interested. At first they will just mouth the ration with their lips, but as they grow they will learn to relish their little taste of grown-up feed. After the kids are weaned, gradually work up to 1 pound (500 g) of concentrate per day. When feeding any goat 1 pound (500 g) or more per day, divide the concentrate into two feedings, morning and evening. Feeding too much concentrate at once upsets the rumen's balance.

The goat's level of production. Mature goats that are not pregnant or lactating require a maintenance ration that provides just enough nutrients to maintain the animal's health and body weight. A maintenance ration for wethers and open dry does on good browse or pasture need not include concentrate. A supplemental feeding of ¼ to ½ pound (125 to 250 g) or 1 to 2 cups (236 to 473 mL) of concentrate per day, however, increases the growth rate of a meat goat and keeps all goats easier to handle because they look forward to your regular visits with the feed can.

Open dry does or wethers raised in confinement can be fed up to 1 pound (500 g) of concentrate per day. The same applies to open dry does and wethers that normally browse or graze but the feeding pattern has been disrupted by bad weather or the forage supply has been curtailed by drought. Since open dry does and wethers have low nutritional requirements, you can save money by feeding them shelled corn, barley, oats, wheat, sorghum, or milo instead of commercial concentrate. Whole grains that are dry and hard do not digest as well as grains that have been rolled, crimped, cracked, or flaked.

Feed a dry doe about 1 pound (500 g) of concentrate per day. During the last two weeks of pregnancy, gradually increase the concentrate to about 3 pounds (1.5 kg) a day by the time she gives birth. During early lactation, when her milk production is increasing, feed her a minimum of 1 pound (500 g) of concentrate per day plus an additional ½ pound (250 g) for each pound (500 g) of milk she produces over 2 pounds (1 kg). During late lactation, when her production has leveled off, feed her ½ pound (250 g) of concentrate per pound (500 g) of milk. After the doe has been bred, gradually decrease her concentrate to 1 pound (500 g) per day, and start the feeding cycle again.

Above all, remember that concentrate is a supplemental ration and not a goat's main diet. No matter how little or how much concentrate a goat gets, it should have access at all times to as much hay as it wants.

Concentrate Feeding Guidelines*

Stage	Status	Concentrate per day
Kid	nursing	nibble
	weaned	1 to 2 lbs. (0.5-1 kg)
Wether or open dry doe	fresh forage available	¼ to ½ lb. (250-500 g)
	no fresh forage available	1 lb. (0.5 kg)
Doe	pregnant dry	1 lb. (0.5 kg)
	2 weeks before kidding	increase to 3 lbs. (1.5 kg)
	lactating	½ lb. per lb. milk (1 lb. [1 kg] minimum)

* Reduce amounts by half for miniature goats.

Note: Adjust all amounts to each goat's condition.

Concentrate Feeders and Storage

Some herd owners feed concentrate from a communal trough, but doing so can mean the timid goats don't get their fair share. Feeding concentrate to each animal separately means each goat gets an appropriate portion, and the portion is tailored to the individual goat's needs.

Placing feeders in a manger the goats can access through the usual feed hole helps ensure that each animal eats its own ration. To keep fast eaters and bullies from stealing feed from others, you may find it necessary to run a chain across their feed holes to lock in their heads until everyone is finishing eating.

Lactating does are often fed on the milk stand to keep them from getting restless while being milked. However, if you train your does to stand calmly without being fed, they won't get restless if they finish eating before you finish milking.

Concentrate comes in 50-pound (23 kg) bags. Store unopened bags away from moisture, out of the sun, and off the ground. Your farm store may let you have a wooden or plastic pallet to store sacks on. Otherwise, raise sacks off the ground by using bricks, concrete blocks, or pieces of lumber.

After opening a bag, pour its contents into a clean trash can with a tight-fitting lid. A 10-gallon (38 L) can holds 50 pounds (23 kg). Storing concentrate in a can keeps it from getting stale or absorbing moisture from the air. Moisture causes concentrate to become moldy; never feed moldy grain to goats. Empty the storage can completely before pouring in another bag so stale concentrate won't build up at the bottom of the can.

Store concentrate where your goats can't get into it. Otherwise, they will jump on unopened sacks until they tear a hole. They will work on the lid of a storage can until they get it open. The resulting feast will speed up rumen fermentation, and the goats will become ill, like kids who eat too much candy the day after Halloween. In the case of goats, however, all that fun could prove fatal. As a safety measure, in case a goat should manage to get out, secure the lid of the storage can with a bungee cord.

Besides preserving the feed, a storage can keeps out munching mice, which can consume an astonishing amount of concentrate while fouling what they leave behind. When you transfer feed from the sack to the can, take care to avoid spillage, since mice are attracted by spilled grain. Sweep the floor of your storage area regularly. In spring and fall, when barn mice are most active, reduce the population by using traps baited with peanut butter. Better yet, keep a cat in your barn and reward it with a daily saucer of warm goat milk.

Soda and Salt

The rumen ferments best within a narrow range of acidity. Feeds that ferment rapidly increase the rumen's acidity. If the acidity goes up too fast, the microorganisms that cause fermentation multiply too fast. As a result, the rumen's balance is upset, and the goat becomes sick. Your goat's health depends a good deal on proper rumen acidity.

Alkaline substances, such as sodium bicarbonate (baking soda), reduce acidity. A goat eats soda to keep its rumen acidity within the proper range. The goat knows when it needs soda and how much. All you have to do is make sure your goats can get soda when they need it. Feed-grade baking soda can be purchased at any feed store and is less expensive than baking soda from the grocery store.

Each day a goat will lap up an average of 2 tablespoons (30 mL) of soda. Lactating does, and all goats on summer forage, eat more soda than at other times. The choice should be theirs.

Sodium chloride, or common salt, helps control rumen acidity, aids digestion in other ways, and helps keep a goat's body tissues healthy. Besides regular salt, goats require many other minerals in minute amounts. They obtain some of these trace minerals from good hay, fresh forage, and concentrate. To make sure your goats get all the minerals they need, give them free-choice trace mineral salt, which is a combination of trace minerals and salt and is sold at any farm store. It comes in loose form, like table salt, or compressed into a block. Loose salt is easier for many people to handle than a heavy block and is easier for goats to lick, especially at times when their salt need is high.

Your farm store may sell a trace mineral mix formulated specifically for goats. If not, get the mix for horses or cows. The mix must contain copper, iodine, and selenium. Copper is essential for dairy goats. If your goats browse on trees and other deep-rooted plants, or if their drinking water flows through copper pipes, they already have some copper in their diet and need less from the trace mineral mix. Iodine and selenium are both essential for a doe to give birth to healthy kids. Iodine and

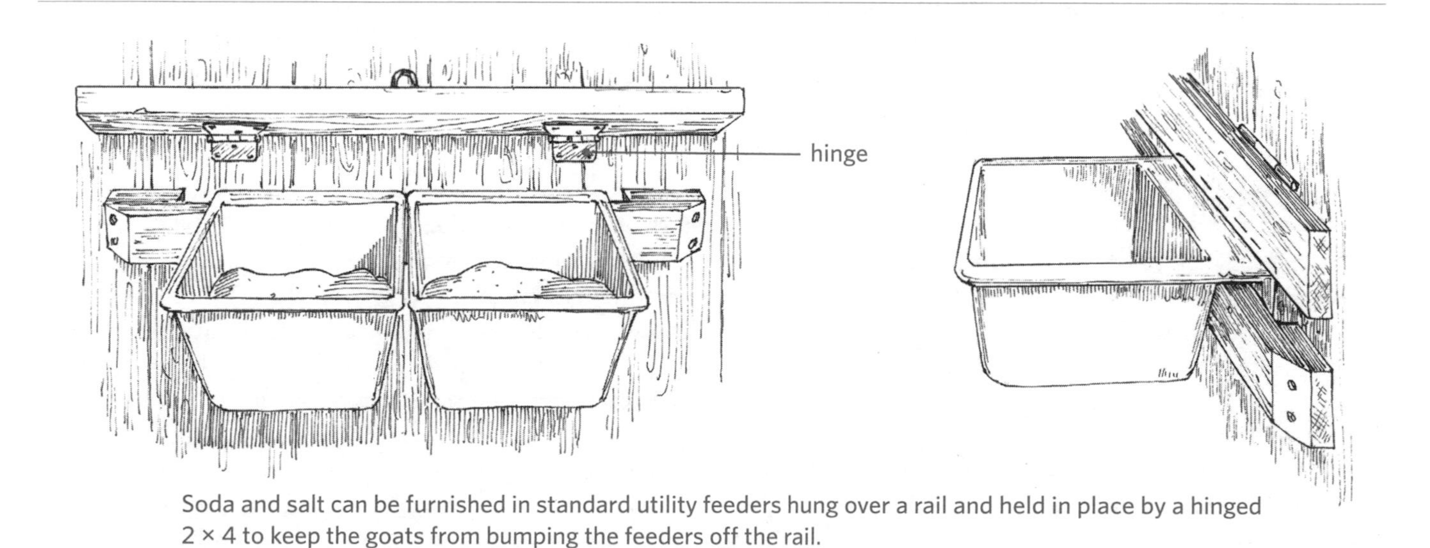

Soda and salt can be furnished in standard utility feeders hung over a rail and held in place by a hinged 2 × 4 to keep the goats from bumping the feeders off the rail.

selenium do not occur in plants grown in regions where these minerals are lacking in the soil.

Be careful not to overfeed trace minerals. Excess iodine in a doe's diet makes her milk taste unpleasant. And because excess selenium in the diet can be toxic, do not give your goats trace mineral mix containing selenium if the soil in your area is high in selenium. The soil tends to be high in selenium in Colorado, Kansas, Montana, Nebraska, North Dakota, South Dakota, Utah, and Wyoming. The soil is deficient in selenium in some areas of the Northeast, the southern Atlantic seaboard, and the Pacific Northwest in the United States and in the Maritimes and parts of British Columbia in Canada. Your local veterinarian or county Extension agent can tell you if the soil in your area is high or low in selenium, and quite likely your farm store will sell the appropriate formulas accordingly.

Two-part soda-and-salt stations are available for a few dollars from nearly any farm store, but standard utility feeders are easier to remove from the wall for cleaning. Clean and refill the feeder often. Salt attracts moisture that causes the surface to crust over. Soda and salt both turn lumpy from the water dripping off the chin of a goat that has just had a drink. And your goats will periodically delight in backing up to the feeder to fill it with droppings.

Water

The most important and least expensive item in a goat's diet is water. Goats should have access to clean, fresh water at all times. Water aids digestion, controls body heat, and regulates milk production. The more water a doe drinks, the more milk she gives. Lactating does drink more water than dry does. All goats drink more water in warm weather. They drink less when they graze on spring pasture because fresh grass has a high water content.

A 5-gallon (19 L) plastic bucket works well as a water container for a small herd. Place the bucket outside the stall, where it can be accessed through a head hole. Keeping the water bucket outside the stall means your goats can't knock it over and wet the bedding, fill it with droppings, or accidentally drop a kid into it while giving birth. Neither will a goat be able to get its head through the handle and spend the day wandering around with a bucket hanging around its neck.

Goats will not drink water that has been contaminated. Items that typically fall into drinking water include hay, hair, manure, insects, and mice (often drowned). Empty and refill the bucket at least once a day, and scrub it with a brush and bleach at least once a week.

To encourage your goats to drink, fill the bucket with cool water in warm weather and warm water in cool weather. In cold weather, you must keep the water from freezing. If the goat house has electricity, the best way to keep the water from freezing is to use a plug-in bucket with a self-contained heater, available at most farm stores. Otherwise check the water at least twice daily and break ice as needed, or keep a spare bucket so you can bring the frozen one inside to thaw while carrying out warm (not hot) water in the thawed bucket.

Breeding Does

A doe that gives milk steadily for more than a year is the exception. Most dairy goats produce more milk than their kids need, but over time, the milk production gradually lessens. So part of a doe's normal annual cycle is giving birth to renew her milk production. The female kids can be sold to help pay for the doe's upkeep, and the male kids can be raised as meat for your family.

When to Breed

In deciding when to breed each doe, consider these factors: her age and size, when she was last bred, and the season.

Her age and size. Do not breed a young doe until she reaches at least 75 percent of her mature weight, which usually happens when she is about 8 to 10 months old. A doe may think she is ready to breed when she is 6 months old or less, but breeding her before she is large enough may stunt her growth. A doe that is bred early will produce fewer kids that are smaller than normal and may be too weak to survive.

When she was last bred. Renewing a doe's milk production by breeding is called freshening. Most dairy goats are milked for 10 months of the year, then are given 2 months off before they freshen again.

The season. A doe must come into heat, or estrus, before she can be bred. Some goats come into estrus every three weeks year-round, but for most goats the breeding season is August, September, and October. Does that are bred during September and October will give birth when spring's green pastures provide the extra nutrition a freshened doe needs.

Estrus

Throughout the breeding season a doe comes into estrus every few weeks. Estrus lasts for 2 to 3 days. The time between the start of one estrus and the start of the next is called the estrous cycle. Different does have different estrous cycles, ranging from 17 to 23 days. The average doe has a 19-day cycle. Keeping accurate records on each doe lets you track her exact cycle.

Most does show some signs of heat, but each has different signs or different combinations of signs. As you record each doe's estrous cycle, note the signs she displays so you'll know what to look for next time.

Signs of estrus are as follows:

- The doe may vocalize more than usual. She may bleat so loudly that you think she is in pain. Don't worry, she isn't.
- She may urinate more often than she usually does, especially when you enter the stall.
- The area under the doe's tail may become dark or swollen and wet. You may see sticky mucus that will be clear early in estrus and white toward the end of estrus.
- The doe may get restless. She may flag, or wag her tail. She may let you handle her tail, or she may move away if you try to touch it.
- The doe may give more milk than usual just before coming into heat, then give less milk than usual for a day or two
- She may mount another doe as if she were a buck, or let other does mount her.
- If a buck lives nearby, you'll have no doubt when a doe is in heat. She'll get as close to the buck's yard as she can. The buck will wag his tongue, slap a front hoof against the ground, urinate on his own face, and otherwise act the fool.
- If no buck is around, you might trick the doe into displaying signs of estrus by using a buck rag, which is a piece of cloth rubbed on the forehead of a mature buck and placed in a sealed container. When you open the container near a doe in estrus, she will show clear signs of interest.

Occasionally a doe shows little or no signs of estrus, a phenomenon known as silent heat. If you have a doe with silent heat, it's good to keep in mind that does living together tend to synchronize their heat cycles. So if you have one doe showing signs of heat and another showing no signs, chances are pretty good the second doe will be in heat at approximately the same time as the first one.

Flushing

A doe gets pregnant more easily and has more kids if she gains weight starting one month before she is bred until one month afterward. When a doe gains weight, more eggs flush from her ovaries during estrus. Getting a doe to gain weight at breeding time is therefore called flushing. Flushing is more important for a thin doe than for a well-conditioned doe. To get the maximum benefit from flushing, deworm your does before breeding season.

You can flush a doe in one of three ways:

- Move her to fresh pasture
- Feed her an extra 1 to 1½ pounds (0.5–1.5 kg) of alfalfa cubes, pellets, or hay each day
- Feed her an extra ½ to 1 pound (0.5–1 kg) of grain or concentrate each day

The Buck

For a small herd of does, keeping a buck is generally not economical. However, by the time you recognize the signs of estrus, make an appointment to have the doe bred, and transport her to the buck's location, you may be too late. To avoid missing your chance, you might arrange to lease a buck during breeding season, especially if none is available close by.

If you find a suitable buck at stud nearby, make breeding arrangements with the owner in advance. Find out whether the owner will be reachable on short notice and what the breeding fee will be. Some stud owners like to keep a doe overnight to make sure she is bred. Others expect you to wait while

Sound, strong, well-trimmed feet are important to a goat's health.

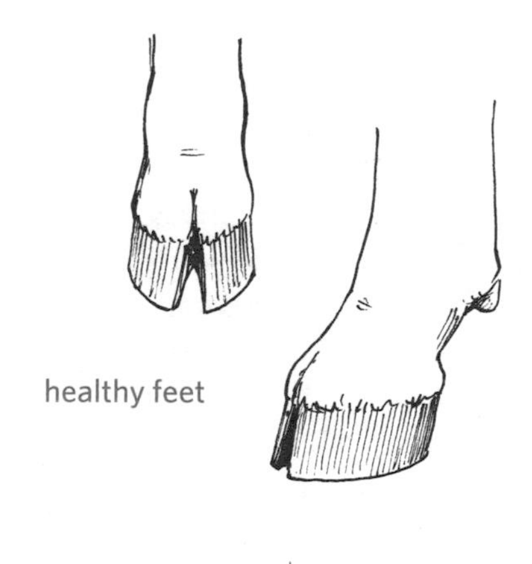

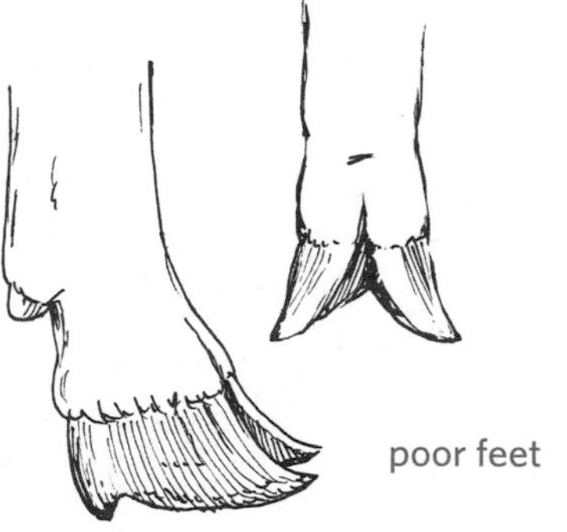

When to Breed Your Doe

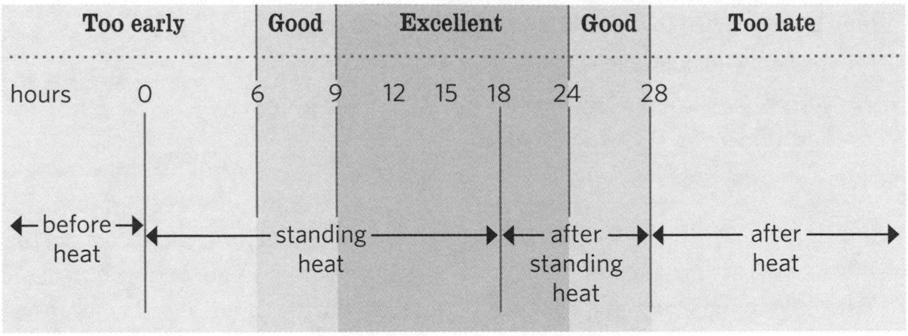

Too early	Good	Excellent	Good	Too late

hours 0 6 9 12 15 18 24 28

← before heat → | ← standing heat → | ← after standing heat → | ← after heat →

the doe is bred so you can take her back home with you.

Seek a strong, healthy buck that's been well cared for. He should have sound, strong, well-trimmed feet. A buck with poor feet may have a hard time mounting does.

If you plan to sell or show the offspring, the buck should be the same breed as your doe. The buck's dam should have a good record in milk production. To improve the udders of your doe's offspring, select a buck whose dam has a good udder. If the buck has been bred previously, try to look over his offspring as well. If he has produced quality kids in the past, he will probably continue to do so.

If you plan to raise or sell the kids for meat, the buck need not be the same breed as the doe. Mating your doe to a buck of a different breed, in fact, usually produces larger, faster-growing kids, especially if the buck is one of the meat breeds.

If you cannot find a suitable buck within a reasonable distance, you may wish to breed your does by artificial insemination (AI). To find an AI practitioner, contact the nearest goat club, check the ads in one of the goat magazines (listed in the appendix), or ask your veterinarian. The AI practitioner will help you select a buck through descriptions in a catalog and arrange to purchase that buck's semen for local storage until your does come into heat. When your doe is in estrus, the AI practitioner may come to your place or may ask you to take your doe elsewhere for insemination. You will be charged for both the semen and its placement. Artificial insemination can be costly; be sure to find out in advance how much it will cost.

Mating

A doe is ready to breed when she is in standing heat. A doe that is not in standing heat will move away when the buck tries to mount her, whereas a doe in standing heat waits patiently to be mated. Some bucks have to go through their whole clownish routine before getting down to business, but the act of mating itself takes just a few seconds. You can distinguish a trial run from the real thing by the way the buck arches his back before uncoupling from the doe.

By remaining on hand for the mating, you can be sure the act has been completed and you can make a close estimate of the kidding date. If you are away a lot, you may prefer to run a buck with your does during breeding season. It's a lot less work for you, but you never know for sure when each doe has been bred and, therefore, when she is likely to kid. In these circumstances, chances are greater you won't be on hand to help out if something goes wrong at kidding time, such as the doe having trouble during labor, the newborn kids suffering frostbite in freezing weather, or kids getting trampled by anxious herd mates or battered to death by a suspicious first-time mother.

As each doe is bred, note the date and the names of the doe and the buck. You might jot this information on a calendar in the goat barn and transfer it to a notebook or computer database at the end of breeding season. If you have your doe bred by a registered buck and you plan to register the kids, you will need a formal service memo signed by the buck's owner that shows the date, the names of the doe and the buck, their registration numbers, and the owners' names. The organization that registers your breed supplies pads of preprinted service memos to buck owners. Since the owner will be busy with the buck and may forget to furnish this record, be sure to ask for it at the time your doe is bred.

When a doe has been successfully bred and becomes pregnant, she is said to have settled. A doe sometimes has a personality reversal a couple of weeks after settling; if she was shy she may become friendly, and if she was friendly she may become standoffish. It's a hormone thing.

A doe that does not settle usually comes back into heat on her next cycle. If she settles, she will not come back into typical heat. She may show some signs of estrus when her next cycle is due, but they won't be as strong as usual. If you put her together with a buck at that time, she will display little interest in him. For the next 150 days, the doe will spend much of her time sleeping. You haven't lived until you hear a pregnant doe snoring.

Gestation & Kidding

The gestation period for goats is approximately 150 days, give or take two weeks. First-time mothers often kid early, especially if they are carrying more than one kid. A doe that kids late may give birth to extra-large kids and may therefore have trouble in labor, so become familiar with assistant methods.

Keep a record of the gestation period for each doe; you will find that each follows her own pattern, which will help you predict pretty precisely when she is likely to kid next time. Note the time of day as well as the date, since each doe tends to be fairly consistent whether she gives birth in broad daylight or in the wee hours of the morning.

Gestation Checklist

Date of breeding ______________

Signs of return of estrus (17–23 days) ______________

Dry off (90 days) ______________

Start increasing feed (100–120 days) ______________

Inject selenium and vitamins A and D (135 days) ______________

Kidding day (150 days, more or less) ______________

Managing the Expectant Doe

A dairy doe can be milked up to two months before she is due to kid, then should be dried off so her body can rest. Drying off consists of discouraging milk production by putting the doe on a maintenance diet and by no longer milking her. Some does dry off on their own. Others will stop producing milk within a few days after you stop milking. A really good milker may continue producing milk. Relieve the doe of the excess milk about one week after you stop the regular milking. Do not continue milking her regularly, because removing the milk from her udder encourages further milk production.

As kidding time draws near, you will need to attend to a few management chores.

Adjust the doe's diet. Using the guidelines in the section on feeding goats (see page 184), adjust the doe's diet to ensure she gets hay of the proper quality and the right amount of concentrate for her age, breed, and condition.

Give vitamins and minerals. Unless you live in an area where the soil is high in selenium, give each doe a selenium injection one month before she is due to kid. If your herd does not have access to fresh forage, also give each doe an injection containing vitamins A, D, and E. Obtain the doses from your veterinarian or the owner of a large goat herd. If you have never given an injection, ask a veterinarian or experienced goat handler to show you the proper procedure.

Crotch the doe. Before the due date, use electric clippers to trim the hair from the doe's udder and tail and beneath her tail. This procedure, called crotching, makes the doe more comfortable and makes kidding cleaner. Crotching lets you more easily watch the doe's udder development and helps newborn kids find her teats, especially if the doe has a shaggy coat. Crotching also makes it easier to clean postkidding fluids from the doe's hindquarters and udder.

Clip your own nails. As kidding time gets close, keep your fingernails cut short. If you have to help out, long fingernails may scratch and injure the doe internally.

Prepare a kidding stall. Have a clean stall ready with fresh bedding where the doe can kid in a relatively sanitary environment. Straw or waste hay makes

Before your doe kids crotch her by trimming hair from her tail and udder.

better bedding than shavings, which tend to stick to a wet newborn kid and can scratch the eyes.

Make sure the water bucket is positioned where a kid can't be accidentally dropped in and drowned. The best place for the water bucket is outside the stall, so the doe has to drink through a head hole.

Arrange for help. If this kidding will be your first, try to find an experienced goat or sheep keeper who is willing to be on hand at kidding time. That person can at least reassure you everything is going fine (which it usually is). If you can't find someone to be on hand in person, find one who is willing to offer advice over the phone at a moment's notice (and at any time of day or night) and come right away should you feel you need help.

Assemble kidding supplies. Before kidding starts, gather the following supplies and keep them in a handy place. After each doe kids, replenish your supplies in preparation for the next kidding. You don't want to be running around looking for things when a doe starts to kid; you need to be there to watch. If you keep your supplies in a clean container with a lid, you can leave them at the goat barn. Store medical items that might freeze in a separate smaller container, such as a lunch bucket, and leave it where you can grab it as you hurry out the door.

- Paper and pencil to write down the date and time of kidding, order of birth, and any unusual events or problems
- A pair of overalls and a washable jacket (kidding can be messy)
- A snack for yourself, in case you're at the goat barn a long while and can't or don't want to leave to get something to eat
- A large box to put the newborn kid in
- Plenty of old towels or large absorbent rags to line the bottom of the box and dry the newborn kid
- A hair dryer to dry the kid fast in cold weather
- A pet heater panel in case a kid is weak or sickly (needed only in freezing weather)
- Tamed iodine (such as Betadine) to coat the kid's navel
- Soap and a container to hold warm water to wash your hands
- Long surgical gloves (available in drugstores) in case you have to reach inside the doe
- Water-based lubricant (such as K-Y jelly) to lubricate your hands in case you have to reach inside
- Two uterine boluses of antibiotic (available from a veterinarian or farm store) per doe to prevent infection in case you have to reach inside
- A stack of old newspapers to wrap the afterbirth for disposal and, if there's a second kid, to keep it from landing in the mess from the first kid
- A scale so you can weigh each kid as soon after birth as possible
- A washcloth to clean the doe's udder after she kids
- A bottle and rubber nipple in case you need to hand-feed a kid with a weak sucking instinct
- Molasses to mix with warm water to feed the doe after she gives birth
- A bowl or bucket to hold the warm molasses water
- A handful of raisins or roasted peanuts to reward your doe

It's Almost Time

When a doe is due to kid, separate her from the other goats. The stall where the herd lives is probably not clean enough for a newborn kid. The other goats may panic at the sight of a baby and butt or trample it. And the other goats may become curious when birthing starts, crowding around and causing the doe distress. The new mother needs space to bond with her new kids without being disturbed.

Signs that kidding is near may be distinct or undetectable. A few days before a doe kids, her udder may swell and look shiny; however, some does won't swell with milk until after the kids are born. The hairless area beneath the doe's tail may redden and swell. The doe may show little interest in eating.

A doe carries kids on her right side. When her time for kidding draws near, you can put your hand on her side and feel the kids moving. As long as you can still feel them, they probably won't be born for at least 12 hours.

Just before birth, the kids will move back toward the birth canal. This shift causes the areas around the doe's tail and hips to look bony. Her muscles may loosen so much she can't hold her

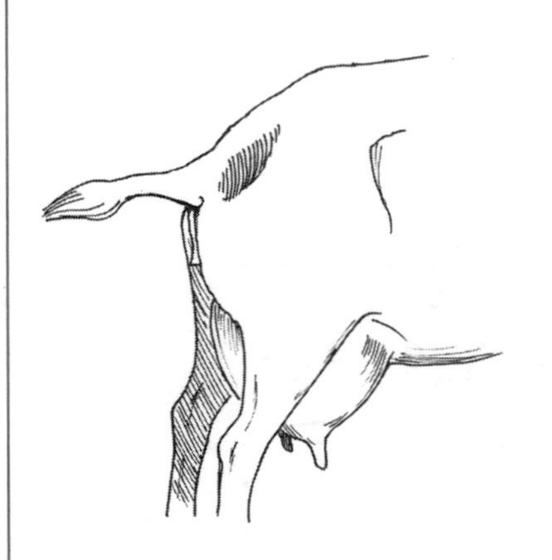

The ligaments in a doe's hindquarters become increasingly relaxed as kidding approaches.

tail down, so it remains in an upward position. The doe will probably become extremely affectionate toward you, perhaps nickering when you approach her and trying to lick your hands and face. She will discharge thick mucus. She will get restless. She will paw the ground and repeatedly lie down, then stand up.

Pregnancy Toxemia

If a doe stops eating late in pregnancy, she may have pregnancy toxemia, also known as ketosis. This condition occurs when a doe draws energy from her own body to feed her developing kids. It is seen most often in first-time young mothers, in does carrying multiple kids, in extremely fat does, and in does that have not received sufficient nutrition throughout pregnancy.

As a result of pregnancy toxemia, the doe becomes progressively weaker, may wobble when she walks, has trouble getting up, and may appear lame. A good preventive is to keep a bag of feed-grade dry molasses on hand to sprinkle on top of the concentrate of any pregnant doe that shows signs of going off feed. If a doe becomes weak, stronger measures are needed to keep her alive. Give her 2 to 3 ounces (59 to 89 mL) of propylene glycol twice a day until she kids. Pregnancy toxemia can be prevented through proper nutrition.

It's Time!

Most kids are born without human help. Chances are good, in fact, that you will go out to check on a doe only to find she has already given birth. It's a good idea to try to be there during the event, in case the doe needs help. However, unless a problem occurs, let your doe kid on her own. The most she will likely need from you is moral support. Some does seem to wait for the comfort of your arrival before they start giving birth. Others seem to deliberately wait until you leave.

When a doe lies down and starts straining and groaning, she is in labor. Soon the water sac will emerge; do not break it. When it breaks on its own and fluid spills out, a kid will soon follow. Usually the first thing you see is a pair of tiny white hooves. Then you will see a little nose resting on the hooves. If you see two hooves and no nose, don't worry. The kid could be coming out back legs first, which is perfectly normal — the birth will just take a little longer.

Once you see the hooves, the doe will strain a few more times and out will come a newborn kid. The doe should lick her kid, which stimulates it to breathe and creates a bond between mother and offspring.

How many kids a doe has depends on her age and breed. An older doe usually has more kids at a time than a doe giving birth for the first time. A Pygmy usually has twins. Most other breeds have either twins or triplets. Nubians and Kinder goats may have four or five kids at a time. A doe that is herself a twin, triplet, or quadruplet is more likely to have twins, triplets, or quads. Flushing a doe before breeding increases her chances of producing multiple kids.

A goat sometimes experiences an early labor pain.

Signs That Kidding Is Near

- The areas around tail and hip look sunken.
- The udder swells with milk and looks shiny.

If the doe starts straining again after the first kid emerges, dry the first kid and put it into a clean box out of harm's way. You don't want it to be trampled or rolled on while the second one is being born. Scatter fresh bedding over the mess left by the first birth, or spread newspapers behind the doe to give the next kid a clean place to fall.

Stay with your doe until you are certain she has had all her kids. You don't want to come back later to find that a kid has died because the doe was so busy taking care of the first one she didn't clean off the second one's face to help it breathe.

The birth of kids is exciting but can cause anxiety the first time around. Talking to your doe and repeating her name in a reassuring tone will help keep both of you calm. No matter how nervous you are, stay as calm as possible so the doe will remain calm as well.

When a Doe Needs Help

If the doe is straining without producing a kid, and you see two hooves but no nose, the kid's head may be folded back. You might see a nose and only one hoof, or no hooves, because one or both legs are folded back. In these cases, the doe needs your help. Here's what to do:

1. Wash the area under the doe's tail with soap and warm water.

2. Scrub your hands and arms with soap and water and put on surgical gloves, if you have them. Lubricate your hands with water-based lubricant (K-Y jelly) or liquid soap.

3. Reach inside the doe and feel for the kid's feet and nose.

- If the kid's head is turned back, push the feet back until you have room to move the head forward.
- If one or both legs are folded back, push the kid back and try to bring both legs forward, one at a time. Cup your hand around each hoof so it doesn't tear the doe.
- Take care to ascertain exactly what it is you are feeling. If the doe is carrying more than one kid, they may be tangled together. Follow each leg to the body so you know both legs belong to the same kid. If two kids are coming through at once, push one back to give the other more room.

4. When you have one kid in proper position, take hold of both of the kid's legs and pull gently when the doe strains. Pull *only* when the doe strains. Do not attempt to pull out the kid when the doe is not straining.

5. To prevent infection due to your intrusion, after all the kids are born, scrub your hands, take two uterine boluses, reach inside the doe again, and deposit the boluses.

If you fail to get things sorted out or your doe strains for more than 45 minutes without giving birth, immediately call for help. A doe that has trouble giving birth may die, her kids may die, or all may die.

After the Birth

Thirty minutes to 12 hours after a doe gives birth, she will pass the afterbirth — a mass of bloody tissue. Most does produce only one afterbirth, but occasionally a doe will pass two. The doe may try to get rid of the afterbirth by eating it. Even though goats are vegetarians, eating the afterbirth is a normal and instinctive act designed to avoid attracting predators to the birth site. If you find the afterbirth before it is eaten, wrap it in newspapers and dispose of it.

Never pull on afterbirth while it is hanging from the doe. It may still be attached, and you could cause the doe to bleed to death. If the afterbirth does not fall out on its own within 24 hours, call the vet.

Giving birth causes a doe to lose a lot of fluids, and she will probably be thirsty afterward. Give her a bucket of warm water with a little molasses mixed in. Besides replacing lost fluid, warm water will help her relax and pass the afterbirth, and the molasses will provide needed energy.

As a reward for a job well done, offer the doe a handful of raisins or roasted peanuts. Feed her as usual, but don't worry if she doesn't eat right away. Depending on how hard she labored and how many kids she produced, she may be tired and prefer to rest. Some goats, however, have their heads in the hay manger the moment kidding is over.

The new mother will discharge bloody fluid for up to two weeks after kidding. Keep her comfortable by cleaning her tail and udder twice a day. Use warm soapy water, pat her dry with a paper towel, and coat the hairless parts with dairy balm (such as Udder Balm or Bag Balm) if they appear dry and chafed.

Breeding and Birth Record

No. ______

On (date) ______________ Buck (name of buck) ______________
Registration No. ______________ Registry ______________
Was Bred to Doe (name of doe) ______________
Registration No. ______________ Registry ______________
Owned by ______________
Address ______________
____

Signed (buck owner) ______________
Date of Birth ______________
Name of 1st Kid ______________ Sex ______________
Description ______________
____
______________ Tattoo: L ______ R ______
____
Name of 2nd Kid ______________ Sex ______________
Description ______________
____
______________ Tattoo: L ______ R ______

Normal Kidding Positions

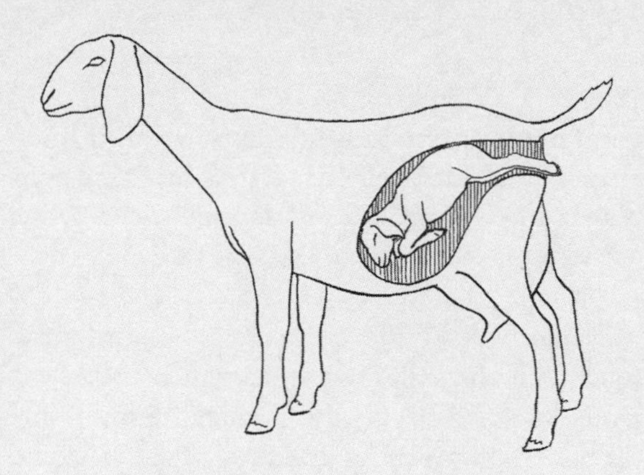

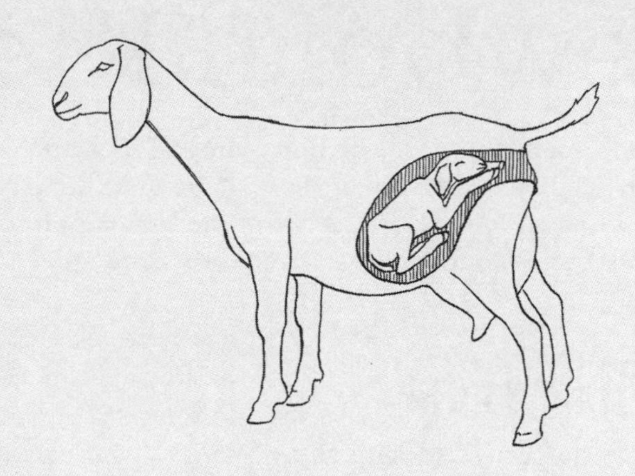

Problem Kidding Positions

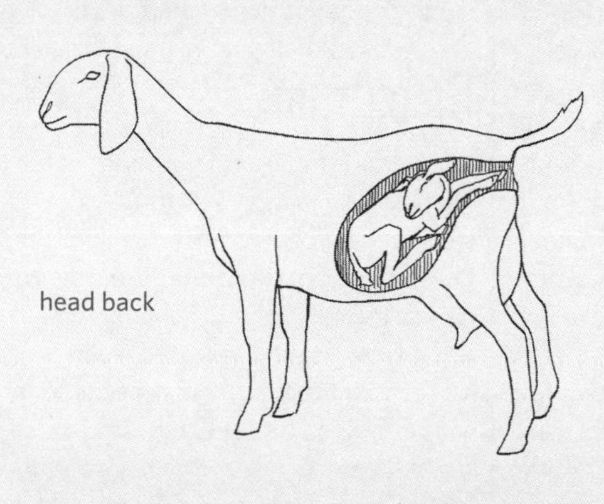

head back

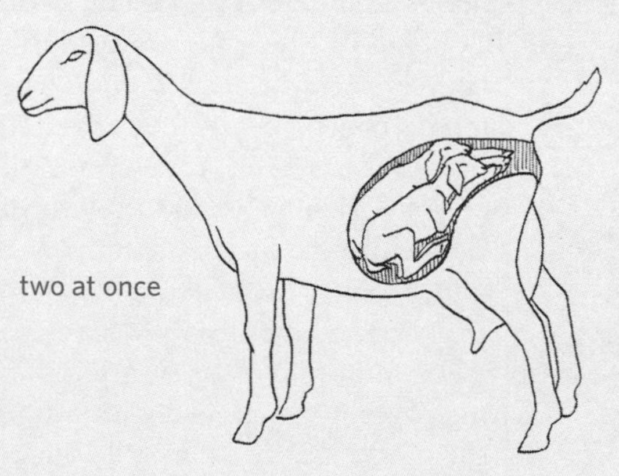

two at once

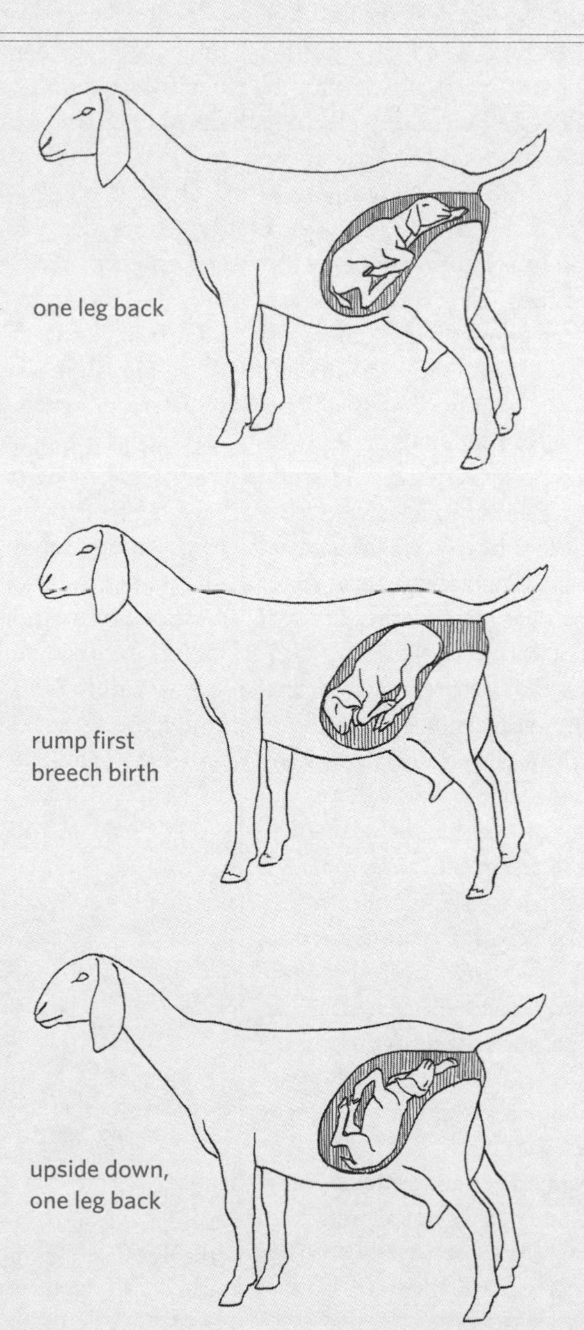

one leg back

rump first
breech birth

upside down,
one leg back

Newborn Kids

As each kid is born, note the date, time of day, and duration of labor. Weigh the kid and jot down its birth weight and sex. When a doe has two or more kids, note any identifying marks that will help you remember their order of birth. If two kids look exactly alike, put temporary marks inside their ears with a marking pen until you can identify some distinguishing feature. You may choose to name them at this time.

First Things First

A kid is born covered with slimy mucus. The doe licks the kid's face to remove the mucus so it can breathe. If the doe is distracted by another kid on the way, help her out by cleaning the first kid's nose and mouth. If the kid does not start breathing immediately, tickle the inside of its nose with straw. If you hear rattling sounds when the kid breathes, lift it by its back legs so mucus can easily drain from its airways. You will know the kid is okay when it bawls, an indication that it has gotten enough air into its lungs.

In cold weather, dry the kid rapidly with old towels or clean rags so it won't become chilled. If the weather is freezing, complete the job with a hair dryer or towels warmed in a clothes dryer. If you find a newborn kid that is not moving or feels cold, warm it up in a hurry. A kid that is shivering but otherwise seems fine can be warmed by holding it inside your jacket. A kid that is not breathing or moving can be warmed by soaking it in warm (not hot) water for 15 minutes, then drying it thoroughly and placing it on a heating pad or under a pet heat panel while it revives. Even a kid that appears to be dead can be brought around, so don't give up too easily. A kid that has been cold or wet for too long, on the other hand, may not revive.

Do not put a kid under heat unless it is premature, sickly, or suffering from extreme cold. As soon as a newborn kid is dry, its body should adapt to normal temperatures. Using a heater prevents healthy kids from adapting and therefore does more harm than good. As long as the kids have a draft-free place to curl up and sleep, they should be fine. In the coldest weather, a small doghouse provides the kids a place to sleep where their own body heat will keep each other warm.

A good, thick layer of clean bedding helps kids conserve body heat. At temperatures of 50°F (10°C) or less for newborn kids (32°F [0°C] or less for month-old kids), the bedding must be thick enough that when a kid curls up to sleep it is burrowed in so well that its legs are hidden by the bedding.

A newborn kid has a bloody umbilical cord hanging from its navel. If the cord is long enough for the kid to step on, reduce its size by cutting it with a clean, sharp knife or scissors a few inches from the body, where the cord starts to thin. Do not cut through the cord close to the body, where it is thick and pink, and do not pull the cord off. To prevent bacteria from invading through the navel area, coat the cord and navel with a tamed iodine solution, such as Betadine, or chlorhexidine (Nolvasan) solution.

Within minutes of entering the world, the newborn kid will struggle to its feet and search for its first meal.

Keep the doe and her kids apart from the herd for a few days while they learn to respond to one another's calls. The kids need time to learn where the teats are and to grow strong enough to scamper away when they get butted for trying to steal milk from the wrong doe.

Colostrum

The first milk a doe produces after kidding is called colostrum. It is thicker and yellower than regular milk because it contains extra nutrients. It also contains antibodies that help protect a newborn kid against disease.

Some kids are ready to nurse as soon as they can stand. Others want to rest first. A newborn kid can absorb antibodies from colostrum for only about 24 hours, by which time it should have taken in several helpings. Don't let more than 2 hours go by without ensuring that a kid gets its first colostrum.

The easiest way to feed colostrum is to let the kid nurse. Clean the doe's udder with a clean cloth and warm, soapy water, and dry it with a paper towel. Make sure the teats are functioning by gently milking a stream from each side. Getting the milk started also removes the waxy plug that sometimes seals each teat, making it easier for the kid to start nursing. If the doe is a heavy milker, the pressure from her milk-swollen udder may already have caused the plugs to pop out. A really heavy milker may be so swollen with milk that she feels uncomfortable and kicks when her teats are touched by you or her babies. In such a case, milk out just enough to relieve the pressure and help her relax. You may need help holding the doe still, but after she gets the idea that milking makes her more comfortable, she'll learn to stand still on her own.

With the teats cleaned and dried, squirt a bit of colostrum into each kid's mouth. Once the kid gets a good taste, it should start sucking right away.

Bottle feeding is time consuming, but some goat owners prefer it and sometimes it can't be avoided.

Bottle-Feeding

Bottle-feeding can be a nuisance, since kids need to eat every few hours around the clock. The milk must be warmed before each feeding, and the bottles and nipples must be scrupulously cleaned afterward. These conditions may not sound intimidating until it's 2 a.m. on a cold winter morning.

Sometimes you have to bottle-feed kids because the doe won't let them nurse. She may try to get away from the kids or butt them, injuring or killing her own offspring. A doe that will not accept her kids was probably herself raised on a bottle. Think about the long-term ramifications before starting your kids on bottles.

Despite the drawbacks, some goat owners choose to bottle-feed their babies. One reason is to keep their goats friendly and people-oriented. Kids raised by hand are like puppies — always happy to see you. Kids raised by a doe, on the other hand, tend to be shy unless you take time *every day* to handle each one.

Another reason to bottle-feed kids is to keep a share of the doe's milk for yourself. This reason is popular among makers of cheese. An easier alternative is to let the kids nurse at will for the first two weeks, then house the doe at night separately from the kids. Milk the doe in the morning for yourself, then put the kids in with her so they can nurse the rest of the day. Dam-raised kids grow better and are usually healthier.

Yet another reason to bottle-feed kids is to avoid spreading caprine arthritis and encephalitis (CAE), a virus for which no cure and no preventive measures are known except to break the disease cycle by preventing its spread from one generation to the next. The best way to avoid this virus is to purchase certified CAE-free goats. A doe that is not certified may be infected with the virus and pass it on to her kids. Kids usually get CAE by nursing from an infected doe, although they

Weak Kids

Sometimes a kid is born too weak to stand or has a weak sucking instinct. If a kid can't or won't suck, milk the doe and put the colostrum into a small flexible bottle, such as a clean single-serving plastic soft drink or water bottle. Fit the bottle with a nipple, such as the Prichard Teat designed for this purpose and available at some farm stores and all mail-order goat supply outlets. Warm the filled bottle in hot water for a few minutes until a squirt of colostrum placed on your wrist feels neither hot nor cold.

Put your finger into the kid's mouth and rub gently back and forth along the length of its tongue. As soon as you feel the kid attempt to suck, squeeze a little warm colostrum onto the back of its tongue and stroke the throat until it swallows. Keep at it until the kid drinks ½ cup (118 mL) of colostrum. After a nap, the kid should be stronger and ready to nurse on its own. If not, keep at it and eventually the kid should come around.

may be infected before they are born or through contact with infected goats after birth. The virus generally causes arthritis (inflammation of the joints) in older goats and, rarely, encephalitis (inflammation of the brain) in kids, but does not affect humans.

Bottle-feeding kids colostrum and milk known to be CAE free helps break the disease cycle among noncertified goats. Colostrum and milk from cows, as well as milk replacer available from the farm store, are virus free and can be fed to kids instead of goat milk. Kids raised on cow milk or milk replacer, however, don't grow as well as kids raised on goat milk. CAE is more likely to be transmitted through colostrum than through milk. Goat colostrum and milk from noncertified does must be heated to destroy the CAE virus.

Each bottle-fed kid should get milk or milk replacer amounting to not less than 15 percent nor more than 25 percent of its body weight each day. Kids kept at the maximum of 25 percent of their body weight will grow fast and sleek. Kids kept closer to the 15 percent minimum will start nibbling on hay and other forage earlier and be easier to wean.

To make sure kids are getting the right amount of milk, weigh them at least every other day and increase their ration accordingly. Divide the total daily amount into four or five evenly spaced feedings for the first couple of days. By the time the kids are 1 week old, they can be fed three or four times daily. Gradually work down to two feedings per day at evenly spaced intervals.

Kids can be fed from individual bottles or from a community bucket with several nipples. Using separate bottles ensures that each kid gets the right amount. If you have more kids than you can feed with a bottle in each hand, you can make or purchase a rack that holds multiple bottles, allowing all kids to nurse at the same time. A bucket is easier to handle than multiple bottles and reduces the amount of time needed to clean the bottles, but slow eaters may not get their full share.

To keep track of feeding times and amounts, make a chart showing what time the kids are scheduled to be fed and how much each should get. Update the chart each time you weigh the kids. Place a check mark on the chart after each feeding — an important procedure if you are busy and forgetful, or if more than one person helps feed the kids.

Heat-Treating Colostrum and Milk

Colostrum is sensitive to heat. If it gets too hot, it turns into a thick pudding and its antibodies are destroyed. To avoid that eventuality, heat the colostrum to 135°F (57°C) and hold it at that temperature for an hour.

The easiest way to properly heat colostrum is to put it into a double boiler (to prevent scorching), clip on a candy thermometer, and stir steadily. Have a thermos bottle filled with hot water ready. While heating the colostrum, do not remove the thermometer or stirring spoon and then return it to the pot, or you may reinfect the colostrum. When the colostrum reaches 135°F (57°C), pour the water out of the thermos and pour in the colostrum. Screw on the lid and wrap the thermos in towels.

After one hour, open the thermos and test the colostrum with a clean thermometer. The temperature must be no less than 130°F. If the temperature dips below 130°F before one hour is up, you'll have to start over.

Milk is far easier to treat than is colostrum, since pasteurization destroys the CAE virus. (See page 208 for directions for pasteurizing milk.) If you opt to buy an electronically controlled pasteurizer, the modern ones are equipped to also heat-treat colostrum.

Cool heated colostrum or milk to 100°F (38°C) before feeding it to the kids. The temperature is just right when you can't feel a drop placed on the inside of your wrist.

Kids can be fed from a community bucket with several nipples.

Weaning

A kid is born with only one of its four stomach chambers functioning — the abomasum. A newborn kid therefore digests milk like a puppy, kitten, or human baby with a single-chamber stomach. By the time the kid is 1 week old, it begins nibbling on hay, grain, and grass. The more solid feed it eats, the more quickly the three other chambers develop. Keeping bottle-fed kids on the hungry side encourages early eating of solid foods, provided free-choice hay or browse is available. As soon as your kids start eating solid foods, make sure they have access to clean drinking water at all times. They may not drink any at first, but they should have the option to do so.

How much concentrate kids need is influenced by the quality of their hay; the lesser the quality, the more concentrate they need for proper nutritional balance and growth. The proper amount for each kid also depends on the animal's size and growth rate. Cut back on the concentrate if a kid is getting too fat — if you can't feel the ribs of a large-breed kid or you can grab a handful of flesh from behind the elbow of a miniature breed. Like milk, concentrate should be divided into two evenly spaced feedings. As a general guideline, by weaning time, regular-sized kids can be gradually worked up to 1 pound (500 g) of concentrate per day, miniatures to ½ pound (250 g).

Most kids no longer need milk by the time they reach three times their birth weight or 8 weeks of age, whichever comes first. Weaning a bottle-fed kid encourages early rumen development and frees up your time for other things. To start weaning bottle-fed kids, substitute water for a small portion of the milk. Gradually decrease the amount of milk and increase the amount of water. The kids will be weaned without even noticing.

If your kids nurse, weaning can be upsetting for the kids, the doe, and you. When you separate a kid from its mother, both parties will carry on dramatically. They won't get so upset if you put them in side-by-side stalls where they can see each other. After a few weeks, the kids will be weaned and can be put back with their mother.

Unless you need every drop of milk a doe produces, you do not have to wean nursing kids. As they grow and their need for milk lessens, the mother becomes less interested in feeding them. Eventually the kids will wean themselves without any distress to all parties concerned. Best of all, kids raised for meat will grow faster and weigh more than kids that have been weaned early.

Tracking Weight

The average newborn kid weighs about 7 pounds (3 kg). Doelings may weigh less, bucklings more. Triplets and quads weigh less than twins or singles. Miniature kids weigh about half as much as full-size kids. Weigh each kid at birth, as soon as it is dry but before it has its first meal. Track each kid's growth by weighing it every other day for the first four weeks, then once a week until it reaches maturity. Record weights on a calendar, or make a chart to record the dates and weights.

After the first week, a kid should gain ¼ to ½ pound (125 to 250 g) per day. Some grow faster than average; some grow slower. Except briefly during weaning, at no time should a kid lose weight.

If a kid fails to gain weight or loses weight, look for a reason. Perhaps it is not getting enough milk. If the kid is nursing, make sure the mother is producing sufficient milk. Check her udder: Perhaps a teat is plugged, or the udder is infected and sore, causing the doe to push the kid away when it tries to nurse.

If the doe is nursing more than two kids, she may not have enough milk for them all. A strong kid may push a weaker kid aside at nursing time. In that case, you may have to bottle-feed the slow-growing kid.

If a kid you are bottle-feeding grows slowly, you may not be feeding it enough. Bottle-feeding charts apply only to the average goat. In real life, no average goats exist. Make adjustments to suit the needs of your individual animals.

Weighing Kids

You can weigh your kids on your dairy scale. Place a kid in a grocery bag with handles and hang the bag from the scale. Be sure to keep the kid's head outside the bag so it can breathe.

When the kids get too big for a grocery bag, use your bathroom scale. Pick up a kid and hold it while you stand on the scale. Then weigh yourself without the kid. Subtract the second number from the first; the difference is the kid's weight.

Scours

A kid's first bowel movement is black and sticky. Then, for the next few days, the kid passes yellow, pasty material that may stick to the hair around its

Kids should gain weight steadily and start nibbling pasture plants and other solids by 1 week of age.

hind legs and rear end and need to be cleaned off. When a kid is about 1 week old, it starts dropping small brown pellets that resemble the droppings of mature goats, only smaller. Droppings that are loose and white, light yellow, or light brown indicate diarrhea.

Diarrhea, or scours, is a fairly common problem, especially during a kid's first few days of life. It more often strikes bottle-fed kids than kids that nurse naturally. Scours may be caused by chilling, erratic feeding, dirty bedding, dirty milk bottles, overeating, and milk that is too rich. If the problem is not corrected immediately, the kid will die.

Avoid scours by washing bottles and nipples after each feeding and rinsing them in warm water mixed with a splash of household chlorine bleach. Feed kids small amounts at a time, and space feedings evenly throughout the day so they don't get too much milk at one time.

If a kid gets diarrhea, stop feeding milk. Substitute an electrolyte fluid in an amount equal to the amount of milk the kid would otherwise drink. Some goat supply outlets sell electrolytes mixed specifically for goats. In a pinch, you can mix up a homemade electrolyte drink from the following ingredients:

- 2 tablespoons (30 mL) table salt (sodium chloride)
- 2 tablespoons (30 mL) Lite Salt (potassium chloride)
- 2½ gallons (7.5 L) water

The scours should clear up within two days of substituting electrolyte fluid for milk. If you have to continue with the electrolyte for more than 24 hours, add ¾ cup (177 mL) corn syrup (*not* sugar) to the above formula. After 48 hours, if the scours don't clear up, consult your veterinarian. The kid may have a disease that requires treatment.

Scouring in a 3- or 4-week-old kid is probably due to coccidiosis, caused by microscopic parasites called coccidia that are always present in the soil. Properly managed kids are exposed to coccidia gradually and develop immunity. Kids that live in filthy conditions or that must drink water with manure in it are exposed to too many parasites at once. The main sign of coccidiosis is diarrhea, which is sometimes tinged with blood. Treatment is with a coccidiostat; this medication is available from farm stores, goat supply outlets, and veterinarians. Even if a kid recovers from coccidiosis, it may never grow to be fully productive. Since coccidia flourish in warm, humid weather, careful scheduling of breeding helps prevent infection. Manage your does to give birth during the cold days of winter or early spring, which gives the kids time to grow and develop immunity before the warm weather arrives.

Take care not to go overboard by feeding your kids too much milk at once, or they will get diarrhea. Diarrhea, also known as scours, is a dangerous cause of weight loss.

Identification

Even if your goats are not registered, each animal should have a unique means of identification to help you sort out the look-alikes and help your veterinarian keep track of health tests. Federal regulations require the identification of goats in certain categories (see the box above). If you plan to pedigree or register your goats, you need a way to positively identify each animal. Different registries recommend different systems of identification, so before you mark your goats, check with the appropriate breed organization. Some registries recommend using two methods to ensure that at least one endures for the animal's life. The most common forms of identification for dairy and meat goats are tags, tattoos, and microchips.

Identification Regulation

The United States Department of Agriculture (USDA) requires certain categories of goat to bear some means of identification, especially if they are involved in interstate travel. The purpose of this program is to eradicate a relatively rare disease called scrapie.

Ear tags are the preferred method of identification, although some states allow tattoos. For details, contact your state's USDA Animal and Plant Health Inspection Service office.

No system of identifying the goats in your herd makes sense unless you keep careful records for each animal. Your records should include information and identification details not only for the goats currently in your herd but also for those you no longer have, including how they were disposed of, such as to whom they were sold and when.

Ear Tags

Ear tags are commonly used for meat goats. Tags are sold at farm supply stores and come in plastic or metal and in a variety of shapes and sizes. Goats are so active that some styles are easily torn off, injuring the goat's ear. The safest tags for goats are round button tags that attach to the center of the ear, where they are less likely to be torn off than tags that attach to the edge. Another option that is easier to read than a button tag is a small rectangular tag that clips to the front of the ear.

Ear tags come in various colors and may be preprinted or blank. Depending on the manufacturer, preprinted tags may be numbered sequentially (you assign the next available number to the next kid born) or may include other

Ear tags come in a variety of shapes and types. Avoid large blood vessels and ridges of cartilage when inserting a tag of any type.

information, such as your name or initials, your herd or farm name, or your city and state. If you use blank tags, you must mark them yourself with an indelible marker. Over time, the information wears off; unless you check often, the original information may be lost before you get around to renewing the markings.

For your goats to comply with federal and state scrapie regulations, you must use USDA-approved ear tags. Along with any other information you wish to include, each tag must bear your premises number (which you will be assigned) and the animal's unique identification number.

Neck Tags

Neck tags are preferred over ear tags for dairy goats, as they tend to be more active than meat goats and therefore more likely to tear an ear. Neck tags are similar in appearance to ear tags, but instead of attaching to the goat's body they attach to the collar. If the goat loses its collar, the tag is gone, which generally isn't a problem unless two or more look-alike goats lose their collars at once, which is possible but not likely.

Tattooing

Tattooing as a means of permanent identification is used for dairy goats and sometimes in combination with ear tags for meat goats. Tattoo kits with goat-size numbers and letters are available from any goat supply catalog. Instructions come with the kit. Additional instructions can be obtained from registry associations. All dairy breeds, except LaMancha, are tattooed in the ear. Since LaManchas have no external ears, they are tattooed in the tail web, the hairless area underneath the tail.

A common system for dairy goats is to tattoo the right ear with three letters designating the herd name. If you join the American Dairy Goat Association, three letters will be assigned to you. If your herd designation is XYZ, for example, all your kids will have "XYZ" tattooed in their right ears.

The left ear has a letter indicating the year of birth and a number indicating the kid's birth sequence. All kids born in 2015, for example, would have the letter *F*. The first kid born in your herd in that year would be "F1."

If your goats must comply with federal and state scrapie regulations, you may be allowed to identify them with tattoos. Regulations vary from state to state, so find out what regulations pertain in your state.

If you expect few kids each year, you may not wish to purchase your own tattoo kit or ear tag applicator. Goat clubs sometimes hold demonstrations where you can take your kids to be ear tagged or tattooed. A neighboring goat keeper may be willing to ear tag or tattoo your kids for a small fee.

Tattoo Year Code

A = 2010	N = 2021
B = 2011	P = 2022
C = 2012	R = 2023
D = 2013	S = 2024
E = 2014	T = 2025
F = 2015	V = 2026
H = 2016	W = 2027
J = 2017	X = 2028
K = 2018	Y = 2029
L = 2019	Z = 2030
M = 2020	

(To avoid confusion, G, I, O, Q, and U are not used.)

A finished tattoo identifies a goat for life.

Microchipping

Embedding a microchip can be done instead of tattooing, or in addition to tattooing, with the chip number corresponding to the tattoo. The microchip is inserted with a needle (like giving the goat a shot) and can be placed on the goat's head near an ear or in the goat's tail web. If the goat has a tattoo on the tail web, the chip is inserted into the end of the tail. As an alternative to placing the microchip under the goat's skin, some ear tags come with embedded microchips.

The microchip system has two parts, the transponder or chip itself, and a scanner or reader. This system is also called radio frequency identification (RFID) because the reader generates a low-power electromagnetic field that activates the chip to send a radio signal back to the reader. When the reader receives the goat's identification code from the chip, it converts it into digital form and displays it on an LCD screen. One of the problems with microchipping is that scanners are not cheap and not all scanners read all chips.

Disbudding

All kids are born without visible horns. Some kids have horn buds that develop into horns. Other kids are polled, meaning they have no horn buds. You can tell the difference between a horned kid and a polled kid from birth. The wet hair on the head of a polled kid lies smooth, whereas the hair on a horned kid is twisted in a whorl at the two spots on its head where horns will grow.

Wild goats use their horns for protection. When goats are raised in a barn, their horns can become dangerous weapons. Without meaning to, a goat with horns can injure herd mates, or you, simply by lifting or turning its head at the wrong time. A kid butts its mother's udder while nursing, and

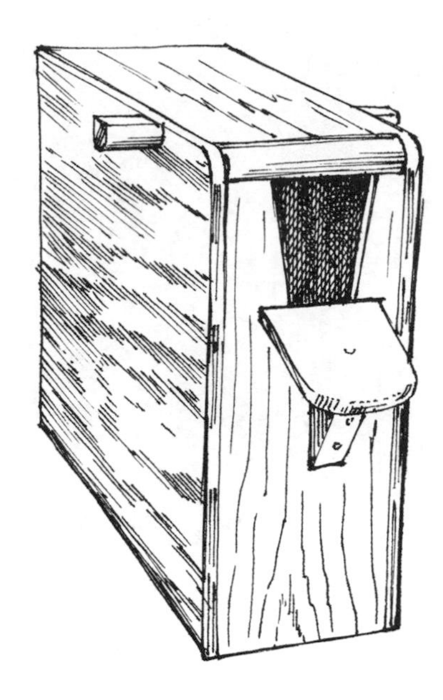

A disbudding box is used to confine a kid for tattooing and disbudding.

pointy emerging horns can bruise or otherwise injure the udder, perhaps leading to mastitis (inflammation of the udder). For these reasons, a kid born with horn buds should be disbudded (unless you plan to bottle feed it and butcher it young).

Disbudding should be done as soon as you see little horns, since the procedure becomes more difficult as the horns grow. Different breeds grow horns at different rates. Most dairy goat kids are ready for disbudding between the ages of 1 and 2 weeks. Older goats can have their horns surgically removed by a veterinarian, but the procedure is painful and the resulting wound takes a considerable amount of time to heal. Disbudding is painful in young kids as well, but the kids forget so quickly afterward that within a few minutes they may resume butting heads in play.

Disbudding requires a disbudding iron and a box to hold the kid. For just a few kids each year, you may wish to invest in these items jointly with another goat keeper, or you might seek a nearby goat keeper who is willing to disbud your kids for a small fee. Often the same person who tattoos kids will also disbud them.

Do not attempt to disbud a kid without first having an experienced person show you how. Use the following tips to help the kids through the ordeal:

- Give each kid a baby aspirin, or half an adult aspirin, to dull the pain.
- Place a small bag of crushed ice on the kid's head immediately after both buds have been burned.
- If you are disbudding bottle-fed babies, comfort each one with a small amount of warm milk as soon as you remove the baby from the disbudding box.
- To prevent tetanus infection, give each disbudded kid an injection of tetanus antitoxin — 500 IU under the skin. One bottle holds 1,500 IU, or enough for three kids. Tetanus antitoxin is available from any farm store or veterinarian.

Castrating Bucklings

Unless you have a prime buckling you plan to keep or sell for breeding, any buckling kept past weaning age should be castrated. Because a castrated buck is called a wether, castrating is sometimes also called wethering. A wether raised for meat grows faster and tastes better. Castrating your bucklings simplifies management, since wethers need not be separated from does or doelings to avoid unwanted pregnancies.

Castrate a buckling as soon as his testicles descend into the scrotum, usually between the ages of 1 and 3 weeks. To do the job, you will need a 9-inch (23 cm) lamb elastrator. Before trying it for the first time, have an experienced person show you how to use it. If you raise only a few kids each year, you might seek a nearby goat or sheep owner willing to castrate your bucklings for you.

Goat Milk

Most people in the United States drink cow milk, yet around the world more people drink goat milk than cow milk. The milk from a properly cared for doe tastes exactly like milk from a cow, although occasionally you may run across a goat that naturally gives off-flavor milk (but the same is true of cows) or a person who claims to be able to tell the difference between goat and cow milk.

Goat milk, like all milk, contains solids suspended or dissolved in water. Goat milk is made up of approximately 87 percent water and 13 percent solids of the following type:

- Lactose (milk sugar), which gives you energy
- Milk fat, which warms your body and gives milk its creamy smooth texture
- Proteins, which help with growth and muscle development
- Minerals, for your general good health

Milking Equipment

Goats are milked in a milk room or milk parlor, which can be built into a corner of your dairy barn or in a separate building. Some people milk their goats in their garage or laundry room. Wherever your milk room is, it should be easy to clean and big enough to hold a milk stand and a few necessary supplies.

Equipment. These items are a one-time purchase:

- Milk stand
- Spray bottle (for teat dip)
- Strip cup
- Stainless steel milk pail
- Dairy strainer
- Milk storage jars
- Pasteurizer (optional)
- California Mastitis Test kit
- Milk scale

Supplies. These items must be replaced as you use them:

- Baby wipes
- Teat dip
- Bag Balm or Corn Huskers lotion
- Milk filters
- Chlorine bleach
- Dairy acid cleaner

A milk stand, homemade or purchased from a dairy goat supplier, gives you a comfortable place to sit while you milk. At the head of the milk stand is a stanchion that locks the doe's head in place so she can't wander away while you're milking. Most people feed a doe her ration of concentrate to keep her from fidgeting during milking, but it's better to train your does to be milked without eating. A doe that's used to eating while she's being milked tends to get restless if she finishes eating before you finish milking.

Your milk pail should be made of quality stainless steel and be seamless for easy cleaning. Seamless stainless steel pails of various sizes are available from many farm stores and pet suppliers (sold as feed buckets) and gardener suppliers (sold as compost buckets) as well as dairy suppliers. Ideally the pail should hold all the milk from one goat at one milking, allowing for extra volume to accommodate the fresh milk's head of foam. In actuality, the size of the goat will determine the size of the pail. A 6-quart (5.5 L) pail is suitable for most full-size does, but a 2-quart (2 L) pail fits better under a Nigerian Dwarf. It may also be easier to maneuver under a full-size goat with a pendulous udder or long, low hanging teats, although you'll have to stop and empty the pail once or twice before you finish milking the doe.

Once the milk is in the pail, it needs to be strained through a milk filter (to remove any bits of hair or chaff that may have fallen in during milking), put into clean jars, and cooled as rapidly as possible. To hold the milk filter and funnel the milk into the jars,

Milking equipment is a one-time purchase.

A milk stand may be used for hoof trimming as well as for milking.

you'll need a dairy strainer, which like the pail should be stainless steel and seamless. For one or two does, a canning funnel of the style known as busy lizzy works fine. For more does, a larger dairy strainer goes faster. Milk strainers and filters are available from dairy supply outlets and some farm stores; use regular dairy filters, as coffee filters don't work well with milk. Quart or half-gallon (1 L or 2 L) canning jars make handy milk storage jars.

Keep all your equipment scrupulously clean to ensure your milk is healthful and good tasting. Every time you use your pail and strainer, rinse them in lukewarm (not hot) water to melt milk fat clinging to the sides. Then scrub with hot water mixed with liquid dish detergent and a splash of household chlorine bleach. Use a stiff plastic brush — not a dish cloth (which won't get your equipment clean) or a scouring pad (which causes scratches where bacteria can hide). Rinse your pail and strainer in clean water, then in dairy acid cleaner (which you can obtain from a farm or dairy supply store), then once more in clean water.

Cleaning your pail is so much simpler if it fits into your dishwasher. Then all you need to do is rinse it out and pop it into the dishwasher. If you never leave the pail and strainer sitting around with milk residue in them before running them through the washer, chances are you'll never need chlorine or acid cleaner.

Milking a Goat

When a doe gives birth, her body begins producing milk for her kids, a process called freshening. A dairy goat is designed to give more milk than her kids need and continue to produce milk long after the kids are weaned. The amount of milk a doe gives increases for the first 4 weeks after she freshens, then levels off for about 15 weeks, after which production gradually decreases and eventually stops until the doe freshens again to start a new lactation cycle.

A doe's milk is produced and stored in her udder. At the bottom of the udder are two teats, each with a hole at the end through which the milk squirts out. The two most important things to remember when you milk a goat are to keep her calm and not pull down on her teats, both of which can be tricky when you're first learning. Keep the doe calm by singing or talking to her and by remaining calm yourself. Not pulling her teats takes practice. The doe will kick the milk pail if you pull her teat, pinch her with a fingernail, or pull a hair on her teat. To avoid pulling a doe's hair during milking, and to keep hair and dirt out of the milk pail, use clippers to trim the long hairs from her udder, flanks, thighs, tail, and the back part of her belly.

To get milk to squirt out the hole, you must squeeze the teat rather than pull it. The first time you try, chances are milk will not squirt out but will instead go back up into the udder. To force the milk downward, apply pressure at the top of the teat with your thumb and index finger. With the rest of your fingers, gently squeeze the teat to move the milk downward. If you are milking a miniature goat — or a full-size goat with small teats — her tiny teats may have room for only your thumb and two fingers.

After you get one squirt out, release the pressure on the teat to let more milk flow in. Since you will be sitting on the milk stand and facing the doe's tail, work the right teat with your left hand and the left teat with your right hand. Get a steady rhythm going by alternating right, left, right, left. Aim the stream into your pail beneath the doe's udder. At first the milk may squirt up the wall, down your sleeve, or into your face, while the doe dances a little jig on the milk stand. Keep at it, and before long you will both handle the job like pros.

When the flow of milk stops, gently bump and massage the udder. If more milk comes down, keep milking. When the udder is empty, the teats will become soft and flat.

If you milk more than one doe, always milk them in the same order

When milking a full-size goat, gently squeeze with all your fingers to move the milk downward.

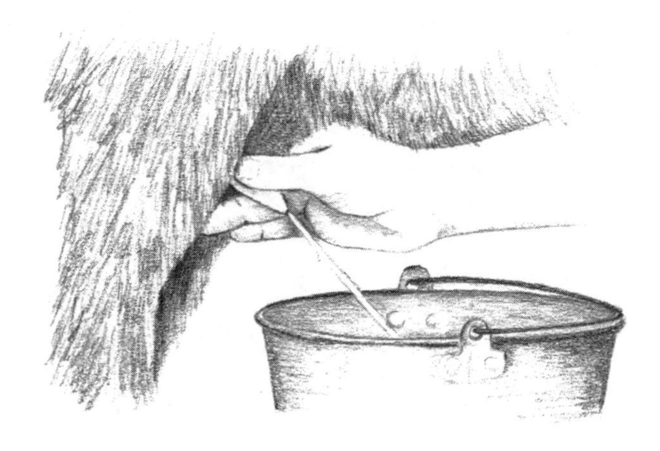

When milking a miniature goat, you may only have room to squeeze the teat with your thumb and two fingers.

every day, starting with the dominant doe and working your way down to the meekest. Your goats will get used to the routine and will know whose turn is next.

As you take each doe to the milk stand, brush her to remove loose hair and wipe her udder and teats with a fresh baby wipe to remove clinging dirt. While you clean the doe's udder, watch for signs of trouble — wounds, lumps, or unusual warmth or coolness. Squirt the first few drops of milk from each side into a strip cup, so-called because it is used to examine the first squirt or stripping of milk. A standard dairy strip cup has a screen on top to catch clots and make them easier to see, but any cup or small bowl will do. Check the stripping to see whether it is lumpy or thick, two signs of mastitis.

When you finish milking each doe, spray both teats with teat dip to minimize the chance of bacteria entering the openings. Use a brand recommended for goats — some dips used for cows are too harsh for a doe's tender udder. In dry or cold weather, prevent chapping by rubbing the teats and udder with Bag Balm (available from a dairy supplier) or Corn Huskers lotion (available from a drugstore).

How to Milk

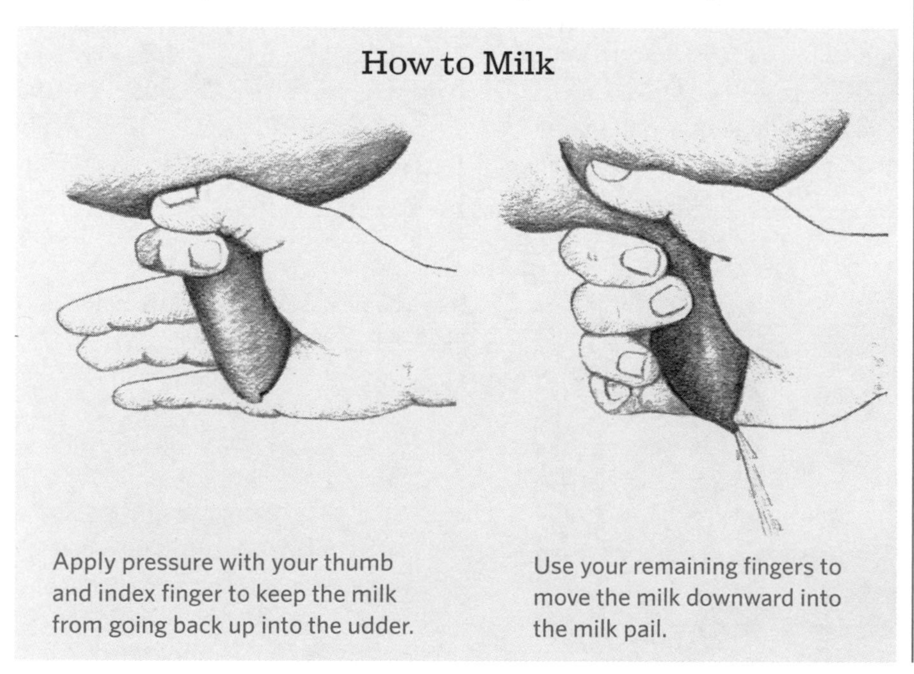

Apply pressure with your thumb and index finger to keep the milk from going back up into the udder.

Use your remaining fingers to move the milk downward into the milk pail.

Milk Output

Exactly how much milk a doe produces in each cycle depends on her age, breed, ancestry, feeding, health and general well-being, and how often you milk. The more often you milk, the more milk the doe will produce. Most goat keepers milk twice a day, as close to 12 hours apart as possible. If milking twice a day gives you more milk than you can use, milk only once a day. Do it every day at about the same time. If you don't milk regularly, your doe's udder will bag up, or swell with milk. Bagging up signals the doe's body that her milk is no longer needed, and the doe begins drying off.

During the peak of production, a good full-size doe in her prime may give as much as 8 pounds of milk (3.5 kg), which is about 1 gallon (4 L) per day. She will then gradually taper off to maybe 2 pounds (1 kg) or about 1 quart (1 L) per day by the end of her lactation cycle. Although a gradual decrease is normal, a sudden decrease may mean the doe is unhealthy, is not getting enough to eat, or is in heat. Weighing each doe's output helps you stay on top of things.

Milk is weighed on a dairy scale, which has two indicator arms. Set the arm on the right to zero. With your empty pail hanging from the scale, set the left arm to zero. When you hang a pail full of milk on the scale, the left arm automatically deducts the weight of the pail. Use the right arm to weigh other things, such as newborn kids.

If you don't have a scale, you can keep track of each doe's output by volume. This method is not as accurate as weighing, because fresh milk has foam on top and it's hard to tell where the foam stops and the milk starts. But it still basically helps you track a doe's output.

Keep a record of each doe's milk output, noting not only the amount of milk obtained from each milking but also anything that might affect output, such as changes you've made in the doe's ration, rainy weather that has kept your does from going out to graze, or the time of day you milked (whether earlier or later than usual). At the end of each month, add up each doe's milk output. At the end of her lactation cycle, add up each doe's total output.

A standard cycle lasts 10 months, or 305 days. To accurately compare the annual output from each doe, or to compare the output of one doe against another, adjust production to a 305-day cycle: Divide the total output by the actual number of days in the cycle, then multiply by 305.

Can you save money by producing your own goat milk? That depends on the price of feed, how much milk your goat produces, and how much you're paying for milk now. Using averages (and for simplicity's sake, ignoring the cost of housing, water, electricity, and so forth) do the math: A milker eats ½ pound of concentrate per pound of milk produced. If she produces 1,800 pounds (900 quarts) of milk in a yearly cycle, she'll eat approximately 900 pounds of concentrate for 305 days plus 60 pounds maintenance ration for the remaining two months, or a total of 960 pounds of concentrate. At the current rate of 26 cents per pound for one brand of all-natural concentrate, that comes to about $250.

She'll also eat about 4 pounds of hay each day, or 1,440 pounds in a year. At 40 pounds per bale, that's 36 bales. The cost of hay varies considerably with location, season, and type of hay, but let's say you pay $4.50 per bale, your outlay for hay comes to $162.00. Total feed cost is $412.00 for 900 quarts (225 gallons) of milk, or about 45 cents per quart. Of course, you have incidental expenses such as for dewormer, shots for pregnant goats, milk filters, and so forth; but if your doe is grassfed the cost of feeding her will be less, and you'll also have a kid or two or three to raise for meat or sell to offset the cost of keeping the doe.

Whether or not you come out ahead financially, and how much, ultimately depends on what you are now paying for milk. Current prices vary with location, ranging from $2 per gallon (50 cents per quart) for cow milk at a discount store up to as much as $20 per gallon ($5 per quart) for goat milk at the farm gate.

Pasteurizing destroys any harmful bacteria that may be in milk, but also destroys some of the vitamins, reduces the availability of the minerals and proteins, and destroys all the natural enzymes.

Pasteurizing Milk

Pasteurizing destroys any harmful bacteria that may be in milk, but also destroys some of the vitamins, reduces the availability of the minerals and proteins, and destroys all the natural enzymes, making the milk more difficult to digest. Pasteurized milk doesn't keep as well as raw milk, and when it starts to turn it tastes terrible.

Despite all this, Americans are taught by promoters of the industrial dairy industry to be afraid to drink milk that isn't pasteurized. Milk is pasteurized by making it hot enough for a long enough time to destroy bacteria. It can be done using a pasteurizer designed for the purpose and sold through dairy supply catalogs. Pasteurizers are expensive but

Use a home pasteurizer if you feel the need to pasteurize goat milk.

easy to use, because they automatically control the time and temperature.

If you have only a few quarts (liters) of milk to pasteurize at one time, you can do it on top of the stove. (Milk cannot be safely pasteurized in a microwave oven.) All you need is a candy thermometer, a spoon to stir the milk, and a double boiler (or a clean pail set in a large pot of water) to keep the milk from scorching.

Heat the milk to 165°F (74°C), stirring to distribute the heat evenly. Do not take the thermometer or stirring spoon out of the milk and put it back in during pasteurization, or you will recontaminate the milk. When the milk reaches 165°F (74°C), continue heating it for 30 seconds more.

Cool the pasteurized milk quickly for best flavor; otherwise it may have an unpleasant cooked taste. Set the pan or pail of milk in a basin of ice or cold water and stir until the milk is cool. Pour the milk into clean jars with tight-fitting lids. If possible, store the jars on the bottom shelf of the refrigerator, where the temperature is coolest and milk keeps the longest.

Whole and Skimmed Milk

After goat milk has been refrigerated for a day or two, its milk fat rises to the surface. Milk fat thinned with a little milk is cream. The milk fat content of goat milk ranges from 2 to 6 percent, depending on genetics, diet, and other factors. An average of 4 percent will give you about 5 tablespoons (74 mL) of milk fat per quart. The milk from Nubians and Nigerian Dwarfs contains more fat than other milk, and the milk from all does varies in fat content during the lactation cycle. Milk fat content is important in making ice cream, butter, and certain kinds of cheese.

If you are trying to limit the amount of fat in your diet, you can remove the milk fat to create skimmed milk. Store your fresh milk in a widemouthed container. In about two days, most of the milk fat will rise to the surface, and you can easily skim it off with a slotted spoon.

Milk Sensitivities

Milk gets its sweet taste from the complex sugar lactose. For your body to digest lactose, it needs the enzyme lactase to break down the lactose into two simple sugars: glucose and galactose. About 75 percent of all adults are lactase deficient. They cannot digest lactose, and when they ingest it, the result is bloating, cramps, gas, nausea, and diarrhea. The problem can be resolved by taking a lactase concentrate, available at most drugstores. Fermentation reduces lactose content by as much as 50 percent, so if you have a problem drinking milk because of lactose intolerance, you may not have a problem eating yogurt.

Not all problems associated with milk are caused by lactose deficiency. About 5 percent of the population is allergic to milk protein. In children, symptoms of milk protein allergy are eczema and digestive problems, including diarrhea, vomiting, and colic. In adults, milk protein allergy causes a feeling of being bloated and gassy. Since the protein in goat milk is not the same as the protein in cow milk, a person who is sensitive to cow milk protein may have no trouble drinking goat milk. Besides having a different protein makeup, goat milk has proportionally more small fat globules, making it easier to digest than cow milk and therefore leaving less undigested residue in the stomach to cause gas and cramps.

Making Yogurt

Yogurt is easy to make from fresh goat milk and tastes better than any yogurt you can buy. To make yogurt, simply add a culture to the milk and keep it warm for several hours. Your yogurt incubator might be an electric yogurt maker designed for the purpose or a glass casserole dish wrapped in towels and placed on a heating pad or an electric hot plate set on warm. Yogurt makers and yogurt culture are available from dairy and cheesemaking suppliers.

You could also culture your goat milk yogurt using unflavored store-bought yogurt, provided it is a live culture yogurt — meaning it contains the living organisms needed to promote fermentation. Once you get your own yogurt going, you can use a spoonful from the previous batch to make the next batch, until the culture eventually weakens and plays out. Then you'll need to start again with a fresh packet of culture or a container of store-bought live-culture yogurt.

To make yogurt, measure out a quart (1 L) of strained milk fresh from the milk room, add a packet of yogurt culture or 2 tablespoons (30 mL) of previously made yogurt, and stir. If you want the yogurt to be thick, add unflavored gelatin at this time. Sprinkle one packet of unflavored gelatin over a little cold water. Allow the gelatin to soften for 1 minute. Heat the water and gelatin in a small saucepan, or in the microwave oven for 30 seconds on high, until the gelatin thoroughly dissolves.

Incubate the cultured milk until it tastes right to you; the longer it ferments, the more tart it will be. About eight to nine hours makes a nice tangy yogurt that isn't overly tart. Store the yogurt in the refrigerator. Fermentation slows to a virtual standstill when yogurt is refrigerated.

Once you get the hang of making yogurt, you may want to go on to make simple soft or hard cheeses. Two good books to get you started are *Home Cheesemaking* by Ricki Carroll and *The Home Creamery* by Kathy Farrell-Kingsley.

The Other Red Meat

Although goat meat is delicious, it has never caught on in mainstream America, perhaps because the English language has no single word for it. "Goat meat" is a more direct term than the words we use for other kinds of meat, such as beef and pork, neither of which names the animal the meat comes from. People who raise and market goat meat therefore borrow words from Spanish and French.

Cabrito is the meat of a 1- or 3-month-old kid weighing less than 50 pounds (23 kg). It is usually cut into serving size pieces for grilling, or into smaller pieces to make shish kebab.

Chevon is the meat of an older kid, generally 6 to 9 months of age, and weighing less than 80 pounds (36 kg). It is usually baked, grilled (as on a spit), or stewed.

Mutton or **chivo** is the meat from older goats, which is not as tender as the meat from younger goats. It is usually ground up and used in spicy foods like sausage and chili.

Goat meat is lower in calories, total fat, saturated fat, and cholesterol than other meats and is easier to digest than other red meats. It makes a good dietary alternative for the many people who eat mostly chicken and fish because they have difficulty digesting other types of meat or because they want to reduce fat intake.

Comparing Goat Meat with Others

(per 3 ounce serving)

Nutrient	Goat	Chicken	Beef	Pork	Lamb
Calories	122	162	179	180	175
Fat (g)	2.6	6.3	7.9	8.2	8.1
Saturated fat (g)	0.79	1.7	3.0	2.9	2.9
Protein (g)	23	25	25	25	24
Cholesterol (mg)	63.8	76.0	73.1	73.1	78.2

Source: Alabama Cooperative Extension UNP-0061, October 2008

Raising Kids for Meat

When raising kids for meat, your goal is to get the fastest weight gain at the least cost. The meat that is the least expensive to produce and has the mildest flavor comes from milk-fed kids that are 6 to 8 weeks old and weigh less than 35 pounds (16 kg).

Raising grain-fed kids is more costly, but their meat is more flavorful. In addition to their milk ration, these kids get a small amount of grain from the age of 6 weeks until they weigh at least 50 pounds (23 kg), which may take up to six months.

The least expensive way to get full-flavored meat is to wether your bucklings and let them nurse as long as they like (and the doe will allow), and then put them on pasture. By the time a wether is 1 year old, it should weigh 80 pounds (36 kg) or more.

The conversion factor from live weight to dressed weight varies with the type of goat being raised. A meat breed, or meat cross, has a better meat-to-bone ratio than most dairy breeds. On average, you can expect a goat to dress out to between 45 and 52 percent of its live weight, including the liver, heart, and kidneys.

Can you save money raising goats for meat? That's not an easy question to answer, because your cost depends on a lot of variable factors including how you acquire the goats and the amount of land you have available for forage. For instance, if you were to go out and buy a bottle baby of a meat goat breed and raise it on a small lot to the age of slaughter, you will have had to not only pay the purchase price but also everything the kid eats starting with milk replacer. Such an endeavor can be quite costly. And when The Day comes, you most likely will have a hard time thinking about eating what has become a family pet. However, if you raise dairy goats for milk, and you have an annual crop of kids to deal with, and they are nursed by their dams and have access to quality grazing and browsing, your capital outlay will be primarily for whatever amount of hay and concentrate they eat from the time of weaning until the time of slaughter.

Using averages (and for simplicity's sake, ignoring the cost of housing, water, electricity, and so forth) do the math: By the time a kid reaches the live weight of 50 pounds it will have eaten about 150 pounds of concentrate and 120 pounds of hay. At 45 percent of live weight, a 50-pound goat dresses out to approximately 22 pounds. At current rates, a whole goat in this weight range can be purchased for approximately

$5.50 per pound, or about $120.00 for a 22 pounder. The current rate for one brand of all-natural concentrate is about 26 cents per pound, or $39.00 for 150 pounds. The price of hay varies with location, season, type of hay, and bale weight, so let's say the bales weigh 40 pounds and cost $4.50 each, for a total of $13.50 for hay. Total feed cost is about $52.50 for 22 pounds of goat meat, or a little less than half the price of the meat at retail market value. Not a bad deal at all. And if the kid is grass-fed, therefore requires less bought feed, the deal is even better.

A good source for information on butchering goats, cuts of meat, and meat storage is a book that covers dressing venison, such as *Basic Butchering of Livestock & Game* by John J. Mettler, Jr., DVM. In size and shape, a goat is similar to a deer, and in flavor, goat meat is virtually indistinguishable from properly dressed venison. If you prefer to take your animals to a slaughterhouse, often the same shops that cut and wrap venison during the hunting season will take in goats at other times of year.

Keeping Goats Healthy

Each goat herd is unique in its combination of breed, age, local climate, prevalent pathogens, and other elements that influence health. Your health care plan must therefore fit your particular situation and cannot be developed by blindly following a formula established for another herd elsewhere.

The most important health care measure is disease prevention, which must start when you acquire your first goats. Purchase only healthy animals. If you feel unsure about how to tell a healthy goat from a sick one, take along an experienced person to coach you. You'll avoid a lot of future expense and heartbreak.

Sickness in goats is usually caused by poor management. Good management includes protecting your goats from illness and watching for early signs of problems. Take care not to allow your goats to become crowded, because crowding causes stress and stress decreases resistance to disease. If your goats regularly come into contact with others outside your herd, you will need a stronger routine health care program and greater attention to keeping vaccinations current than someone whose herd is isolated from other goats that may potentially be unhealthy.

The health care routine for all goats should include weighing, hoof trimming, and regular deworming, as well as vaccinations and booster shots as determined by diseases that are prevalent in your area. Consult your veterinarian and local goat owners to develop a suitable health care program for your herd.

Weighing

Each time you have a goat confined for hoof trimming, weigh it and record the date and weight on a chart. Your weight records will tell you if a young goat is growing properly, a young doe is big enough to breed, or a pregnant doe is getting enough to eat. Loss of weight is often the first sign something is wrong.

Measuring heart girth is a way to estimate your goat's weight.

When a goat gets too heavy to lift onto a scale (see box on page 201), the best you can do is estimate its weight. Even though the estimate is not 100 percent accurate, it will tell you whether the animal is gaining or losing weight. To estimate a weight, measure the goat's heart girth — the distance around the goat's middle, just behind its front legs, over its heart. Have the goat stand on a level surface with its legs solidly beneath it.

Use a dressmaker's tape measure or a weight tape from a goat supply catalog. The weigh tape automatically converts heart girth to estimated weight. If you use a dressmaker's tape, the chart (see page 212) will help you convert inches to pounds.

Hoof Trimming

Hooves are made of keratin, the same material your toenails are made of. Like toenails, hooves grow uncomfortably long if they aren't trimmed. Wild goats generally live in rocky areas, where abrasion wears down their hooves as they move around seeking fresh forage. By contrast, when a goat spends its days in a barn or on

Estimating Your Goat's Weight

Heart girth (in./cm)	Weight (lbs./kg)	Heart girth (in./cm)	Weight (lbs./kg)	Heart girth (in./cm)	Weight (lbs./kg)
10.75/27.25	5/2.25	21.75/55.25	37/16.75	32.75/83.25	105/47.5
11.25/28.5	5.5/2.5	22.25/56.5	39/17.75	33.25/84.5	110/50
11.75/30	6/2.75	22.75/57.75	42/19	33.75/85.75	115/52.25
12.25/31	6.5/3	23.25/59	45/20.5	34.25/87	120/54.5
12.75/32.25	7/3.25	23.75/60.25	48/21.75	34.75/88.25	125/56.75
13.25/33.5	8/3.5	24.25/61.5	51/23	35.25/89.5	130/59
13.75/35	9/4	24.75/62.75	54/24.5	35.75/90.75	135/61.25
14.25/36.25	10/4.5	25.25/64.25	57/26	36.25/92	140/63.5
14.75/37.5	11/5	25.75/65.5	60/27.25	36.75/93.25	145/65.75
15.25/38.75	12/5.5	26.25/66.75	63/28.5	37.25/94.5	150/68
15.75/40	13/5.75	26.75/68	66/30	37.75/95.75	155/70.25
16.25/41.25	14/6.25	27.25/69.25	69/31.25	38.25/97.25	160/72.5
16.75/42.5	15/6.75	27.75/70.5	72/32.75	38.75/98.5	165/74.75
17.25/43.75	17/7.75	28.25/71.75	75/34	39.25/99.75	170/77
17.75/45	19/8.75	28.75/73	78/35.25	39.75/101	175/79.5
18.25/46.25	21/9.75	29.25/74.25	81/36.75	40.25/102.25	180/81.75
18.75/47.5	23/10.5	29.75/75.5	84/38	40.75/103.5	185/84
19.25/49	25/11.25	30.25/76.75	87/39.5	41.25/104.75	190/86.25
19.75/50.25	27/12.25	30.75/78	90/40.75	41.75/106	195/88.5
20.25/51.5	29/13.25	31.25/79.25	93/42.25	42.25/107.25	200/90.75
20.75/52.75	31/14	31.75/80.75	97/44		
21.25/54	35/15.75	32.25/82	101/45.75		

pasture, its hooves keep growing. Unless they are trimmed, eventually the goat will be unable to walk properly. Hooves left untrimmed for too long fold under, trapping dirt and moisture that create an ideal environment for bacteria that can cause foot rot. Excessive growth also alters an animal's stance, eventually crippling the animal permanently.

To learn what a properly trimmed hoof looks like, study the feet of a newborn kid. Note how its hooves are flat on the bottom and look boxy. When a hoof has been properly trimmed, the bottom is parallel to the growth rings and the two toes are of equal length.

How often you need to trim hooves depends on how fast they grow. Some hooves need trimming every two weeks; others may not need to be trimmed more often than every two months. For many goats, the rear hooves need trimming more often than the fronts. The rate of growth of a particular goat's hooves is influenced partly by genetics and partly by environment. Like wild goats, domestic goats can wear off some hoof growth if they have access to an abrasive surface, such as a rock outcrop or a concrete pad on which to play.

A goat that is not used to having its hooves trimmed may struggle and kick. Avoid this problem by training kids to the procedure early and by continuing to check and trim hooves often throughout each animal's life. Frequent trimming of small amounts at a time is easier than infrequent major trimming.

You will need a pair of sharp shears, such as good garden pruning shears or shop snips, or a hoof trimmer designed specifically for the purpose and available from any livestock supply store or catalog. Keep the shears sharp by periodically rasping the blades with a file.

You'll need to confine the goat so it doesn't attempt to wander away while you're trimming its hooves. You can use the milk stand to hold each goat during trimming, or fasten the goat's collar to a wall or a fence and crowd the animal with your body so it can't move around. Face the goat in one direction to trim the two hooves away from the wall, then turn it around to trim the other two hooves.

While a kid is still small, you can trim its hooves by turning it on its back and holding it firmly on your lap. A kid's hooves shouldn't need much trimming, but starting early is a good idea to get the animal accustomed to having its hooves handled.

Trim when hooves have been softened by rain or dewy grass. To prevent injury to the goat, work in good light and trim away a small amount at a time. Here's the procedure:

1. Grasp one leg by the ankle and bend it back to see the bottom of the foot.

2. Use the point of the shears to scrape away dirt.

Hoof Trimming, Step 2.

3. Cut off long growth at the front of the hoof.

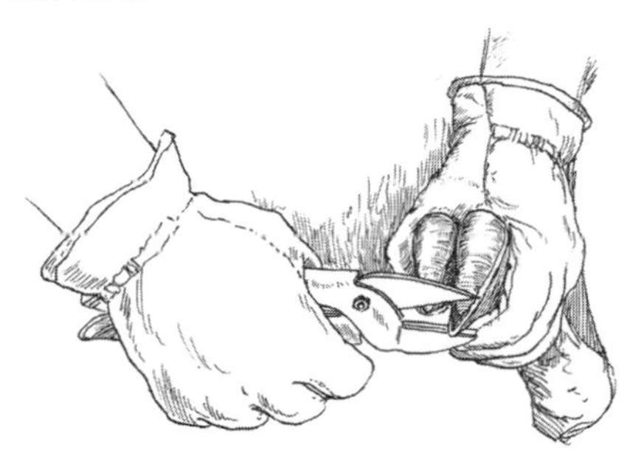

Hoof Trimming, Step 3.

4. Snip off any flaps that fold under the hoof.

5. Trim the bottom, one tiny slice at a time, cutting toward the toe. Stop trimming when the hoof looks pink, indicating you are getting close to the blood supply. Do not keep cutting, or the foot will bleed. If that happens, pour tamed iodine (such as Betadine or chlorhexidine) over it.

6. If you reach the blood supply before an overgrown hoof is properly trimmed, avoid causing damage by continuing the trimming a few days later, allowing time for the blood supply to recede.

Signs of Illness

Take time to study your goats while they are healthy so you will readily notice any changes in the way they look, eat, move, or smell. The sooner you realize a goat is getting sick, the quicker you can do something about it, and the greater the chance it will recover.

Notice the size, shape, firmness, color, and smell of your goats' droppings. Any change may indicate a dietary imbalance, the beginning of a disease, or an infestation of parasites. Parasites may be internal, such as worms and coccidia, or external, such as lice and ticks.

Listen for teeth grinding, a sign the goat is in pain. Look for changes in the color of the gums and the lining around the eyes. These areas should be bright pink. If a goat is in shock or has lost blood (for instance through a parasite overload), these areas may turn pale. A purple or blue color may indicate damaged airways or other breathing problems. If the color is pale gray or blue and the goat has a hard time breathing, call your vet immediately.

Before calling a veterinarian about any goat's health, first take the animal's temperature and be prepared to relay the information to the vet. Inexpensive digital thermometers, complete with instructions for their use, are available through veterinary supply catalogs. Restrain the goat as you would for hoof trimming and insert the thermometer into its rectum. If you know the animal's normal temperature at rest, you will have a better idea when its temperature is abnormal. A good idea is to take each goat's temperature whenever you weigh the animal, and record the temperature along with the weight. Any time you note a change from normal, look for the reason.

Before calling the vet, you might also take the goat's pulse by using one of the following two methods:

- Place your fingertips on both sides of the lower rib cage and count the beats for 1 minute.
- Place your finger on the big artery on the upper inside part of one of the rear legs and count the beats for 1 minute.

Signs of Health in Adult Goats*

- Pulse rate: 70 to 80 beats per minute
- Breathing rate: 12 to 20 breaths per minute
- Rumen movements: 2 or 3 every 2 minutes
- Rectal temperature: 101.5°F to 105°F (39–41°C)

*Kids have higher pulse and breathing rates than adult goats.

Common Goat Problems

If one of your goats becomes sick, don't try to treat it yourself unless you know for certain what the problem is. Giving medications incorrectly can do more harm than good. The first time you give your goat a shot or a drench (liquid medication given through the mouth), have an experienced person show you how.

The following list describes the most common problems in goats. Alternative names are included to help you discuss the problem with others or look it up online.

Signs of Illness in an Adult Goat

Behavior
- Inactivity
- Teeth grinding
- Complaining noises
- Frequent coughing
- Quick, shallow breathing
- Any change in normal behavior

Digestion
- Reduced or no appetite
- More (or less) frequent urination
- Change in manure color or consistency

Milk
- Inexplicable drop in production
- Change in color, odor, or consistency

Coat
- Rough or dull
- Hair falls out or scabs appear
- Goat scratches or bites itself

Body Temperature
- Above or below normal

Abscesses

Also called: Cassous lymphadenitis, pseudotuberculosis
Sign: Firm swellings under the skin
Cause: Bacterial infection of lymph nodes by *Corynebacterium pseudotuberculosis*. These bacteria most likely enter through skin wounds.
Prevention: Provide a safe environment that minimizes the risk of skin wounds. Isolate or remove affected animals.
Treatment: Antibiotics are not effective. Surgical removal can be attempted for valuable animals.

Bloat

Also called: Acidosis or ruminal tympany
Signs/effects: Swelling on left side, kicking at stomach, grunting, slobbering, lying down and getting up; can lead to death
Cause: Excess gas in rumen
Prevention: Feed balanced rations. Make any dietary changes gradually. Prevent overeating of concentrate or lush pasture.
Treatment: Keep the goat on its feet (propped between hay bales, if necessary). Rub its stomach to eliminate gas. Drench with 2 cups (500 mL) of mineral oil followed by ¼ cup (60 mL) of baking soda dissolved in 1 cup (250 mL) of water. Call veterinarian immediately.

Caprine Arthritis and Encephalitis

Also called: CAE or CAEV
Signs: Weak rear legs in kids; stiff, swollen knee joints in mature goats
Cause: Infection by caprine arthritis-encephalitis virus
Prevention: Purchase certified CAE-free goats. Do not feed kids raw colostrum or milk from infected does.
Treatment: No effective cure

Coccidiosis

Also called: Cocci, coccidia
Signs/effects: Loss of appetite, loss of energy, loss of weight, diarrhea (sometimes bloody); can cause death
Cause: Infection by *Eimeria* species of protozoal parasites
Prevention: Keep bedding, feeders, and waterers scrupulously clean.
Treatment: Coccidiostat (sulfa drug), used as directed on the label

Foot Rot

Also called: Hoof rot
Signs/effects: Lameness, ragged hoof, grayish discharge, smelly feet, foot deformity, loss of weight, tetanus; can cause death
Cause: Infection by *Bacteroides nodosus* and *Fusobacterium nodosum* bacteria
Prevention: Keep bedding clean and dry. Trim hooves regularly.
Treatment: Trim hoof to healthy tissue. Soak foot in copper sulfate solution (½ pound per gallon [226 g per 4L] of water) for two minutes. Severe cases may require antibiotics, as recommended by your veterinarian.

Ketosis

Also called: Pregnancy disease, pregnancy toxemia (the disease occurs in does just before or after kidding)
Signs/effects: Sweet-smelling breath, urine, or milk; loss of appetite; doe lies down and can't get up; can cause death
Cause: Metabolic disorder triggered by a sudden change in diet, overfeeding during early pregnancy, or underfeeding during late pregnancy
Prevention: Proper feeding
Treatment: Drench with 1 tablespoon (15 mL) of baking soda dissolved in ¼ cup (60 mL) of water, then drench with 2 ounces (60 mL) of propylene glycol (or, in a pinch, 1 cup [236 m] of honey or corn syrup) twice daily. If the doe is weak or can't swallow, call vet immediately.

Lice

Also called: Pediculosis, body lice, external parasites
Signs: Scratching, loss of hair, loss of weight, reduced milk production
Cause: Contact with infested animals, living in damp housing
Prevention: Avoid contact with infested animals
Treatment: Powder, dip, spray, injectable, or pour-on insecticide approved for dairies. Repeat in two weeks to kill newly hatching lice eggs. If infestation is severe, repeat again in two weeks.

Mastitis Test

At least once a month, test the first stripping of milk from each teat for mastitis. The California Mastitis Test (CMT) kit is sold by most farm stores and dairy suppliers. The test is designed for cows, which have four teats, so you will use only two of the four cups in the kit, which comes with directions.

Mastitis

Also called: Infected udder
Signs: Doe stops eating; may have a fever; udder is unusually hot or cold, hard, swollen, or painful; milk smells bad or is thick, clotted, or bloody (slightly pink milk at freshening is normal).
Cause: Various bacteria, often following injury or insect sting to the udder
Prevention: Keep bedding clean. Remove objects from housing and yard that could damage the udder. Disbud nursing kids. At each milking, check udder and milk for signs.
Treatment: Isolate the doe. Apply hot packs four or five times a day. Milk three times a day. Milk the infected doe last. Contact your veterinarian about antibiotic treatment.
Human health risk: Do not drink infected milk or feed it to goat kids. Do not drink milk from treated does until the time span specified on the drug label or by your veterinarian has passed.

Overeating Disease

Also called: *Clostridium perfringens* infection, enterotoxemia
Signs/effects: Twitching, swollen stomach, teeth grinding, fever; can lead to death
Cause: Infection by *Clostridium perfringens* bacteria
Prevention: Avoid abrupt changes in diet. Vaccinate.
Treatment: No effective cure

Pinkeye

Also called: Conjunctivitis, infectious keratoconjunctivitis
Signs/effects: Red-rimmed, watery eyes; squinting; may cause blindness
Cause: Various bacteria; less commonly viruses or other microorganisms
Prevention: Avoid dust, eye injuries, and contact with infected goats.
Treatment: Antibiotic eyedrops or ointment under the eyelids. Treat all goats, even those without signs.

Plant Poisoning

Also called: Toxic reaction
Signs/effects: Frothing at the mouth, vomiting, staggering, trembling, crying, rapid or labored breathing, altered pulse rate, convulsions; can lead to death
Cause: Ingestion of plants sprayed with pesticides or toxic plants such as black nightshade, bracken fern, death camas, hemlock, horse nettle, laurel, Japanese yew, milkweed, oleander, and rhododendron, as well as the wilted leaves of trees that produce stone fruit — cherry, peach, and plum. Limp leaves that are still green or are partially yellow are the most dangerous; fully dried leaves are no longer toxic. Ask your county Extension agent for an illustrated list of poisonous plants in your area.
Prevention: Feed a balanced diet, including free-choice hay. In autumn, keep goats away from stone-fruit trees that are dropping their leaves.
Treatment: Place 2 tablespoons (30 mL) of salt on the back of the goat's tongue to induce vomiting, followed by ¼ to ½ pound (113 to 226 g) activated charcoal powder mixed with ½ cup (226 g) mineral oil and 3 to 4 quarts (up to 4 L) of water or as much as the goat will drink. Call the veterinarian.

Pneumonia

Also called: Lung sickness
Signs/effects: Coughing, runny eyes and nose, loss of appetite, fast breathing, fever; may cause death
Cause: Various bacteria and viruses (usually after exposure to drafts and dampness), parasites, allergies
Prevention: Provide dry, draft-free housing with good ventilation. Do not heat housing.
Treatment: Contact your veterinarian about antibiotic treatment.
Human health risk: None

Ringworm

Also called: Dermatophytosis
Signs: Circular hairless patch, usually on the head, ears, or neck; sometimes on the udder
Cause: Various fungi in the soil
Prevention: Avoid contact with infected animals.
Treatment: Scrub area with soapy water, and coat with iodine or fungicide (be careful not to get any in the goat's eyes).
Human health risk: Can infect humans. Wash your hands after handling an infected animal.

Scours

Also called: Diarrhea
Signs/effects: In kids, watery, bad-smelling loose manure, loss of appetite, loss of energy; can cause death
Cause: Various bacteria
Prevention: Feed kids properly. Disinfect containers after each feeding, especially when feeding milk. Keep housing clean.
Treatment: Substitute electrolyte fluid for milk ration for 2 days (see page 202). Call your veterinarian if diarrhea does not clear up right away.

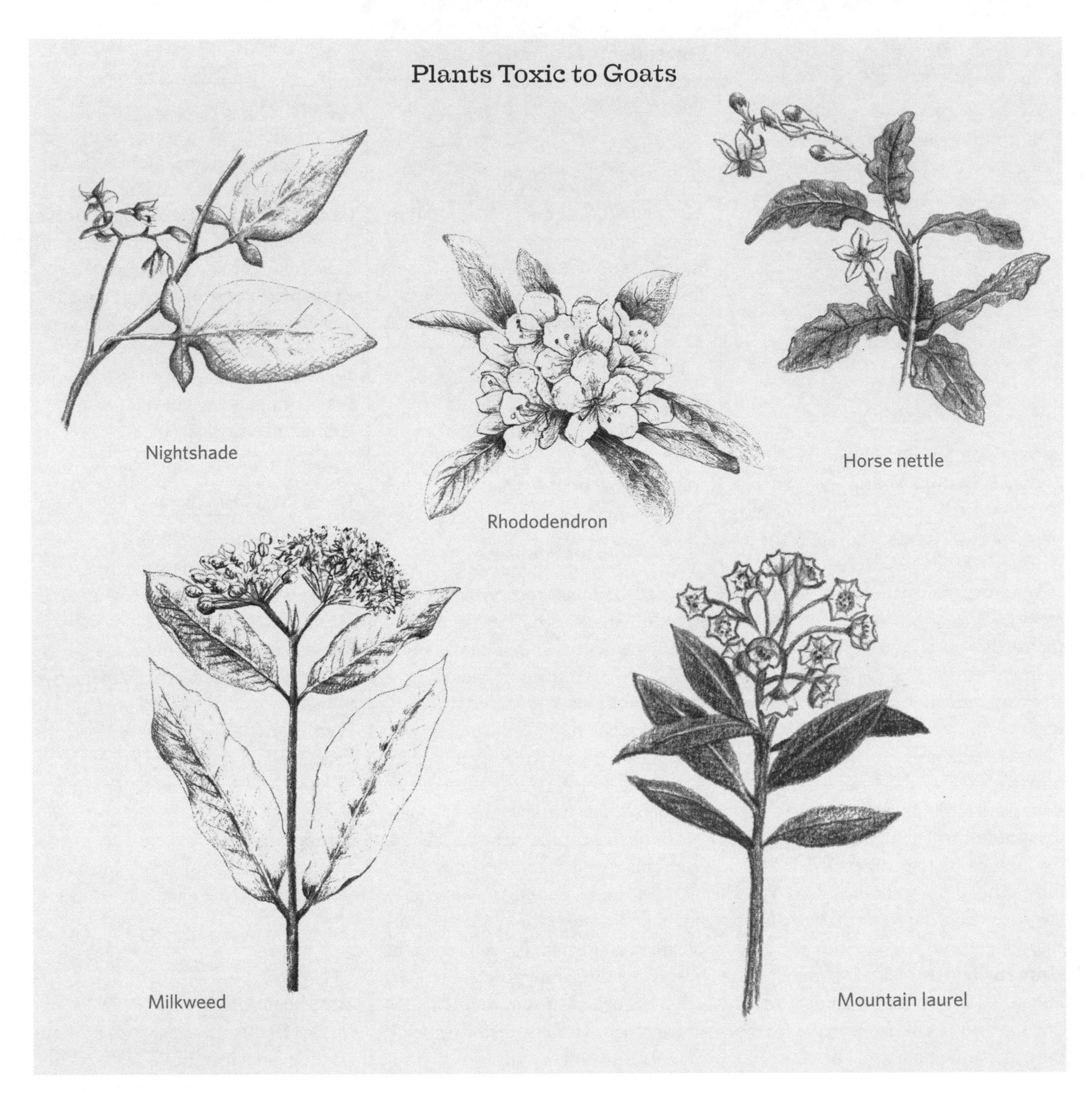

Scrapie

A neurologic disease called scrapie is not a significant problem in goats, but the USDA has imposed regulations related to scrapie because the disease can be transmitted from goats to sheep or sheep to goats. Exactly how the disease passes from one animal to another is not known, and its incubation period of one to seven years makes tracking difficult. Scrapie is similar to mad cow disease in cattle and has no known treatment or cure.

In an attempt to eradicate scrapie, the USDA requires herd certification and identification of individual goats within each herd, hoping to trace any outbreaks back to the herd of origin. Most states have additional regulations, some of which exceed federal requirements. Since these regulations may change from time to time, the best way to find out if, or how, you comply, as well as locate your state regulatory agency, is to contact the USDA Animal and Plant Health Inspection Service office by calling 866-873-2824 (toll free) or visiting their website at www.aphis.usda.gov.

Sore Mouth

Also called: Contagious ecthyma, orf
Signs/effects: Crusty sores on the lips and muzzle of young animals can cause

difficulty nursing and eating and loss of condition.

Cause: A pox virus that can survive for years in the environment

Prevention: A vaccine is available but should be used only in infected herds under a veterinarian's supervision.

Treatment: Sores will heal by themselves in one to four weeks. Give extra milk or feed to kids that aren't eating well.

Human health risk: This virus will also cause skin infection in people. Use gloves when handling affected kids and sterilize nipples, bottles, and feed tubs after use.

Tetanus

Also called: *Clostridium tetani* toxemia, lockjaw

Signs/effects: Stiff muscles, spasms, flared nostrils, wide-open eyes; results in death

Cause: *Clostridium tetani* bacteria entering a wound

Prevention: Vaccinate kids with ½ mL of tetanus toxoid at 4 weeks of age (or before disbudding). Repeat in 30 days and annually thereafter.

Treatment: No effective cure

Human health risk: You cannot get tetanus from your goats, but you can get it from the same sources that they do. Talk to your family doctor about immunization.

Ticks

Signs: Rubbing, scratching, loss of hair, loss of weight

Cause: Browsing in wooded areas

Prevention: Check goats daily during tick season.

Treatment: Dust or spray with pesticide approved for dairies. To remove an attached tick, grasp the tick carefully with tweezers near the point of attachment and lift firmly. Wrap the tick in a tissue and flush it down the toilet.

Human health risk: Ticks transmit diseases such as Rocky Mountain spotted fever and Lyme disease. During tick season, check your body for ticks after handling your goats

Urinary Stones

Also called: Calculosis, bladder stones, kidney stones, urinary calculi, urolithiasis, water belly

Signs/effects: In a wether, difficulty urinating, kicking at abdomen, loss of appetite; can lead to death

Cause: Sandlike crystals (calculi) in the urinary tract due to dietary imbalance or drinking too little water

Prevention: To acidify the urine and prevent stone formation, add ammonium chloride at the rate of 0.5 percent to the wether's total diet. Keep a wether's diet low in concentrate and the leaves from beets, mustard, and Swiss chard, and keep it high in free-choice hay, salt, and clean water. Feed grass hay, never a legume hay.

Treatment: Surgery

Worms

Also called: Internal parasites

Signs: Paleness around eyes, loss of weight or failure to gain weight, loss of energy, weakness, poor appetite, diarrhea, coughing, rough coat, reduced milk production, strange-tasting milk

Cause: Eating worm eggs from manure, as may happen when feed is thrown onto the ground or pasture is grazed without frequent rotation. Worm egg numbers skyrocket after a particularly mild winter, in areas with a mild climate, and during warm, rainy weather.

Prevention: Keep feed and water free of manure. Move goats to fresh pasture often. Isolate and deworm new animals.

Treatment: Take a fecal sample to your vet and obtain a dewormer.

How a Goat Gets Worms

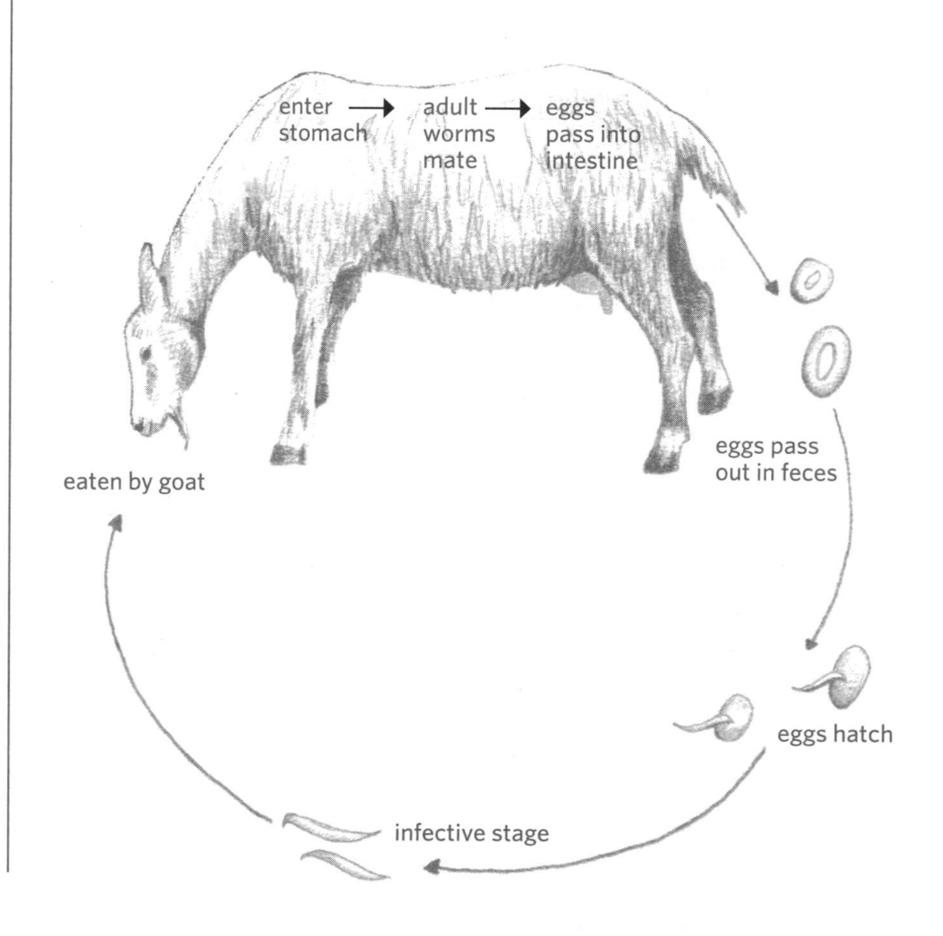

Wounds

Curiosity can get a goat into all sorts of trouble, including serious injury. If one of your goats is wounded, clean the wound with saline solution wound wash so you can see how serious it is. Cuts on the udder usually look worse than they are because they bleed a lot.

Stop the bleeding by pressing a folded clean cloth, towel, or disposable diaper against the wound; if possible, tape it on tight. If the bleeding does not stop, call your vet. Keep the goat quiet and continue to apply pressure until the vet can take a look.

If the wound is a simple shallow cut or scratch, clip away the hair around the wound. Wash the area with warm, soapy water and rinse with clean water. Pour wound wash over the cut, dab it with a clean tissue, and coat the area with an antibiotic ointment (such as Neosporin). Clean the cut daily and coat it with tamed iodine or antibiotic ointment until it heals completely.

Watch for signs of infection, such as redness, swelling, tenderness, and oozing. If infection occurs, call your vet. A shallow cut is not likely to become infected if you keep it clean while it heals.

Health Records

Keep a health maintenance chart for each goat so you can track its medical history, recognize recurring problems, and keep worming and vaccinations up to date.

Taking a Fecal Sample

Place a self-sealing (zipper) plastic bag, inside out, over your hand. With the covered hand, grasp a handful of fresh droppings. With your other hand, turn the bag right side out to enclose the droppings. Seal the bag and take it to your veterinarian as soon as possible. Your vet will examine the sample through a microscope. Ask if you can take a look, too. If your goats have worms, the vet will tell you what kind they are and recommend the proper treatment.

Take a fecal sample in fall around breeding time, and then again in spring after kidding. On the basis of these samples, your vet can recommend a suitable routine deworming program.

First-Aid Kit for Goats

If you assemble a first-aid kit ahead of time, you will be ready to handle most emergencies. Keep the items listed below clean and dry in a large lunch box, tackle box, ammunition case, or any sturdy plastic or metal container with a tight lid. Inside the cover, tape the names and phone numbers of at least three goat-oriented veterinarians. In an emergency, you may not be able to reach your regular vet.

Include the following items in the first-aid kit:

- 1 rectal thermometer, to take temperature
- 1 quart (1 L) isopropyl alcohol, to sterilize thermometer
- 6 disposable syringes (3 mL and 5 mL), to give shots
- 6 needles (18 gauge), to go with syringes
- 3 clean towels or diapers, to stop bleeding
- 1 bottle tetanus antitoxin, in case of wounds
- 1 pint (500 mL) saline solution wound wash
- 1 tube antibiotic ointment, to dress wounds
- 1 container tamed iodine (such as Betadine), to treat wounds
- 1 quart (1 L) mineral oil, to treat bloat
- 1 quart (1 L) propylene glycol, to treat ketosis
- 1 package powdered electrolytes, to treat scouring kids
- 1 jar udder balm, for chapped udders (and hands)
- Deworming medication, as recommended by your vet
- Dairy-approved pesticide spray or powder, if needed to treat lice and ticks

Health Maintenance Chart

Date	Name of Goat	Medication Used	Dosage	Remarks

CHAPTER 7

Sheep

For thousands of years, people have raised sheep for three critical reasons: milk, meat, and wool. Of course, other barnyard animals are also able to provide humankind with these items, but sheep have many advantages: They are much easier to handle than larger farm animals, such as cows and pigs. They require little room, they're fairly easy to care for, and they can be trained to follow, come when called, and stand quietly.

Sheep are also earth-friendly. Land that cannot be used to grow vegetables, fruits, or grains is fine for sheep. They eat weeds, grasses, brush, and other plants that grow on poor land, and their digestive systems are designed to handle parts of food plants, such as corn, rice, and wheat, that people cannot eat. Even just an acre, subdivided into four to six paddocks, will support a small flock of three or four ewes and their lambs during the spring, summer, and fall. Many of the world's most popular cheeses are made from sheep milk. Sheep manure fertilizes soil.

Our domestic sheep, also known by the scientific name *Ovis aries*, were first domesticated about 9,000 years ago in an area that corresponds to modern Iraq, Iran, and Turkey. The wild Mouflon sheep, which is considered vulnerable to extinction, is the forebear of our domestic sheep.

Sheep Q & A

Sheep rely on their owners for food, protection from predators, and regular shearing, but they require less special equipment and housing than most other livestock. One or two lambs or ewes can be raised in a backyard with simple fencing and a small shelter. Here are some answers to common questions.

Are children safe around sheep? Sheep are among the safest four-legged farm animals for children to handle. Most sheep are small and docile. Rams can be aggressive at times, but sheep, especially those that are around people every day, are usually gentle and even-tempered.

What do sheep eat? Sheep don't need fancy food. In summer, they can live on grass; in winter, they can eat hay that is supplemented with small amounts of grain. Fresh water, salt, and a mineral and vitamin supplement complete their diet.

Are sheep dumb? Sheep are anything but stupid. They learn quickly and are among the smartest of all farm animals. Many sheep recognize and respond to their individual names. Sheep are often thought to be stupid because of the way they react to perceived danger. They have no way to defend themselves; if an enemy threatens them, they cannot kick like horses, butt like goats or cattle, or bite like pigs. They can only bunch together and run away. Sometimes, when sheep are frightened, they run headlong into obstacles, which makes them seem stupid.

Do sheep stink? Definitely not. All farm animals have their own distinctive odor. The natural odor of sheep and their manure is not as strong as that of cattle, and is comparable to that of rabbits and goats.

How many sheep should I keep? No sheep should be raised alone. Sheep have a built-in social nature and a flocking instinct. They are happiest when they have companions. Orphaned, or bummer, lambs, however, are often just as happy around humans as they are with other sheep. An orphaned lamb quickly becomes attached to the person who feeds it.

How much land do I need to graze sheep? Lambs are a good choice if you are interested in raising meat animals but lack much land. You can raise feeder lambs (lambs fed for butcher) in a pen, allowing about 25 feet (7.6 m) per animal. Although they can be raised in a pen, lambs will do better with access to a well-fenced pasture area. You could raise four or five lambs on just one-quarter acre of grass.

How old should sheep be when I buy them? Most people start with weaned lambs, which are 2 to 3 months of age. You can also buy an older ewe that has been bred to lamb in the spring. These ewes can often be purchased at a low cost, and you'll get two sheep for your money.

How much do sheep cost? Prices for lambs and adult ewes vary widely. Shop around at several local farms to get a sense of the average market prices in your area. A good crossbred (a sheep whose parents are of different breeds) will be much less expensive than a purebred or registered animal.

Sheep Speak

buck. A mature male; also called a ram.

bummer. A bottle-fed lamb; maybe an orphan, a lamb abandoned by its dam, or a lamb whose dam is not producing enough milk to raise all of her lambs from a multiple birth.

club lamb. A feeder lamb raised as a 4-H, FFA, or other club project.

docked tail. A tail that has been cut short, usually when a lamb is 2 to 3 days of age, for health reasons.

ewe. A mature female.

ewe lamb. An immature female.

feeder lamb. A weaned lamb raised specifically for butcher.

flock. A group of sheep.

hothouse lamb. A lamb born in fall or winter and butchered at 9 to 16 weeks of age.

lamb. A newborn or immature sheep, typically under 1 year of age.

ram. A mature male; also called a buck.

ram lamb. An immature male.

wean. To accustom a lamb to obtain nourishment from feed other than milk; usually takes place around 2 or 3 months of age.

wether. A castrated or neutered male.

Choosing a Breed

Your reasons for wanting sheep will help you determine the best breed for your purposes. Choose a breed whose characteristics are most important to you, whether for meat, wool, or mowing the lawn. Sheep can also be raised for milk, and some famous cheeses — such as feta, pecorino, and Romano — are traditionally prepared from ewe's milk, but you'd need a lot of sheep to get a sufficient amount of milk for cheesemaking. For this reason, a goat or a cow is a much more appropriate choice for backyard milk production.

Your climate will influence the breed you select, particularly if you live in a really wet area or hot climate. Look around and see what breeds of sheep are being raised locally — these breeds may also be the best ones for you.

Sheep come in so many breeds that it would take a whole book to describe them all. This section describes some of the most popular breeds, as well as a few minor ones. These brief descriptions will help you decide which breed is right for you.

Columbia. Developed in the United States, Columbia sheep are large animals that produce heavy, dense fleeces and fast-growing lambs. Columbias have a calm temperament and are easy to handle. They have an open (or wool-free) white face and are polled. Mature ewes weigh 150 to 225 pounds (70 to 100 kg); rams weigh 225 to 300 pounds (100 to 135 kg).

Corriedale. Coming from Australia and New Zealand, Corriedales are noted for their long, productive lives and ability to thrive in a wide variety of climates. These large, gentle-tempered sheep have been developed as dual-purpose animals, offering both quality wool and quality meat. A strong herding instinct makes them excellent range animals, as well. They have an open, white face and are polled. Mature ewes 130 to 180 pounds (60 to 80 kg); rams weigh 175 to 225 pounds (80 to 100 kg).

Dorset. One of the earliest breeds of British sheep, the Dorset is considered one of the best choices for a first sheep for both wool and meat. Their lightweight fleece is excellent for hand spinning, and they have large, muscular bodies and gain weight fast. Dorset ewes are good mothers, and Dorsets are one of the few breeds that can lamb in late summer or fall. A Dorset has little wool on its legs and belly, which makes lambing easier. Its

Quick Guide to Sheep Breeds

Breed	Size	Horns	Foraging capability	Best known for	Status
Columbia	large	naturally polled	good	fast growth; dense fleece	common
Corriedale	medium-large	naturally polled	good	high-quality meat and wool	common
Dorset, horned	large	impressive curling horns	good	fast growth; heavy muscling; lightweight fleece	uncommon
Dorset, polled	large	naturally polled	good	fast growth; heavy muscling; lightweight fleece	common
Hampshire	large	naturally polled	poor	fast growth; heavy muscling	common
Katahdin	medium-large	usually polled, occasionally horned	excellent	meat production; no wool	uncommon
Polypay	medium-large	naturally polled	excellent	high-quality meat and wool; high twinning rate	common
Romney	large	naturally polled	excellent	high-quality meat; long, soft fleece	common
Suffolk	large	naturally polled	poor	fast growth; heavy muscling	common
Tunis	medium-large	naturally polled	excellent	tender meat; lustrous fleece; heavy milker	uncommon

Columbia

Corriedale

Dorset

Hampshire

Katahdin

Polypay

Romney

Suffolk

Tunis

face is usually open, and it is white on both the face and the legs. Dorsets are medium size, have a gentle disposition, and may be either polled or horned. Mature ewes weigh 150 to 200 pounds (70 to 90 kg); rams weigh 225 to 275 pounds (100 to125 kg).

Hampshire. Developed in England, the Hampshire breed is among the largest of the meat types, and the lambs grow fast, making this breed popular for commercial production and 4-H projects. Its face is partially closed; the wool extends about halfway down. It has a black face and legs and is polled. Hampshires have gentle temperaments that make them popular with children. Mature ewes weigh 200 pounds (90 kg) or more; rams weigh 300 pounds (135 kg) or more.

Katahdin. Originating in Maine, the Katahdin is an easy-to-raise meat sheep with hair instead of wool. It does not require shearing, because it sheds its hair coat once a year. Katahdins can tolerate extremes of weather. Except that Katahdins do not produce wool, they possess all the ideal traits for a small flock: They are gentle, with mild temperaments; have few problems with lambing; are excellent mothers; and have a natural resistance to parasites. They have an open, white face and are polled. Mature ewes weigh 120 to 160 pounds (55 to 75 kg); rams weigh 180 to 250 pounds (80 to 115 kg).

Polypay. Developed in Idaho, the large and gentle-tempered Polypay is a superior lamb-production breed with a high twinning rate, a long breeding season, and good mothering ability. Polypays are known for having strong flocking instincts, quality meat and wool, and milking ability. They have an open, white face and are polled. Mature ewes weigh 150 to 200 pounds (70 to 90 kg); rams weigh 240 to 300 pounds (110 to 135 kg).

Sheep Attributes

White or black face. These terms describe the color of the wool on the sheep's head and face. Normally, the wool on the lower legs is the same color as that on the face.

Open or closed face. These terms are used to describe how much long wool is on the sheep's face. An open-faced sheep has only short, hairlike wool on its face. A closed-faced sheep has long wool on its face. On a closed-faced sheep, the wool may grow all the way down to the animal's nose. Too much wool around the eyes causes the sheep to become wool blind. The excess wool must be clipped away so the animal can see.

Prick or lop ear. Just as a German shepherd's ears stand up and those of a cocker spaniel hang down, a sheep's ears can stand straight up (prick ear) or hang floppily down (lop ear). Some sheep's ears stand out to the side.

Horned or polled. Horned sheep have horns. Polled sheep are born without horns.

Open and black face

Closed and white face

Romney. Coming from England, Romneys are gentle-tempered sheep and excellent around children. They are polled and have an open, white face, black points (noses and hooves), and a long, soft fleece that is ideal for hand spinning. They also produce good market lambs. Romney ewes are quiet, calm mothers. Romneys are best suited to cool, wet areas. Mature ewes weigh 150 to 200 pounds (70 to 90 kg); rams weigh 225 to 275 pounds (100 to125 kg).

Suffolk. Originating in England, Suffolks are by far the most popular breed in the United States for commercial production of meat and fiber and a common club lamb for 4-H and Future Farmers of America. They are similar to Hampshires, being large and having fast-growing lambs. They have an open, black face (unlike the Hampshire's partially closed face) and are polled. Suffolks are usually gentle, but some can be headstrong and difficult for younger children to manage. Mature ewes weigh 180 to 225 pounds (80 to 100 kg); rams weigh 250 to 350 pounds (115 to 160 kg).

Tunis. Believed to have roamed the hills of Tunisia and parts of Algeria prior to the Christian era, this breed has been around for more than 3,000 years, making it one of the oldest sheep

Qualities of a Good Sheep

- Good conformation, with no obvious defects
- Rear legs plump and well muscled
- Teeth meet dental pad well; jaw not overshot or undershot
- Strong pasterns (ankle joint just above the hoof)
- Rams: good fertility (have it checked by a veterinarian)
- Ewes: good udder with no lumps or damage

breeds. It is considered a minor breed because relatively few Tunis sheep may be found in the United States. They are medium size, hardy, docile, and good mothers. The reddish tan hair covering their legs and closed faces is an unusual color for sheep. They have long, broad, free-swinging lop ears and are polled. Their medium-heavy fleece is popular for hand spinning. The Tunis thrives in a warm climate, and the rams can breed in hot weather. The ewes often have twins, produce a good supply of milk, and breed for much of their lifespan. Mature ewes weigh 125 to 175 pounds (60 to 80 kg); rams weigh 175 to 225 pounds (80 to 100 kg).

Top 10 Tips for Health and Happiness

1. Have regular feeding times when you are supplementing your sheep's diet with grain or hay. Feedings are usually given once in the morning and once in the evening.
2. Stick to the regular diet. If you are going to change feed, do so gradually. Start adding the new feed and reducing the old feed over a period of about 10 days.
3. Provide plenty of fresh water. Water is an important part of your sheeps' good health, and they will drink more if it is fresh. If manure gets into the water, empty the container and clean it well before refilling it.
4. Lock down the grain supply. Sheep love grain and will try to figure out how to get into the grain storage area. Overeating grain can cause severe stomach upset and possibly death, so keep grain in a tightly lidded container the sheep cannot reach or knock over.
5. Never allow your dog or any other dog to play with your sheep. Sheep are afraid of dogs, and a dog barking at them through the fence or running in the pasture causes sheep severe stress, which can lead to overexertion, heat stress, and heart failure.
6. Watch for external parasites, and treat them immediately if you see any sign of keds, lice, mites, or fly-strike. Seek treatment advice from your veterinarian, county Extension agent, or an experienced local shepherd.
7. Keep hooves trimmed. Sore, overgrown hooves are not comfortable. For more on hoof trimming, see page 243.
8. Treat for worms. Consult your veterinarian or Extension agent for instructions on type and use of deworming medications. For more about deworming, see page 245.
9. Provide shade. Sheep need shade in the summer. An open-sided shed, shade trees, or a canopy roof will keep them cool.
10. Train your sheep to be calm and cooperative when routine handling is necessary. Peanuts and small bits of apple make great rewards and treats.

Heritage, Rare, and Minor Breeds

The breeds that have fallen out of favor with industrialized agriculture are referred to as rare, heritage, or minor breeds. Many of these breeds were major breeds just a generation or two ago, but as agriculture has focused on maximum production regardless of an animal's constitution, these old-fashioned breeds have begun to die out. The loss of heritage breeds can have an especially grave impact on homesteaders, who are usually interested in low-input agriculture, which means less work on the part of the farmer.

These breeds, although not the most productive in an industrialized system, have traits that make them especially easy to maintain. Some are dual-purpose, able to produce both meat and fiber. Others are acclimatized to regional environments, such as hot and humid or dry and cool conditions. Many perform well on pasture with little or no supplemental feeding. Others resist disease and parasites. Some have such strong mothering skills that they leave you with little to do during lambing season.

Interest today in preserving heritage breeds of livestock, including sheep, is increasing. A driving force in this movement in the United States is the American Livestock Breeds Conservancy (ALBC) and in Canada, Rare Breeds Canada (RBC). For more information about heritage breeds of sheep, contact these organizations or visit their websites.

Buying Sheep

Try to buy a sheep directly from the person who raised it, because you can ask questions about the sheep's history and see the flock it comes from. Avoid buying a sheep at an auction; you usually can't talk with the owner, you may be rushed into making your decision, and you can't be sure of the animal's state of health.

If you're buying a lamb for meat, look for a weaned lamb. Should you buy, or accept the gift of, a bummer lamb or a bottle baby? That depends on how much time you have, and what you intend to do with the animal in the long term — butchering a bottle lamb will be emotionally tough and some people simply can't do it.

In looking for the ideal sheep, above all it should be healthy. You can gauge a sheep's condition by examining it carefully for the following characteristics:

- It should seem alert and thriving.
- It should come from a flock with no major medical problems. If the whole flock is healthy, your sheep has probably been well cared for, is of sturdy family stock, and has not been exposed to disease or parasites.
- It should be the normal size and weight for an animal of its age and breed. A large-boned sheep will have more meat, and the ewes will handle pregnancy and birth more easily.

Physical Condition

Examine the sheep's eyes, teeth, feet, and other body parts, including its fleece. Look for signs of good health and conformation, and avoid purchasing a sheep with any of the following problems.

Eyes. Runny, red, or damaged eyes may mean the sheep is diseased.

Teeth. Worn or missing teeth will interfere with eating.

Jaw. The lower jaw should not be undershot or overshot (see the illustrations on page 228).

Head and neck. Lumps or swelling under the chin may be related to bottle jaw, indicating an untreated worm infestation, or to abscesses caused by a bacterial infection.

Feet. A sheep with untrimmed or overgrown hooves has not been cared for properly. If the sheep is limping, it may be injured or have foot rot (see page 243). Even if the sheep you're considering seems healthy, notice whether other sheep in the flock are limping. If a sheep in the flock has foot rot or other foot problems, your sheep may also be infected.

Body conformation. Sheep with narrow, shallow bodies tend to be light muscled. Sheep with wide backs and deep bodies are more desirable.

Body condition. Don't choose a ewe that is extremely thin. However, it's normal for a ewe that has just raised lambs to be thin, which does not mean she has a health problem. Don't choose a fat ewe, either. She may have trouble lambing.

Potbelly. A thin lamb with a potbelly usually has a heavy infestation of worms.

Udder. If you are buying an adult ewe, check the udder. If it is lumpy, she may have had mastitis, which sometimes causes a ewe to have no milk for any future lambs.

Proper Sheep Conformation

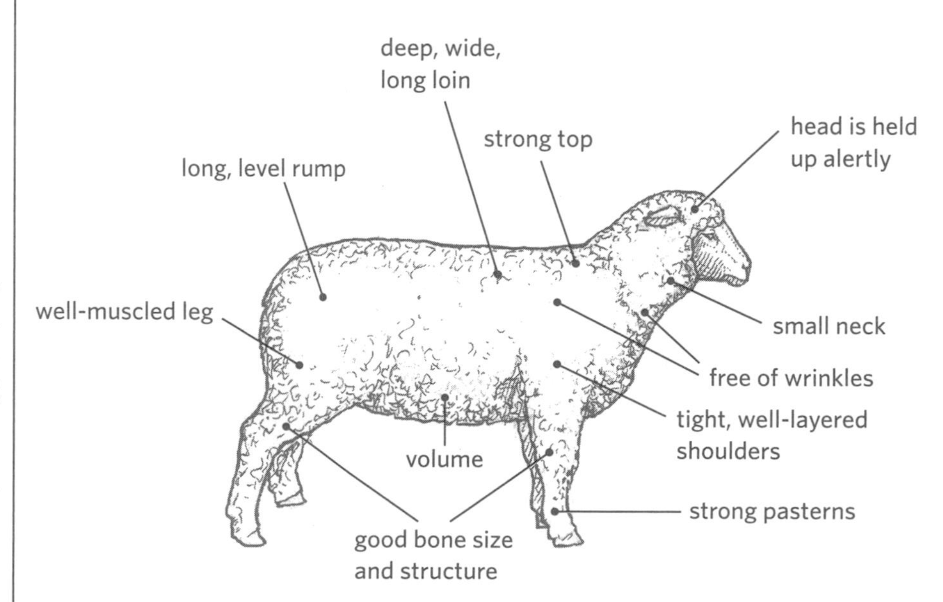

Tail. The tail should be cut short, or docked, which for health reasons should be done to all lambs at 2 to 3 days of age. An extremely short tail, however, is a serious defect in a breeding ewe, as it weakens the tail area and may create problems during lambing.

Manure. Runny droppings or a messy rear end can mean the sheep is sick or has worms. However, a lamb that has been grazing on new, lush spring grass may naturally have runny droppings.

Wool covering. Each breed has a certain amount of wool on its face or legs. Your sheep should have the correct amount for its breed. Avoid sheep with excessive wool around the eyes, which can lead to wool blindness.

Fleece. A ragged, unattractive fleece may be an indicator of disease, keds, or lice (see page 245).

External parasites. Look for any signs of external parasites, including keds and lice.

Leg Conformation

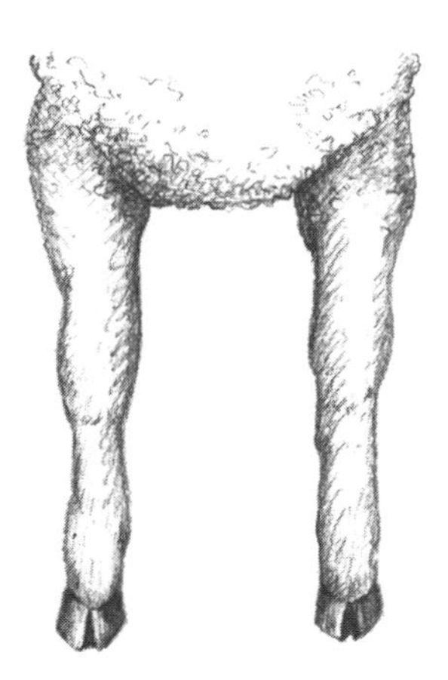

Correct

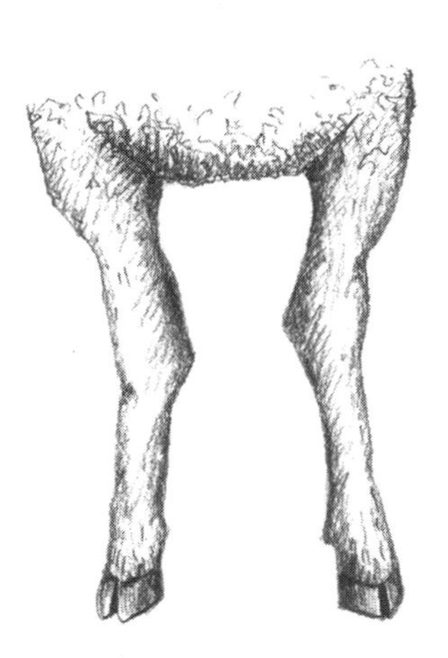

Knock-kneed and splayfooted

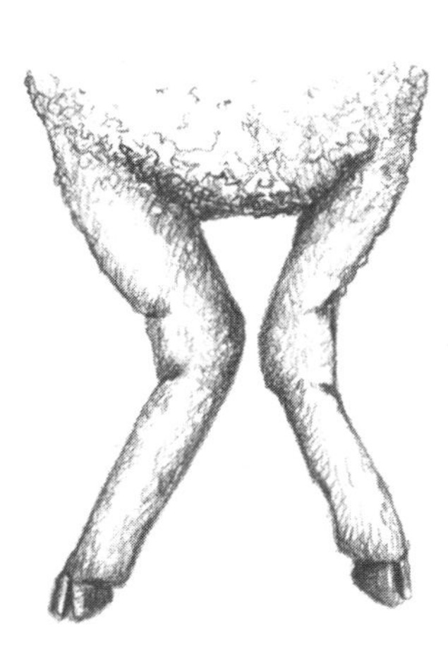

Bent leg

Jaw Conformation

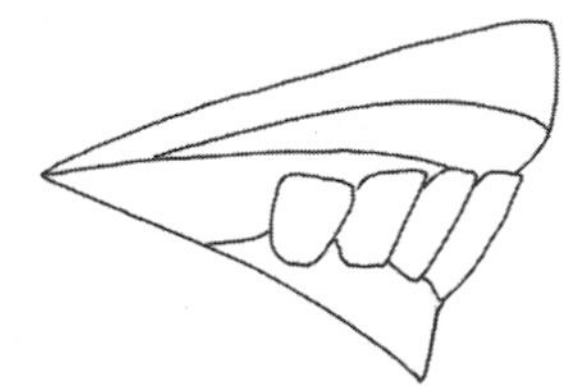

Good jaw conformation

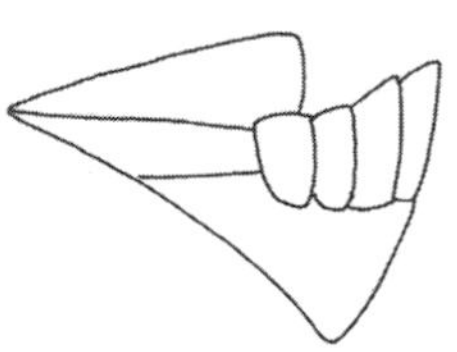

Undershot jaw

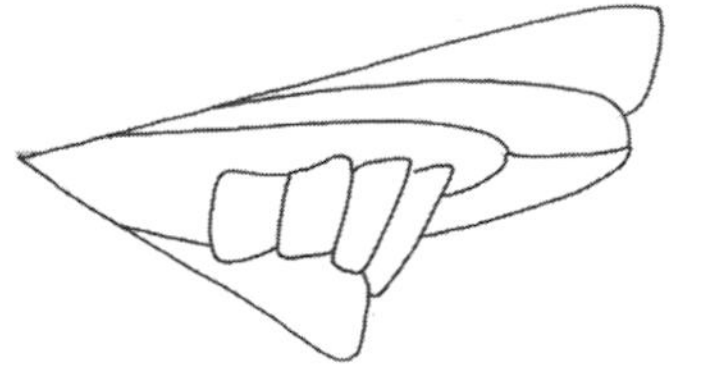

Overshot jaw

How Old Is My Sheep?

Until the age of 5, you can determine the age of sheep to some extent by their teeth. A well-cared for backyard ewe may remain productive to the age of 10 or 12 years, and some hardy individuals live 20 years or more.

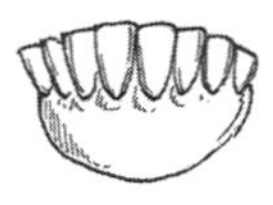
lamb

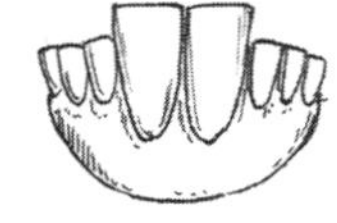
yearling

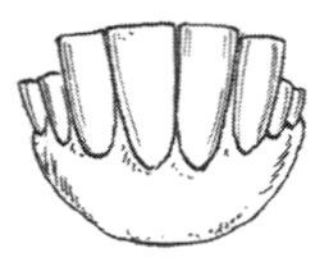
2-year-old

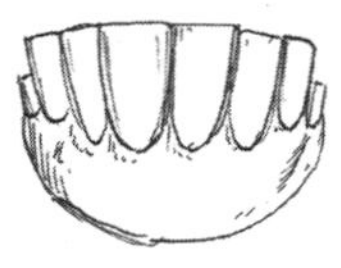
3-year-old

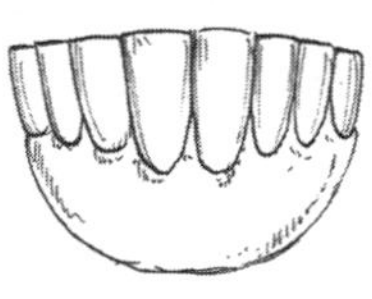
4-year-old

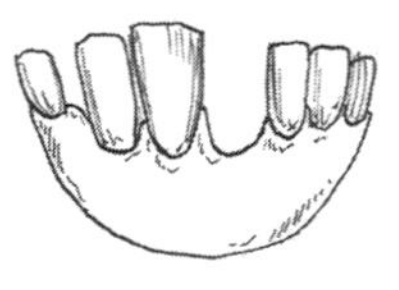
older than 4 years

Judging the Age of a Sheep

The front teeth of sheep are similar to the front teeth of humans, with one big difference: Sheep have front teeth only on the bottom. Where you would expect to find upper teeth, a sheep has a hard gum line called a dental pad.

You can estimate the age of a sheep by examining its teeth. The front teeth of sheep begin to show at 2 to 3 weeks of age, starting with one pair in the center. This pair is followed by three more pairs a few weeks later. The lamb will have eight teeth in all. Sheep lose their baby teeth, and permanent teeth grow in. When the sheep is about 1 year old, the first pair of permanent teeth appears. Each year after that until the sheep is 4 years old, it gains another pair of permanent teeth. When the sheep is 4 years old, all four pairs of baby teeth have been replaced with permanent teeth.

Past the age of 4, the incisors start to spread, wear, and eventually break. When a sheep loses some of its teeth, it's called broken mouthed. When an aged sheep loses all its teeth, it's called a gummer.

What to Ask the Sheep's Current Owner

Don't be afraid to ask these important questions only the seller can answer.

What vaccinations or treatments has the sheep had, and when? Young lambs may have received some, but not all, of their primary vaccinations. The sequence and timing of the vaccinations are important. You may be purchasing a lamb in the middle of its immunization schedule. You will need to know what shots were given and when so you can get the rest of the shots on schedule and with the same product.

When was the sheep dewormed, and with what? You will need deworming information so you can give scheduled maintenance dewormings using the same product the animal has already received. (For more about deworming, see page 245.)

What kind of feed is the sheep being fed? Any sudden change in the kind of feed you give your sheep can make it sick. Digestive upset can also slow down your sheep's growth or even

Top Tips for Buying Lambs

- Buy a healthy lamb. Ask your veterinarian to examine the animal for signs of disease.
- Learn to recognize the finer points of conformation. Exceptional conformation may be worth the money.
- Price does not always reflect quality. The most expensive lamb is not necessarily the best one.
- Most faults don't get better with time. If the lamb has faulty conformation, its defects will become more pronounced with age.
- Don't buy what you cannot see. Learn how to discern desirable traits, such as a thick leg muscle, by feeling the lamb.
- Spend only 80 percent of what you think you can afford. Expenses you haven't planned for will always come up.
- An inexpensive lamb with poor conformation is not a bargain. Keep shopping until you find affordable quality.

cause its death. If you are not sure what to feed your sheep, continue to give it the same kind and amount of feed it has received up to now.

Was the sheep a twin? Was its mother a twin? Twinning is a highly desirable genetic trait. The possibility of twinning is mostly determined by the ewe. If the ewe is a twin, she is more likely to give birth to twins. However, even if you are buying a ram, ask if he was a twin. If so, his daughters are also likely to give birth to twins.

If the sheep is an orphan lamb, why is it an orphan? A lamb may have been pushed away by the mother because she had triplets or quadruplets, and this one was just one too many for her to care for. But if the ewe was simply a poor mother — some ewes are not as interested in mothering as others are — or had no milk, her daughter may have inherited these traits. The female bummer will therefore be less valuable for breeding when she grows up.

Has the flock had a history of medical problems? Diseases spread quickly in a flock. A sheep from a sick flock may have been exposed to a number of diseases, especially foot problems. Notice whether any sheep in the flock limp or kneel when grazing. If their feet show signs of neglect, such as overgrown hooves, be cautious about buying from the owner of this flock. Along the same lines, don't buy sheep that show signs of having lice, keds, or problems with internal parasites.

Is there a record of this sheep's growth? Ask for the date of the lamb's birth, its weight at birth, the date of its weaning, and its weight at weaning. You can use this information to figure out its rate of growth. A lamb's rate of growth is more important than its size. Two lambs of equal weight may be two or more months apart in age. If you are raising a lamb for meat, do not purchase a slow-growing runt, because it won't gain enough to be economical. After you buy your lamb, weigh it frequently and write down the weight on your sheep record.

Handling Sheep

To be a successful shepherd, you must learn to take advantage of the sheep's way of thinking. A sheep's most powerful instinct is, at all costs, to avoid being trapped. A cat has claws, a porcupine has quills, and a skunk has scent, but a sheep has only one defense from danger — she can run to escape.

This drive to avoid being trapped is why dogs in the pasture can cause so much trouble. When the sheep see the dog, they run to escape. When the sheep begin running, the dog thinks it's fun and begins chasing them. It's a vicious circle.

Sheep will come to you if you don't do anything to startle them or make them think you're trying to chase them. For instance, if you try to drive sheep into a barn, they will avoid going in if they think you are trying to trap them. But sheep love to eat and will do almost anything to get their favorite treat. They especially love peanuts, apples, and grain. If you show them a bucket of grain and use it to coax them to follow you, you can lead them into the barn easily. In fact, they will be so eager to follow you they will get pushy.

Allow the sheep time to eat the grain before you try to pen them or catch one. The grain is their reward for coming in, and eating it should be a good experience for them. Once the sheep are indoors or in a pen, use lambing panels or a gate to squeeze them together so they can't run away.

Basic Behavior Characteristics

Understanding your sheep's behavior will be easier if you remember the following:

- Sheep fear noise, unfamiliar surroundings, or unfamiliar items in their surroundings (a jacket hanging on the shovel against the wall, for example), strange people, dogs, and water.
- Sheep move readily from a dark area to a light area, from a confined area to an open area, from a lower area to a higher area, and toward food.
- Sheep like to follow one another and move away from people, dogs, and buildings.
- Sheep will cling to a wall in a pen but will bunch up in sharp corners and stay there.

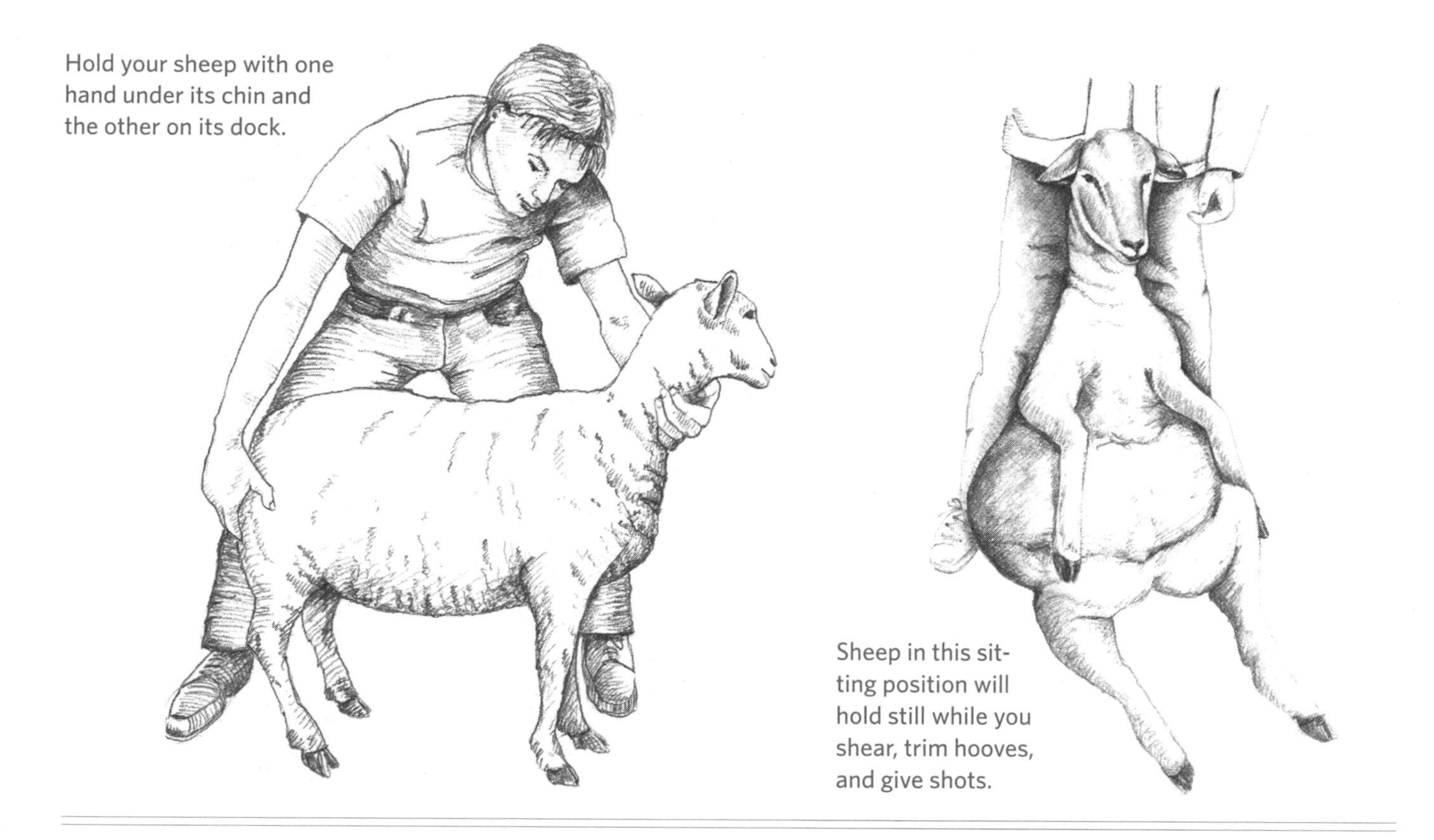

Hold your sheep with one hand under its chin and the other on its dock.

Sheep in this sitting position will hold still while you shear, trim hooves, and give shots.

Then walk up to the group, catch one sheep, and move it to an area where you can handle it.

If you say a sheep's name and offer it a treat at the same time, it will soon come when you call its name. You'll be surprised how quickly it will learn. If you sit quietly on a small stool or box in the middle of the sheep pen, your sheep will be curious about you and begin to approach. Remain still and talk softly; your sheep will soon come close enough to smell your hair and nibble at your clothing.

As your new sheep become tame, they will probably want to be scratched and petted. Never attempt to scratch a sheep on the top of its head or nose; sheep don't like it. They prefer to be stroked under the chin and scratched on their chest between their front legs. Some sheep like to have their back or rear scratched.

The time will come when you will need to make a sheep do something it doesn't want to do. By training a sheep properly, you can get it to do such things without becoming frightened or losing its affection for and trust in you.

Although you can train calm adult sheep that are used to people, it's easier to begin training as early as possible. A lamb that has not been handled much is practically impossible to catch in a pasture. Training should start once the lamb begins to nibble at grass at about 2 weeks of age.

Never hold a sheep by its wool; that hurts. To make a sheep stand still, place one hand under its chin and the other hand on its hips or dock (tail area) and slightly to the rear. When held this way, it can't go forward or backward, and it can't swivel away from you. With your hand gripped firmly under the chin, walk your sheep forward by giving it a gentle squeeze on the dock. As the sheep starts moving, all you need to do is keep up with it. When you want it to stop, hold it back with the hand under its chin.

Training should start once the lamb begins to nibble at grass at about 2 weeks of age.

The Chair Hold

To give a sheep shots, trim its feet, or shear its wool, take advantage of a reflex every sheep has: Once all four of its feet are off the ground, it can be placed on its rump and it will sit still.

To get your sheep into this sitting position, slip your left thumb into the sheep's mouth in back of the incisor teeth and place your other hand on the sheep's right hip. Bend the sheep's head sharply over its right shoulder, and swing the sheep toward you. Lower it to the ground as you step back. From this position, you can lower it flat on the ground or set it up on its rump for foot trimming.

Withhold feed for several hours before handling or shearing sheep. This sitting position is uncomfortable for a sheep with a full stomach.

Halters

When a lamb is about 1 month old, begin to train it with a halter. Your sheep will not like its halter at first, but with practice and repetition, it will soon settle down. After the halter is in place, grip it close to the head and place your other hand over or behind the sheep's hips, just as you would without a halter. Start leading by pulling on the halter and pushing on its rear. Be patient. A lamb may buck and fuss a bit, but it will soon get used to working with the halter.

Making a Halter

To make a halter for training and tying up, you will need an 8- to 10-foot (244–305 cm) length of ⅜-inch-diameter (0.95 cm), three-ply nylon rope. Cut the rope by holding it over a flame at the desired length. Slowly rotate the rope over the flame. Once the nylon rope has melted apart, and while the melted nylon is still hot enough to stick together, use pliers to squeeze the ends to seal the nylon strands together. Do not use your fingers; the ends of the rope will be hot. Loop the rope together, as shown in the following steps.

To create holes in the rope, twist the ply open.

Step 1

Step 2

Step 3

Step 4

Feeding Sheep

Sheep, like goats and cows, are ruminants. They have four compartments in the upper part of their digestive systems that work together to break down plant material such as fresh grass and hay. (See pages 184-85 for a detailed description of the ruminant digestive system.)

Ewes and Rams

Feed for mature ewes and rams consists of grass pasture, salt, a mineral/vitamin supplement, and water. If you are going to breed your sheep you will need to increase their level of nutrition. From the day you bring your sheep home, you will need to provide a constant supply of fresh water. For a few sheep, a washtub works nicely for watering. Do not use a bucket, which is easily tipped over.

Sheep feed themselves much of the year from pasture grass. In periods of drought, when grass becomes short, or in winter, supplement your sheep's diet with hay and/or grain. Feed sheep a green, leafy grass or alfalfa hay. Never throw the hay onto the ground, where it will become dirty and wet and contribute to worm infection. Use a feeder.

A supply of salt is important for good health. The salt should be loose or granulated. Never feed sheep a salt block intended for cattle. The sheep may attempt to chew the block and harm their teeth. A good mineral and vitamin supplement is also necessary, and should be of a sort intended for

Suggested Feeding Schedule for an Orphan

Age	Amount of milk
1-2 days	2-3 ounces (59-89 mL) six times a day, with colostrum
3-4 days	3-5 ounces (89-148 mL) six times a day (gradually changing over to lamb milk replacer)
5-14 days	4-6 ounces (118-177 mL) four times a day, and start with leafy alfalfa and crushed grain or pelleted creep feed
15-21 days	6-8 ounces (177-237 mL) four times a day, along with grain and hay
22-35 days	Slowly change to 1 pint (0.5 L) three times a day; after the lamb is 3 months old, feed whole grain and alfalfa or pelleted alfalfa containing 25 percent grain, but change rations very gradually

sheep. Cattle minerals often contain levels of copper that can be toxic to sheep. These supplements are usually placed in sturdy wooden boxes that won't tip over, or in a hanging feeder.

If you need to change their feed, do so gradually. Sheep cannot adapt quickly to big changes in their diet, such as a change from feeding on pasture grass to feeding on grain. The microorganisms that help digest grass differ from those that digest grain. Therefore, make dietary changes slowly, to give the appropriate microorganisms time to grow in numbers. A grassfed animal needs a week or more before it can digest a large amount of grain.

Bummer Lambs

Bottle lambs take a tremendous amount of effort, but raising them successfully is a truly rewarding experience. They need to be fed frequently, day and night, but in small quantities (see chart above for schedule), and you can't skip feedings or the lamb may fade away amazingly fast. It is very important that you only feed small quantities of food: overfeeding can cause scours (a severe form of diarrhea). Almost half of all lambs that die are the victims of scours, and bottle-fed lambs are particularly prone to this affliction.

Keep bottles, nipples, and milk containers clean, and the milk refrigerated between feedings. Warm the milk immediately before feeding, but don't get it too hot or it will scald the lamb's mouth.

If scouring occurs, substitute a day's feeding with oral electrolytes — give no milk. In a pinch, you can make a homemade electrolyte solution (see recipe on page 202), but ultimately it pays to keep some powdered electrolyte on hand that is specifically prepared for livestock. Such products (those labeled for calves work fine for lambs) contain not only electrolytes but also vitamins, minerals, and energy components. A few teaspoons of Pepto-Bismol or diarrhea medication for human infants can also be used for baby lambs.

Pasture Rotation

Pasture rotation is good for grass growth as well as for sheep health. Use woven-wire fencing or portable electric fencing to divide the area into four to eight smaller paddocks. While the sheep spend a few days grazing one paddock, the grass in the empty paddocks has time to regenerate. With the sheep rotated through smaller pastures, the grass they eat is always fresh.

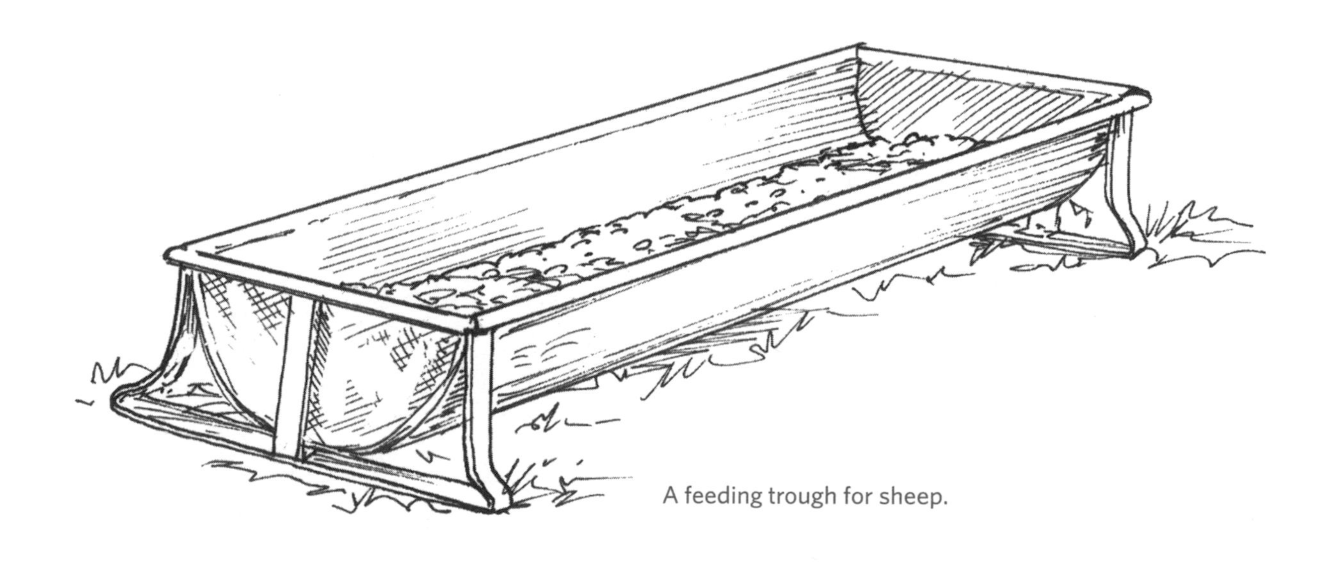

A feeding trough for sheep.

Plants Toxic to Sheep

Sheep are curious and love to nibble at strange plants, so you must prevent them from contact with any poisonous plants or shrubs in their grazing area. Your county Extension agent can tell you what poisonous wild plants grow in your area, but you'll need to investigate potentially dangerous ornamental plantings. The most common plants that are poisonous to sheep are American or Japanese yew, lupine, milkweed, nightshade, oak, ragweed, rhododendron, sheep laurel, and tansy ragwort.

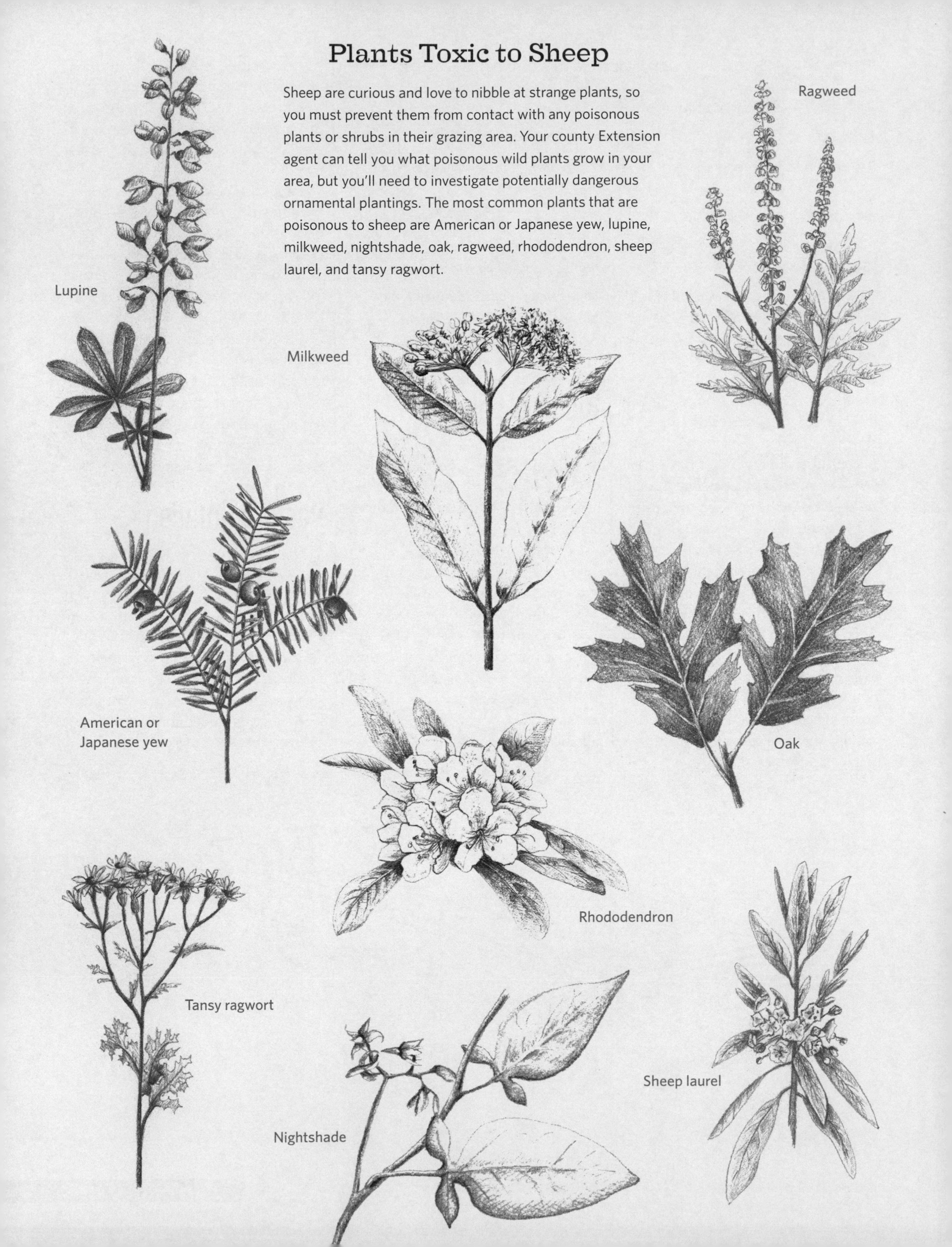

Fencing & Shelter

Before you bring your first lamb home, fencing around your sheep pasture should be your first priority. Build a fence if you have none, or check existing fences to be sure they are secure. Fences keep in sheep and keep out predators, such as dogs and coyotes. A dog playing with a sheep in the pasture can kill or seriously injure it, and coyotes prey on sheep. Sheep and lambs need shelter only in bad weather — heat, cold, rain, and snow.

Fences. Of the many types of fence, the best for sheep is a smooth-wire electric or a nonelectric woven-wire fence. Some nonelectric wire fences have six or more tightly stretched wires and heavy posts, but woven-wire sheep fence is better. This fencing is designed not only to keep sheep in, but also to keep predators out.

The strength of the electric fence is in the shock. Electric fence chargers, or energizers, are available at farm supply stores and fence supply dealers. Most of these chargers operate on a few dollars' worth of electricity per month.

Fences permit you to rotate your pasture. Cross-fence, or divide, your large pasture into smaller paddocks by using inner fences. The outer perimeter fence must be permanent and sturdy to keep dogs out of the pasture, but the inner fences can be made of inexpensive woven-wire or portable electric fencing, which is quick and easy to construct.

Shelter. When they must seek shelter, sheep are happiest in a south-facing three-sided shed; it's well ventilated and offers adequate protection from the weather. If your sheep shelter is an enclosed shed, it must be well ventilated. You can build or buy a small portable shed that can be moved easily to whichever paddock the lambs and ewes are currently in. Allow 15 to 20 square feet (1.5 to 2 sq m) per adult animal.

Whatever sort of shelter you have, keep it clean. If you think it smells bad, so will your sheep — and the dirty shelter will be a potential breeding ground for disease.

Even good fencing can't always provide enough protection for sheep at night. The shelter should be located within a corral or small pen constructed of tightly woven wire or cattle panels that will positively keep dogs and other predators from getting to the sheep. If you place feed in the shed each evening, the sheep will naturally head toward it every night.

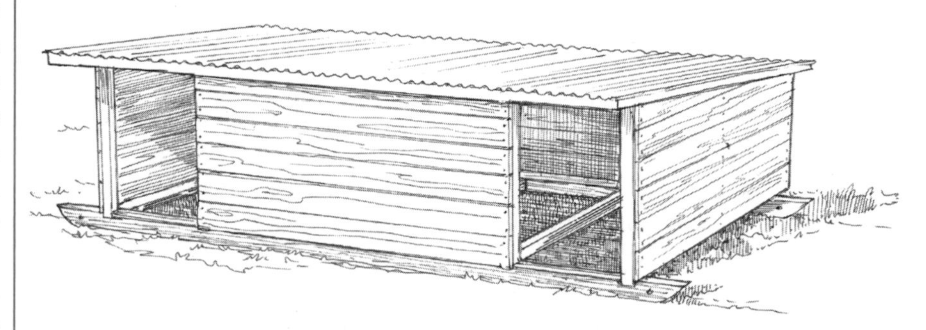

A portable structure provides flexibility and inexpensive shelter for sheep and lambs on pasture. Because sheep use the shelter only in bad weather, up to six ewes and their lambs can share one shelter.

Of the many types of fence, the best for sheep is a smooth-wire electric or a nonelectric woven-wire fence.

Sheep Manure

Use sheep manure as fertilizer in your garden. If you don't have a garden, you can sell packaged manure to gardeners in your area. Sheep manure is dry, has little odor, will not burn plants, and is an excellent substitute for commercial fertilizer. Sheep manure also has more nitrogen, more phosphorus, and more potassium than cow manure.

Predators & Protection

Predators are a potential problem for all shepherds. But not all predators kill sheep, and predators are important members of the food chain, creating a balance that nature depends on. Predators keep populations of wild herbivores, such as deer and elk, from overpopulating their ecosystems, and they feed on lots of small rodents and rabbits. They'll also eat insects and carrion, which is often abundant along highways.

Predator species tend to be opportunistic animals, seeking the easiest target to meet their needs. In other words, they usually go for young, old, weak, or sick animals first, though some have been known to attack mature, healthy animals in their prime. All predators become more aggressive as their hunger increases — during a drought, for instance — and may attempt to take anything they can sink their teeth into.

As a shepherd, you can learn to manage your flock so a predator will decide that eating at your place is a lot more difficult than chasing mice and rabbits. The box on page 237 will give you some ideas on how to minimize predation. Keep in mind that your sheep are less likely to suffer from predation if they are strong and healthy; good feed and adequate health care pay in more ways than one.

Identifying Predators

Although coyotes kill more sheep than dogs do, more shepherds experience predation by dogs. Bears and wildcats can also create nightmares for a shepherd, and occasionally birds of prey (eagles and hawks) and carrion birds (vultures and ravens) may be the culprits.

Sometimes predators get a bum rap: If the corpse of a dead sheep has obvious bite marks, it's natural to think a predator was the perpetrator. But sheep die from a number of causes, and unless you see a predator attacking a live animal, the sheep may have died of natural causes and then been fed on by scavengers.

Coyotes. Wile E. Coyote may have looked the fool in all of his encounters with the Road Runner, but he's not a good example of the species. Coyotes are intelligent, curious, and adaptable and they have been expanding their range into urban and suburban areas of the United States.

Coyotes generally select lambs over adults, unless hunger has made them desperate. Smaller lambs and those born to young, old, or crippled ewes are more commonly victims than those of middle-aged and healthy ewes. And a coyote is more likely to take the smallest lamb from a set of twins or triplets than to take a larger, single lamb.

Dogs. Dogs are a special class of predator for shepherds. An estimated 62 million or more owned dogs in the United States — not counting countless feral and abandoned dogs — means that

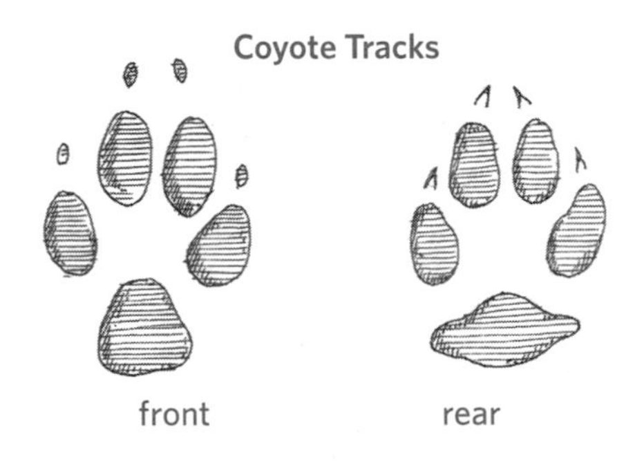

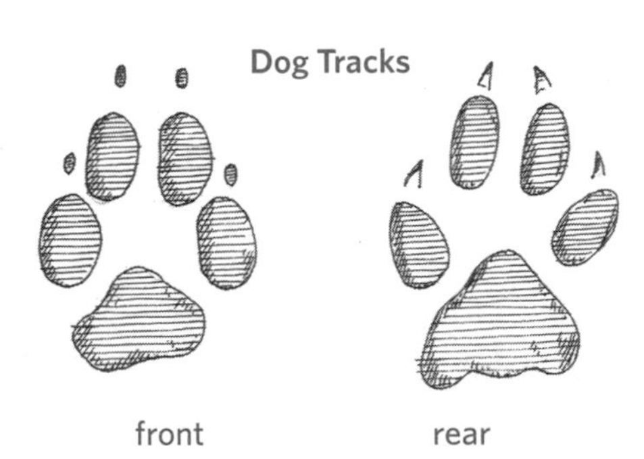

The coyote track is about 3 inches (7.5 cm) long for the front foot, and 2½ inches (6.3 cm) for the rear. Dog tracks can vary from slightly larger than a coyote track to quite a bit smaller, depending on the size of the dog. Notice that the pads vary. If you live where wolves present predator problems, look for a doglike track that is about 5 inches (12.7 cm) long in the front.

Discouraging Predators

Predators can be discouraged by the following techniques:

- Keep guardian animals, like dogs, donkeys, or llamas.
- Use a lighted night corral with a high, predator-tight fence.
- Put high-frequency bells on some of your sheep; you can hear the bells if the sheep are being chased, and predators are less likely to chase sheep that are wearing bells.
- Keep your sheep in an open field in sight of your house.
- Use coyote snares along fence lines to catch both dogs and coyotes. First check the legality of using snares in your area.
- Have a gun and know how to use it. Even a pellet gun can drive off an attacking dog. A dog running through a flock of sheep is not an easy target, but like most predators a dog will spook at the sound of a gun shot into the air.
- Use a live trap, which is a cage that allows the trapped animal to be set free. Such traps work well with dogs but are of little value with coyotes because they are too wily to be caught. State wildlife officers may supply live traps for bears or wildcats that are repeat offenders.
- Use a propane exploder (which produces a loud explosion), a radio, or other noise-making device and frequently change its placement, volume, and timing to prevent predators from getting used to the device and losing their fear of it.
- Use a combination strobe light and siren, a device developed and tested by the USDA that seems to significantly reduce predation.

for every sheep more than six dogs are roaming around. And they can prove more dangerous than coyotes, as one or two dogs can maim or kill dozens of sheep in one night. Many people have been driven out of the sheep business because of dogs.

Fido and Spot don't have to be wild, vicious, or even brave to chase sheep. When dogs chase sheep, they're following their natural impulse to chase whatever runs. Unfortunately, sheep run at the slightest disturbance. The offending dogs are not as much at fault as owners who don't keep their pets at home.

Dogs attack rather indiscriminately — they grab at any part of a sheep they can. Sheep often survive dog attacks but may be badly injured. Even dogs that are too small to kill or maim a sheep can still cause heart failure in older ewes and abortion in pregnant ewes. Broken legs may also result when dogs chase sheep.

Other predators. For most shepherds, wolves and foxes are less problematic than coyotes, though both are similar to coyotes in their style of predation and feeding. Wolves tend to hunt in packs of two to four animals and can easily take down adult sheep. Because of their small size, foxes take only fairly small lambs.

Bears and wildcats (mountain lions, lynx, or bobcats) are common in more remote areas, especially in the West. These animals can kill perfectly healthy, strong adult sheep as easily as they can a young or an old sheep. They often kill more than one animal in a single attack and then feed on their favorite parts of each kill.

Cats generally attack by biting the top of the head or neck. Bears usually use their massive paws to strike a sheep down. Eagles and other birds of prey occasionally kill lambs. They attack by dropping out of the air at high speed and closing their talons into the head.

Laws

Many wild predators are protected or controlled by federal or state laws and regulations. If you have, or suspect you have, a problem with wild predators, call the Wildlife Services office of the United States Department of Agriculture or your state's wildlife office to learn about specific remedies and laws in your area. In Canada, the Livestock, Poultry and Honey Bee Protection Act provides for compensation for sheep producers who lose animals to predators; contact the Ministry of Agriculture in your area to learn more.

Your county Extension agent should be able to tell you the county's dog laws or, better yet, give you a copy of the county or state laws. You may find they are strict and well spelled out but lack enforcement.

Some states permit the elimination of any trespassing dog that is seen molesting livestock. Others require the owner to have the dog destroyed or be charged with a misdemeanor. Most states allow the livestock owner to recoup payment from the dog's owner (if known) for both damage and deaths to livestock. If a dog is chasing or killing sheep, promptly contact your local sheriff or animal-control officer. They can assist you in determining the dog's owner and can impound the dog and press charges against the owner on your behalf.

Guardian Animals

Sheep have been bred for thousands of years to be docile, a trait that makes them easy victims. However, other species become quite aggressive when predators invade their territory, and shepherds have harnessed this trait to defend their sheep for almost as long as sheep have been domesticated.

Attacks usually occur at night or early in the morning when you're normally asleep. A guardian animal is on duty 24 hours a day and is most alert and protective during the hours of greatest danger.

Few guardian animals actually kill predators, but their presence and behavior can reduce or prevent attacks. They may chase a trespassing dog or coyote but should not chase it far. Chasing for a prolonged distance (or time) is considered faulty behavior because the guardian should stay near the flock, between the sheep and danger. The best guardians balance aggressiveness with attentiveness to the sheep.

Dogs, ponies, llamas, donkeys, mules, and cows are all used as sheep guardians. Even geese have been used to guard sheep, although they're more effective against domestic dogs than against wild predators. Whichever type of guardian you're considering, remember the following:

- The guardian needs to bond with the sheep, and the bonding process can take time.
- Guardians should be introduced slowly, across a fence; it's usually easiest to make the introduction in a small area rather than in a large pasture.
- One guardian is generally sufficient except on open range, where at least two are required. In a large pasture or on open range, bigger animals (such as donkeys and llamas) may be more effective than guardian dogs — though dogs can significantly cut down on losses, even in large areas.
- Each animal is an individual and wil react differently in different situations. Some individuals don't make good guardians.

Guardian Pros and Cons

Although guardian animals can be a great help to a shepherd, keeping them may have some drawbacks as well.

The benefits of using a guardian animal include these:

- Reduced predation
- Reduced labor and fencing costs
- Increased utilization of pastures
- Environmentally benign predator control

Some of the potential problems include the following:

- Playfulness, which can be deadly to sheep
- Lack of guarding ability — some guardians don't guard but rather are uninterested or roam away from the flock
- Aggressiveness with people
- Interference with your need to work or move the flock
- Destructive behavior — some guardians destroy property by chewing, digging holes, and so on

A donkey will usually bond with sheep and protect them from predators.

Different breeds of dog are suited for different circumstances, requiring careful selection to best suit your sheep-guarding needs.

Harvesting Wool

With the exception of hair sheep, which have little or no wool, sheep must have their wool sheared off annually. Shearing is usually done in the spring so the sheep are free of their heavy wool coats for the hot summer months. Shearing is important for keeping your sheep healthy and comfortable. This annual haircut has the additional benefit of producing wool, which you can either spin yourself or sell to handcrafters.

Many people raise sheep for the single purpose of supplying themselves with wool. You can dye and spin the wool yourself to make yarn with which to knit sweaters, socks, mittens, hats, and many other projects. Or, you may decide to felt your wool. By employing the technique of wet felting, you can make scarfs, hats, mittens, and many other products; or, use needle felting to create toy animals and other felted objects. For more information on the type of wool produced by each of the wool sheep breeds, refer to *The Fleece and Fiber Sourcebook* by Carol Ekarius and Deborah Robson.

Even if you are not a spinner or felter you can profit from your fleeces. You might sell them to a wool pool, which combines and processes fleeces from many sources. However, selling good hand-spinning fleeces directly to individual spinners can bring in considerably more money. If you earn a reputation among spinners for having prime-quality fleeces, your fleeces will be in demand far ahead of shearing time.

Fleece Quality

Several characteristics determine the quality of a fleece, including wool type, strength, and lack of vegetation. The type of wool — silky, curly, or coarse, for instance — is a characteristic of the breed. The strength of the wool is determined by health and nutrition. But even if your sheep has a strong fleece of a desirable type, it will be less valuable if it is not kept clean. To keep a fleece clean, use a sheep coat to protect it from vegetation, such as seeds, hay, burrs, and bedding. (See page 240 for how to make your own sheep coat.)

Preparing for Shearing Day

The first shearing you observe may be quite a surprising scene — a sheep must be held in awkward positions while its fleece is removed with sharp clippers or shears. While sheep don't enjoy the process, they feel a lot better after shearing. They're happy to be rid of their heavy winter coat of wool, and shorn sheep are easier to keep free of external parasites. To reduce the stress for both sheep and shearer:

- Don't give sheep food or water for about 12 hours (or overnight) before shearing. A sheep with an empty stomach is more comfortable during shearing.
- Be sure your sheep is dry. If rain is predicted, keep the sheep indoors the night before.
- Get your sheep into a pen where each one can be caught easily. A shearer must not be expected to chase or catch your sheep.
- Prepare a good shearing floor. A smooth plywood platform is better than a dirt floor or grassy area. Spread a tarp or an old wool rug on the platform to make it less slippery.
- When shearing is done indoors, provide sufficient light.
- Have cold water or another beverage available for the shearer.
- Arrange a skirting table, which is a raised table with a slatted top. When you skirt the fleece (remove a strip about 3 inches (7.5 cm) wide from the edges of the fleece), the skirtings (the trimmed-off edges) will fall through the table to the floor. Sanded lath, long dowels, or 1-by-2-inch lumber set on edge makes a tabletop suitable for skirting.
- Have first-aid supplies on hand in case a sheep or the shearer is severely cut. Don't worry about little nicks on sheep. They may look bad, but the lanolin in sheep's wool helps them heal quickly. In hot weather, spray each shorn sheep with a fly repellent before releasing it to prevent infection in the tiny cuts.
- Handle the fleece carefully, so it doesn't become contaminated with manure, straw, or dirt. Sweep the floor after each sheep.
- Shearing is a good time to trim hooves, check udders, deworm, treat for parasites (after the wool is removed), and check the teeth of oldsters. Have all the necessary supplies on hand.

Skirting

After each sheep is shorn, set the fleece sheared side down on the skirting table. (This position will shake off any second cuts — undesirable short clumps of hair resulting from imperfect shearing.) Then sweep the floor of the shearing area so that the shearer can get going on another sheep.

Skirt the fleece, allowing the cuttings to fall through the slats of the skirting table onto the floor. Roll up the fleece and secure it with paper twine (available from sheep supply stores) or place it in a paper feed sack (turned inside out) or a cardboard box.

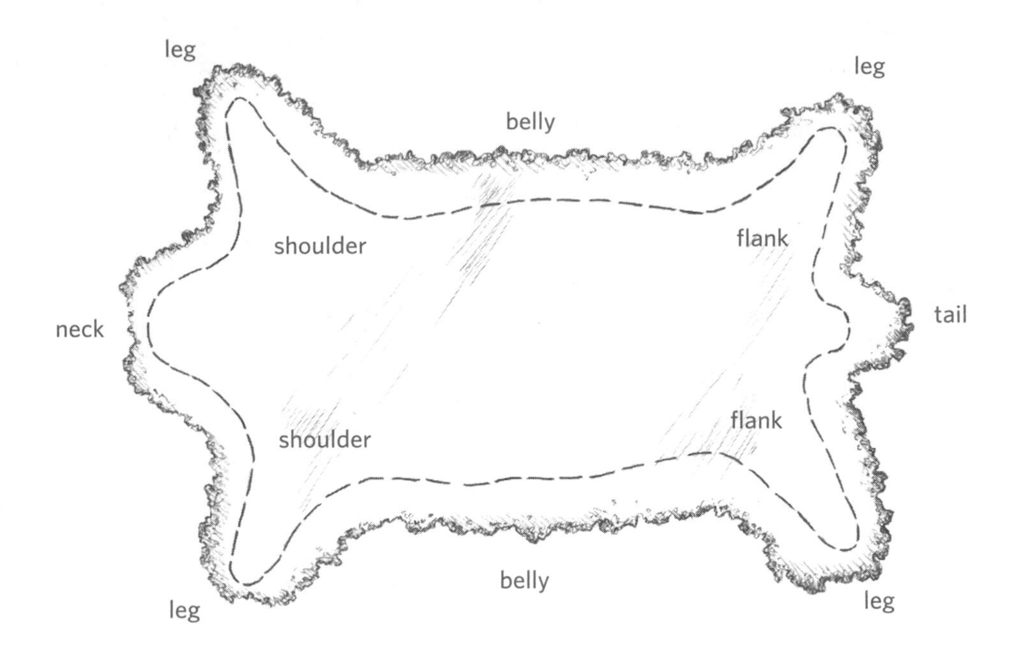

Skirt a fleece at the dotted line.

Making a Sheep Coat

A sheep coat — also called a blanket, cover, or rug — is used to keep a fleece clean while it is still on the sheep. In areas with severe winters, coats help sheep stay warmer, and the energy that might have gone toward keeping them warm is expended on increased wool growth and heavier lambs. Many owners of small flocks who depend on the sale of choice fleeces use coats on all their sheep.

You can make inexpensive sheep coats from woven-plastic feed sacks, which allow air circulation and hold up well. The edges must be well hemmed so that they don't fray. The easiest material to work with, however, is #10 cotton duck. Prewash it before constructing the coat to prevent shrinkage after the coat is made.

Use the dimensions suggested on page 241, or measure your sheep and make a custom-fitted garment. For the best protection, the coat should extend 3 inches (7.5 cm) below the belly. If your material isn't wide enough to cut on the fold, cut the two pieces and make a seam along the center at the backbone. To custom fit, determine the coat length by measuring from the center of your sheep's breast to the end of the back thigh. Make the hind leg loops of soft pajama elastic, which is about 1 inch (2.5 cm) wide and not as stiff as most elastic. Although the garment itself should last more than one season, the elastic will probably have to be replaced each season. Check the sheep often to be sure the elastic isn't rubbing its leg and causing injury.

To get the right fit, first make a rough model out of an old sheet and try it on your sheep. The front edge should be as close to the head as possible for maximum protection against hay and weed seeds. If you make a center back seam, you can get a closer fit by shaping the seam to conform to the curve of the sheep's back.

1. To make neck openings, match section A-B on both sides (see illustration on next page) and overlap ½ to 3 inches deep (1.3 to 7.5 cm), depending on the size of the sheep and stitch.

2. Make a ½-inch (1.3 cm) hem on all outside edges (shown by dotted line).

3. At points C and D on both sides, attach loops of soft elastic, each 24 to 27 inches (61 to 69 cm) long, depending on the size of the sheep.

Making Your Own Sheep Coat

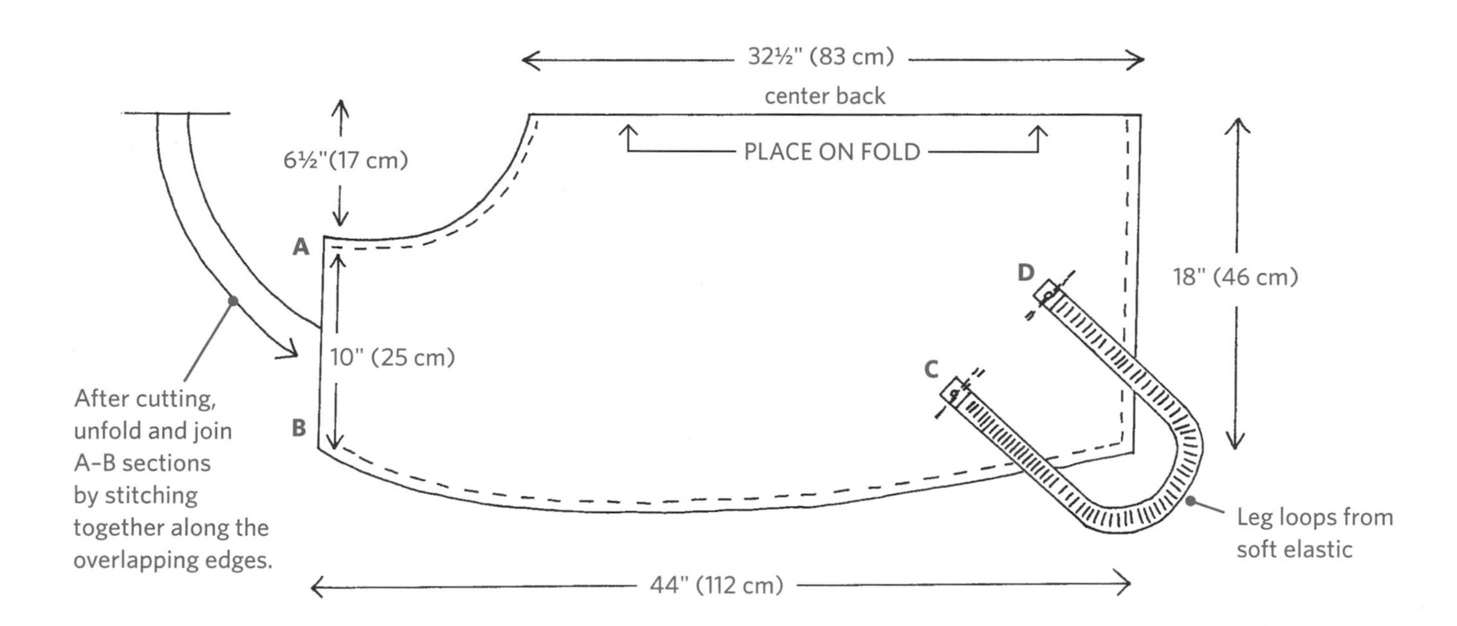

Using these measurements, make a pattern to size (from brown paper), then place on fold of material and cut. Finish following instructions on opposite page.

Raising Sheep for Meat

If you live in an area that has lots of shepherds you should have no trouble finding a weaned lamb, but in many areas of the country finding lambs may be a challenge. The prices for weaned lambs vary across a pretty wide spectrum — you might pay less than $50, or you might be counting out hundreds of dollars.

Club lamb producers specialize in raising lambs to weaning, and then selling them to kids participating in 4-H and FFA (Future Farmers of America) programs at school. Club lambs are typically sold to the young person 120 days prior to their first major show (often the county fair), weighing 60 to 80 pounds (27 to 36 kg). Depending on the breed, they will weigh 110 to 140 pounds (50 to 64 kg) at show time. Your Cooperative Extension office should have information on club lamb breeders in your area.

It may be worthwhile to regularly check Craig's List on the Internet, as sometimes you can find really excellent animals. If possible, buy from someone close by. Ask lots of questions — if the seller doesn't want to answer all of your questions, you may want to keep looking.

Can you save money buying a couple of lambs, raising them through the summer, and putting them in the freezer in the fall? Since a weaned lamb will thrive on pasture and water, your main expense (after housing and fencing) is the cost of the lamb. If you buy a 60-pound (27 kg) weaned lamb for $50 and butcher it at 120 pounds (54 kg), you can expect about 30 pounds (14 kg) of meat, giving you an approximate cost per pound of about $1.65, which is a good deal no matter what you're paying for lamb now. But if you purchase higher-end lambs from the club market, or purebred lambs that are aimed at the breeding-stock market, you could end up losing money.

Dressed Weights and Yields

Dressing percentage represents the comparison of carcass weight to live weight, with carcass weight being taken from an animal with hide, internal organs, and head removed. Locker lambs are often sold based on their dressed weight. Yield, or the amount of packaged meat you receive after the carcass is cut and wrapped, typically represents 50 to 60 percent of the

Relative Percentages of Various Lamb Cuts*

Cut	Percentage dressed eeight	Percentage live weight
Loin	8	4.4
Rib (track)	7	3.9
Leg (boned/rolled)	24	13.2
Shoulder (boned)	20	11.0
Ground lamb	10	5.5
Stew meat	7	3.9
Bone, waste, etc.	24	13.2

**Example*: If you have a 115-pound (52 kg) lamb butchered, you'll get about 5 pounds (2.3 kg) of loin cuts (115 × 0.044, or 52 × 0.044).

dressed weight. The actual percentage for both dressed weight and yield varies based on the condition of the animal, its breed, its bone structure, and the amount of fat it is carrying. For example, if you take in a 100-pound lamb for butcher, typical dressed weight is around 50 pounds, and typical yield is 25 to 30 pounds.

Here are some important terms to know if you are interested in raising lambs for slaughter: **market lambs** have a live weight of 90 to 120 pounds (41 to 54 kg) and their average dressing percentage is 50 percent; **hot house lambs** have a live weight of 40 to 60 pounds (18 to 27 kg) with a 55 to 70 percent dressing percentage (they're popular for Christmas and Easter); and **cull sheep** have a great variety of live weights, widely based on age and condition, and they have a 37 to 52 average dressing percentage.

Causes of Tough Meat

To get the best quality and best-tasting meat you need to keep the lambs growing at a steady and strong pace. If they are set back by poor feed or health problems the meat's quality will suffer. Young lamb is naturally expected to be tender, but several factors, one at a time or combined, can reduce tenderness:

- Stress imposed on animals prior to slaughter, such as rough handling when they are caught and loaded.
- Slow growth rate — one good reason to feed your lambs grain in a creep feeder if your pastures aren't in top form.

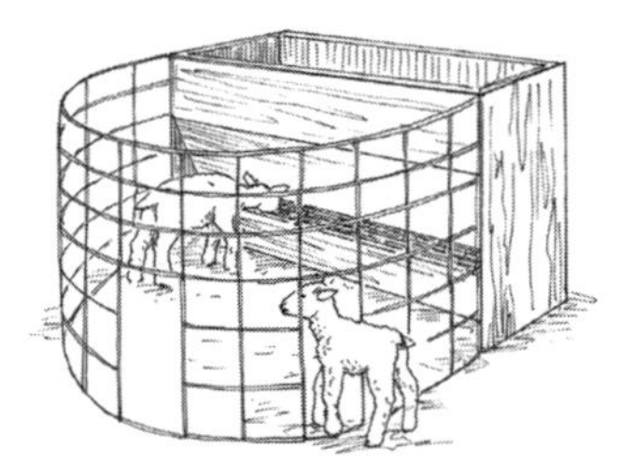

A creep feeder allows lambs to enter and eat extra feed, while keeping larger animals out.

- Slow freezing of meat, causing it to dry out; most cut-and-wrap facilities freeze meat faster than you can do it in your home freezer.
- Length of time in freezer storage; one year is the maximum lamb should be stored.
- A lamb can be cut and wrapped quickly after slaughter, but a yearling or mature animal must be tenderized by hanging it for at least one week in a chilling room prior to cutting and wrapping.

Butchering a Sheep

You can take your sheep to a custom packing plant to be slaughtered and butchered or you can do it yourself. If you want to do it yourself, get a copy of *Basic Butchering of Livestock and Game* by John Mettler, DVM, which provides excellent slaughter and cutting instructions, with lots of illustrations to help along the way. Your county Extension agent may also have a booklet available on the topic.

If you're going to work with packers, you will have to give them some directions. To get the maximum use and enjoyment from your sheep, give these instructions:

- Cut off the lower part of the hind legs for soup bones.
- The hind legs can be left whole, as in the traditional French-style leg of lamb, or cut into sirloin roasts or steaks and leg chops or steaks.
- The loin can be cut as tenderloin into boneless cutlets or as a loin roast.
- Package riblets, spareribs, and breast meat into 2-pound (0.9 kg) packages. The spareribs and breast can be barbecued or braised; for tips on cooking riblets see sidebar on page 243.
- The rack, or rib area, can be cut into that favorite, lamb chops, or left as a rack roast. The shoulder can be cut into roasts or chops, and the neck and shank can be used as soup bones. Stew meat or ground lamb can also come from these front cuts.
- If you want kebab meat, have it cut from the sirloin or loin.

Riblets

Riblets, which are sometimes referred to as short ribs, are almost inedible when prepared by most cooking methods but when prepared in a pressure cooker for about 45 minutes, with an inch (2.5 cm) of water in the bottom to start, the meat will fall off the bone. You can substitute barbecue sauce, curry sauce, or your favorite sauce or marinade for the water.

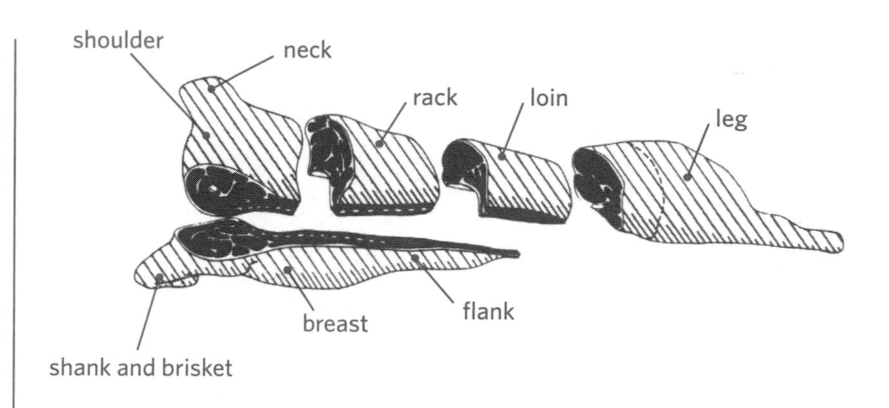

Correctly breaking a lamb carcass can be done in one of several ways, and no one method can be considered best. However, for many purposes, the method shown here is ideal. (From *Lamb Cutting Manual*, American Lamb Council and National Livestock and Meat Board)

Health Care

All sheep need grooming, health care, and medical treatment. The basics for keeping sheep in good health include hoof trimming and guarding against hoof problems, vaccinating sheep against diseases prevalent in your area, regular deworming, and protecting your sheep from external parasites.

Hoof Trimming

Wild sheep and those that graze on mountainous pastures wear down their hooves by traveling over rocky ground. Domestic sheep walk primarily on soft soils. Consequently their hooves become long and overgrown, which makes walking painful. Sheep need to have their hooves trimmed so they can walk properly and to help prevent hoof disease, such as scald and foot rot (see below).

Before buying an adult sheep, take a good look at its feet. Ask when they were last trimmed. If they show any need for trimming, the owner should be able to demonstrate the procedure for you. If the owner can't show you how, don't buy the sheep — the sheep's hooves have probably not been properly cared for.

How often to trim your sheep's feet depends on your pasture, paths, and barn floor. Many shepherds trim twice a year, whereas others need to trim only once a year. If you notice limping sheep or sheep on their knees, check their feet. Whenever you take care of other needs, such as deworming or shearing, make a habit of trimming the animal's hooves as well.

Wear leather gloves while trimming hooves — a kicking sheep that is not held firmly or properly can injure you with the sharp edges of the freshly trimmed hoof. Set the sheep or lamb on its rump as shown on page 231. Trim the rear feet first, and then the front feet.

Hoof Troubles

Foot problems are some of the most common problems a new shepherd needs to look for. Fortunately, they are generally easy to spot.

Foot rot, a bacterial condition that causes hard-to-cure infections, is one of the worst diseases that can infect your sheep. Active outbreaks of foot rot occur during warm, wet weather. Foot rot is a painful disease; the bottom of the sheep's hooves literally rots off, exposing the soft tissues beneath.

Before buying a sheep, look at the rest of the flock. If you see any limping sheep or sheep with severely overgrown, misshapen hooves, go somewhere else. Foot rot can be introduced into a clean flock only by an infected animal or a carrier animal (one that shows no symptoms but spreads the infection to others). Once foot rot is introduced into your flock, it is almost impossible to get rid of.

When bringing a new sheep into your flock, a good precautionary measure is to disinfect its feet. To disinfect a sheep's feet you must walk the sheep through a germicidal footbath — the same as used to treat foot scald (see page 244).

Plugged toe glands occur when the deep gland between the two toes of a

Trimming Hooves

Many styles of hoof shears are available, but few are as easy and safe to use as Felco No. 2 pruning shears. Many farm and garden stores sell them. If you have many sheep to trim, the investment is well worth the money.

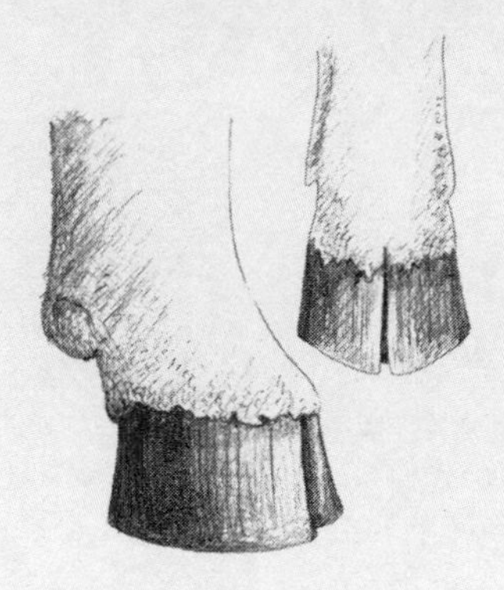

properly trimmed hooves

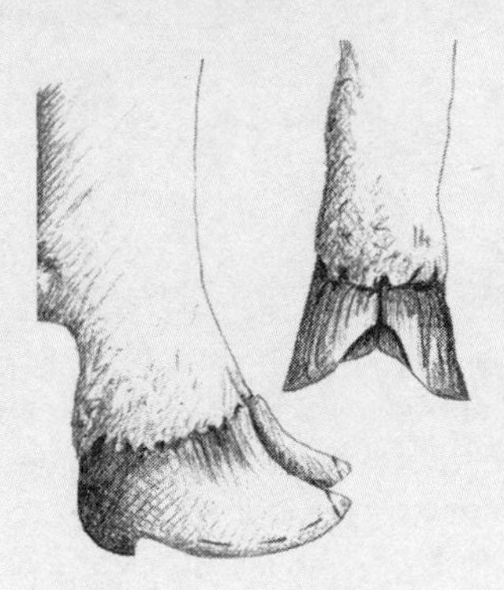

overgrown hooves

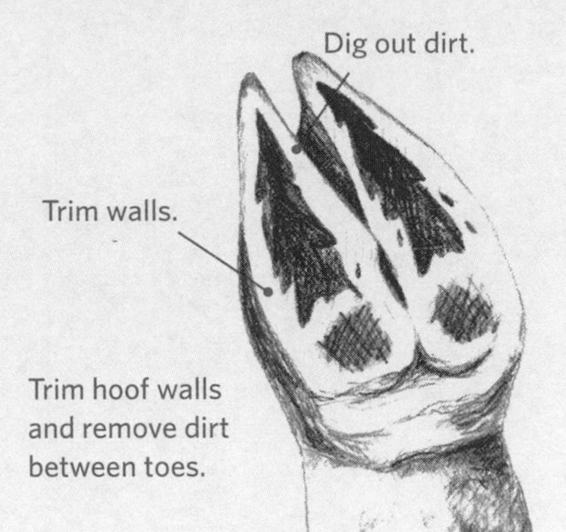

Trim hoof walls and remove dirt between toes.

front foot becomes plugged, causing swelling and lameness. Look for a small opening at the top of the front of the hoof. The gland secretes a waxy substance that has a faint odor; its purpose supposedly is to scent the grass to reinforce the herding instinct. When the gland becomes plugged with mud, the secretion is trapped, and the toe swells and causes lameness. To unplug a toe gland, squeeze it so the plug pops out. Then clean the area with warm soapy water and disinfect it with a povidone-iodine (Betadine) topical antiseptic.

Foot scald is a skin infection that occurs between the toes that occurs most commonly after a long, wet period. The first sign of scald is limping. Usually the front feet get sore before the hind feet do. The skin between the toes is moist, hairless, and red from irritation. Begin treatment by moving the affected animal to a clean, dry area. Trim hooves and then spray them with hydrogen peroxide. If you see no improvement within 72 hours, treat with a zinc sulfate footbath, which can be done using either a large fruit-juice can or booties made for the purpose and available from your veterinarian or any source of sheep and goat supplies. The procedure is as follows:

1. Pour the footbath solution to a depth of 2 inches (5 cm) into the fruit-juice can, or place a bootie on the affected foot and fill it with the footbath solution.

2. Soak the affected foot for 30 minutes.

3. Repeat once or twice a week until the foot scald is resolved.

Possible Causes of Lameness

- Overgrown, untrimmed hooves
- Mud, stone, or other matter stuck between the toes
- Plugged toe gland
- Abnormal foot development (an inherited defect)
- Foot abscess
- Foot scald
- Foot rot
- Thorns, punctures, bruises, or other injuries

Vaccinations

Lambs should be vaccinated at an early age. Consult your veterinarian, a county Extension agent, or an experienced shepherd about vaccinations for your sheep.

Sheep are commonly vaccinated against diseases that infect the lungs, the digestive system, and the reproductive tract. For example, lambs (as well as ewes and rams) can be vaccinated to protect them against pneumonia, and all lambs should receive immunizations against enterotoxemia, a clostridial disease caused by overeating. Ewes should be vaccinated or given booster shots for clostridial diseases before lambing. This vaccine not only protects the ewe but also increases the disease-fighting antibodies she passes on to her newborn lambs when they nurse. Take this opportunity to vaccinate your ram, too; you will be less likely to forget any sheep if you do them all at once.

Disease control programs can be somewhat complicated. The important thing to remember is that most vaccinations must be given either before breeding or before lambing. Don't wait until the ewes are almost ready to lamb before you vaccinate them.

Keep regular records of your sheep's health history, including dates and vaccinations given.

Deworming

Sheep have a high resistance to disease but low resistance to internal parasites. All sheep, especially lambs, must be dewormed. Sheep can pick up many types of worms through grazing. The worms are microscopic and live off the blood of the sheep. Lambs can die of severe worm infection.

When you raise sheep for the first time, your pasture is probably uncontaminated by previous use. Cross-fencing and pasture rotation help reduce exposure to worms. Clean feeding facilities further help prevent rapid or serious buildup of stomach worms.

Even with precautions, however, sheep can become infected with worms, and they need to be treated periodically. Deworming is especially important in mild climates or after a particularly mild winter. Several good deworming medicines are available that are safe even for pregnant ewes and small lambs. Obtain deworming medicine from your veterinarian, who can give you information on the proper dose and timing. The medicine may come in the form of a bolus (large pill), which is placed on the base of the sheep's tongue by using a bolus gun, available at farm supply stores. Deworming medications also come in the form of liquid drenches, pastes, feed blocks, and injections.

Most internal parasites have a similar life cycle. Sheep ingest worm larvae while eating and pass eggs in their manure. The eggs hatch in the soil, and the larvae migrate onto the grass.

Symptoms of worms. A lamb with worms may be underweight for its age and have a potbelly, prominent hip bones, a scruffy wool coat, runny or loose manure and manure buildup on its rear, and anemia, which makes it weak. Anemia can be prevented by deworming lambs when they are 2 or 3 months old. If you have only one or two sheep on clean pasture, you may need to deworm your lambs only once more during the first year. In a larger flock that has limited or contaminated pasture, the lambs may have to be dewormed monthly. Adult sheep with heavy worm loads will sometimes have a swelling under their throat or chin called bottle jaw. Ewes should be dewormed before breeding, before lambing, and before going out on fresh spring pasture.

After deworming. Keep your sheep in the last pasture they grazed prior to being dewormed, then transfer them to a clean pasture 24 hours later. The sheep will expel the worms and eggs in the old pasture and won't contaminate the new pasture as rapidly; worm control is one important reason to cross-fence pasture. The eggs and larvae of many worm species can survive out on pasture for as long as three months in cool, damp weather but may die within three weeks during hot, dry weather.

External Parasites

Sheep are subjected to a number of external parasites, including keds, lice, mites, and maggots from fly-strike. If they seem to be bothered by flies, act itchy, or have broken and patchy wool, they are suffering from external parasites. Your veterinarian, county Extension agent, or an experienced local shepherd can offer helpful advice on causes and treatments.

Other Health Problems

Sheep can be affected by other health problems, such as abscesses, copper toxicity, Johne's disease, sore mouth, pneumonia, and scrapie. Some illnesses show no signs, and others present symptoms fairly early. If you notice changes in the appearance of a sheep's skin or wool, difficulty breathing, or intense itchiness, contact your veterinarian.

Keds

Keds are wingless flies that cause severe itching and discomfort. Your sheep will stay free from keds unless you bring in an infected animal or let your sheep come into contact with infected animals.

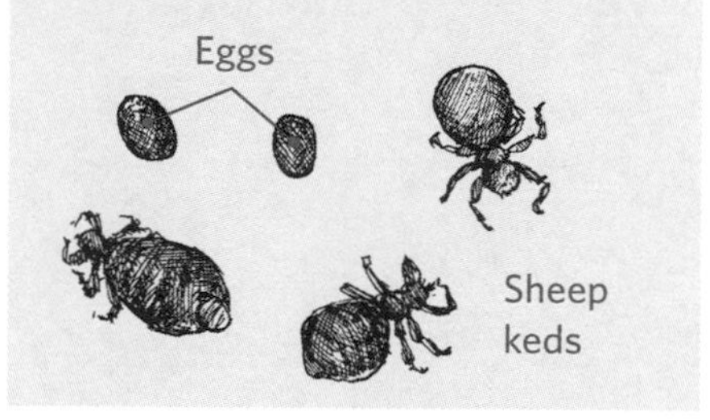

CHAPTER 8

Pigs

Of all of the major livestock species, none is more misunderstood or less appreciated than the hog. No other major livestock species is more durable, adaptable, or productive. The hog is one of the livestock animals with the longest association with man. Its hardiness, vigor, and grit have to be admired.

Several mistaken beliefs circulate regarding hogs. Let's address some of them here:

Hogs are inherently dirty animals. Not true. Hogs do lack the ability to sweat and sometimes run into problems in hot and humid weather because they can release heat only through panting and the evaporative cooling process. So they wallow in muddy water to cool off, but otherwise have no affinity for mud. In small lots and around feeding and watering equipment, their sharply pointed hooves tear the soil surface and can cause muddy areas to develop.

Hogs are inherently greedy or gluttonous. Hogs are mistakenly assumed to be gluttonous because growing hogs and sows nursing young are fed to consume roughly 3 percent of their body weight daily in order to foster rapid growth and heavy milk production. It is also part of their herd instinct for all the hogs in a group to rise up and want to eat at the same time. Hogs are accustomed to rooting about for at least a portion of their diet, but they are not eaters of swill and garbage as is so often depicted.

Hogs are more vicious or treacherous to work with than other livestock species. The tales of mean hogs date back to the times when they were truly free range, when you were as likely to encounter feral hogs as domestic ones. We worry about this far less today; for a great many generations now hogs have been bred selectively for a more docile nature and quiet temperament.

Choosing a Breed

All of today's breeds are selectively bred for leanness, feed efficiency, and meatier carcasses, and any full, purebred herd can do a creditable job producing basic meat hogs. Purebreds are becoming fewer, as the commercial pork industry favors crossbreds for their increased growth, fertility, and disease resistance. But by raising one of the purebreds, you help further their survival, and some of the older breeds are favored by connoisseurs. Here are three groups of purebred hogs to consider.

Heritage breeds have a long association with North American farms; among them are the Berkshire, the Spotted, the Poland China, and the Chester White. Many of the heritage breeds have fallen to near minor-breed status and had their gene pools corrupted as some unscrupulous sorts have sought to make them more extreme — heavier muscled, excessively lean — and similar in type to the confinement breeds.

Rare endangered breeds once had roles of economic importance and were considered true farmers' hogs, although they were often popular only in certain regions of the country. They include the Tamworth, the Hereford, and the Large English Black. A true, rare old-timer is the Mulefoot, which even in texts from early in the twentieth century is referred to as "the old breed."

Porcine outsiders is a wide-open group that includes the Red Wattle. This group encompasses truly minor breeds that are supported by a limted number of breeders. A small group of Red Wattle hogs was recently processed and submitted to a group of chefs as part of a Slow Food group study, and they tested favorably for both taste and cooking qualities.

In simple and straightforward terms, all pigs are pork. The various breeds

Quick Guide to Hog Breeds

Breed	Size of mature sow/mature boar	Best attributes	Pasturing ease	Status
Berkshire	450-550 lbs./550-650 lbs. (204-250 kg/250-295 kg)	durability; well muscled and lean	good	common
Chester White	450-550 lbs./550-650 lbs. (204-250 kg/250-295 kg)	hardiness; good mothering; large litters	good	common
Duroc	450-550 lbs./550-650 lbs. (204-250 kg/250-295 kg)	durability; fast growth; efficient feed conversion	fair	common
Hampshire	450-500 lbs./550-650 lbs. (204-227 kg/250-295 kg)	durability; rapid growth; well muscled and lean	fair	common
Hereford	600 lbs./800 lbs. (272 kg/363 kg)	adaptability; fast growth; efficient feed conversion	excellent	uncommon
Large Black	600-700 lbs./700-800 lbs. (272-318 kg/318-363 kg)	large size; good mothering; efficient feed conversion	excellent	uncommon
Poland China	450-550 lbs./550-650 lbs. (204-250 kg/250-295 kg)	durability; well muscled and lean	good	common
Red Wattle	600-800 lbs. (272-363 kg) for sows and boars	hardiness; rapid growth; lean meat	excellent	uncommon
Spotted	450-550 lbs./550-650 lbs. (204-250 kg/250-295 kg)	durability; fast growth	good	common
Tamworth	500-600 lbs. (227-272 kg) for sows and boars	vigor; good mothering; large litters; lean meat	excellent	uncommon
Yorkshire	450-550 lbs./550-650 lbs. (204-250 kg/250-295 kg)	good mothering; large litters	fair	common

Talkin' Hog

Visit even briefly the sale-barn alleys, show barns, or hog lots, and you will find the people there speaking a language uniquely their own. Not only will they talk of "gilts" and "barrows," but you will also hear about pigs with "daylight" that are "blown apart" or "coon footed."

Each new trend in swine type seems to create at least half a dozen new words. Still, the language has ancient roots. Where else would you find ancient words like *farrow* and *sow* used in the same sentence with *sonoray*?

A feeder pig generally weighs 40 to 70 pounds (18 to 32 kg) and is between 8 and 12 weeks of age.

What separates the swine pro from the tenderfoot is how the word *pig* is used. To be country correct, it is the term for a young pig. A *shoat* is a young swine from weaning age to 100 pounds (45 kg); a *hog* is a swine weighing more than 120 pounds (54 kg). One farmer might tell another that the 400-pound (181 kg) boar standing before them is a "pretty good ol' pig," but the greenhorn who then strolls up and says "Wow, look at that big pig!" has shown himself to be, well, green.

A *butcher* or *market* hog is one weighing 220 to 260 pounds (100 to 118 kg) and ready for sale for slaughter. These swine generally are 5 to 7 months of age. A *feeder* pig is an animal weighing 40 to 70 pounds (18 to 32 kg) that is sold to a farmer/feeder to be *fed out* (raised and fed) to market weight. Such pigs are generally between 8 and 12 weeks of age and may be a bit heavier or lighter in weight. A *finishing hog* is a hog that weighs between 100 pounds (45 kg) and market weight.

A pig with *daylight* has good leg length, because you can see some daylight underneath the animal. A *coon-footed* hog walks with a flat and flexible foot and sloping pasterns that indicate it should stand up well on concrete or other unyielding surfaces. Its foot, in many ways, is akin to the foot of a raccoon.

You have to have earned your parlance by time in the mud and muck to be able to tell the difference between "pig," "hog," and "shoat."

were developed to function in different environments and under different sets of economic circumstances; over time their roles have evolved. Some breeds are more suitable for outside survival than others. They have the vigor and form that provides natural methods of protection. And certain breeds — including the Duroc, the Berkshire, the Tamworth, and the Chester White — produce more distinctive and flavorful meat than the pork that comes from the confinement industry.

How Long Does a Hog Live?

Most hogs don't live much past the age of 6 months, at which time they become hams and bacon. Commercial breeding hogs are usually productive for about 4 to 5 years, after which they become uneconomical and are turned into sausage. In a backyard situation, a breeding hog may live to the age of 10, and a well-kept, healthy, hardy hog may live as long as 15 years.

Colored Breeds

A distinction of sorts is made between colored and white purebred breeds of swine. The colored breeds are considered to have the stronger economic traits. As a group they are noted for vigor, faster yet leaner growth, and meatier carcasses. They are the meat-type counterpart to the white mother breeds.

The Berkshire is of English origin, as are all breeds with *shire* in their name. It is an old-line, hardy, and durable breed now enjoying a resurgence in popularity.

Berkshire

Duroc

Hampshire

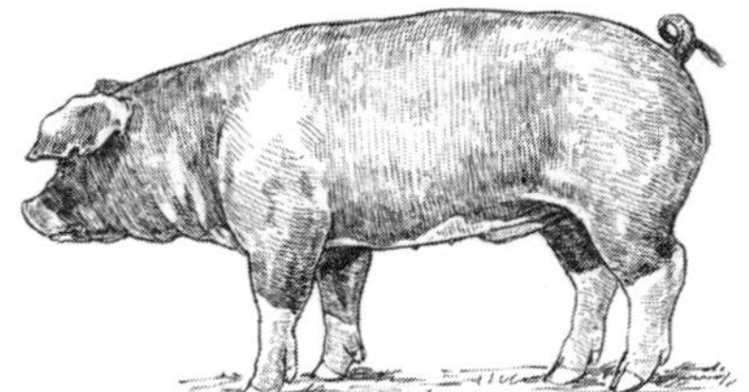

Poland China

Spotted

The reason for its surge in popularity is its exceptional meat qualities. The Berk is a well-muscled breed noted for both leanness and high-yielding carcasses. Pork from pure- and high-percentage-bred Berkshires is noted for quite large loin eye areas and finishing qualities that give the pork something akin to the marbling in high-grade beef.

The Duroc is a truly American breed. It was originally known as the Duroc-Jersey, which arose from two types of hog: the Jersey Red of New Jersey and the Duroc of New York. The Duroc ranges in color from dark, brick red to several shades lighter; it has smallish, drooping ears. Because of its excellent muscling, outstanding meat quality, and hardiness, the Duroc is often used as a terminal-cross male in a crossbreeding program. The Duroc is an ideal outdoor pig and is bred in relatively large numbers.

The Hampshire is a black hog with a distinctive white belt encircling the shoulder region. This color pattern dates back centuries, to a fad for breeding all manner of domestic animals for this striking color effect. Animals without a white belt encircling the whole body or white splashes in the upper part of the body are not issued a pedigree document. They are often termed "off belts" and are generally sold at discounted prices; however, they are not impaired physically by their lack of belting and still produce quality meat animals.

Hampshire boars are often used as terminal sires, and the Hampshire × Yorkshire gilt is the stuff of legend in the swine industry, known for its hardiness and productivity. Hampshires can impart rapid growth and excellent leanness to their offspring.

When seeking purebred Hampshires, beware of animals with excessively heavy muscling and too little fat cover. These animals are often too extreme and unnatural in form, and may not be genetically pure.

The Poland China, sometimes known as the Black Poland China, was developed in Ohio and is nearly identical to the Berkshire in its color pattern but has drooping ears. This hog has a flatter and longer topline than the Berkshire does and produces exceptional meat qualities: The carcasses are lean and well muscled and yield a good amount of meat, which includes excellent hams.

The Poland China is widely known as one of the most durable of all of the swine breeds. At one time this breed's association handbook could lay claim to more females in their eighth litter and beyond than any of the other major swine breeds.

The Spotted, once known as the Spotted Poland China, has a black and white spotting pattern that gives it its name. This breed also has drooping ears and is known for its good mothering abilities.

Meet the Hog

Pigs belong to the Suidae family, which evolved from prehistoric species that date back more than 40 million years. About 33 million years ago, the forebears of the domestic pig (*Sus domesticus*) and the wild boar (*Sus scrofa*) also roamed the planet. Wild boars were quite adaptable, living in a wide range of climatic conditions. Archaeological evidence suggests that Turkey was the first place where pigs were domesticated, about 9,000 years ago.

Terminal Cross

A *terminal* cross is one made to maximize growth and muscling when all pigs produced are to be harvested for meat.

Chester White

Yorkshire

The carcass is long and trim yet is as heavily muscled as the Hampshire or the Duroc.

Spotteds share a common ancestry with the British heritage breed known as the Gloucestershire Old Spot. In the United States, both breeds are tracked by the same breed association. In England, where it is also known as the orchard hog, the Spotted has enjoyed the support of the royal family. The British claim Spotteds have been bred outside confinement for more than a hundred generations.

White Breeds

The white breeds are generally considered to be stronger than the colored breeds in the traits needed for successful pig raising: They produce milk and nurse and nurture their litters better than the colored breeds and tend to farrow pigs in larger numbers. They also have the docile nature needed to raise and wean large litters. The white breeds therefore remain useful in both their pure state and in various crossbreeding rotations. When selecting one of the white breeds, look for animals with structural soundness, true vigor, moderate muscling, and internal dimension (body width and depth; the distance between the front legs should extend the length of the body). You need a big barrel of a body on these hogs; you may also wish to seek out individuals with older, deeper pedigrees.

The Chester White is an American-developed white breed that originated in Chester County, Pennsylvania. It has a medium-sized frame and drooping ears. It is particularly hardy, making it good for people who raise hogs mostly outdoors with simple facilities.

The sows are good mothers and become pregnant soon after weaning the previous litter. Purebred Chesters are a bit smaller at birth than some other purebreds, and a bit slower growing — trailing the norm by 3 to 8 days to harvest weight — but still achieve 230 pounds (105 kg) in a respectable number of days (160 to 170 days).

The Yorkshire lays claim to the title of the mother breed in its advertising and, in most years, leads the nation in total registrations. Large litter size at both birth and weaning is a breed standard and evidence of its superior mothering and milking abilities. Yorkshires are among the most widely available of all of the purebred swine breeds.

The Yorkshire is all white, with sharply erect ears and dark eyes. It shares its origins with the English Large White — both are recorded with the same breed association in the United States. Although the Large White is the definitive confinement breed, American Yorkshire breeding lines are particularly good for the backyard raiser because they are more selectively bred to live outdoors.

Rare, Minor, and Heritage Breeds

Interest in the minor breeds has grown remarkably in recent years for two reasons. Heritage breeds are old breeds valued by consumers and raisers for the same reasons people value heirloom poultry or produce. Heirloom tomatoes and heritage hogs are reminiscent and representative of days of yore, and they are often tasty and suited for small farms. Demand is also growing for local and regional food items and food supplies, a market for which the minor breeds were developed.

Crossbred Feeder Pigs

In shopping for feeder pigs you are apt to encounter crossbred pigs. Most crossbreds result from one of the colored boars being bred to a white female or one with a high percentage of white breeding. It is generally a good thing, as the heterosis or hybrid vigor they receive from a well planned cross results in larger and more vigorous pigs at birth. Some popular crosses are Duroc × Spotted, Hampshire × Yorkshire, Duroc × Yorkshire, and Hampshire × Duroc.

Crossbred

Hereford

Large Black

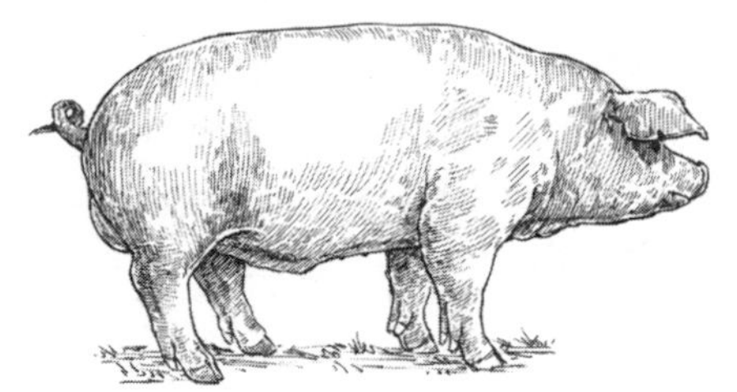

Red Wattle

Heirloom breeds are valued for a number of reasons, including the simple fact that they represent some of the hardiest of all the swine genetics. As a group they are quite naturally lean, adapt readily to a wide range of environments, and are among the best choices for pasture- or range-based production.

The Hereford was developed in the American Midwest and is an especially attractive breed with dropping ears and a red and white pattern mimicking the distinctive color pattern of the Hereford breed of cattle. Although its frame is a bit smaller than that of some other swine breeds, young pigs grow rapidly and can reach as much as 250 pounds (113 kg) in just six months. This is a pretty but practical breed that has a gentle disposition, is easy to pasture, and does well in a variety of climates.

Pedigree and Purity

Pedigree papers are no assurance of genetic purity. An old adage in the livestock sector holds that a pedigree is only as good as the man or woman who hands it to you. Tamworths with Duroc type or color, for example, should raise all sorts of red flags. When buying purebred swine, be thoroughly familiar with the conformation and character of the breed that interests you, or bring along an experienced person who is knowledgeable about your chosen breed.

The Large Black is another minor breed and is the rarest of the British breeds. It has a strong Canadian presence and is occasionally found in the United States as well. A prolific breeder and good pig raiser, this breed is all black and has large, floppy ears. Although it isn't as long or trim as other breeds, it is prized for its innate vigor. Like other minor breeds, the Large Black has been neglected by the industry and supported only by a few breeders who value its ability to thrive on pasture.

The Red Wattle was developed in Texas and suffered a bit of a scandal some years ago when the wattle trait (consisting of fleshy attachments hanging from both sides of the neck) was bred onto hogs with a far different background. If you're looking for purebred Red Wattle hogs, you must be conscious of selecting for true breed type and character. Red Wattles have a gentle disposition, thrive on pasture, produce large litters that grow rapidly, and do well in a variety of climates.

The Tamworth is of English origin. It is a light-red breed with erect ears and is probably the hardiest and strongest forager of the swine breeds. It has an unearned reputation for having a bad temper.

With a frame that is slightly smaller than some other breeds, the Tamworth has remained hardy and retains many of the characteristics of what was once called the range hog — it is eminently self-sufficient. Because it thrives on low-energy foods while growing slowly to produce lean bacon with a fine grain, it also was once called the bacon breed. This hog is so lean that some lines are in fact too thin. Unfortunately, like many of the minor breeds, the too-lean type is a bit stuck in time genetically. Tamworths need more people to take them up and selectively breed them — to maintain the traditional strengths of the breed but also to improve growth and muscling, which translate into better carcass yield and more cuts that are prime hams and loin.

Tamworth

Wild and Feral

Russian or wild boars are the bona fide, naturalized variety; they are real-deal pork — straight from nature. Some purebred wild boar herds exist in the United States, but they must be maintained in tightly enclosed pastures. Today's feral hogs sprang from domestic stock that have somewhat reverted to wild type and have existed in the United States since colonial times. Given a moderate climate and access to an array of potential food sources, a hog is a self-sustaining animal.

Buying a Pig or Two

The old nursery rhyme goes, "To market, to market, to buy a fat pig. Home again, home again, rig-a-jig-jig." Well, buying a pig or two may not be quite as easy as this rhyme implies. Still, it is pretty simple and straightforward once you have a little background and know what to look for.

A meat hog of good type will yield about 85 percent of its live weight in various cuts and meat products. Unfortunately, hogs are not all ham and chops, but one or two hogs fed out each year will go a long way toward meeting the protein needs of a typical family of four. Hogs are herd animals, so it is best to raise at least two. The second animal can always be sold or used by other family members or friends if your family doesn't need it. If sold, the money will help offset some of the costs of buying the animals and feeding them.

Pig raising is simplest and most efficient in climates without great extremes in temperature or precipitation. The grow-out period will normally be between 90 and 120 days, depending on the starting weight of the pig, and in many places fits into the spring or fall season, to avoid the weather extremes of a Missouri summer or a Maine winter.

Where to Buy

The best place to buy feeder pigs is at their farm of origin. There you can often view the sire (father), dam (mother), and siblings, which will give you an idea of what you can expect from the pigs as they grow out and reach finished weight.

Pigs at an auction are often stressed from the transport and handling. They also have the real possibility of having been exposed to disease organisms by way of sick pigs.

When to Buy

Early spring may be the most expensive time of the year to buy feeder shoats because fewer sows farrow in the cold months of December, January, and February, when the early feeders are born. Still, these pigs are desirable because they will reach a good slaughter weight before the weather becomes excessively hot and humid.

Late summer through fall is also a good finishing period, as the pigs will usually be grown out before the late-autumn rains and winter cold begin, when feed utilization is impaired due to temperature and weather extremes. These pigs are classic for furnishing the winter's meat.

Pigs are herd animals and will be more content if fed out at least two at a time.

Barrows versus Gilts

While barrows (castrated males) may grow a bit faster than gilts (young females), some people prefer to feed out gilts. Not only do gilts hang a leaner carcass, but their growth rate can be boosted with richer feedstuffs. Barrows tend to pack on a bit more weight late in the feeding period.

Castrated males are preferred over uncastrated males because the latter can produce a boar taint in the flavor of the meat. Some folks describe boar taint as a strong cooking odor and a slightly off flavor.

Cost

Expect to pay a bit more per head for a single pig or two than the going rate for pigs in groups (droves) because the farmer needs to be compensated for the added bother and stress of working a whole group to sell just a couple of pigs. Also, he or she may find that by reducing the size of the group to be sold, the value of the remaining pigs is reduced.

A couple of rough rules of thumb: A 40-pound (18 kg) feeder pig should sell for between 1.75 and 2.25 times the going per-pound rate for butcher hogs; a 40-pound (18 kg) pig should cost as much as the going price for 100 pounds (45 kg) of butcher hog. Add a bit more if you are going to buy just one or two head. Heavier pigs should cost correspondingly less per pound, and pigs at 25 to 30 pounds (10 to 15 kg) a bit more per pound.

Recognizing a Good Feeder Pig

A feeder pig is an animal generally weighing 40 to 70 pounds (18 to 32 kg) that is sold to a farmer/feeder to be fed out (raised and fed) to slaughter weight. Such pigs are generally between 8 and 12 weeks of age.

A good pig for feeding is not simply a scaled-down model of a finished hog. It should show indications of good muscling and potential growth. Muscling is demonstrated by roundness or flaring in the areas where the primary pork cuts are to be found: the ham and the loin. Growth potential can be seen in free and easy movement and a large, stretchy frame.

You can judge frame size by looking at the width between the front and back legs, the bone diameter in the legs, and the width and depth of side. In making your final selections, focus on details such as width between the eyes, size of the jaw, or foot size. Where these dimensions are larger-than-normal size, the rest of the animal will "grow to them."

In selecting pigs, it is nearly always best to go with first impressions. Even the best animal can be picked apart if you go over it hair by hair.

Signs of a Healthy Pig

A healthy pig will display an alert and curious manner. As you approach their pen, pigs should rise up and move to the far side of the pen away from you. They will then be lured back by their own curiosity to check out this new feature in their space. If they appear to act sluggish, they generally are ill or lame.

Listen for labored breathing, sneezing, coughing, or other sounds of respiratory distress. Certain diseases, such as transmissible gastroenteritis — an infectious disease of the stomach and intestines — create fecal material with a distinctive odor. A cloud of dust or excess mud can indicate a stressful environment that might impair pig health or performance.

Any pig you buy should have these signs of vigor and good health:

- **A bright and clean hair coat** is a must in feeder-pig selection.
- **Free and easy movement** is vital if the pig is to remain sound and grow well.
- **Good growth for its age** as indicated by selecting from among the largest pigs in a pen.

Farm Environment

Note closely the environment in which the pigs have been raised. Buying pigs from a farm environment that most closely matches what you have back home goes a long way toward reducing stress on the newly arriving pigs.

Tips for Buying Feeder Pigs

Here are a few tips to help you with feeder-pig selection.

- The greatest variety and highest quality of pigs are available in the 40- to 60-pound (18–27 kg) weight range.
- A pig lighter than 40 pounds (18 kg) is a young pig and more prone to stress and setbacks. It may not hold up to the rigors of life in the finishing pen or be old enough to cope with transport.
- Crossbred pigs are generally more vigorous and faster growing than purebreds.
- Pigs should be crosses from a planned breeding program and not simple mongrels.
- If you are seeking barrows, buy pigs that have already been castrated and healed.
- Gilts may grow more slowly than barrows but will generally produce leaner carcasses and can be pushed to grow faster with more nutrient-dense rations.
- Buy pigs that have been treated recently for both internal and external parasites.
- Buy pigs that are well past the stress and strain of weaning.

Parts of a Feeder Pig

When checking out pigs, analyze them by parts: the feet and legs, the topline, the bottom line, the front half, and the back half.

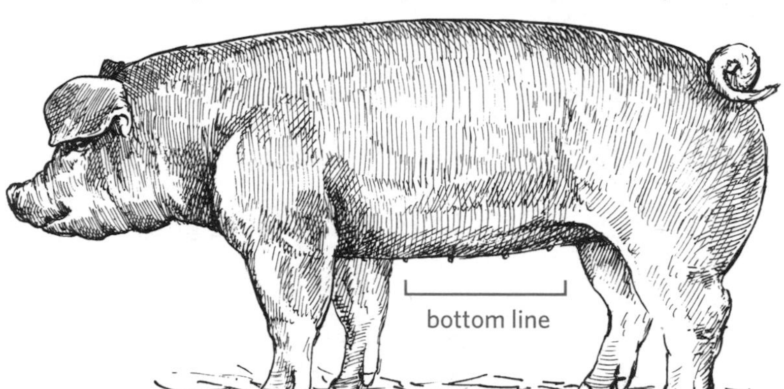

Pigs to Avoid

Here are some easily detected indications of bad pig health and, in some cases, the probable causes. These signs should immediately raise a red flag:

- **Exceptionally long and coarse hair coat and a head that appears too large** for the pig's body appear on an animal that is generally overage and badly stunted.
- **Twisted, swollen, or misshaped snout** is an indication of advanced atrophic rhinitis, a disease that reduces growth rate and performance and can lead to death. Do not buy any pig from a group where the disease is present.
- **Sniffling sounds as the pigs rise up and move about** may be a symptom of respiratory ills.
- **Discharge from the nose and eyes or slight bleeding from the nose** can indicate rhinitis.
- **Nodes or swellings (called knots) in the jawline** and other places are likely to be abscesses. They sometimes can be lanced and drained, but such procedures can introduce the abscess-causing organisms to your premises. Steer clear!
- **Dull, sunken eyes, or a listless manner** are indicators of an emerging health problem.
- **Dull hair coat, hair on end** (a sign that the body is trying to trap heat), or an excessively dirty hair coat are signs of stress. Pigs with hair on end and piled closely together are obviously chilled and undergoing severe stress. If you note a greasy appearance or a spotty hair coat, or an excessive amount of scratching, look behind the ears for lice or lice eggs attached to individual hairs.
- **Swelling in foot and leg joints** can indicate injury or arthritis. Pigs with healthy legs and feet can do fine with little or no bedding. Be on guard where straw bedding is piled deeply, as it may conceal serious foot and leg conditions.

And don't take a pig with a stiff gait, front legs that are too straight, hind legs that are tucked too far under the body, knocked knees, front legs that both appear to be coming out of the same hole, or hooves with toe points of uneven size. These signs indicate inferior body type, poor growth, and being prone to respiratory ills.

Bringing Your Pigs Home

With your feed and facilities ready at home and your pigs selected, you are ready for hands-on pork production. This adventure begins with getting your animals safely home and off to a good start. It's important to remember that a stressed pig is vulnerable to sickness, so keep it happy.

If you're hauling them in a pickup or stake-sided trailer, cover the front of the racks with a tarp or sheet of plywood to keep chilling drafts off the pigs while in transit. Bed the racks with straw to a depth of 4 inches (10 cm) in damp or cold weather. In hot weather haul early in the morning or late in the day; use damp sand or sawdust for bedding.

At home, unload the pigs into a well-bedded sleeping area. You may need to block the small animals away from a portion of the sleeping area to discourage them from developing the habit of dunging inside the house. Wetting down a far corner of the pen for the first couple of days may encourage them to begin dunging in that area.

Managing Stress

Moving pigs stresses them, and one sign of stress is going off feed. It is not at all uncommon for newly moved pigs to go off their feed within a few hours. But it shouldn't last for more than several days to a week, and the pigs should continue to drink water even while off their feed.

To help animals through this stressful time, many farmers add a vitamin/electrolyte product to the pigs' drinking water. This supplement helps them maintain both health and body condition should they go off feed or have trouble adjusting to a new ration. Adding flavored gelatin to drinking water will further increase water consumption and can also be used to mask any unpleasant tastes from medications that are administered through the water.

Sprinkling the same gelatin powder on top of feed will draw the pigs to the feed and increase consumption there, too. Still, never forget that abundant drinking water is a hog's most important feedstuff.

Young, newly moved, and stressed pigs should not be offered anything in the way of new or exotic feeds. It's a good idea to buy some of the feed the pigs were previously eating or match it as closely as possible to further help them through those first stressful days at their new location.

Buy some of the feed the pigs were previously eating or match it as closely as possible to further help them through those first stressful days.

Parasite Control

Unless you have been assured otherwise — and have some sort of documentation — assume that all pigs have at least been exposed to all of the major internal parasites or worms, as well as to external parasites and any localized problem parasites. Your new pigs will need to be dewormed. Talk with your veterinarian about the best products and their application for your situation.

Managing Flies and Manure

A couple of adult ducks of any variety make an effective, natural method of fly control. Some hog owners additionally introduce red worms under their units to hasten the breakdown of manure for future use in the garden or wherever else composted wastes are needed.

Beware of Hot Weather

The first hot spell of the summer, especially if it comes on quickly and is accompanied by high humidity, can take quite a toll on finishing hogs. In such conditions, hogs have had no time to become acclimated to the weather and thus react poorly to it. When it's hot out, provide extra drinking water and keep it fresh and clean. You can even sprinkle your hogs with water from above in the hottest part of the day (if your hog has hyperthermia, don't water it down; see page 268). These measures will add greatly to hog comfort and survival.

Housing Hogs

Some people keep their hogs confined to a small structure with a pig patio, while others allow their pigs to range on the ground, with a small shelter to which they can retreat in inclement weather. Whichever you decide, make sure the finishing pen — the place where you bring your hogs to slaughter weight — is oriented so the sleeping area opens away from prevailing winds. This will protect your animals against storms.

Many hog owners place the pen at the foot of the family garden, where garden wastes can be thrown over the fence. An alternative is to maintain two garden sites, using only one each year while penning a hog or two in the fallow site, where the animals naturally till and fertilize the ground.

Small Finishing Unit

A small finishing unit can be made of native oak lumber and sheet metal with a capacity for two to four head. It can provide shelter in inclement weather, is easy to move if necessary, will not appear on property tax lists as a permanent structure, and will keep the hogs clean and easily accessible. It keeps the hogs off the ground, which eliminates or reduces mud and helps keep parasites in check. Basically, it is made up of two parts: a small house and a slatted, floored pen fronting it. Some call the latter part of this unit a pig patio or sunporch. Both parts can fit into a space as small as 60 square feet (5.6 sq m).

Width. Width is determined by the sleeping-bed option chosen. A three-sided hog hut of sheet metal, or sheet metal and wood construction, can be bought in dimensions from 4½ feet by 6 feet to 5 feet by 7 feet (1.5 × 1.75 m to 1.5 × 2 m) for a modest price. Used ones are even cheaper. Such a hut is light enough and of the right size to use as a feeding unit.

The foundation. Begin with two or three treated 4 inch by 6 in (10 × 15 cm) runners of 12 to 16 feet (4 to 5 m) in length. For a unit 4½ to 5 feet (1.5 m) wide, two runners should be adequate; for one that extends to 6 or 7 feet (2 m) in width, three will be needed. Elevate the runners at the corners and center points with two to four 1-inch concrete blocks and at a height just above the top of the ground. You don't want them to sink into the mud and freeze down there. The runners will form the foundation for the entire unit; elevating them on blocks will facilitate cleaning the pen and make loading and unloading the hogs easier.

The floor. Place a solid floor at least 2 inches (5 cm) thick on the half of the unit where the house or sleeping bed will be. Two-inch (5 cm) full-cut native hardwood lumber, such as white oak, works well. It is durable and, unlike some treated wood products, will not be a skin irritant to lighter-color hogs. Solid sheathing such as plywood may form a surface too slippery for hogs to stand and walk upon.

The pig patio. The outside pen or patio floor should be made from 2-inch- (5 cm) thick hardwood planking, with

This two-stall finishing unit on runners is fitted with lift gates that can be lowered to confine pigs as needed.

Insulate the Roof

You can choose to use plywood or sheet metal and 2-inch (5 cm) framing lumber to build sleeping beds. Just be sure to insulate the roof to prevent condensation from forming on the inside of the roof and falling on the pigs and their bedding.

Swine Pasture Crops

The pasture crop that would absolutely stand up to the wear and tear of grazing hogs would be a natural variety that has a great deal in common with Astroturf. Alas, such a species does not exist. Still, hardiness and durability are important factors to take into consideration when selecting plantings for swine pasture. Some of the hardier alfalfa varieties, clovers, and various grasses can be used to pasture swine. Multiple varieties create a more enduring pasture; mixes also are often less costly to establish and maintain than pure stands, especially pure stands of some of the legumes.

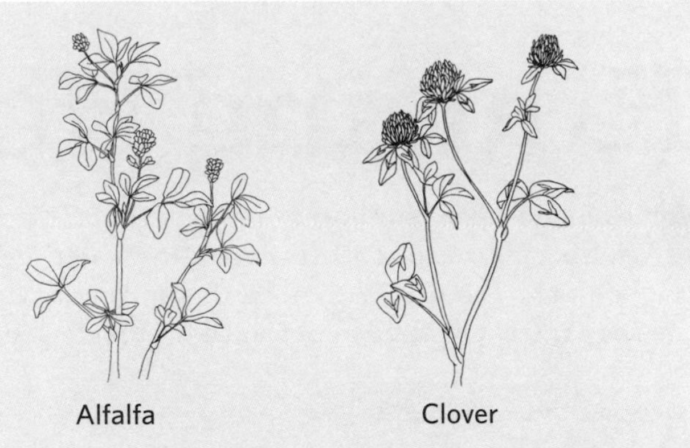

Hardier species of alfalfa and clover can be mixed in with various grasses when raising hogs on pasture.

a 1-inch (2.5 cm) space between the boards. The slots allow wastes to work through the floor and away from the hogs. At 1 inch (2.5 cm), however, they will not catch the feet or legs of young pigs nor encumber larger hogs. The pen's sides can be made of 1-inch- (2.5 cm) thick planking or 34-inch- (85 cm) high hog panels trimmed to fit the pen's sides and end.

Provide a 2-inch (5 cm) space between the bottom plank or panel bottom and pen floor to facilitate cleaning. In cold weather, the base of the unit can be insulated with bales of straw or sheet metal to prevent chilling drafts from coming up through the floor. Insulate only around the floor; leave the slots open.

Keeping Pigs on the Ground

Some owners prefer to keep their hogs on the ground. If this is your choice, you'll need to provide at least 150 square feet (14 sq m) of pen space per shoat to keep mud problems from developing. In a wet or low area, that amount of space may have to be tripled.

Hogs can fare well even on total legume pastures — something few ruminants can do — but hogs are not the most efficient users of pasturage. They are omnivores and have a single gut. They do not totally utilize the browse — twigs, leaves, and shoots — and they need richer sources of energy.

Hog lots become muddy not from the rooting activity of the pigs, which can be controlled with the use of humane nose rings, but from excessive foot traffic and those sharp, pointed hooves. In a drylot, feeders and waterers should be placed on concrete pads or hardwood platforms to prevent hogs from slipping and mud holes from forming around these well-used sites.

As an environmental safeguard, maintain a strip of sod 10 to 20 feet (3 to 6 m) wide at the foot or bottom end of a drylot. This strip naturally filters the runoff from the lot and thus serves as an environmental safeguard. A hard rain clears the lot to what is essentially a new surface, and the sod maintains the lot in a natural way, with minimal runoff getting past the strip and little or nothing by way of erosion in the lot.

You'll still need a house for shelter, such as the small unit described above, since it will be the animals' only dry retreat in wet or raw weather. It should have a step up to help keep muck out of the bedding area. Of course you'll eliminate the pig patio, since you'll be raising your hogs on the ground.

A house in a lot or pasture should still have flooring atop the runners or, if floorless, be placed atop a 4-inch- (10 cm) high pad or stone, packed earth, or cull lime (lime not good enough to use as fertilizer). It's a good idea to block between the runner ends with planking and dig shallow trenches around the range house to help carry runoff water away from the hogs' sleeping area. In raw weather a 4-inch- (10 cm) deep layer of clean straw will increase the comfort level of the sleeping area by at least 10°F (5.5°C).

Numerous variations on these housing themes are possible, but remember your primary goals: to keep sleeping hogs dry, protect them from drafts, and get them up out of the muck.

Hogs on pasture should have a humane ring on their nose to keep down rooting damage to pasture crops and the soil surface. Humane rings are applied through the cartilage and stay in place much longer than regular rings.

Fencing Options

To contain your growing hogs within a lot, use steel hog panels, wooden gates, or electric fence. These fencing choices are easy to take down when rotating lots or to enclose an idle garden plot.

Hog panels are all metal, can be easily moved by one person, and attach quickly to steel posts. They can be attached with long-shanked staples, clips, or short lengths of tie wire. They are 34 inches (85 cm) high by 16 feet (5 m) long. To erect a pen for growing and finishing hogs from such panels, 5-foot- (1.5 m) long steel posts are adequate.

Wooden gates are nonconductive. They're simple and inexpensive to build and are typically 12 to 16 feet (4 to 5 m) long and 34 to 40 inches (about 1 m) high. They span openings in the fence line large enough to enable a tractor or truck to enter the pen.

Electric fencing requires only two charged strands; one 4 inches (10 cm) above the ground, the other, 12 inches (30 cm). The lower wire will contain small pigs, while the higher wire is needed for larger hogs weighing about 80 pounds (36 kg) or more.

Critical to the successful operation of an electric fence is having the controller properly grounded. A 6-foot- (2 m) long copper rod driven at least 4 feet (1.2 m) into the ground will serve admirably to ground the charger. The rod must be driven deep into the ground to ensure it is constantly in contact with moist soil. In some areas a 6-foot (2 m) rod may have to be driven nearly its full length into the ground. In dry weather or climes, regularly dousing the area around the ground rod with several buckets of water will ensure it remains effective.

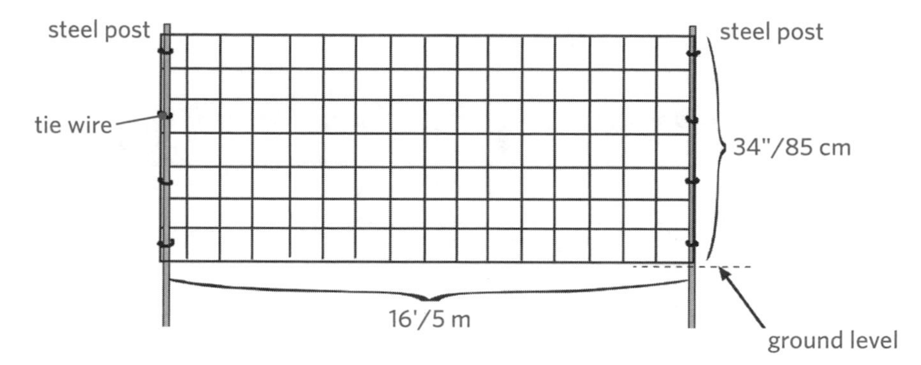

Steel hog panels attach quickly to steel posts.

Electric Fencing

Here are a few bits of wisdom on how to operate and maintain an electric fence line and charger for best results:

- Hogs that have never been exposed to charged wire tend to run right through it on their first encounter. To train them to respect hot wire, expose them to it inside a woven-wire or panel-enclosed lot. Run a short strand through the pen, place a bit of feed adjacent to and beneath it, and they will soon learn to give it a wide berth.
- Following a strong wind or storm, walk the line completely. Weeds, grass that's damp and heavy, and limbs across the fence line can short it out or greatly reduce its ability to contain hogs.
- Install a battery-operated fence monitor where you can easily see it. The monitor clips to a fence wire and emits a bright red flash if it detects low voltage on the wire, alerting you to a problem that must be found and repaired.
- Monitor the charger and fence line several times a day. A good practice is to walk the line and check the charger while doing chores.
- Invest in one of the simple testing tools for measuring the strength of the charge being carried by the fence line.

A single strand of charged wire suspended 12" (30 cm) above the ground will contain most hogs 80 lbs. (36 kg) and up. For smaller pigs, a second charged wire about 4" (10 cm) above the ground will be needed.

Essential Equipment

Besides facilities to house hogs and fencing to keep them in, you will need the following materials to help you easily and safely handle hogs, no matter how few. The feeling of comfort these items provide for the hogs will promote rapid and efficient hog growth. They will also ensure the safety of both you and your animals.

Loading Chutes

A chute is used to load hogs onto a truck. A simple loading chute 22 to 30 inches (55 to 75 cm) wide and long enough to extend from the ground or pen floor to the truck bed can be made from little more than scrap lumber. Such a chute generally runs from 3 to 9 feet (1 to 2.5 m) in length. It should be kept narrow enough so hogs cannot turn around inside it but wide enough that someone can walk behind the hogs while loading.

Solid sides will make the hogs feel more secure in the chute and will make them easier to handle. Solid sides also prevent shadows, shifting light patterns, and outside activities from upsetting the hogs. Cleats in the floor 8 to 12 inches (20 to 30 cm) apart will give them more secure footing. The chute decking should be made of material at least 2 inches (5 cm) thick. A few holes or small slots in the floor decking will allow rain and urine to run through, speed floor drying, and protect it from rot.

A bit of straw may encourage a hog to climb up an unfamiliar chute, but be sure to clean the chute floor after every use to extend its life. The best way to keep hogs moving up a chute is to walk behind them slowly, blocking any retreat with a short gate or hurdle (see the illustration on page 261).

Essential Equipment for Raising Hogs

- Housing unit and pig patio; or ground space and housing unit
- Fencing
- Loading chute
- Handling panels
- Snare or gripping tongs
- Waterer
- Feed trough

Hog Movers

The most difficult and time-consuming task is getting hogs from point A to point B while maintaining a modicum of dignity and your religion. Options depend on the size of the animals.

Small pigs can be moved in a plastic trash can on wheels. Two or three young shoats will travel quite comfortably in this unconventional but quite inexpensive and simple-to-use trailer. It also will hold a whole litter of baby pigs.

For a larger hog use a farm boy hog mover made from one 16-foot- (5 m) long hog panel and a 5- to 6-foot- (1.5 to 2 m) long segment cut from another. Bow the longer panel into a horseshoe shape, then securely wire the shorter panel across one open end (you'll be left with an open bottom and one open end). This simple unit can be used as a small holding pen, a sorting aid inside a larger pen, and with the help of a second person, a tool for moving even a large sow or boar over short distances.

The best way to keep a hog moving is to block the way behind it. Also, as much as possible, allow the hog to move along at its own pace and in the company of other animals to help it remain confident and at ease. Driving two or three hogs back off a trailer is often easier than driving just one hog onto it.

A plastic trash can on wheels can be used to move two or three shoats quite comfortably for a short distance.

A farm boy hog mover can be used for moving a large sow or boar over short distances.

Handling Panels

Invest in an assortment of short, solidly made wooden gates, or hurdles. *Hurdle* is an old term for a short, solid gate formed from 4 feet by 4 feet (1 × 1 m) sheets of plywood, often called pig boards. Hogs can't see through the hurdles, and they are useful for sorting, loading, and restraining the animals.

When snugly pressed between a barn wall or gate and one of these handling panels, a hog can be held safely for brief health treatments, sorting, or evaluation and measurement with a weight tape. These short gates are also useful for matching a loading chute mouth to various truck and trailer racks to pro-

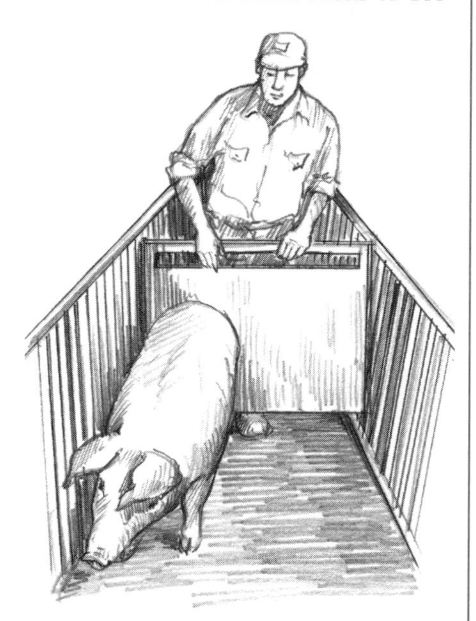

A hurdle is formed from 4 ft. by 4 ft. (1 × 1 m) sheets of plywood. Hogs can't see through them, and they are useful for sorting, loading, and restraining.

Restraint Devices

A snare or set of gripping tongs is good for restraining hogs of market weight and larger. The snare surrounds the hog's snout; the tongs apply pressure to the back of the neck. Both are pressure points and restrain the hog in such a way that you can administer any necessary health-care treatments.

A restraining device, either gripping tongs or a head gate, must be used while a nose ring is being placed to prevent rooting activities. Ring pliers clip the ring across the end of the nose or the end of the gristly tissue around the nose.

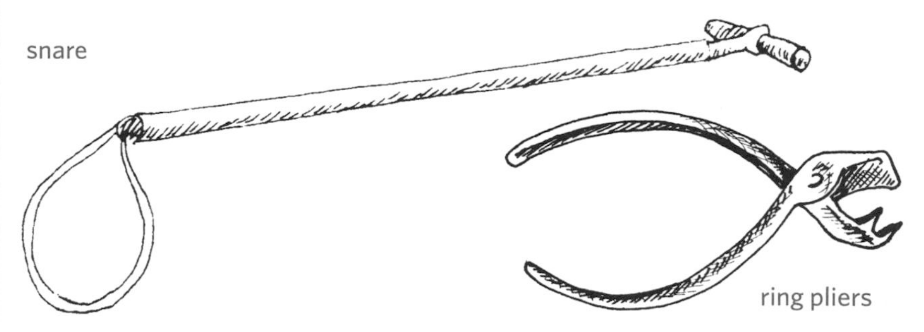

Beware of Light

Whenever you're moving hogs, think ahead and try to think a bit like a hog. Changing light patterns, intensities, shadows, and contrasting colors in their line of sight can be most disturbing to animals being driven from one place to another. Especially difficult can be driving a hog from a dark barn environment into bright sunlight and vice versa.

In an area where hogs will be regularly worked or along frequently used lanes of travel, line the fencing or enclosure with plywood or sheet tin to completely block out shifting light patterns and other outside distractions. A solid-floor loading chute also encourages hogs to move more easily than a chute with a slatted floor.

A hog is wary of shifting light patterns and contrasts in flooring and other footing surfaces when being moved and may balk or turn back from them.

Handle with Care

Just as with any other large animals, when handling hogs in close quarters, plan ahead and leave yourself at least two ways out. The hog has an undeserved fearsome reputation, but its size does dictate thoughtfulness and due consideration when you are working in close proximity to it.

Move about your hogs with a quiet and assured manner. Never shout, and never try to work them when angry or stressed yourself. Never back them into a corner — they then have no choice but to strike out by lunging past you or biting.

When handling young pigs and shoats of up to 70 pounds (30 kg), the hands-on approach has no substitute. Handy grab points are the ears and hind legs. That tail makes an awfully tempting handle, but never attempt to lift a hog of any size by the tail. It is a part of the animal's spinal system.

Feeding and Watering Equipment

Feeding and watering equipment can be as elaborate or as simple as your desire — or pocketbook — allows. The two essentials are keeping the drinking water clean and fresh and keeping the feedstuffs palatable so none are wasted.

Provide Adequate Trough Space

Provide at least 12 inches (30 cm) of trough space for each hog in the pen to accommodate the animals' herd instinct to eat at the same time. One space or hole in a self-feeder can accommodate three to five head of growing/finishing hogs because it keeps the feed before the hogs at all times. A water container should be large enough to hold at least 12 hours' worth of drinking water.

Homemade containers. Feed and water pans can be made by sawing a 30-gallon (115 L) plastic barrel in half down the center line, to make two small tubs.

Store-bought containers. Rubberized pans and tubs are inexpensive and easy to move, and in cold weather they can be flipped over and stomped on to pop out ice. A one- or two-hole self-feeder of sheet-metal construction bought at farm supply stores will hold 1 to 5 bushels (35 to 175 L) of feed. The lifespan of these feeders hinges on your efforts to maintain and keep them clean. Tie the feeder lids up or open for a few days, until you are sure the pigs are large enough and able to operate them. The same holds true for lids or covers on drinking fountains.

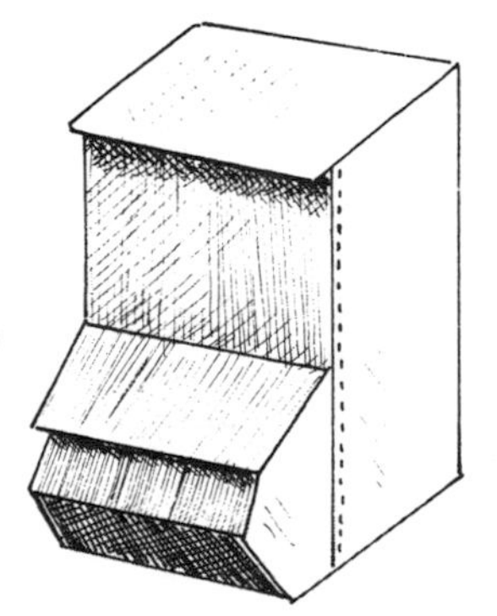
A 1-hole self-feeder is suitable for a single animal or a small group of hogs.

Fountain-style waterers. A number of fountain-type drinkers can be attached to 30- and 55-gallon (115 and 200 L) drums to provide drinking water in volume during warm weather. They can be set into a pen corner and will hold enough drinking water for several days. They are durable and quite simple in design.

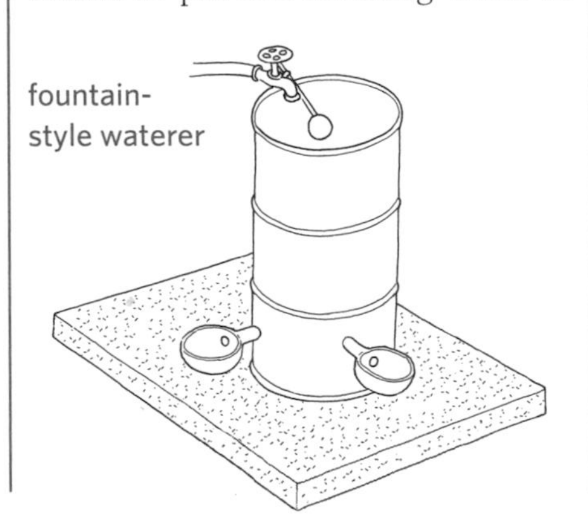
fountain-style waterer

A simple waterer that will work for one or two shoats and that takes up little in the way of pen space can be made from a 4- to 5-foot (1 to 1.5 m) piece of 6- or 8-inch (15 or 20 cm) polyvinyl chloride (PVC) pipe, one pipe cap, and a nipple-type or Lixit waterer head. Cap one end of the pipe, then affix the nipple through a low hole on that capped end. The nipple must be a gravity-flow model rather than a pressure type. When all of the gluing is thoroughly dry, the tube waterer can then be wired into a pen corner at the height appropriate for the pigs you are holding. This unit will hold a good amount of drinking water, is quite inexpensive, and is easy to store when not in use.

Feeding Hogs

Hogs are not garbage disposals and have a digestive system similar to our own. They are true omnivores and require a well-rounded, basic diet. To build a truly balanced ration from ever-changing table scraps and garden wastes is nearly impossible. It is best to consider these extras as treats, something over and above a pig's regular ration, and never to feed them in amounts that will disrupt the hog's digestive system.

The old rule of thumb was that in growing from 40 pounds (18 kg) to a handy market weight (230 to 240 pounds [105 to 110 kg]), a hog would roughly consume 10 bushels (350 L), which is roughly 685 to 750 pounds [310 to 340 kg] of corn and 125 pounds (55 kg) of protein supplement. The math still holds up pretty well, although many farmers now feed rations that have higher crude-protein levels or are more complex in their formulation.

Be on constant guard for freshness; be concerned if bags are stained or dusty or the contents are caked, have a bad smell, show mold, or are excessively dusty in their composition. With small numbers to feed, it is nearly always the best practice to plug into your local supplier's complete feed program, which should be readily available in easy-to-handle 40- or 50-pound (20 kg) bags.

Daily Consumption

To give you an idea of how much hogs eat, the following chart details the average daily consumption of hogs on full feed.

Age	Weight	Daily consumption	Percent crude-protein feed
8 weeks	40 lbs. (18 kg)	1.2 lbs. (0.5 kg) pelleted starter-grower	17–18
12 weeks	75 lbs. (34 kg)	2.5–3 lbs. (1–1.5 kg) grind & mix	11–15
16 weeks	125 lbs. (55 kg)	4–5 lbs. (1.8–2.3 kg) grind & mix	15
To market	150–230 lbs. (68–105 kg)	6–7 lbs. (2.7–3.2 kg) grind & mix	14–15

The Basics: Corn and Soybeans

Yellow corn is the basic feed grain for hogs and most other livestock species. To improve digestibility, and to make the nutrients more available to the hog, feed corn in a meal form. When milled, the corn is sometimes combined in a mixing tank with all the ingredients of a balanced ration.

If yellow corn is the foremost feed grain in use, its counterpart in protein supplements is soybean oil meal, commonly abbreviated SBOM. It is a meal-type by-product of the manufacture of soybean oil. Varying in content from 44 to 48 percent crude protein, it is blended with corn or other feed grains then supplemented with vitamins and minerals to create a complete growing/finishing ration. It is also often the base protein supplement in young-pig and breeding-stock rations.

Alternatives to Corn

Alternatives to corn as the primary grain in swine rations include grain sorghum or milo, barley, and wheat. For best results, all should be ground to a meal form and should probably replace no more than 50 percent of the corn in the ration mix. The so-called birdproof grain sorghums should not be used, because the factors that deter birds from feeding upon sorghum as it ripens also greatly reduce its palatability.

Soybean meal can be found in two other forms: hog forty, a 40 percent crude-protein content that is already supplemented with the more commonly required vitamins and minerals to form a complete swine ration, and extruded soybean meal, made by running whole soybeans through an extrusion screw that processes them with both heat and pressure.

Specialty and Not-So-Special Feeds

You may wish to put a local stamp on your feeding practices, such as having your hogs glean in a local apple orchard and feeding them some of the by-products of apple processing. Hogs could also be fed locally available by-products from gardening or food

Pellets or cubes are top feed choices for hog owners with small numbers of animals.

processing. These practices will favorably affect the flavor of the pork produced.

On range, or when providing food items with a lesser-known nutrient content, keep your hogs on a complete dietary ration appropriate for their age and role as sow, finisher, grower, and the like. Consider your additions as a little something extra and offer them in modest amounts that the animals will clean up quite quickly. Hogs are like big kids and can easily render their diets completely unbalanced by choosing items they favor in taste over less palatable, healthier options.

Meat Scraps and Animal Protein Supplements

Skimmed milk and meat scraps were once about the only protein supplements to be fed to hogs. However, the threat of mad cow disease, spread by the consuming of meat scraps from the spine and brain of prion-infected animals, has changed the thinking on this practice.

Some sources of animal protein, generally fish and dairy, are still to be found in some of the more protein- and nutrient-dense, complex rations such as pig starters and early-stage growers. Humane and chemical-free early-age rations are supplemented with whey or egg-based products. If the animal-based protein source is from a species distinct from the species being fed, the chances of any potential health problems' being transferred are greatly reduced.

A number of totally vegetable-based rations are now available for hogs at nearly every stage of life, but always bear in mind that hogs are true omnivores in their natural state.

Pellets and Cubes

You could grind and mix your own feed (or buy it from someone else), but pelleted feed is the best option when dealing with only a few hogs. Under heat and pressure and with a binder such as alfalfa, the feedstuffs are formed into pellets or cubes that range from ⅜ inch (1 cm) long up to about the size of a man's thumb. The heat and pressure may free as much as 5 percent more of the nutrients within the feedstuffs, and when fed in a trough on the ground, pellets and cubes create less waste. Pellets and cubes can be handled and fed easily, are widely available, and deliver nutrition more efficiently than other types of ration.

Feeding Methods

Feeding out pigs to market weight can be accomplished through one of two approaches.

Free choice or full feeding is the first of these approaches. Simply put, the hogs have access to all the feed they might want 24 hours a day. By the numbers, a hog on full feed will eat about 3 percent of its weight daily in feedstuff. By the end of the feeding period, a finishing hog will be eating 7 pounds (3 kg) or a bit more of feed daily.

Limit feeding, the second approach to feeding out an animal to about 90 percent of its appetite, will produce a slightly trimmer hog and will somewhat reduce daily feed costs. It will also extend an animal's time on feed; as a result, you will probably realize no overall savings on feed costs. Still, it will produce some extra-trim pork.

A metal or plastic trash can is a good way to store small amounts of feed. The container should have a tight-fitting lid to keep moisture down. Clean feed-storage bins often, and if you use ground feed, use it rapidly so it doesn't have a chance to get moldy.

As a finishing hog ages, both its growth rate and feed efficiency slow, and much of the weight gain in the late stages of the grow-out period is often finish (fat cover) rather than lean or muscle gain. To help cope with this natural pattern, you can reduce the crude-protein content of the ration as the hog grows.

If you do opt to make changes in swine rations, do it gradually over three to seven days. Sudden changes can cause gastric upsets and stress.

Finishing Weight

The last thing you'll have to determine when finishing hogs is the weight at which you want to butcher them. You can butcher a hog at 250 to 260 pounds (115 to 120 kg) and still have a quite lean and high-yielding carcass.

A hog will normally yield about 70 percent of its live weight in pork and pork products. A typical family can feed out two hogs a year for their needs: one for slaughter in late winter and another for the fall. Fresh and cured pork is also a most appreciated gift item during the holiday season.

Home Butchering & Pork Processings

***Slaughter* and *butcher* are hard words for many people, and they are processes you may not want to take on as a do-it-yourself project. They are traditional farming activities and can result in measurable cash savings, but they are also time-consuming and call for some special skills. If you wish to process your own hogs, see *Storey's Guide to Raising Pigs* by Kelly Klober for step-by-step how-to information.**

Can you save money by raising your own hogs for meat? It's a simple question that has no simple answer. The current cost of a 40-pound feeder hog is about $40.00. To reach a good weight, it will eat 10 bushels of corn (at $3.50 a bushel, that's $35.00) and 125 pounds of protein supplement (at 11 cents a pound, that's $13.75). Out-of-pocket costs are then $88.75, with additional overhead costs of around $10.00, for a total of $98.75. Custom processing on a 240-pound hog will run around $75.00 and special processing, such as curing or links, will add roughly 30 cents for each pound handled.

From a 240-pound hog, you should get about 170 pounds of pork for a total cost of $173.75, including basic processing. Roughly this comes to $1.00/pound and quite a good price for sausage, fresh ham, and pork chops. You are also getting, however, tail, feet, head, heart, and a few other extras. If you take tenderloin, you won't get chops; you must choose between shoulder steaks or roast; and not everyone likes fresh ham. If you do your own butchering, you can get chops cut to your exact thickness, sausage seasoned to your taste, and bacon in any size package you desire, but then you must put in the labor to produce such things. Many factors are involved, but for most people the benefits of raising their own pork go beyond economic considerations; to know the meat is truly fresh and has been produced to your own exacting health and humane standards has its own immeasurable value.

Freezing the Pork

Pork products can be frozen, canned, cured, or smoked. Freezing is the most common means of home meat storage. It is also the simplest and least time-consuming.

Freezing simply slows the changes that affect food quality. Bacteria are not killed by freezing; they are simply halted from multiplying. How the meat is prepared for freezing, how the freezer is maintained, and how foods from the freezer are thawed prior to preparation all affect eating quality and wholesomeness. Freezer management is an ongoing chore.

To preserve their quality, you should freeze pork and other meats as quickly as possible. Set the freezer control to –10°F (–23°C) or slightly below. Ice crystals break down meat cells and fibers and adversely affect juiciness and texture in the meat when it is later prepared for the table. Rapid freezing, on the other hand, causes smaller ice crystals to form in the meat and thus preserves better eating quality. When the meat is completely frozen through, you can return your freezer to the higher temperature setting of 0°F (–18°C).

Never attempt to freeze an amount of meat that will not freeze thoroughly within 24 hours. Also, never try to freeze more than 1⁄15 of the freezer's capacity at a time. A good guideline is to freeze only 2 to 3 pounds (1 to 1.5 kg) of meat for each cubic foot

Tips for Freezing and Thawing Pork

While freezing hastens color changes in red meat, pork's eye appeal, flavor, and juiciness are little affected when the meat is properly handled and wrapped, and then used within the approved storage times. Do not, however, season ground pork before freezing. Most seasonings — especially sage — are intensified by freezing. Lard, after complete cooling, can also be stored in the home freezer. Freezing times that best maintain the table qualities of various pork items are as follows:

Cured bacon	1 month or less
Cured ham	1–2 months
Sausage	1–2 months
Fresh pork chops	3–4 months
Organ meats	3–4 months
Pork roasts	4–8 months

The best way to thaw frozen pork is in the refrigerator. For small roasts and steaks, allow three hours of thawing time per pound; for large roasts, four to five hours per pound.

(30 L) of freezer space. When you have your meat cut and wrapped professionally, it will have been properly frozen when you pick it up.

To save space in the freezer, debone the cuts of pork and trim them into pieces as uniform and compact as possible before wrapping them and packing them into the freezer. Frozen meat is best when consumed as soon as possible following thawing, so wrap it in portion-sized amounts that can be used at one time or for a single meal.

Grinding the Pork

Before grinding, all animal heat must be gone from the meat; it should also be chilled, to firm it up. Cut the chilled meat into cubes or strips for easiest feeding into the grinder.

An electric grinder will speed up the pork-grinding chore, but a hand-turned grinder is adequate for small amounts of well-chilled or partially frozen trimmings. All-metal grinders offer the greatest durability and ease of cleaning. They will either bolt to a large board or table or clamp to the edge of the work surface. They come with grinding plates of different sizes that you can use to double-grind the meat to the desired texture.

For the best taste, some fat should be ground with the lean meat to enhance the juiciness and flavor of the sausage. The fat content of pork sausage normally ranges from 20 to 30 percent; the latter is typical of many supermarket sausage products. Whole-hog sausage is generally leaner (80 percent lean or better), as it includes all of the leaner cuts of pork.

Pork for sausage should first be run through the coarse blade of the meat grinder. Then mix it thoroughly with your hands and spread it out thinly. Season it evenly (choose any of a variety of spices according to your own taste and preference), and mix it thoroughly again by hand. Finally, regrind it through a finer plate. Sausage to be frozen is best left unseasoned, and sausage for canning should receive no sage seasoning.

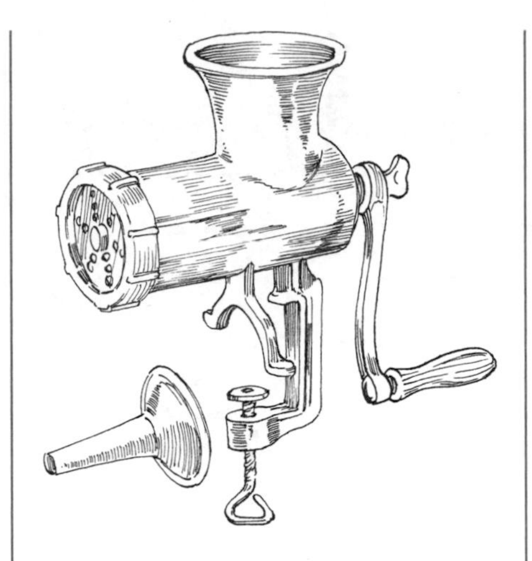

A metal hand-turned grinder is sufficient for processing small amounts of meat.

Preserving Pork

Drying, smoking, and curing are the oldest known methods for preserving meat. Pork-preserving methods are time-consuming by today's standards, and the finished products may require refrigeration for safest storage. Still, their taste, aroma, and texture can be produced in no other way.

Most cures are quite often closely guarded family secrets and may include such widely varied ingredients as red pepper and cloves, but the two basics of most cures are sugar and noniodized salt. Salt is both a good preservative and a flavor enhancer. It moves through the meat by that process we all read about way back in high school: osmosis. It also helps inhibit bacterial growth. Sugar helps counter some of the salt's harsh edge, brings out further flavor in the meat being cured, and lowers the pH of the curing solution.

The third ingredient common to many pork cures is saltpeter (nitrate), which is different from curing salt. It fixes the desired taste and color in cured meat and protects it from the often-fatal botulism organisms. The nitrate question is a touchy one for many — nitrate has been associated with some forms of cancer, but it is the best protection the home processor has against botulism. If used, saltpeter should be stored and handled with great care and added only in the exact amounts set down in whatever curing recipe or formula you are using. Some people delete the saltpeter and make sure to use the cured pork quickly.

Quality Assurance at Home

For the assurance of wholesomeness and quality, home processing of pork and pork products cannot be surpassed. Processing pork may at first appear to be beyond your skills or desires, and allowing a custom processor to do some or all of the pork-processing tasks is certainly a reasonable and economically viable alternative. Still, the savings, assurance of quality, and freshness and taste of home rearing and processing of a meat hog or two do make it a worthwhile project even if you have just a small bit of acreage.

Swine Health

The best thing you can learn about swine health is how to monitor your animals' well-being and healthful appearance several times each day, carefully noting any changes from the norm, and seeking professional help when things appear to be going awry. Hogs are tough animals, and as long as they are managed properly — they are well fed, well watered, and well housed — not much is apt to go wrong.

The typical backyard is not likely to have a great deal of trafficking with other swine populations, and thus you are unlikely to introduce health problems to your animals. The important thing is to start with sound and healthy animals, then keep them as free from stress as possible, and provide them with high-quality, age-appropriate feedstuffs.

A good vet will know of any disease problems in your area, can advise on management changes needed as the seasons change, offer health products in the smallest amounts needed, and can advise on preventive care and basic health management. Once you establish a good relationship with a veterinarian, a lot can be done over the phone.

Watch your pigs closely, observe and note (time and degree) when any behavioral changes and responses occur, and call your vet for help before the condition worsens. The hard truth is that the longer you wait to call for help, the more it will cost and the less likely it is to succeed.

Supply the veterinarian with a good health history: how the animal is being fed, the age of the animal, and the disease signs and time of onset. The veterinarian will need an area where the animal can be confined and easily accessed, and where the vet can have good footing. Also be prepared to provide warm water and some old-fashioned manual assistance. The small, growing/finishing unit described on page 257 is a good place for a vet to tend to animals.

Health Products

Many health products are available for each major livestock species. Some cost $100 or more per vial, all are invaluable when needed, and most will never be needed by someone raising just a few head of growing hogs.

It is a good idea to keep on hand a small bottle of epinephrine to treat shock and a small bottle of a long-lasting antibiotic. Also, rotate deworming products after three or four uses to prevent parasites from developing immunity to a particular product.

Veteran hog producers have raised literally hundreds of boars that receive absolutely no other injections than their iron shot at birth. Still, injections are necessary to some extent. Injections can be given in one of three ways: subcutaneous, intramuscular, and intraperitoneal. Ask your veterinarian to teach you

Normal Temperature for Pigs of Different Ages

Age	Rectal temperature
Newborn	100.2°F (37.9°C)
1 hour old	98.3°F (36.8°C)
24 hours old	101.5°F (38.6°C)
Unweaned pig	102.6°F (39.2°C)
Growing pig (60–100 lbs [27.2–45.4 kg])	102.3°F (39.1°C)
Finishing pig (100–200 lbs [45.4–90.7 kg])	101.8°F (38.8°C)
Gestating sow	101.7°F (38.7°C)
Boar	101.1°F (38.4°C)

Adapted from the text *Diseases of Swine*, 7th edition (Iowa State University Press, 1992).

Taking a Hog's Temperature

In many other types of animal, the heart and respiratory rates are commonly checked as well as the temperature. But in pigs and hogs, the temperature, coupled with the way the animals act, is a pretty good indicator of illness.

A fever develops because the body's systems are rising to its own defense to fight off an infection. But a below-normal temperature can indicate trouble, too. It can mean the body systems are shutting down, which in swine occurs with kidney failure.

To take a hog's temperature, use a standard rectal thermometer with a ring in the end. A short piece of stainless wire attaches to an alligator clip, which then attaches to the animal's hair coat; your hands are thus left free for other chores while you wait. Insert the thermometer about 3 inches (75 mm) into the rectum and leave it in for at least two minutes.

how to administer each, then practice on something harmless like an orange.

Keep your health products in a separate refrigerator or in their own plastic container with a sealable lid within your primary refrigerator. Store all products with their complete instructions, check them often to ensure they have not expired, and buy in the smallest amounts possible. Veterinarians can sometimes provide drugs in small units that can be used quickly for specific needs. And keep a regularly updated inventory sheet as to what animal health products you have on hand and their uses.

Here are a few more tips on properly storing and using health products:

- Administer a product only when an exact diagnosis has been made. Follow completely the directions given to you.
- Store products correctly. Drugs in dark vials or nonopaque packages are light sensitive. When the label says keep refrigerated, it means it.
- Keep the tops of injectable bottles clean and wipe them with alcohol before and after each use.
- Use different needles to fill the syringe and inject the animal.

Swine Ailments and Illnesses

A number of problems in swine are easily managed yourself; for others, you'll need to enlist the help of your veterinarian. Most important, however, is that you familiarize yourself with the signs of common illnesses in your hogs. If you have any doubt about what's wrong, or are concerned that you may not be able to handle the problem yourself, consult your veterinarian. In hogs, prompt treatment of a disease can prevent a slowdown in weight gain and may even be lifesaving.

Dehydration

Dehydration is usually caused by an illness that results in scouring, or diarrhea. Oral glucose and electrolytes are administered as antidotes for dehydration, and a number of good ones — such as RE-SORB — are available just for pigs. In a pinch you can use Pedialyte, a product for children available at your local drugstore, or even Gatorade, which can be administered orally with a syringe (without the needle attached). Keep dehydrated animals warm and comfortable.

Hyperthermia and Hypothermia

Hyperthermia, or a body temperature above normal, means the animal is too hot. A hog that gets hyperthermia goes into shock. Do whatever you can to calm it. Get it away from anything that might upset it. *Do not hose the hog down.* This could kill it. At most, brush a bit of cool water across the bridge of the nose. Administer an intraperitoneal injection of epinephrine, and keep your fingers crossed.

Hypothermia, or low body temperature, means the animal is too cold. It isn't a common problem in hogs. Little pigs, however, are susceptible to chilling, so keep them warm.

Lice and Mange

Lice and mange are common external parasites of swine. Treat with a product that kills the parasites.

Lice cause pigs to rub themselves to relieve itching. Suspect lice if pigs act itchy, and especially if you see small white eggs on their hairs — which are lice nits. You can see the lice, too, around a pig's head, ears, and flank area. The pig louse *(Haematopinus suis)* is more likely to be seen in animals that are not well managed. It is transmitted via pig-to-pig contact, and the lice suck blood from the pig's skin, which can result in anemia. The primary problem, however, is skin irritation and itching. Infested pigs may grow slowly and not efficiently utilize their feed.

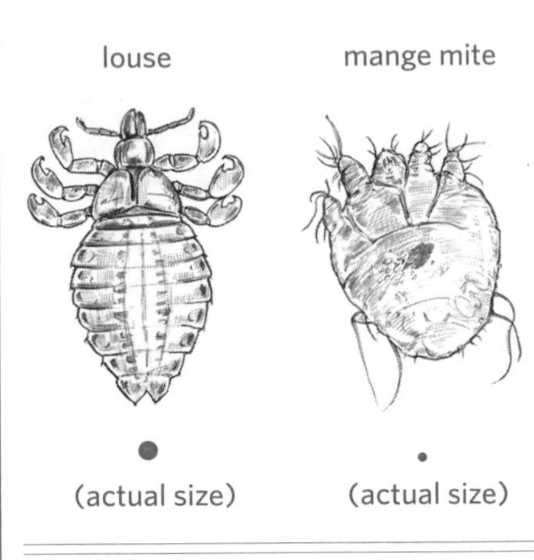

Mange, like lice, is not uncommon among swine. It is an infection caused by a mite called *Sarcoptes scabiei*. This external parasite spreads via pig-to-pig contact, and may be transmitted from an infected dam to her litter. Mites favor the head and ear area. Hogs may develop scabs from rubbing their itches. Some young pigs are allergic to mites, so their response to this infection will be more

Signs of Illness

- Getting off the bed slowly
- Eating less or not eating at all
- Looking drawn (dehydrated and excessively thin; sides appear hollow and pinched)
- Suffering from diarrhea
- Acting depressed
- Vomiting
- Experiencing abortions or stillborns
- Lame or walking stiffly
- Acting uncoordinated
- Among nursing pigs, seeming to be poor doers (animals that are smaller and slower growing than their pen mates)

severe; their itching will be worse, and skin irritation can involve their hind ends and abdominal areas. Young pigs with mite infestations grow more slowly.

Scouring

Scouring, or diarrhea, can be caused by a variety of factors, ranging from dietary problems to infectious disease. It can be lethal in an amazingly short time, and unless you are absolutely sure of the cause, consult a veterinarian to determine the cause and suitable course of action. Prompt treatment will not only increase pig survival but also help maintain reasonable growth in the face of a health problem. Pigs with scours usually become dehydrated and need rehydration (see Dehydration on page 268).

Shock

Shock occurs when the body tissues experience an inadequate blood flow. Numerous causes include severe stress, trauma, serious infections, and dehydration. Of these, trauma is the most common cause.

A hog in shock may be prostrate and have a rapid pulse, rapid breathing, and low body temperature. An animal in shock needs to be kept warm and dry, and should receive treatment for dehydration.

If you suspect a hog is in shock, *immediately* administer an intraperitoneal injection of epinephrine and back away. It usually works fast. The epinephrine will not harm the animal if it turns out the animal is not in shock. The cause of shock also needs to be addressed — a wound, for example, must be treated.

Swine Influenza

Swine influenza has frightened many potential new swine raisers. While the disease is deadly and can possibly be passed from pigs to humans, the chances of the pigs in your backyard catching the flu are remote, and the chances of them spreading it to you are even more remote.

Swine influenza is usually caused by the type A influenza virus. It can manifest as a deep, dry cough; loss of appetite; discharge from eyes and nose; and as a secondary bacterial infection in sows that can hurt their reproductive performance. Treatment involves reducing stress, keeping the animals dry and comfortable, and possibly treatment with an antibiotic.

The virus can pass from hogs to humans and even from humans to hogs. However, basic cooking practices and proper food sanitation will prevent the spread of the virus through consuming pork or poultry. Generally, the virus moves to humans when humans and livestock live in quite close proximity, where an exchange of blood, nasal spray, or other body fluids is likely. Where livestock are housed separately and basic sanitation rules are followed, the chance of it spreading to humans is greatly reduced.

And animals in good housing, that have access to fresh air and sunshine, are not crowded and stressed, and are well fed, are far less vulnerable to any and all health concerns in the first place. So, if you practice good sanitation and treat your hogs properly, you will most likely never see swine influenza.

Trichinellosis

Trichinellosis is a parasitic infection. It can be transmitted via undercooked meat to humans, in whom the disease is called trichinosis. It isn't nearly the problem some people believe it to be — the fear really is needless. Improved swine management has greatly reduced incidents of trichinellosis and increased public awareness of the importance of thoroughly cooking meat has effectively eliminated trichinosis.

Whipworms

Pigs from about 2 to 6 months of age that develop bloody diarrhea with mucus in it likely have whipworms. And pigs that ingest whipworm eggs in feces also become infected.

An infection with whipworms, also known as *Trichuris suis*, leads to dehydration and wasting. Pigs with a whipworm infection may go off their feed and become anemic. They can die.

As with most types of parasitic infections that affect swine, good sanitation and an effective parasite-management program will control this problem.

Wounds

In hogs the most common wounds are scrapes, cuts, and punctures. They tend to be shallow and will often heal with little or even no treatment. But an aerosol wound spray, such as Blue Lotion, or tamed iodine can be applied quickly. It disperses over the wound site, dries rapidly, and can be used without restraining the hog. But some hogs become skittish when they hear the hissing sound of an aerosol. Fortunately, many wound sprays are now available in pump and squeeze bottles.

With deep puncture wounds, some products may not penetrate fully and the wound may heal over at the surface but leave an infection underneath. For such wounds, obtain from your veterinarian and keep on hand a prescription product made to combat infection.

When using any spray around a hog's head, be careful not to get the product into the eyes or ear canals. Watch out for your own comfort, too. Tamed iodine can also sting you if it finds a nick or cut.

Tracking Treatments

To keep track of treatment, mark each animal as you administer the drug. Use different colors for multiple doses. Mark little pigs on the top of the head and adults on the topline with a stock crayon.

CHAPTER 9

Dairy Cows & Beef Cattle

Raising a milk cow, or a calf or two for beef, is an efficient way to produce food, since cattle can mow the hillside behind your house that is too steep for a garden or survive on a back forty that has too much brush, rocks, or swamp to grow any crop other than grass. Cattle provide us with meat or milk while keeping weeds trimmed, which is a good measure for fire control and yields a neater landscape. Getting set up to raise cattle often does not involve much expense, except for the initial fencing to keep them where you want them. And if you don't want to bother with purchasing hay and grain, you can use weaned calves to harvest your grass during the growing season, then send them for butchering when the grass is gone — thus making seasonal use of pasture and creating a harvest of meat.

Raising cattle can be a soul-satisfying experience. They are fascinating and entertaining animals; working with cattle is never boring. It can be physically challenging at times, as when delivering a calf in a difficult birth or trying to catch an elusive animal. But for those of us who enjoy raising cattle, the chores of caring for them are not really work. Our interaction with these animals is part of our enjoyment of life.

When you raise cattle, you participate in one of the oldest known human activities. Humans have lived with cattle since prehistoric times. Wild cattle were the main meat in the diet of Stone Age people. Our prehistoric ancestors began to domesticate cattle about 10,000 years ago to have a supply of meat for food and hides for leather clothing. Later humans discovered they could hitch these animals to a cart or a plow. In fact, oxen were used for transportation before horses were; cattle were domesticated sooner than horses, probably because they were easier to catch.

Getting Started

You can raise a steer in your backyard in a corral or on a small acreage, or you can raise a herd of cattle on a large pasture, on crop stubble after harvest, or on steep, rocky hillsides. Cattle can be fed hay and grain or can forage for themselves. Economics and individual circumstances will dictate your methods. If you have pasture, all you'll need is proper fencing to keep the animals in, so they won't visit the neighbors.

You will need a reliable source of water and a pen to corral the animals when they need to be handled for vaccinations or other management procedures. If you have a milk cow, you may want a little run-in shed or at least a roof, so you can milk her out of the weather if it's raining or snowing. Most of the time, cattle don't need shelter; their heavy hair coat insulates them against wind, rain, and cold. In hot climates, however, shade in summer is important. A simple roof with one or two walls can provide shade in the hottest months and protection from wind and storm in winter.

As a novice cattle raiser you may need advice from time to time from a veterinarian, cattle breeder or dairyman, or your county Extension service agent. Don't be afraid to contact an experienced person to ask questions or request help.

Meet the Cattle Family

Cattle are of the order Artiodactyla, consisting of hoofed animals with an even number of toes. Most large land mammals are Artiodactyls, including goats, sheep, and pigs. Cattle are in the family Bovidae, consisting of animals with hollow horns. This family includes antelopes, goats, and sheep. The genus is *Bos*, the true cattle, of which there are two species:

B. taurus: descendants of British and European breeds, as are most cattle in North America

B. indicus: cattle with large ears and humped backs, which are especially adapted to hot climates, such as the Brahman

Choosing the Right Kind of Animal

Your choice in a calf will depend on how much space you have and whether you want to raise a steer to butcher or a heifer that will grow up to be a cow. A calf can be raised in a small area, even in your backyard if your town's ordinances permit livestock. But if your goal is to have a milk cow that someday will have a calf of her own, she'll need more room.

If you are raising a calf to butcher, you should probably raise a steer. Steers weigh more than heifers of the same age. However, heifers are more flexible — you can raise them as beef or keep them as breeding or dairy animals. Mature dairy heifers are worth more money than beef cattle. If you want to eventually raise a small herd of cattle, choose a heifer to start with. Her calves will become your herd.

Bull. When a male calf is born, he is considered a bull because he still has male reproductive organs. Most bulls are castrated and become steers. Bulls can be unpredictable and dangerous.

Steer. A bull becomes a steer when he is castrated. The steer may still have a small scrotum or his scrotum may be entirely gone, depending on the castration method used. Beef calves are sold as steers.

Heifer. A heifer is a young female animal. Between her hind legs, she has an udder with four teats on it that will grow as she matures. A bull or steer calf also has small teats, just as a boy has nipples, but they don't become large.

Cow. Once a female animal becomes older than 2 years of age and has had a calf, she is called a cow. A fresh cow is one that has recently given birth to a calf; a cow freshens once a year.

Your choice in a calf will depend on how much space you have and whether you want to raise a steer to butcher or a heifer that will grow up to be a cow.

Bull

Steer

Heifer

Cow

Miniature Might

Intimidated by the size of a standard cow? Consider a miniature one. Miniature livestock require less housing space, pasture, fencing, and feed than do their full-size counterparts. You can stock two or three miniature Herefords or Lowline Angus to one garden-variety cow. The three kinds of miniature livestock are naturally diminutive breeds that evolved as small animals to better survive the conditions nature handed them (like Dexters); small breeds that retained their original breed character when their parent breeds were selected for greater size (such as Miniature Jerseys and Lowline Angus); and breeds that were deliberately miniaturized by breeders who selected for smaller stature, often through outcrossing to an established smaller breed (like Miniature Highlands and Miniature Longhorns). One warning about selecting miniatures: Be sure you read up on dwarfism before you buy; dwarfs are not miniatures and should mostly be avoided.

Choosing a Dairy Breed

You can be successful with any of the dairy breeds, but you may want to choose one that is popular in your area. The cost will be lower than for a breed that's less numerous, and when time comes to breed your cow, you'll have an easier time finding a bull of the same breed, if you so choose.

Ayrshire. Originating in Scotland, the Ayrshire is a medium-size red and white cow that may be any shade of red and sometimes dark brown. The spots are usually jagged at the edges. Ayrshires are noted for good udders, long lives, and hardiness. They manage well without pampering and can adapt to rocky land and harsh winters. They give rich, white milk.

Brown Swiss. Coming from the Swiss Alps, the Brown Swiss may be light or dark brown or gray. These cows are noted for their long lives, sturdy ruggedness, and ability to adapt to harsh climates. They give milk with high butterfat and protein content — ideal for cheesemaking.

Guernsey. Originating in the British Isle of Guernsey, Guernsey cattle are fawn and white with yellow skin. The cows are small as cows go, have good dispositions, and have few problems with calving. Their milk is yellow in color and rich in butterfat. Heifers mature early and breed quickly.

Holstein. Coming from northern Germany, Holsteins are large black and white or red and white cattle. A Holstein calf weighs about 90 pounds (41 kg) at birth compared to smaller breeds with calves typically weighing 60 to 70 pounds (27 to 32 kg). The cows are bred to turn grass into large volumes of milk that is low in butterfat. Holsteins are the most numerous dairy breed in the United States.

Jersey. Developed on the British island of Jersey, the Jersey is a small breed that may be fawn colored or cream, mouse gray, brown, or black with or without white markings. The tail, muzzle, and tongue are usually black. Jerseys mature quickly, calve easily, and are noted for their fertility. Jerseys produce more milk per pound of body weight than any other breed, and their milk is the richest in butterfat.

Milking Shorthorn. Originating in Britain, Milking Shorthorns may be red, red and white, white, or roan (a mix of red and white hair). These large cows are hardy and noted for long lives and easy calving. Their milk is richer than that of Holsteins but not as high in butterfat as that of Jerseys or Guernseys.

Acquiring a Milking Cow

Buying a mature lactating dairy cow is the easiest way to start with a family milk cow, compared to buying a dairy heifer you must raise for two years, breed, and train before you get your first drop of milk. A mature cow should already be accustomed to being milked, be gentle, and be easy to handle and milk. Check with the person selling her to make sure she's had the proper vaccinations and they are up to date, and she has been tested for brucellosis and tuberculosis, or any other requirements in your state. When in doubt about the required tests or vaccinations, ask your local veterinarian.

Also find out if the cow has been bred back again for her next calf. Unless she calved within the past few weeks, she should be bred again. You can have her checked by a veterinarian to tell if she is pregnant. If she has been milking for several months and is not pregnant, either the farmer didn't rebreed her, or the cow has a fertility problem.

A dairy cow should always be rebred about two to three months after she calves, so she will calve again the next year. A cow may go several years without having a calf and still be milked, but her milk production will decline after the first six months and will gradually keep dropping. She will produce more milk for your family if she has a calf every year and is given a chance to go dry (not be milked) for at least 45 days, or better yet 60, before her next calving.

If the cow has recently calved and is not yet rebred, give her two to three months to recover from calving before rebreeding for her next calf. Some cows will come into heat less than a month after calving if they are well fed, but it's too soon to rebreed them. Just keep

Milk Factory

A top-producing dairy cow gives enough milk in one day to supply an average family for a month. The average milk cow produces 6 gallons a day (23 L), which is 96 glasses of milk. A world-record dairy cow can produce 60,000 pounds (27,216 kg) of milk per year — that's 120,000 glasses of milk.

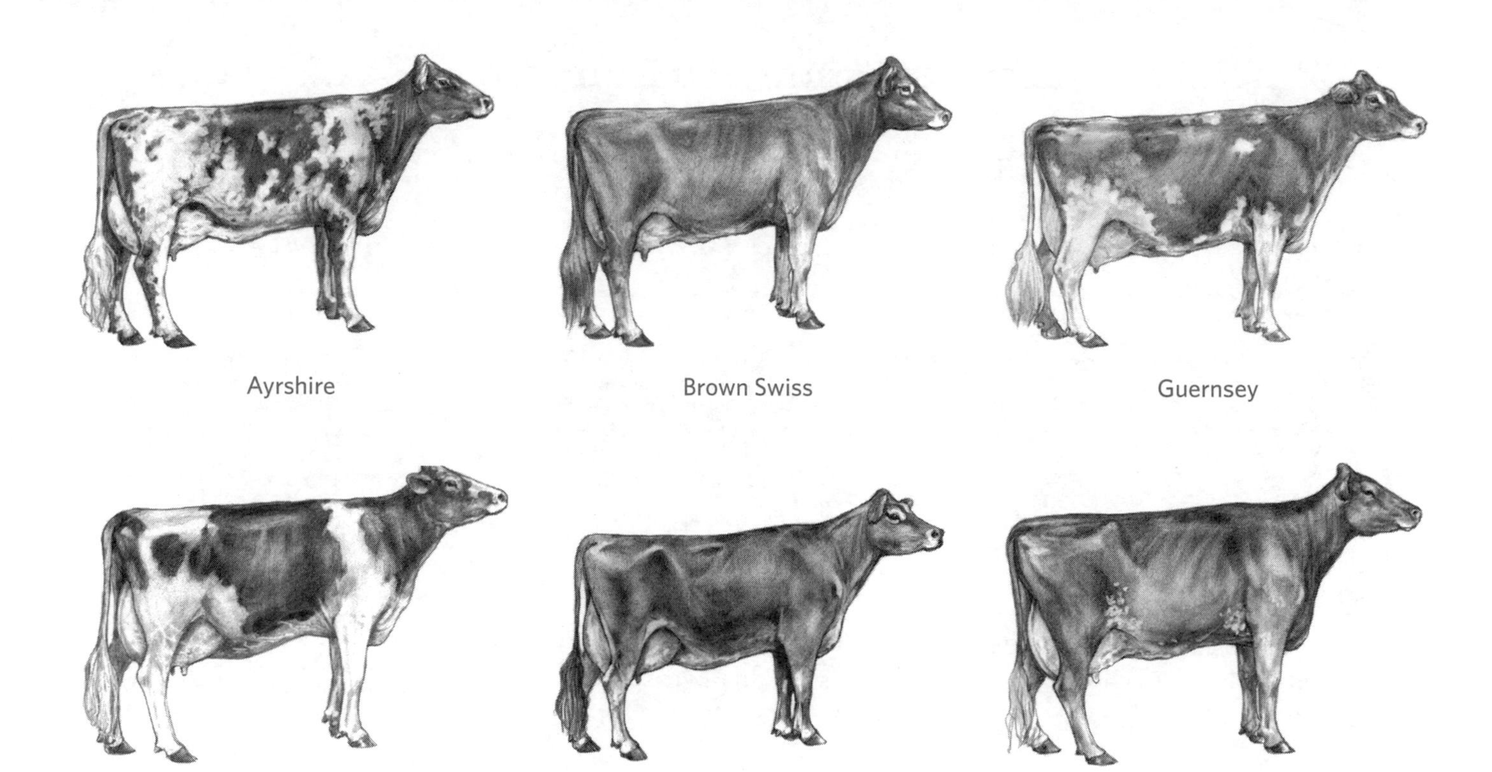

Ayrshire

Brown Swiss

Guernsey

Holstein

Jersey

Milking Shorthorn

Quick Guide to Dairy Cow Breeds

Breed	Weight in lbs. (kg)	Milk/year in lbs. (kg)	Butterfat (%)	Protein (%)	Coat color
Ayrshire	1,200 (544)	17,000 (7,711)	3.9	3.3	red and white
Brown Swiss	1,400 (635)	21,000 (9,525)	4.0	3.5	brown or gray
Guernsey	1,100 (499)	14,700 (6,668)	4.5	3.5	fawn and white
Holstein	1,500 (680)	28,000 (12,700)	3.5	3.2	black and white or red and white
Jersey	950 (431)	16,000 (7,257)	5.5	3.7	fawn, gray, brown, or black
Milking Shorthorn	1,500 (680)	15,500 (7,031)	3.8	3.3	red, red and white, white, or roan

track of her heat cycles when she comes back into heat, so she can be bred at the proper time. (For information on detecting heat, see page 302.)

When your cow produces her annual calf, you might sell the calf to help pay for the cow's maintenance, or you might raise the calf for beef. If the calf is a heifer, you might raise it as a second milk cow or as a replacement for your older cow if she is getting on in age.

Disposition

A cow's disposition is created partly by heredity. She inherits from her parents a tendency toward being nervous or placid, flighty or calm, smart or stupid, kind or mean. Just like humans, some cattle are smarter than others, and some are more emotional. But disposition is also influenced by how a cow is handled or trained. With patient handling, a timid, nervous heifer that is smart will often grow up to be a gentle cow. On the other hand, some wild and nervous animals can be frustrating, and dangerous, because they never learn to trust you.

The cow's attitude will give you clues. She should be mellow and calm rather than nervous and flighty. A milking cow should be accustomed to having people near her and touching her udder. If she

Conformation of a Dairy Cow

Ideal leg conformation (from side)

Post legged (hind leg too straight)

Sickle hocked (hocks too bent)

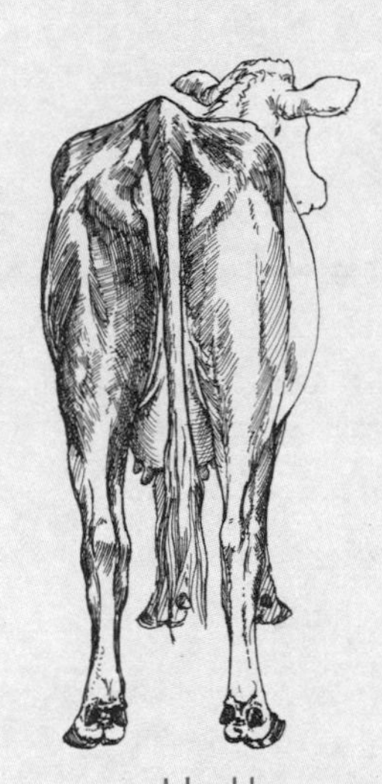

Ideal leg conformation

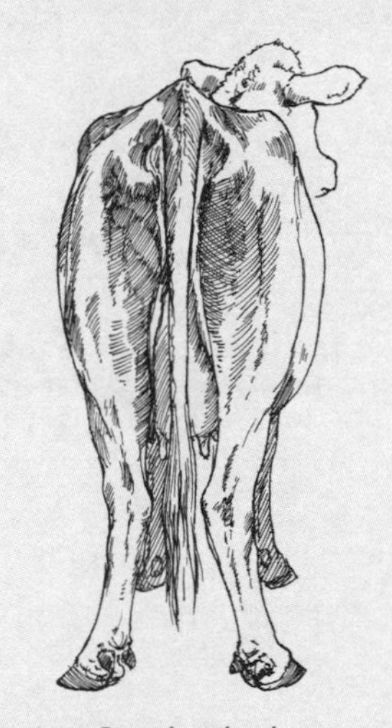

Cow hocked (hocks too close together)

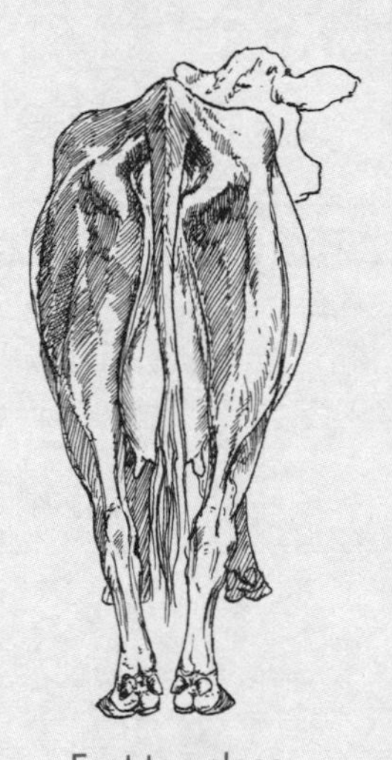

Feet too close together

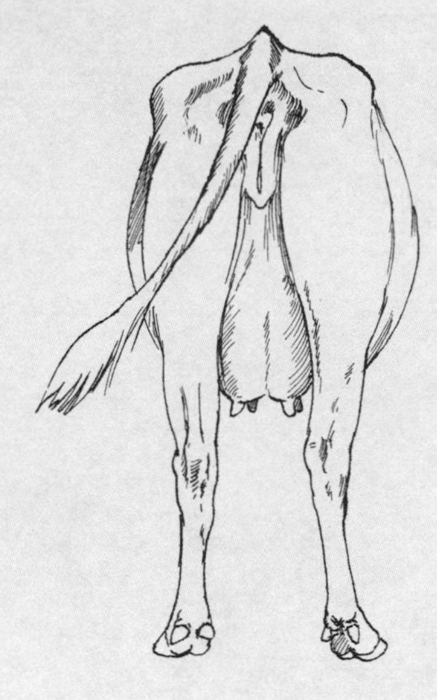

A cow's udder should be well attached between the thighs (top), and have short, small teats and well-balanced quarters.

Teats too long

Teats too fat

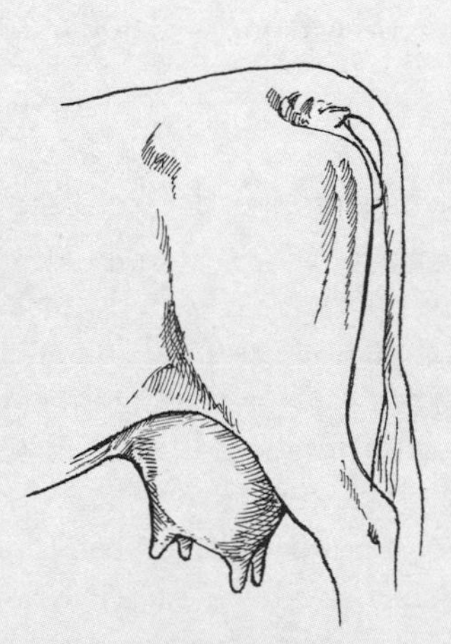

Unbalanced quarters or saggy udder

Parts of a Dairy Cow

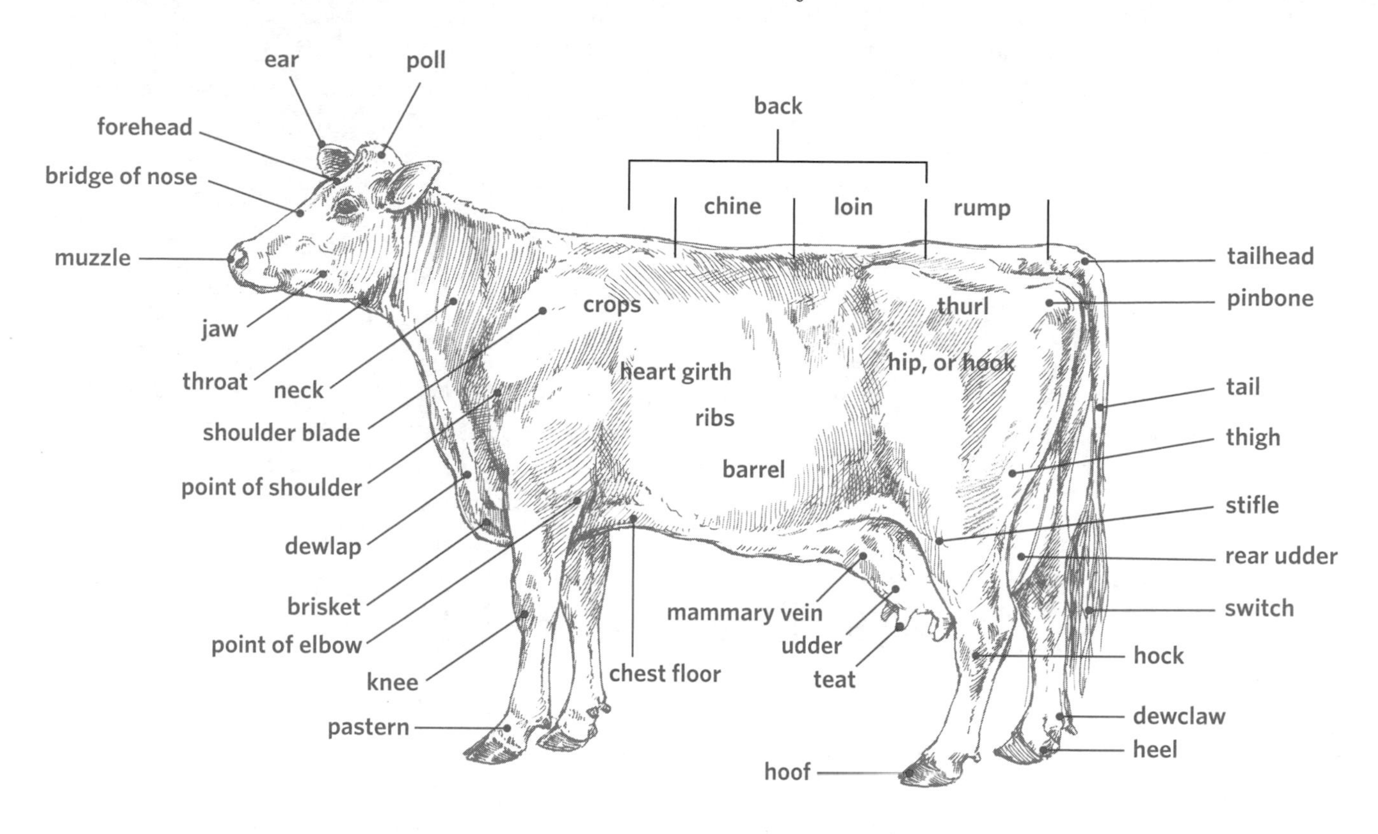

is nervous, doesn't want to stand still, or seems inclined to move away or kick when you touch her udder, she will not make a satisfactory family milk cow.

Purebred versus Grade

The word *purebred* refers to an animal whose ancestors are all of a single breed. A registered purebred has a registration number, recorded in the herd book of its breed association. The association gives the owner a certificate stating that the animal is the offspring of certain registered parents. However, not all purebreds are registered.

If you are thinking of selling your cow's offspring, you'll have to decide whether to buy a registered purebred or a grade cow. You are better off buying a good grade cow than a poor registered one. Registration papers won't guarantee high production or good conformation, and grade cattle can take advantage of the benefits of crossbreeding. So don't select an animal just because it's registered.

If you opt for a purebred cow, join the breed association, which can give you educational materials and information and help you market your cow's offspring. Transfer the registration of your new cow to your name. Be sure the color markings or the ear tattoo on the registration certificate is correct. For the Ayrshire, Guernsey, and Holstein breeds, you may use photos of both sides of your cow or sketches of the markings. For the solid-color breeds such as Jersey and Brown Swiss, an ear tattoo is required.

Performance records and the cow's pedigree are two other documents you will want if you plan to register and sell your registered cow's daughters. A pedigree is a record of all the cow's ancestry. Most successful breeders keep additional detailed records that can be used to compare such factors as birth weight, weaning and yearling weights, milk production, and fertility. If you are not familiar with pedigrees and performance information, have the seller explain the records so you can understand how to use them.

Dairy Character

A dairy cow is judged as much by the way she is built as by any other factors. Conformation is as important as performance records. A good milking cow should have outstanding breed character, which means that she closely resembles her ideal breed type. She should have style — all of her parts come together to create a good-looking animal.

Your cow should have good feet and legs. You don't want a cow that will eventually become crippled because of poorly formed feet and legs. Her hind legs should be straight, not too close together at the hocks or splayfooted, and set squarely under her body. Her front legs should be straight. She should walk freely and smoothly without throwing her feet out to the side or swinging them inward.

The cow must have a feminine head and neck and a long body. A long body gives a cow more room for carrying a calf. She should have a deep, wide rib cage; a long, graceful neck; sharp withers; and a straight back (not humped or swayed) with wide, strong loins. Her rump should be level and square, not tipped up at the rear or slanted downward. If her rump is tipped up and her tailhead is too high, she will have trouble calving. The cow should be well balanced and well proportioned in all of her body parts, not short bodied, shallow bodied, or too short legged.

To give lots of milk, she must have a well-constructed udder. The teats should be evenly spaced, and all the same length and diameter. They should be well shaped and of a size to fit comfortably in your hand — not too long or too fat. Long teats and fat teats are difficult to milk, as are extremely short teats.

The udder should be well balanced, with all four quarters similar in size and shape, well spaced, and level — the front quarters not higher or lower than the rear quarters. The bottom of the udder, between the teats, should be flat rather than bulging. The udder should have strong attachments, meaning the muscles and ligaments holding it up against the cow's abdomen keep the udder relatively high and tight, rather than hanging low and pendulous. Otherwise the udder may break down as the cow ages, swinging back and forth and hitting her legs as the cow walks or trots, making the udder susceptible to bruising and mastitis. Looking from behind, the udder should appear well attached between the thighs, rather than sagging.

Where to Buy a Cow in Milk

You may be able to buy a cow from a neighbor or a local dairy. You can also check the ads in your local or regional farm newspapers or ask at the farm store — or see if the local veterinarian or county Extension agent knows of anyone with a milk cow for sale. If you pass the word along that you are looking for a family cow, you will eventually find one. It's always better to buy directly from the previous owner than from an auction. A cow going through an auction is usually being culled for one reason or another. She may be old or crippled, infertile, or have some other problem. Good cows may go through an auction, but you can't always tell by looking. For instance, a nice young cow may have a uterine infection, making her infertile.

If you buy from the previous owner, you have more chance to examine the cow closely and ask questions. You can learn more about her history and make sure she's had all her vaccinations. A dairy might sell you a young, healthy cow that is just not producing enough milk to pay her way. She would be fine, however, for a family milk cow that doesn't need to be a top producer.

Another place to buy a cow is from a dispersion sale or farm sale, when a dairy is selling out or a farmer is retiring. In such a case you may be able to buy a cow that has just calved, with her new calf at her side.

How Long Does a Cow Live?

In an industrialized dairy — where the cows are crowded, pushed for milk production, and stressed out — the average life span is only 4 or 5 years. By the age of 6 a cow is considered old, and few cows are milked beyond the age of 7. A well cared for family cow, on the other hand, may produce milk to 15 years or older, and may survive to the ripe old age of 20 or even 25. The oldest cow on record was an Irish milker named Big Bertha; she died in 1993 just three months short of turning 49.

Choosing a Beef Calf

Beef breeds are stockier and more muscled than dairy breeds. The latter have been selected for their milking ability rather than for beef production, and the cows are finer boned, are more feminine, and have larger udders so they can give much more milk. The many beef breeds have differences in size (height and body weight), muscle traits (lean or fat), color and markings, hair coat and weather tolerance, and so on.

Your best consideration in selecting from among the many breeds is to determine which ones are available locally — since they are more likely to be adapted to your climate, as well as being easier to locate and transport to your place.

You may also want to consider finish size. A small-framed animal takes less feed and usually matures and reaches finish weight more quickly than a large one. If you have a small place or don't want a huge amount of meat, a small-framed animal like a Dexter or even a crossbred dairy-beef calf from one of the smaller dairy breeds might be just right. If you have a larger acreage and a big family and you want lots of beef, you may prefer to select a calf from one of the larger-framed breeds. A crossbred beef calf is often the best selection, since it will have the advantage of hybrid vigor for better feed efficiency and faster weight gain. Crossbred animals also tend to be hardier and healthier.

Most cattle breeds are horned and some are polled (naturally have no horns). Angus cattle are polled, and some of the horned breeds are infused with Angus genetics so they have black and polled offspring — two traits popular with many stockmen. Now the traditionally red and horned European breeds like Limousin and Simmental have black, polled versions. Angus and Angus-cross cattle have an additional advantage of finishing faster and being ready for butchering in little more than a year, compared to larger-framed breeds that may take two or more years to finish, although the Angus-type cattle will produce less beef.

Dexters are the smallest cattle breed and are used for both milk and beef. The average cow weighs less than 750 pounds (340 kg) and bulls weigh less than 1,000 lbs. (454 kg). Dexter cattle are quiet and easy to handle, and the cows give rich milk.

Beef Breeds

Beef breeds in North America are descendants of cattle imported from the British Isles, European countries, Australia, or India. Many modern breeds are mixes of these imported breeds.

British breeds originated in England, Scotland, and Ireland. They include Angus, Dexter, Galloway, Hereford, Scotch Highland, and Shorthorn.

Continental, or European, breeds are generally larger, leaner, and slower to mature than the British breeds and are popular for crossbreeding with British breeds to add size and muscle (and sometimes milk). These breeds include Braunvich, Charolais, Chianina, Gelbvieh, Limousin, Maine Anjou, Normandy, Pinzgauer, Piedmontese, Romagnola, Salers, Simmental, and Tarentaise.

Some beef cattle breeds originated in places other than the British Isles and Europe. They include the Brahman from India, the Murray Grey from Australia, and the Texas Longhorn, an American breed descended from wild cattle left by early Spanish settlers in the Southwest.

Disposition

The breed you choose for raising a beef animal for your freezer is not as important as the disposition of the individual you select. Some calves are more placid and easygoing than others. If you are getting only one calf, try to select a mellow one, not a skittish one that will get nervous being by himself. Try to choose a smart and gentle one that will readily learn to trust people.

Don't choose a wild calf. A wild, snorty calf is a poor risk, even if he is big and beautiful. You may have trouble gentling him, and he may try to go through fences. He could also be dangerous — he might knock you down or kick you. A wild calf won't gain weight as well as a placid calf. Rate of gain (pounds gained per day) is almost always better with a gentle calf.

Where to Buy a Calf

An auction is the riskiest place to buy a young calf. Even though the calf may have been healthy when taken to the auction, it may get sick after you take it home. Some calves become sick because they are taken from their mother and sold before they have had a chance to nurse enough colostrum, which is loaded with antibodies that protect the calf against disease until it can develop its own immune system.

A sale yard is also a good place to pick up diseases. Cattle come and go, and they spend time in pens before being sold. Some of the cattle brought to a sale may be sick or coming down with an illness. Even if most of the cattle that go through the pens are healthy, germs may contaminate those pens. Don't buy a calf at an auction if you have other options.

A good place to buy a beef calf is at a feeder calf sale in the fall or at a farm or ranch. A local purebred breeder or commercial cattle producer is always the best source. Buying direct from a local farmer or rancher gives you an opportunity to look at calves, ask questions, and determine the personality and tractability of each animal.

Most dairy farms have many newborn calves to sell in the spring. Dairy cows must have a calf every year to produce their maximum amount of milk. A cow makes much more milk after she freshens. Her volume of milk is greater a month or two after calving. From then on, her production gradually declines. Dairy cows are kept at maximum production by being bred every year to have new calves and then being allowed to dry up briefly before the new calves arrive.

Some dairies sell all their calves. Others keep their heifers and raise them to sell to other dairies. Bull calves are usually sold as soon as they are born and are

The Murray Grey is a silver-gray breed from Australia that is gaining popularity in North America because of its moderate size, gentle disposition, and fast-growing calves. The calves are small at birth but often grow to 700 pounds (318 kg) by weaning.

Watch for Signs of Sickness

Before you take home your new cow or calf, make sure it is healthy. The animal should look bright and perky, be lively and energetic, and have a glossy hair coat and a sparkle to the eyes. Bowel movements should be firm but soft, not hard or excessively runny.

If an animal is dull or slow moving, has a dull or rough hair coat, has foul-smelling manure, or has droopy ears, it is sick. Also beware of an animal that stands with its back humped up or has a cough or a runny, snotty nose. If you are in doubt about the health of an animal, have a veterinarian or other person with cattle experience look at it.

After you bring home a calf, pay close attention to detect early signs of illness. Calves easily get pneumonia, and a calf with pneumonia may die. If your calf looks like it doesn't feel well, don't wait to see whether it gets better or worse. Get advice immediately from a person experienced at raising cattle, or call a veterinarian.

Parts of a Beef Animal

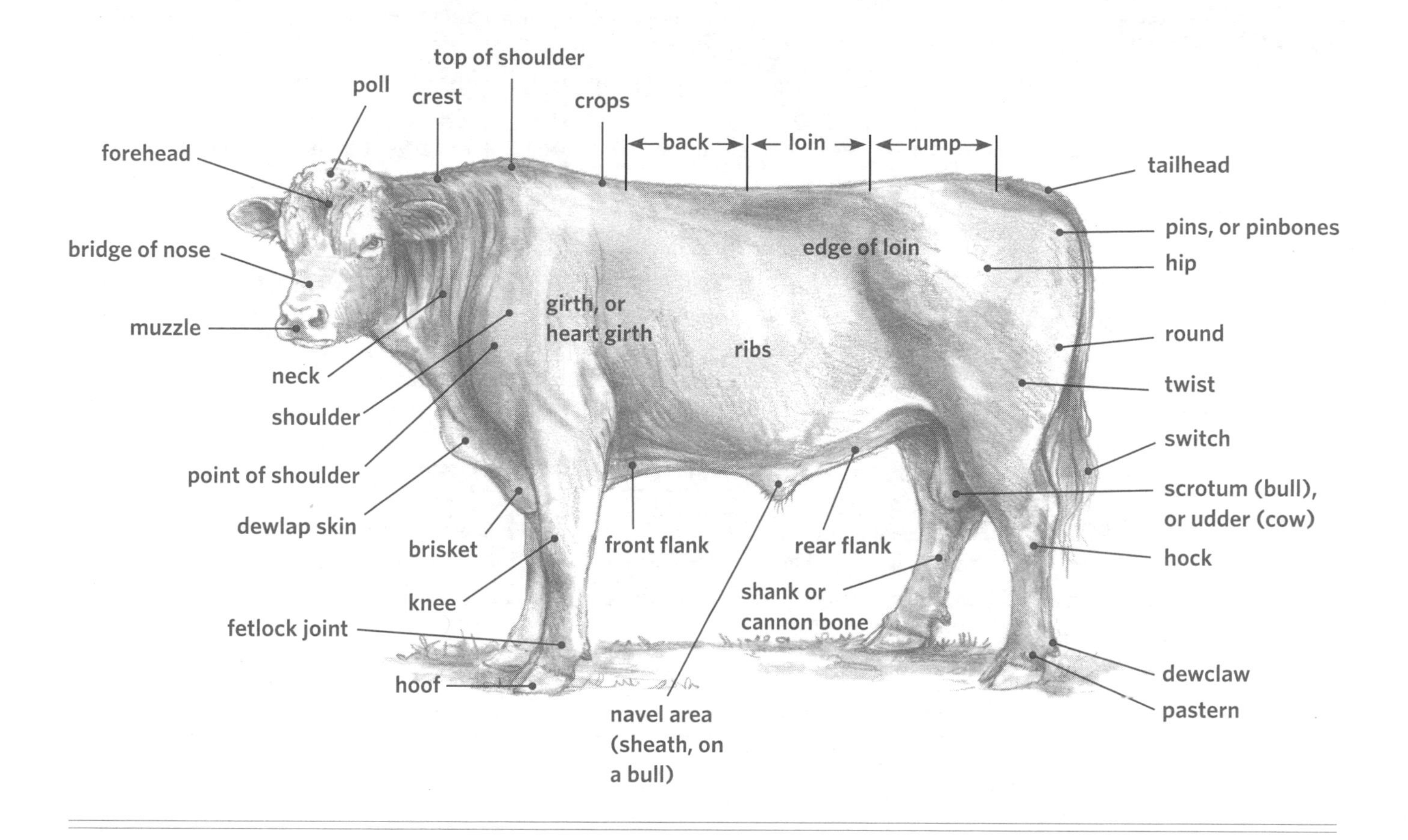

cheap because most dairy people don't want to take the time to raise them.

Some of the calves at a dairy can be crossbred (half beef), if the dairyman breeds his heifers to a beef bull that sires small calves for easy calving. Crossbred calves of either sex can often be purchased cheaply. They make good bucket calves to raise for beef.

Frame Score

The best steer to raise for butcher is a fast-growing, well-muscled animal that will reach a market weight of 1,050 to 1,250 pounds (476 to 567 kg) by the time he is 14 to 20 months of age. Beef cattle are categorized by frame score, which tells you whether they are small, medium, or large bodied. A small-frame, early-maturing steer will not produce as much meat on his small carcass as a larger-frame animal. If you try to get him to grow bigger, he'll just get too fat; he is not genetically capable of attaining a larger size. A large-frame steer will grow too big before he gets fat enough to butcher and will use up more feed than is necessary. The most practical kind of beef steer has a medium frame.

The animal you pick should have a lot of muscle, not a lot of fat. If the calf is already fat, he may not grow well. On the other hand, the animal should not be too thin, either. Thinness may indicate that the animal has been sick or is currently not healthy. He should have nice, smooth lines and should not be swaybacked. He should have a deep body, neither shallow nor potbellied. He should be long and tall, but not overly tall.

A Steer Is No Bull

If you buy a weaned calf, make sure it has been castrated. You want a steer, not a young bull. Although a dairy calf or a dairy-beef cross is usually cheaper than a beef calf, it most likely will not be weaned. You'll have the task of feeding the calf milk replacer by bottle or bucket for a few months until it becomes large enough to thrive on pasture alone. Such a calf will likely still be a bull, so you'll need to castrate it (for details, see page 313). Never try to raise a young bull. Even though he may be friendly and mellow as a baby, he will become more aggressive and unpredictable (and dangerous) as he gets older.

Getting Your Cow or Calf Home

Make arrangements with someone who has a trailer or a pickup truck with a rack to haul your animal home. If you are buying the animal from a farmer, rancher, or dairyman, he may be able to haul it for you; ask what he would charge.

If you will be unloading into a pen or a pasture, a trailer often works best, because it is low to the ground and the calf or cow can step out of it easily. An animal transported on a truck must be unloaded at a loading chute. Even a pickup truck with a rack is often too high for an animal to jump out of without risk of injury, unless the truck can be backed up to a bank or you have a ramp.

Remember that most beef calves have lived with their mothers in large pastures. Some may have seen people close up only during vaccinations or medical treatment, which are scary and painful experiences. Therefore, your calf may be frightened by you, and it may even try to run over you if you get in its way as it comes out of the truck or trailer.

When you get home, make sure the trailer is backed far enough into the pen that the animal has no choice but to enter the pen. The gate of the pen should be swung tight against the trailer. A scared animal may try to bolt through even a small opening. Don't stand in a place where the animal might run over you. If you are unloading from a pickup truck, make a ramp of sturdy boards.

Figuring Frame Score for Beef Calves

To figure a calf's frame score, measure his height from the ground at the hip when he is standing squarely. Then look up his age on this chart and find the hip height on that age line. Look to the top of the column for the frame score. For instance, a 10-month-old calf that is 45 inches (114 cm) tall at the hips would have a frame score of 4. Calves with frame scores of 1 or 2 are small, 3 to 5 medium, and 6 and 7 large.

Age (months)	Frame score (hip height in inches/cm)						
	1	**2**	**3**	**4**	**5**	**6**	**7**
5	34/86.25	36/91.5	38/96.5	40/101.5	42/106.5	44/111.75	46/116.75
6	35/88	37/94	39/99	41/104	43/109	45/114.25	47/119.25
7	36/91.5	38/96.5	40/101.5	42/106.5	44/111.75	46/116.75	48/122
8	37/94	39/99	41/104	43/109	45/114.25	47/119.25	49/124.5
9	38/96.5	40/101.5	42/106.5	44/111.75	46/116.75	48/122	50/127
10	39/99	41/104	43/109	45/114.25	47/119.25	49/124.5	51/129.5
11	40/101.5	42/106.5	44/111.75	46/116.75	48/122	50/127	52/132
12	41/104	43/109	45/114.25	47/119.25	49/124.5	51/129.5	53/134.5
13	41.5/105.5	43.5/110.5	45.5/115.5	47.5/120.5	49.5/125.75	51.5/130.75	53.5/136
14	42/106.5	44/111.75	46/116.75	48/122	50/127	52/132	54/137
15	42.5/108	44.5/113	46.5/118	48.5/123	50.5/128.25	52.5/133.25	54.5/138.5
16	43/109	45/114.25	47/119.25	49/124.5	51/129.5	53/134.5	55/139.75
17	43.5/110.5	45.5/115.5	47.5/120.5	49.5/125.75	51.5/130.75	53.5/136	55.5/141
18	44/111.75	46/116.75	48/122	50/127	52/132	54/137	56/142.25

Understanding Cattle Behavior

Wild cattle were safer in herds. If wolves approached, the cows would bellow and form a tight group. That's why yearlings and young cattle generally travel in a group. If one goes to water, they all do; if the leader decides it's time to graze, they all go. They are not just copycats; they hang together for protection.

Your new calf or cow is probably lonely and scared, unless it has another animal for a buddy. Try not to frighten the animal; it may become upset and crash into the fence. To understand your new animal, try to think like it does. Cattle are herd animals and are happiest when they are in a family group with other cattle.

If your calf was weaned before you bought it, it has already gone through the emotional panic of losing its mother. It will miss the other calves it was with, but it won't be quite so desperate to get out of your pen to find its mother. But if the calf is still going through weaning when you bring it home, it will have several days of stressful adjustment. It may pace the fence and bawl, and it may show little interest in feed or water. A calf being weaned is more susceptible to illness because stress hinders the immune system. In cold, rainy, or windy weather, a weaning calf may be particularly likely to get pneumonia.

Making Friends

Cattle like to be petted and scratched, especially in places that are hard for them to reach. Most love to be scratched under the chin, behind the ears, and at the base of the tail. But don't rub the top of the head or the front of the face. Rubbing these spots will encourage a cow or a calf to bunt at you.

Getting Acquainted

As you begin to get acquainted with your new animal, give it time and space. Don't try to get too close. Until it gets to know you, it may react explosively if it feels cornered.

Speak softly and move slowly around the animal. As you approach the pen with feed, let it know you are there. If its attention is diverted and then it suddenly sees you, it may run off. Talking softly or humming a little tune can help gentle a frightened animal.

When you are in the pen with your animal, don't look directly at it. It will relax more if you act as if you aren't paying attention to it. If you come too close, approach too quickly, and look directly at it, it will see you as a predator. Instead, ignore the animal, but talk softly as you go by. Pretty soon it will come eagerly to meet you when you bring feed.

Your calf or cow has probably not spent much time with people, so don't just turn it out and ignore it. Spend some time walking around in the pasture and let it get used to you. Cattle will let you get closer once they know you are nothing to fear.

Handling and Gentling Cattle

Some cattle are not timid and will be curious about you from the beginning. Use this curiosity to your advantage. If you are patient and quiet, the cattle will come closer to you.

The flight zone. Cattle have a certain personal space in which they feel secure. As long as you don't enter that zone, they feel safe. But if you get too close, they'll get nervous or scared and run off.

Different cattle have different-size zones. A wild or timid calf has a large zone; a gentle or curious animal has a much smaller one. As your cattle get to know you well, the flight zone will disappear.

Use feeding to your advantage. When cows and calves begin to associate you with food, most will lose their fear and come right up to you. They may need a few more days before they will let you touch them, but they will soon stand beside you and eat.

Cattle are good at making associations between things. If you have a special call for feeding time, they'll come to you every time they hear it.

Don't spoil your animal. Don't make the mistake of spoiling a cow or calf. It should trust you, but it must also respect you. Remember that cattle are social animals and accustomed to life in a group, in which they boss other cattle around or get bossed. Cattle will think of you as one of the herd. You must be the dominant herd member; they must accept you as the boss cow. Otherwise, they will try to be too pushy.

If a calf or cow starts pushing you or bunting at you when you are feeding or petting it, discipline it with a swat. Pushing and bunting is the cattle version of play. A calf or cow will naturally want to play fight with you, as cattle do with one another.

If you spoil your animal by letting it do whatever it pleases, you will regret it later. Carry a small stick when you go to feed your animal; if it gets sassy, rap it on the nose. This swat will remind it that you are the boss.

Be careful. Although calves and cows are not likely to attack a person (unless a cow is defending a new calf), they can accidentally hurt you because of their size and weight. Always keep an escape route in mind when trying to corner or work with a cow or a calf. Leave enough room to dodge aside if one backs into you or turns around and runs back out of a corner.

Cattle can be dangerous when handled in a confined area, because they tend to panic. Don't wave your arms, scream, or use a whip. If an animal won't move forward into the catch area, prod it with a blunt stick or twist its tail. Just be careful to not twist the tail too hard. You can twist it into a loop or push it up to form a sideways S curve. If you have to twist the animal's tail to get it to move, stand to one side so it can't kick you.

Be gentle. If you yell or chase your cattle, you may scare them badly. Even if they are stubborn or suspicious and won't go into the catch pen or behind the gate or panel on the first try, don't get impatient. If you lose your temper and yell, you'll confuse or scare them and make things worse. They'll be harder to handle next time.

Housing & Facilities

Before you buy a calf or cow, prepare the place where you'll keep it. A young calf needs shelter from sun, wind, and rain. A mature animal is hardier but still needs protection from driving wind and hot sun. In most climates a three-sided shelter offers sufficient protection from the elements, but if you'll be milking your cow on cold winter days you'll appreciate a more secure structure.

The Pen

If the animal will be living by itself, build a strong pen to put it into for a few days before you turn it out to pasture. Be sure the fence is constructed so the animal cannot jump over or crawl through it. A frantic, homesick animal in a new place may try to escape. If the calf you purchase has already been weaned, it won't be so desperate to get back to its mother.

Make sure your pen or pasture has no hazards, such as nails or loose wires that might injure the animal. Calves are curious, just like little kids, and they often get into trouble. A pole or board on the ground with nails sticking out of it can cause serious injury if stepped on. Nails or bits of wire lying around near the feeder may puncture the animal's stomach if it eats them, causing hardware disease, which is often fatal. Don't leave baling twine hanging on a fence or lying on the ground, and watch for stray garbage: If a calf or cow tries to eat baling twine or chews on a plastic bag that has blown out of your trash, pieces of the material may plug its digestive tract and kill it. Any electrical wires in the barn must be out of your animal's reach, as well.

The pen must be dry and have good drainage. If necessary, put sand in the bedding area or a shady spot where the animal sleeps to make sure the area stays dry.

Pasture Fencing

Build pens and erect fencing on solid ground. Posts set in a boggy, wet area will become loose and wobbly. The postholes can be dug with a shovel if the ground is mainly dirt with just a few rocks. If the ground is really rocky, you'll need an iron bar to loosen the rocks as you dig.

Use metal or pressure-treated wood posts to prevent rot. Set the posts in a straight line. A crooked fence is not as strong as a straight one. Set the corner posts and sight between them to line up your postholes and your posts, or stretch a long string between them to give you an exact line. Set holes around the posts with dirt and rocks. To set the posts solidly, put in a little material at a time and tamp the dirt firmly with an iron bar or a tamping stick before adding more. (For more information on putting up fencing, consult a good reference book such as *Fences for Pasture and Garden* by Gail Damerow.)

Wire fence. A good wire fence will hold cattle that are not being crowded or trying hard to get out. The wire must be tight, without slack, so the cattle won't

Build a Calf Pen

All calves need shelter, but a brand-new calf is especially fragile and needs to be kept warm and dry. If you live in a cold-weather region, you'll need to keep a young calf in a warm barn stall or in your garage or back porch until it is several days old and can live outdoors.

A calf pen can be built with sturdy wood posts. The posts should be at least 8 feet (2.5 m) long, enough to set deeply into the ground but make a fence at least 5 feet high (1.5 m), and should be set 8 to 10 feet (2.5 to 3 m) apart. Use poles, boards, wood or metal panels, or strong woven-wire netting as fencing between the posts. Barbed wire or smooth wire won't work, because a calf can get through it if it tries hard enough. Do not use an electric fence to create the pen. You may need to corner the calf in the pen — or it may corner you — and you don't want it or you to get shocked. Don't skimp on materials; a good pen may be expensive to build but will last a long time.

If the calf will spend all of its time in the pen, it should be large enough to give the calf room for exercise, at least 900 to 1,000 square feet (275 to 300 sq m). You can configure the pen however you like, such as 10 by 100 feet (3 by 30 m), 20 by 50 feet (6 by 15 m), or 30 by 30 feet (9 by 9 m) — whatever fits the space you have. If you are raising more than one calf, add at least 200 square feet (60 sq m) of space to the total area for every additional calf. The pen should offer shade from a building or a tree. A calf needs about 100 square feet (30 sq m) of shade in summer.

You'll need a small catch pen in one corner of your pen and a place where you can restrain the calf for giving shots and medications. A small, enclosed shed or feeding area can be used for cornering and catching the calf. If you make a gated chute at one corner of the pen, you can herd your calf along the fence and into the chute and swing the gate shut behind it. You might include a head catcher or stanchion in the calf's feeding area, which allows you to lock its head in place when it sticks it through to eat. The stanchion will restrain the calf sufficiently for veterinary care.

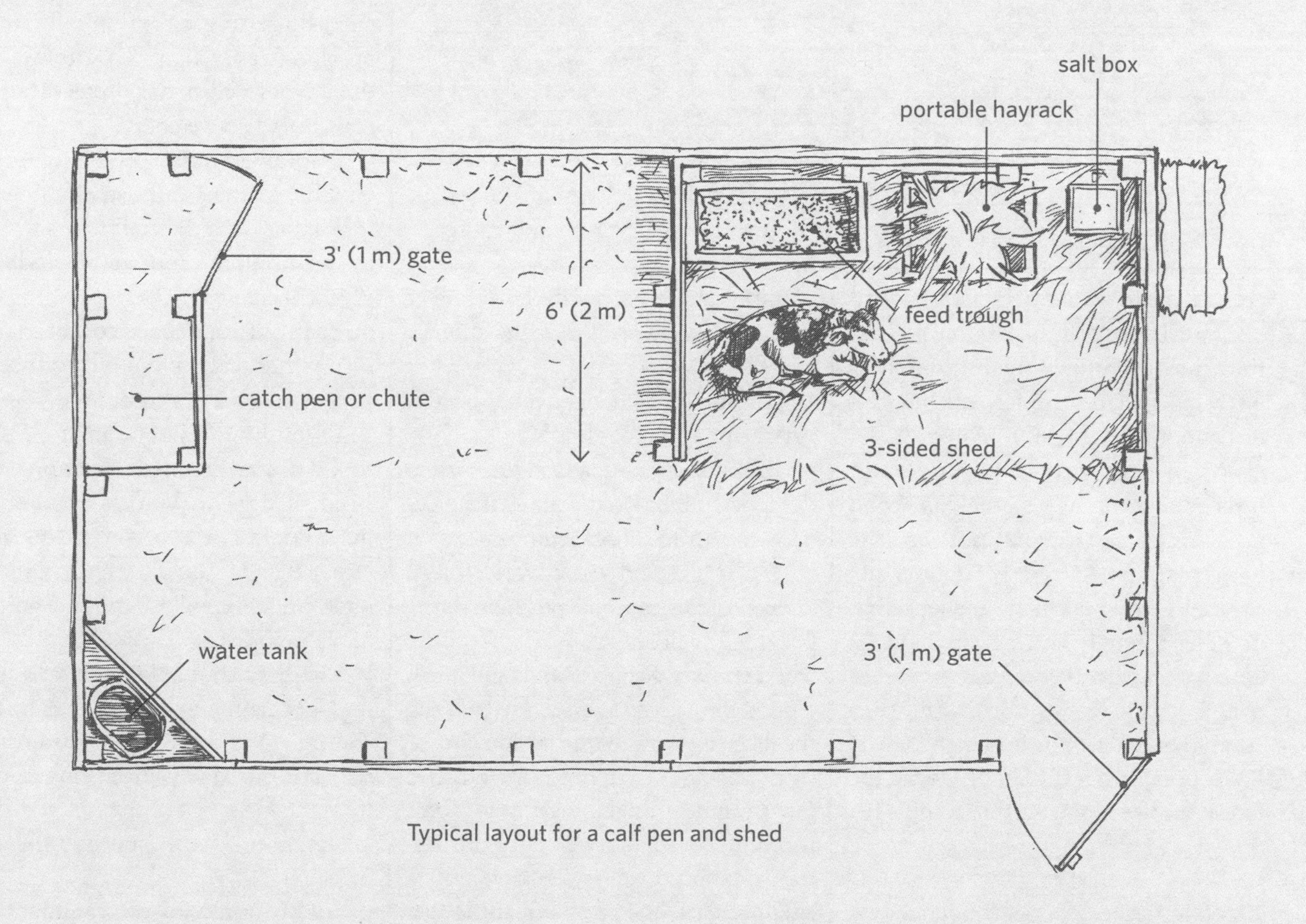

Typical layout for a calf pen and shed

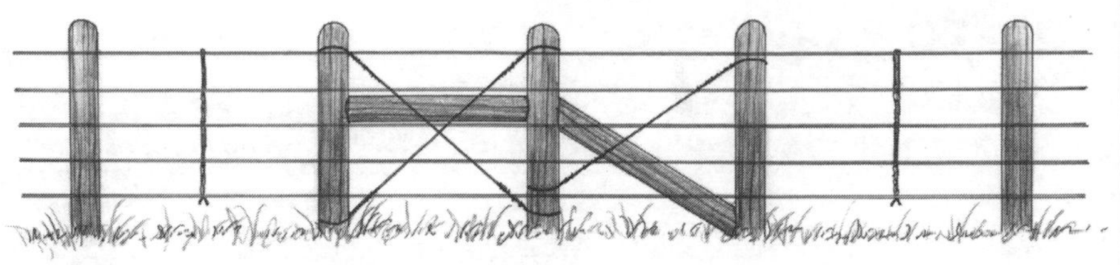
Wire fence with braces and metal stays

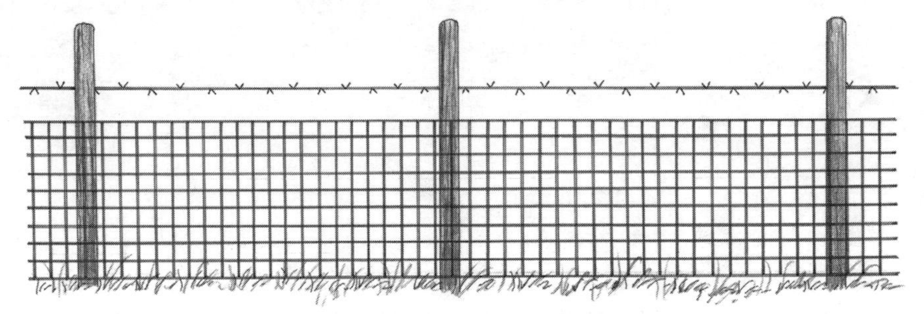
Net wire fence with one barbed wire on top

Corral fence with posts and poles

get into the habit of reaching through it. If they can reach through it to eat grass, they may eventually push through it. Net wire is the best option, because cattle cannot get a nose through it.

Whenever you inspect your fence, replace any missing staples on wood posts or clips on metal posts so the wire is attached properly. Tighten any sagging wires. Make sure the fence has no holes through which a calf might be able to get out. If the fence stretches over a dry ditch, a calf or yearling may be able to walk right under the fence. Put a pole across the ditch, under the fence, and secure it so that a calf can't push it away.

Electric fence. An electric fence will prevent animals from getting through or rubbing against a wire fence. After being shocked a few times, the animals won't touch the fence.

For an electric fence, you'll need a battery-powered or electric fence charger and insulators to attach the wire to the fence posts. Do not allow the electric wire to touch anything metal; it will short out and won't work. It also shouldn't touch wood posts or poles, because it will short out whenever the wood gets wet. Keep all weeds and brush around the electric fence clipped to keep the fence working and to avoid a fire.

A portable or temporary electric fence can be used to divide a large pasture into several smaller ones for pasture rotation. Even if you have only a small acreage, the grass will last longer if you practice rotational grazing

Check That Fence

If you're raising animals on pasture, you'll need to check the fences frequently and carefully to ensure they are in good shape. And be sure to give fences a good once-over before you bring home a new animal. A calf or cow that has never lived alone may be frantic when you first bring it home. It may try to get out of the pasture to rejoin the herd it lived with.

Pasture Shelter

In a mild climate, cattle may need only a small three-sided shed, or a protected fence corner with a roof and some boards or plywood on the sides for windbreak. You can make a simple shed by setting a roof on tall, sturdy posts. A freestanding shed with walls on three sides will better protect the animals from bad weather.

Before you build a shed, figure out which way the wind usually blows in that spot. Place the shed walls to offer the greatest protection from wind. Two sheets of exterior-grade plywood placed on each side of a fence corner make a nice windbreak; add another sheet of plywood to make a roof.

If you live in a hot climate, a shed roof will provide shade, but you'll also need airflow to help keep the animals cool. The roof should be high rather than low, and the shelter should have no walls, which would halt air movement.

The shed should be built on a high, dry spot with good drainage. The roof should slope so rain or melting snow will run off. Make sure it slopes away from the main pen so the runoff doesn't create mud in the pen or flow into the shed.

Add bedding for your animal to lie in. Straw, bark mulch, or wood chips

A simple shed with two or three walls can provide adequate shelter.

scattered into a bed in the corner of the shed will give them a dry place to sleep. Make sure the bedding area is in a high, dry spot. The animal should always have dry bedding. Moist, dirty bedding contains harmful bacteria and also conducts warmth away from the animal's body, causing it to become chilled and more susceptible to disease. Also, ammonia gases given off by bedding that is wet from urine and manure can irritate and weaken the animal's lungs, especially a young calf's, and allow bacteria to become established and lead to pneumonia.

Make a Water Trough

You can make a water trough from anything that will hold water and can be cleaned easily, such as an old washtub or large bucket. You will also need to make a stand or frame to hold the tub. Nail a board across the corner of the pen or stall, leaving room for the tub or bucket to fit snugly between it and the walls. You can easily pull the tub or bucket up out of the corner to rinse and clean it, but cattle can't tip it over.

Manure Disposal

In a large pasture, manure serves as fertilizer. But in a pen or shed, it needs to be cleaned out. If an animal spends much time in its shed, manure and soiled bedding must be cleaned out regularly so it doesn't build up.

A corral may be easiest to clean with a tractor and blade or loader, whereas to clean a shed or bedding area, a wheelbarrow and manure fork will suffice. A manure fork resembles a pitchfork but has more tines, so manure and straw can't fall through it easily.

Manure makes excellent fertilizer. Spread the manure over your pasture or garden, or make a compost pile from manure and old bedding.

Halters and Ropes

You will need a good halter and rope for restraining your calf or cow so that you can tie the animal to the fence or to the side of the chute if necessary. An inexpensive, adjustable rope halter with lead rope can be purchased at a feed store or through a mail-order catalog from a livestock supply company. When putting on a halter, place it on the animal so the adjustable side is at the left. When you pull it tight, the pressure should be mostly on the rope under the chin, rather than behind the ears.

Tying Knots

At some point, you may need to tie up your calf or cow. For your safety and that of your animal, you'll need to know how to make a good knot that will stay tied but can be untied easily, even if the animal has pulled hard on the rope.

Overhand knot. The simple overhand knot is the one you make first when tying your shoes. This basic knot is often the first step in forming more complex knots.

Bowline knot. The bowline knot is probably the most useful nonslip knot for working with livestock. It allows you to tie a rope around the animal's neck or body without the danger that it might tighten when the rope is pulled, and it is relatively easy to untie. An easy technique for remembering how to tie a bowline knot is to think of the following story. The first loop is the rabbit hole, the standing part of the rope is the tree, and the working end of the rope is the rabbit. The rabbit comes out of the hole, runs around the tree, and goes back down its hole.

Double half hitch. The double half hitch knot is quick and easy to tie, acts like a slipknot, and is a handy way to secure the rope around an animal's leg when you are tying a leg back or to secure the end of the rope when no other knot seems appropriate.

Square knot. The square knot is a stronger version of the overhand knot; it consists of two overhand knots, one tied on top of the other. The square knot is perfect for joining two pieces of rope, as when you are joining a broken rope or tying a rope around a gate and gatepost to keep the gate closed. A properly tied square knot will not slip from its position.

Quick-release knot. The quick-release knot (also called a reefer's knot, a bowknot, or a manger tie) is useful for tying your calf to a fence post. Like the square knot, it is a good nonslip knot. The quick-release knot has the advantage of being easily untied even after it has been pulled tight, as will happen if your calf pulls back on the rope.

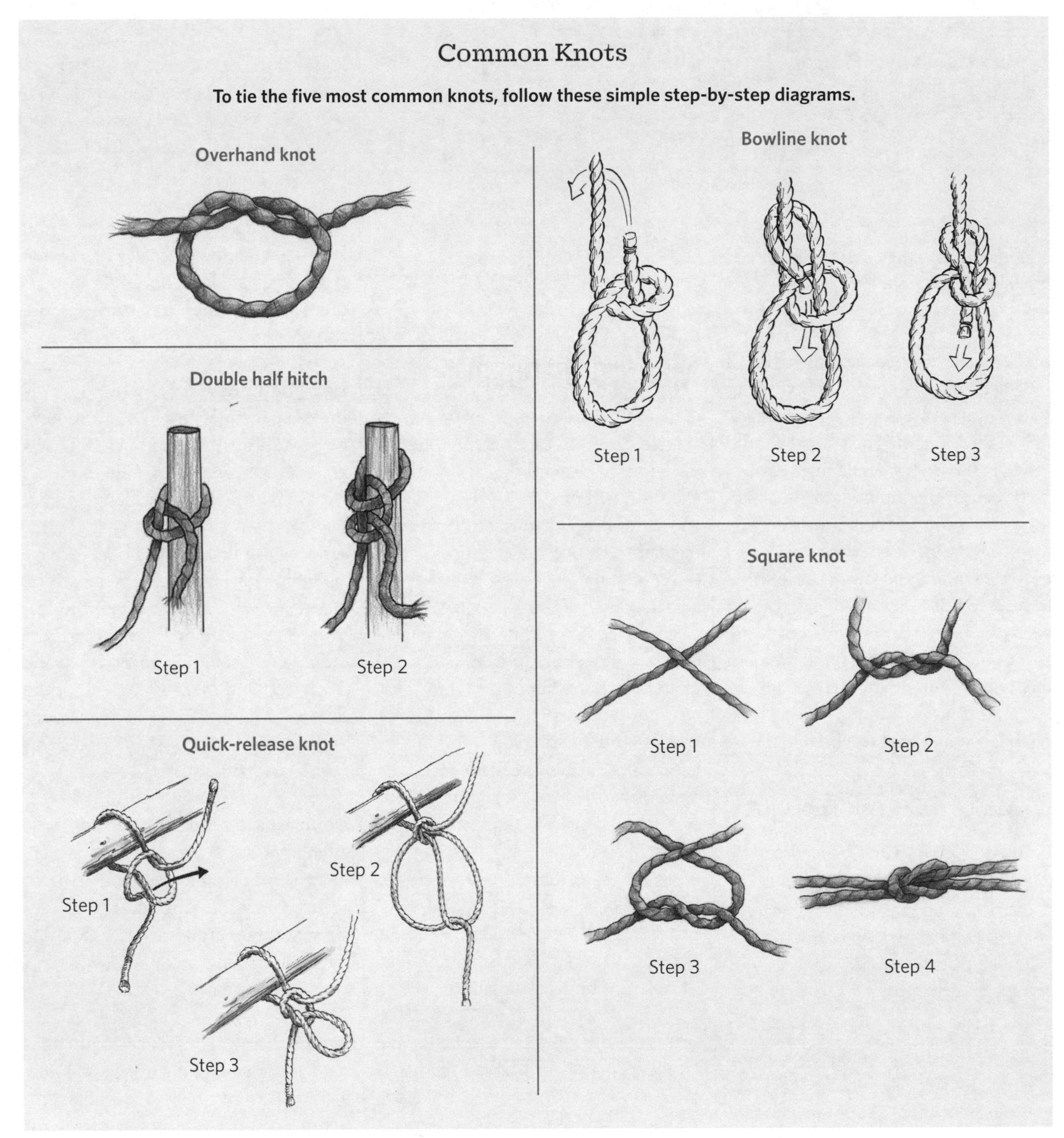

Feeding & Watering

Cattle require a source of fresh water, which can consist of a tub or tank filled twice a day with a garden hose. A water tub for calves should be set up off the ground, but no higher than 20 inches (50 cm); anything higher will keep them from drinking easily. Calves may step or poop in a tub on the ground. Using a feed rack or manger will reduce hay wastage. Cattle won't eat hay that has been stepped on or has manure on it.

Water Supply

You must make sure your cattle have water available at all times. Cattle drink more in hot weather than in cold weather, so check the trough more often in summer. In the winter, keep the water from freezing, even if it means breaking ice every morning and evening. In really cold climates, a rubber tub is a useful water tub, because you can tip it over and pound on it to get the ice out without creating a leak. If you use a hose to fill the trough, drain the hose thoroughly after each use in cold weather to keep it from freezing.

The water should be kept away from the feed rack or feeding area to keep cattle from dragging feed into the water. It should be located far from where the cattle bed. If the cattle have to walk some distance to the water, they will be less likely to stand close to it and defecate in it by accident. Keep the water fresh and clean, even if you have to dump and rinse the tub every day; cattle won't drink dirty water. Use a tub or tank that is easy to dump or has a drain hole at the bottom, and rinse it before refilling. The more cattle you have, the larger the tank you will need, or the more often you'll have to fill it.

Hay Feeder

Clean out any hay that collects in the bottom or corners of the feeder; wet hay may become moldy. Moldy feed may make your animal sick, and the mold spores that are released into the air when the animal eats may make it cough. To help prevent dampened hay, place the feed area inside a weather-resistant shed. If you feed your cattle outdoors, build a roof over the feed manger, hayrack, or grain box or tub.

If you are feeding hay in winter or when pasture gets short, scatter the hay on well-sodded ground. More hay will be wasted if cattle are fed on bare dirt or mud.

Grain Box

Some people like to fatten their beef cattle on grass, without any grain, though grass feeding takes longer to get a calf up to butchering weight. If you prefer grainfed beef, you'll have to feed your calf grain every day after it is weaned.

You'll need a sturdy trough or grain box, mounted off the ground so the calf won't step in it and held securely so the calf can't pull it down. Calves will not eat dirty grain. A rubber tub is easy to wash out and works well if you have only one calf. Build a roof over the tub or feed box to keep the grain dry.

Clean out any leftover kernels before adding new grain to the box. If birds have pooped in the tub or trough, or any old, fermented, or moldy grain remains in the corners, the calf may refuse to eat the next batch you put in.

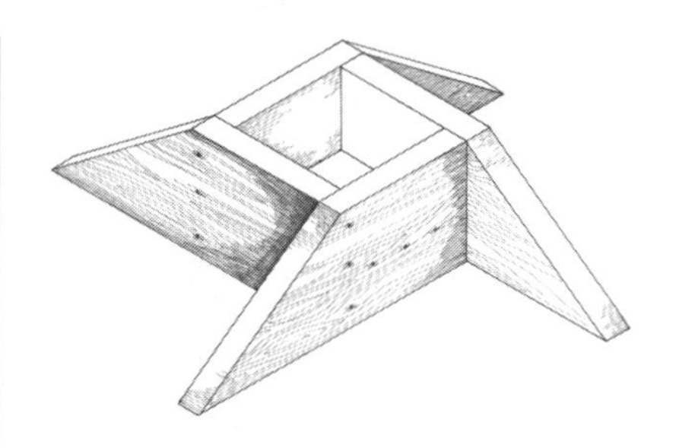

A salt box or grain box can be made of four 1 in. by 8 in. (2.5 × 20 cm) boards and a bottom.

Nutrition Basics

Cattle are ruminants, meaning they have four stomach compartments and chew their cud. The four compartments of a ruminant's stomach are the rumen, the reticulum, the omasum, and the abomasum or true stomach, which is similar to the human stomach.

When a ruminant eats, it chews food only enough to moisten it for

Make a Feed Trough

You can make an inexpensive feed trough with 2-inch (5 cm) lumber cut into lengths. If several calves will be using the trough, allow 3 square feet (1 sq m) per calf. Make the sides of the trough at least 6 inches (15 cm) high. Set the trough no higher than 18 to 20 inches (46 to 50 cm) off the ground.

swallowing. After being swallowed, the food goes into the rumen to be softened by digestive juices. After the animal has eaten its fill, it finds a quiet place to chew its cud. It burps up a mass of food along with some liquid, swallows the liquid part, and then chews the mass thoroughly before swallowing it and burping up some more. The rechewed food goes into the omasum, where the liquid is squeezed out, and then goes on into the abomasum. Ruminants developed this way of eating so that they could cram in a lot of feed while grazing in open meadows and then retreat to a safe, secluded place to chew more thoroughly.

Cattle do well on a wide variety of feeds. To some extent, what you feed your animal depends on whether it is being raised for beef or milk. However, the basic elements of good nutrition are the same for all cattle. Make sure the feeds contain a balance of the basic ingredients for good nutrition: protein, carbohydrates, fats, vitamins, and minerals.

Protein. Protein is necessary for growth. Good sources of protein include high-quality legume hay, such as alfalfa or clover; pasture grasses; or high-quality grass hay. (With alfalfa, care must be taken to avoid bloat — see page 316.) Protein supplements include cottonseed meal, soybean meal, and linseed meal. Cattle that are feeding on good hay or pasture don't need supplements.

Carbohydrates and fats. Carbohydrates and fats provide energy and are used for body maintenance and weight gain. Barley, wheat, corn, milo (grain sorghum), oats, and grain by-products, such as mill run and molasses, contain a high proportion of carbohydrates and a small amount of fat. Extra fat can be fed using a high-fat product designed for ruminants, such as Calf Manna.

Vitamins. Vitamins are necessary for health and growth. Green pasture, alfalfa hay, and good grass hay contain carotene, which the animal's body converts into vitamin A. Overly mature, dry hay may be deficient in carotene. The other vitamins cattle need are either in the feed or created in the animal's gut, except for vitamin D, which the animal's body synthesizes from sunshine. Your cattle will get enough vitamin D unless they spend all their time indoors.

Minerals. Minerals occur naturally in roughages and grain. Cattle don't normally need mineral supplements beyond those found in ordinary feeds. However, if the soil in which their feed was grown is deficient in iodine or selenium, they may require supplements of these minerals. In some regions and situations, they may also need copper supplementation, phosphorus, or some other mineral to prevent deficiency. Check with your local Extension agent or cattle nutritionist for advice on the mineral needs of cattle in your area and always inquire before adding supplements. Some supplements are harmful if overfed.

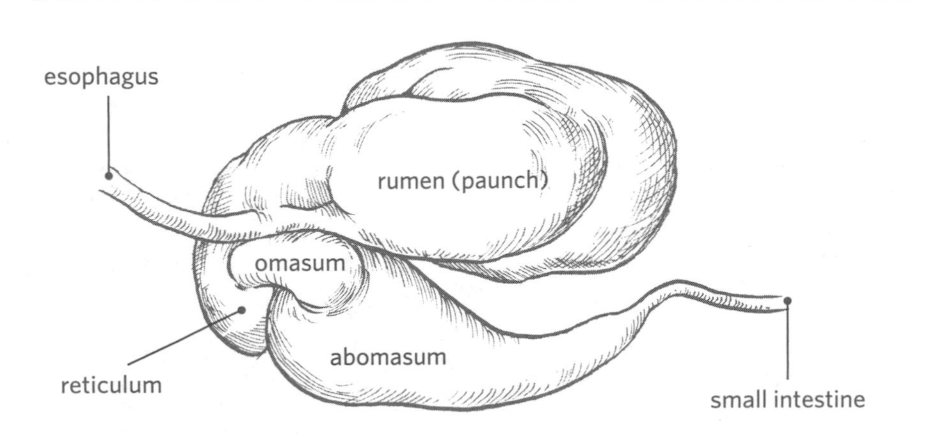

Cattle and other ruminants have four-part stomachs.

Cold-Weather Feeding

In cold weather, cattle need more feed to generate body heat and keep warm. Roughages provide more heat than do grains, because of the fermentation that takes place during digestion. If the weather is cold, increase the ration of grass hay.

Salt is important for proper body functions and for stimulating the appetite. It is the only mineral not found in grass or hay. Always provide salt for your cattle, either in a block or as loose salt in a salt box. Trace mineral salt can be used if feeds in your geographic region are deficient in certain minerals. Trace minerals include cobalt, copper, iodine, iron, manganese, selenium, and zinc. Your veterinarian or county Extension agent can help you figure out which kind of salt to use and whether it should include trace minerals.

If your cow or calf is grazing on lush spring pasture, you may need to feed extra magnesium to avoid grass tetany, which can be fatal. Check with your vet, Extension agent, or local feed store.

Grain

Grain enables a beef calf to reach market weight faster, and a cow needs grain to meet her nutritional needs while producing milk as well. Grains, also called concentrates, include corn, milo, oats, barley, and wheat. In the Pacific Northwest barley is plentiful and can be used instead of milo or corn.

Daily Ration

Cattle need about 3 pounds (1.5 kg) of hay daily for each 100 pounds (45 kg) of body weight. For example, a 500-pound (227 kg) calf needs about 15 pounds (7 kg) of good hay each day. If you wish, you can reduce the hay ration by replacing some of the hay with grain. Since grain is a more concentrated source of nutrients, substitute about a half pound (225 g) of grain per pound of hay. Overfeeding on grain can cause digestive problems or founder, so grain should never constitute more than half the ration by weight.

Wheat is usually too expensive to feed to cattle. Corn is high in energy and is commonly used for calf feed when it is available. Oats also make good feed, as does dried beet pulp with molasses.

Roughages

Roughages, sometimes called forages, are feeds that are high in fiber but low in energy — such as hay or pasture — and are the most natural feeds for cattle. Although cattle do well on them, they don't grow or fatten as fast as they do on grain. If you are raising a beef animal and don't need it to grow quickly, or you are raising a weaned heifer to keep as a cow, feed mostly roughages and little or no grain.

If you don't have pasture for your cattle, you'll have to feed them hay, which is basically pasture plants that have been harvested and dried. The type of hay you feed will depend on what's available in your area. Alfalfa, clover, and timothy are common, but any number of other grass and legume hays are suitable.

Alfalfa hay is richer in vitamin A, vitamin E, protein, and calcium than grass hay, but be careful not to overfeed your cattle on it. Alfalfa hay can cause digestive problems and bloat, meaning the rumen becomes too full of gas. The gas causes the rumen to expand like a balloon and puts pressure on the animal's lungs and large internal blood vessels, causing it to die. Feeding a mix of grass hay and alfalfa hay is safer.

First-cutting alfalfa often has a little grass in it and can be an ideal hay. Second- or third-cutting alfalfa is generally richer and more likely to cause bloat. In addition, alfalfa hay becomes moldy more readily than grass hay if it gets wet or is baled when it is too green. When buying alfalfa, make sure it is green and bright, with lots of leaves and fine stems; it should not be coarse or brown and dry. Alfalfa that is cut early, before it blooms, is finer and more nutritious. Alfalfa that has bloomed has less protein and coarse fiber with larger stems.

Make sure any hay you buy is not moldy or dusty, because mold and dust may create digestive or respiratory problems. Hay for calves should not be stemmy or coarse. Because the protein and nutrition of hay are mainly in the leaves, stemmy hay is not nutritious, and it's hard for a calf to chew. Adult cattle can handle coarser feed than calves can. If you have no experience in buying hay, ask your county Extension agent or another knowledgeable person to help you.

Pasture Management

Pasture should contain several types of nutritious grasses. Cattle won't do well in a weed patch. If you are a pasture novice, have your county Extension agent look at your pasture and offer advice on any needed renovation.

In the early spring, pen up your cattle and feed them hay for a while to let the pasture grow. Otherwise, your cattle will eat the new green grass as soon as it starts to grow, and it won't become tall enough to provide sufficient feed for the summer. Some pasture plants become coarse as they mature, and your cattle will not eat them. Weedy areas may also be a problem. You can improve the pasture by mowing or clipping weeds so they don't go to seed and spread. If the pasture has bare spots, you can seed them by hand scattering a pasture mix when the ground is wet.

If you live in a rainy area, your pasture will grow just fine without much help. But in a dry climate, pasture must be watered with a ditch or sprinklers so it won't dry out by late summer.

Lush green grass has as much protein and vitamin A as good alfalfa hay. For a growing calf or a milking cow, good pasture is hard to beat, but keep a close watch on your grass. Dry pasture is poor feed, because it loses its nutrients. If the grass gets short or dry, feed your cattle some good hay to supplement the pasture.

Pasture Rotation

When cattle stay in the same area all the time, they overgraze short, tender grasses and ignore the mature, coarse grass unless nothing else is available to eat. The grasses and plants in a pasture become less healthy if they are overgrazed or undergrazed. To avoid this problem, confine cattle to one segment of pasture where the grass is at least 4 inches (10 cm) tall and move them to another segment before they graze the first one too close to the ground. The first section will have regrown by the time the cattle get back to it. By dividing your pasture into two or three portions and grazing your cattle on them sequentially, you can improve your overall pasture condition, increase forage production, and extend the grazing season to save money on purchased hay.

Feeding an Unweaned Calf

Whether you purchase a newborn calf from a dairy or the calf is born on your farm, you'll be responsible for feeding it. (If it is born on your farm, you'll be milking its mother.) For the first few days of a calf's life, split the daily feeding into three parts and feed every 8 hours. You can feed the calf early in the morning when you get up, again in the middle of the day, and at night just before you go to bed. Once the calf is 1 week old, you can begin feeding twice a day (every 12 hours, morning and evening), which makes life a bit easier.

Colostrum

Your calf should have adequate colostrum. When buying a calf from a dairy, the calf may have been allowed to stay with its mother until it has nursed once or twice. Some dairymen prefer to take the calf away before it has nursed, put it into a clean pen, and feed it from a bottle. The colostrum from the cow is milked out and saved to feed to calves.

When taking a newborn calf home from a dairy, ask to buy a gallon (4 L) or two of fresh colostrum to take with you. Store the colostrum in scrupulously clean containers in your refrigerator, and feed it as long as it lasts. If the calf was born to your milk cow, it will get plenty of colostrum by nursing naturally. (See page 310 for information on making sure the calf nurses.)

Bottle-Feeding a Calf

Teaching a calf to drink from a bottle is easier if she has never nursed from her mother. A hungry newborn calf will eagerly suck a bottle for her first meal. But the calf that has already nursed from her mother is spoiled, preferring the taste and feel of the udder. These calves can be stubborn and require patience to get them to nurse from a bottle.

If the calf was with its mother awhile, it knows how to nurse from a cow but not from a bottle. You'll need to quickly teach it to nurse from a bottle — you don't want it to go hungry for very long.

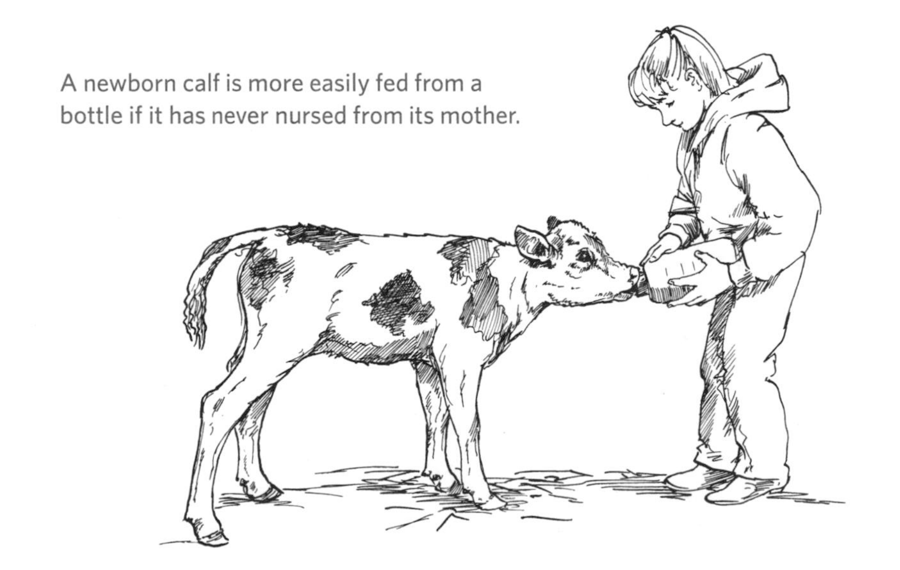

A newborn calf is more easily fed from a bottle if it has never nursed from its mother.

Don't Overheat Milk

Never overheat milk or milk replacer. Overheating damages the proteins.

A young calf needs to eat several times a day.

If a calf's first few feedings with a bottle are colostrum instead of milk replacer, it will more willingly suck the bottle. Colostrum not only tastes better but is also the best food for a calf at this time.

The keys to teaching a stubborn calf to suck a bottle are persistence and the use of real milk (preferably colostrum). The milk should be warm; young calves hate cold milk. Heat the milk so it feels pleasantly warm on your skin but not hot. If it is too hot, it will burn the calf's tender mouth and the calf won't suck.

To feed the calf, back it into a corner so it can't get away from you or wiggle around too much. Straddle its neck and use your legs to hold it still, leaving both hands free to handle its head and the bottle.

Use a nipple that flows freely when the calf is sucking, so it won't get discouraged by having to work too hard. However, the milk shouldn't flow so fast it chokes the calf. Hold the bottle so the milk will flow to the nipple. The calf shouldn't be sucking air. Don't let it pull the nipple off the bottle.

Transitioning to Milk Replacer

If you have no cow to provide milk for your calf, or you wish to use your cow's milk for other purposes, milk replacer is available as a nutritional substitute for feeding young calves. A calf accepts milk replacer more readily when it is introduced gradually.

If you got some colostrum for your young calf, divide it into several feedings to get through the first day or two while you are teaching the calf to nurse from a bottle. If you cannot obtain colostrum, use whole milk, preferably raw milk from a dairy rather than pasteurized milk from a store. The calf will like the taste of raw whole milk better than that of milk replacer.

Before you run out of colostrum or milk, start mixing it with milk replacer to gradually adjust the calf to the taste of what she'll drink from then on. If you switch suddenly to milk replacer, she may dislike the taste of the new stuff and be stubborn about accepting it.

You can buy milk replacer at a feed store. Of the many kinds and brands, some are better than others. Ask a dairyman or your county Extension agent to recommend a good brand, and read the label to find out what the milk replacer contains.

Protein and fat content. The National Research Council recommends using a milk replacer with a minimum of 22 percent protein and 10 percent fat. But calves will do better if the milk replacer contains 15 to 20 percent fat; they will grow faster and be less apt to get scours from inadequate nutrition.

Fiber content. Check the fiber level in your milk replacer. Low fiber content (0.5 percent or less) is ideal because it means the replacer has more high-quality milk products and less filler.

Calf Feeding Program

Age	Ration
Birth to 3 days	colostrum
4 days to 3 weeks	whole milk or replacer; grain mix or starter
3 to 8 weeks	whole milk or replacer; grain mix or starter, with access to good roughage
8 weeks to 4 months	2 to 5 pounds (1 to 2.5 kg) of calf ration (grain mix) with access to good roughage; calves can be weaned as early as 8 weeks but do better if weaned a little later
4 to 12 months	3 to 5 pounds (1.5 to 2.5 kg) of calf ration, with access to good roughage

Protein sources. Check the protein sources in a milk replacer. Are they milk-based or vegetable proteins? Milk protein is the highest quality and best for the calf, because the newborn calf has a simple stomach. Her rumen, for digesting roughages and fiber, is not working yet. She can digest and use protein from milk or milk by-products more easily and efficiently than she can use protein from plants.

Mixing milk replacer. Follow the directions on the bag. The powder is mixed with warm water and fed like milk. The recommended amount varies by brand.

The powder mixes better if you put the warm water into your container first and then add the replacer to the water and stir until it is all dissolved. It won't mix quickly if the water is cool or lukewarm. Start with water that is a little hotter than you want it to be when you feed the calf; the temperature will be just right by the time you mix in the powder and take it out to feed the calf.

Storing milk replacer. Keep milk replacer powder dry and clean. It will spoil if it gets damp. Close the bag immediately after measuring out the correct amount. Keep it in a container with a tight cover. The quality may be reduced and the replacer may become contaminated with germs if the bag is left open and exposed to light, moisture, flies, and mice.

How Much to Feed

It's just as bad to overfeed a calf as to underfeed. Too much milk can upset digestion and cause diarrhea. Feed your calf according to its size: A big calf needs more milk than a little one. Weigh or measure the milk to make sure you are not overfeeding the calf.

Feed 1 pound or about 1 pint (453 g or about 475 mL) of milk daily for each 10 pounds (4.5 kg) of body weight. Thus, a calf that weighs 90 pounds (41 kg) should get 9 pints (4.25 L) daily — 4½ pints (just over 2 quarts) — in the morning and again in the evening, or about a gallon (3.75 L) a day.

Feed at the same time each day on a regular schedule so as to not upset the calf's digestive system. If the calf gets

Keep Feeding Equipment Clean

Always carefully wash your bottle or nipple bucket after each feeding; otherwise, bacteria will grow on it and may make the calf sick. Nipple buckets must be taken apart and cleaned. Use a bottle brush to thoroughly clean bottles.

diarrhea from being overfed, immediately halve the amount of milk for the next feeding. Then gradually increase it to the recommended amount for the calf's size. As it grows, you can increase the amount of milk, but don't feed more than 12 pounds (5.5 kg), or 1½ gallons (5.5 L), of milk daily.

Feeding from a Nipple Bucket

If you are feeding more than one calf, a nipple bucket can save you time. Once a calf learns how to suck a bottle, you can switch it to a nipple bucket. You don't have to hold the nipple bucket while the calf nurses. The bucket can be hung from a fence or a stall wall. Hang it a little higher than the calf's head, where the calf can reach it easily.

Don't enlarge the nipple hole on a nipple bucket. Some people widen it so the milk flows faster, decreasing the time the calf takes to drink the milk. But if the milk runs too fast, the calf may inhale some of it because it can't swallow the milk fast enough. Milk in the lungs can lead to aspiration pneumonia, which can't be cured with antibiotics and will kill the calf.

Feeding from a nipple bucket saves time, because you don't have to hold it while the calf drinks.

A calf needs help to learn to drink from a pail.

Feeding from a Pail

You can teach your calf to drink from a pail instead of a nipple bucket. Put fresh warm milk into a clean pail and back the calf into a corner. Straddle its neck and put two fingers into its mouth. While it is sucking your fingers, gently push its head down so its mouth goes into the milk. Spread your fingers so milk goes into the calf's mouth as it sucks. After several swallows, remove your fingers. Repeat this procedure until the calf figures out that it can suck up the milk. A pail is easier to wash than a nipple bucket or a bottle.

Start Your Calf on Solid Feed

Get your calf to eat dry feed — hay and concentrates — as soon as possible. At first the calf won't consume much dry feed, but it should learn how to eat it.

Hay. A growing calf needs roughages for fiber. A calf uses the bulkiness of roughage — hay, grass, corn silage, straw, and cornstalks — to develop its digestive system so its rumen can begin to function properly. A baby calf that can follow its mother's example begins eating hay or grass at just a few days of age. But when a calf doesn't have its mother to show it how to eat, you have to encourage it to eat hay. Put a little leafy alfalfa hay into its mouth after every milk feeding, until it learns to like it.

Give your calf hay as soon as it will start nibbling on it. Calves have small mouths and cannot handle coarse hay, but they will nibble on tender leafy hay. Fine alfalfa, clover, or grass hays — or a mix of these — are all nutritious. Give the calf just a little bit of fresh hay once or twice a day. Don't give your calf much hay at one time, because the hay will be wasted; baby calves won't eat hay that has been tromped or lain on.

Good green pasture is excellent feed for a calf, as long as it is getting some milk (or milk substitute) and grain. If the pasture is not top quality and lush, the calf may also need a little alfalfa hay, which has more protein and other necessary nutrients for the growing calf than mature or dried-out pasture. A calf that is penned without access to pasture definitely needs alfalfa hay as its roughage source.

Concentrates. A growing calf needs concentrates for energy. Concentrates are feeds that are high in nutrients relative to their bulk, and can be in the form of grains, starter pellets containing grains plus milk products, or a complete starter containing not only grains and milk products but also some roughage. Teach your calf to eat concentrates as soon as possible. Put some into its mouth after each feeding of milk until it learns to like it. You can then feed it in a tub or feed box.

Starter pellets have a high protein level, as well as providing the energy a calf needs for growing. The starter can be offered as early as the first week of life. A calf that will eat enough high-quality dry starter ration won't need milk and can be weaned young. Early weaning can reduce costs if you are buying milk replacer. You can feed the calf alfalfa hay and starter pellets until the calf no longer needs the milk products. At that point you can transition the calf to grain and good hay or pasture.

Grains should be offered to the calf by the time it is 3 weeks of age. At first, about 1 cup or ¼ pound (236 mL or 125 g) of grain is all a young calf can eat each day. Increase the amount gradually until the calf is eating about 2 pounds (1 kg) of grain daily by the time it is 3 months old. Never wean a calf until it is eating about 2 pounds (1 kg) per day. After the calf has been weaned, continue feeding all the grain starter it will finish daily. By the time it is 3 to 4 months old, it may eat as much as 4 to 5 pounds (2 to 2.5 kg) of grain daily. A calf being fed a grain starter also should be given hay, and should be eating hay well for at least 1 week before weaning.

Complete starter works well if you don't have a source of good-quality roughage for the calf; if you have alfalfa hay, you don't need the complete starter. If you do use a complete starter, feed it free choice, meaning leave it available at all times and let the calf eat as much as it wants. Since complete starter includes roughage, the calf won't need hay until it is about 3 months old. Before you discontinue the complete starter, give the calf some hay for at least two weeks, and make sure your calf is eating the hay well.

Grain starter or a complete starter can be fed to a calf until it is 4 months of age to help it through the weaning process. A calf should eat at least 1 pound (500 g) of starter daily for every 100 pounds (45 kg) of body weight before it is weaned. Use a weight tape to estimate your calf's weight.

The Importance of Water

Make sure your calf has fresh, clean water every day and access to trace mineral salt. Calves need water, even though they get fluid with their milk or milk replacer. Water is especially important in hot weather.

Weaning Your Calf

The age at which you wean your calf will depend on several things, including feed sources, the calf's health, and how long it has been eating solid food. A calf can be weaned from milk when it is as young as 8 weeks, but most calves are weaned at the age of 3 months. A calf weaned before it is eating enough grain and hay won't do well. It's better for such a calf to stay on the nursing program longer.

Weaning is easier if the process is gradual. Start by decreasing the amount in its twice-daily bottle or bucket feedings. Cut back to about three-quarters of the amount you've been giving. Feed this reduced amount for a few days, and encourage the calf to eat more grain, feeding it right after it finishes its bottle or bucket. The calf will then be interested in the grain and not as upset with you for shortchanging on milk. Then go to one feeding of milk a day, giving grain at the other feeding time. Then stop the milk feedings. Give grain at the time of day when you used to offer the milk.

The rumen takes a while to enlarge so the calf can eat enough solid food to give it the nutrition it needs. Right after weaning, a calf still doesn't have much rumen capacity. It may eat just a small amount of hay compared with the amount of grain it can handle. The amount of hay consumed will increase as the rumen develops further.

Don't Wean Too Soon

It doesn't hurt to keep a calf on milk or milk replacer for quite a while, but it does hurt to wean a calf too soon. Use your best judgment to decide when you think your calf is ready to be weaned.

Feeding a Weaned Calf

After your calf is 3 months old and weaned, gradually change from feeding starter to a growing ration. A growing ration should contain at least 15 to 18 percent protein. If you have good pasture or alfalfa hay, the necessary protein can be supplied by supplementing pasture or hay with 4 to 5 pounds (2 to 2.5 kg) of daily grain.

Keep some hay in front of your calf all the time in a place it can easily reach. The hay should be fine stemmed and leafy, with no mold or dust. As the calf grows, hay can become a larger part of its diet. After it is 5 or 6 months old, good pasture can be used in place of hay.

A beef calf purchased from a cattle farmer will already have been weaned by its mother and will be accustomed to eating solid feed by the time you acquire it. When you bring the calf home, have feed and water in the pen.

Leave some good hay where the animal can find it easily, but not in a corner or along the fence line where the calf will walk on it every time it goes around the corral looking for a way out. Use really good grass hay or a mix of grass and alfalfa. Don't give rich alfalfa hay to a newly acquired calf; it may make the animal sick or bloated. You can gradually adjust it to good alfalfa hay later.

Give it all the hay it will eat. Then slowly start it on grain, if you wish. Give the calf just a little bit until it learns to eat the grain, and then increase the amount gradually. Too much grain all at once may upset the calf's digestion.

Give water in a water trough or in a bucket that is hooked to the fence or the stall wall so it can't tip over. If the animal has never drunk from a bucket, you may have to put the bucket next to the feed, or feed the animal next to the water trough for the first day or two, so it will find the water when it comes to eat the hay.

Grassfed versus Grainfed

A weaned beef calf can be raised on roughages alone but will grow faster and get fatter sooner if you feed it grain. Some people prefer the flavor of grassfed beef and feel grainfed beef has too much fat. Others find grainfed beef to be more tender and juicy compared to grassfed beef. The eating quality of beef depends on many things besides taste preferences in beef, including the quality of forage during the finishing phase of grassfed beef and the age of the animal at finishing. Other factors include the animal's age, length of time on grain or on high-quality forage, genetic differences in marbling ability, and whether or not the animal was gaining weight at the time of slaughter. An animal that is just maintaining weight or is losing weight, as well as an older animal past the age of 3 or 4, will generally be less tender.

Expected Finish Weight for a Beef Steer (in lbs./kg)

	Angus	Hereford	Shorthorn	Charolais	Simmental	Limousin	Holstein
Angus	1,000/453	1,050/476	1,050/476	1,225/556	1,225/556	1,125/510	1,225/556
Hereford		1,050/476	1,075/488	1,250/567	1,250/567	1,150/522	1,250/567
Shorthorn			1,050/476	1,250/567	1,250/567	1,150/522	1,250/567
Charolais				1,400/635	1,425/646	1,325/601	1,425/646
Simmental					1,400/635	1,325/601	1,425/646
Limousin						1,200/544	1,325/601
Holstein							1,400/635

If your calf is a cross of two of these breeds, look at the figure where their charts meet. For instance, the top line shows the weights of Angus and Angus crosses. Heifers of the same breeds and crossbreds weigh about 80 percent of these values.

Rate of Gain

A beef steer generally weighs 1,000 to 1,300 pounds (454 to 590 kg) and has about 0.25 to 0.45 inch (6 to 11 mm) of outside fat at the time of slaughter. A beef heifer will finish at a lighter weight (about 900 to 1,000 pounds [408 to 454 kg]) than a steer of the same age will. A dairy steer's finish weight depends on its breed. For instance, a Holstein steer might finish at 1,300 pounds (590 kg) while a Jersey steer might weigh 900 (408 kg).

Most beef animals eat about 7 pounds (3 kg) of feed to gain 1 pound (0.5 kg) of weight. To some extent, a calf's rate of gain depends on its genetics; some cattle have better rates of gain than others. On average, a growing steer should gain 2 to 3 pounds (1 to 1.5 kg) per day. Some crossbred steers will gain more. Feed your calf 15 to 20 pounds (7 to 9 kg) of feed daily.

If your beef calf is a heifer, feed her as you would a steer. Remember, though, that she will not finish out as large as a steer of her same age and breed and therefore will not need as much feed. Adjust your feeding figures to fit her target finish weight.

By knowing the calf's weight when you get it, and at what weight you want to butcher it, you can calculate the gain needed to get it ready for butcher. By knowing the total number of days until that time, you can determine the necessary average daily gain.

For example, if you buy a 500-pound (227 kg) steer in November and want to butcher him in August (270 days away), you can use the following formula to figure out how much your calf must gain per day to finish at 1,100 pounds (500 kg): Finish weight, minus present weight, divided by number of days until slaughter, equals the average daily gain. For example, 1,100 pounds (499 kg), minus 500 pounds (226 kg), divided by 270 days, equals 2.2 pounds (1 kg) for his necessary daily gain.

Good average daily gain for a steer is 2½ to 3 pounds (1 to 1.5 kg). The steer in the example should have no trouble meeting his finish weight. Good average daily gain for a young heifer is 1½ to 2 pounds (0.5 to 1 kg). Most dairy steers will gain somewhere between the two, although a Holstein may gain as well as a beef steer.

Butchering Beef Cattle

When you're raising a calf to butcher, you will probably want to let it grow to good size. Some folks like baby beef (from a calf at weaning age), but if you have enough pasture to raise your calf through its second summer, you will get a lot more meat for your money by letting it grow bigger. The ideal age at which to butcher a steer or a heifer is $1^1/_2$ to 2 years. At that age, the carcass is tender and is nearly as large as it will get.

The animal's breed can be a factor in determining when it is ready to butcher. Beef animals generally do not marble until they reach puberty (or in the case of a steer, the age at which he would have reached puberty if he had been a bull). Different breeds mature at different ages. Angus and Angus-cross cattle often reach puberty at a younger age (and a smaller weight) than do larger-framed cattle, such as Simmental, Charolais, or Limousin.

An Angus-type beef calf may finish faster and be ready to butcher when it is a yearling or a little older. If you feed it longer, it may not get much bigger, just fatter. A Simmental calf, in contrast, may still be growing and not fill out (carry enough flesh to be in good butchering condition) until it is at least 2 years old.

Thus, the ideal age at which to butcher your beef animal depends on its breed and on whether it is grassfed or grainfed. Cattle will grow faster and finish more quickly on grain, but at greater cost. Whether you feed grain depends on personal preference (some people prefer grainfed beef to grass-finished beef, and vice versa) and your situation. If you have lots of pasture, raising grassfed beef is usually most economical.

You can take cattle to a custom packing plant to be slaughtered and butchered, or you can do the butchering yourself. A good resource on home butchering is *Raising a Calf for Beef* by Phyllis Hobson.

Cost Considerations

Can you save money raising your own beef? That depends on many factors, including the price of cattle in your area, whether you start out with a weaned or unweaned calf, whether you prefer grass-finished beef or grain-fed beef, the price you're currently paying for beef, and whether you do your own butchering, cutting, and wrapping or have it done at a commercial slaughterhouse.

Weaned and grassfed. If you have adequate pasture for one or two calves, with fences that will reliably contain them, the simplest way to raise your own beef is to buy weaned calves or yearlings in the spring when your pastures are growing, then butcher them in the fall before your pasture is gone or its quality declines after freezing weather. If your pasture is good quality — if you can split the pasture into sections with electric fence and rotate the animals into new sections periodically to allow grazed areas to regrow — calves will gain nicely on grass. A yearling purchased in the spring can be easily finished on grass by fall, you won't have the expense of hay or grain, and — assuming cattle prices don't drastically change during the summer months — the value of the larger animal(s) in the fall will be more than you paid in the spring. On the other hand, a newly weaned calf may not be old enough or big enough to finish by fall and you would have to carry him over winter — feeding hay, or hay and grain — which greatly adds to the expense.

Unweaned dairy calf. If you purchase a day-old dairy or dairy cross calf to raise for beef, you'll have the expense of milk replacer, starter pellets, hay, and grain until the calf is big enough to turn out on grass, and you'll need to keep him through the next winter, feeding hay and possibly some grain. The exact cost of providing milk replacer, grain, hay, and so forth varies greatly, depending on the season, type of grain or hay, your location, and freight costs. The expense of raising a baby calf is partially offset by the fact your purchase price for a day-old calf (sometimes only $100 or less for a male dairy calf) will be much lower than the price for a beef weanling (500 to 600 pounds at $1.00 to $1.40 per pound — more or less, depending on its size and the market at the time of purchase) or yearling (700 to 900

Marbling

Marketplace beef cattle carcasses are inspected by the U.S. Department of Agriculture and judged for quality, using several grades to rate the tenderness of the meat. The main thing that determines the grade is the amount of marbling — flecks of fat in the muscle — which makes the meat more tender, tasty, and juicy. The highest grade is prime, followed by choice, select, and standard. Prime has the most marbling; standard has little.

pounds at $0.80 to $1.05 per pound, again depending on size and the market at time of purchase).

Your cow's calf. The cheapest way to raise beef, if you keep a family milk cow, is to breed her each year to a beef bull and raise the crossbred calf to yearling size for butcher. Your cow then supplies you not only with milk but, after the initial growing period of her first calf to yearling age, she also furnishes enough beef each year to feed your family. By starting the calf with part of the cow's milk, you won't incur any additional feed costs for the calf's first several months of life. After the calf is weaned, it can be grown on pasture and/or hay, and finished on pasture during its yearling summer. Your only expenses are the stud fee for breeding to a bull (or, if the cow is bred AI, the breeding fee for the semen and technician's time) and the cost of winter feed during each calf's first year as a weanling.

Care of the Milking Cow

When a cow starts giving milk at the beginning of a lactation cycle, the amount of milk she produces may increase so rapidly that her feed intake cannot keep up with her increased energy needs. The difference is mobilized from her body, causing the cow to lose weight, and a thin cow doesn't produce much milk. Proper nutrition is therefore important in determining how much milk your cow will give.

Hay and Pasture

About two-thirds of a dairy cow's total nutrition should come from forages — that is, hay or pasture. Make sure the quality is good. Cows will eat a greater total volume of hay if you feed it fresh several times a day. The more good forage you can get a milk cow to eat, the more milk she will give.

Some forage-based dairies rely on good-quality pasture and feed a minimum of grain, preferring to be satisfied with less volume of milk as a trade-off for a longer productive life of the cows.

But grass-based dairies utilize top-quality forage that is higher in nutrients than pasture that's generally available to the backyard cow. Cows in a grass-based dairy are usually grazing across a lush field in strips, being moved to the next strip as often as three to four times per 24-hour period so they always have fresh forage containing optimum protein and energy and moved often so they will eat as much as possible.

> Most backyard cows will more willingly stand for milking if given some grain while being milked and some may not even have to go into a barn but will just stand eating grain.

Grain

Even though a dairy cow has a large rumen, she cannot eat enough forage to supply all of her nutritional needs. For good milk production, the dairy cow needs grain. Grain serves a dual purpose — besides providing the extra nutrition a cow needs, grain gives her a treat while she's being milked. Most backyard cows will more willingly stand for milking if given some grain while being milked, and some may not even have to go into a barn or stanchion but will just stand eating grain.

The amount of grain must be adjusted to fit her needs: more during the peak of her lactation, when she is making the most milk, and less toward the end. How much grain a cow needs depends in part on the quality of her forage; a cow that's expected to milk adequately while living in a weed patch will need more grain than a cow that enjoys good-quality pasture. The amount of grain a cow needs depends also on her breed, milk output, and the cow's body condition. If she is becoming too thin to produce the milk, she needs more grain. If she is holding good flesh and still producing a satisfactory amount of milk she could probably get by with a minimum amount. Just keep her happy while she's being milked, make sure she has good pasture or hay, and increase the grain ration if she loses weight.

Water

Your cow needs a constant supply of clean water. A milking cow needs 3 to 5 gallons (11 to 19 L) of water, including the moisture in her feed, for every gallon (4 L) of milk she produces. A cow eating hay needs to drink more than a cow on lush green pasture, which contains a lot of moisture.

Milking a Cow

When done correctly, milking is a pleasurable experience for both the cow and the person milking her. You — and your cow — may need a few tries before you feel completely comfortable with the task. Just remember to be gentle and observant when milking, and keep the following tips in mind.

Short fingernails, clean hands. Before milking your cow, make sure your fingernails aren't long. If you poke the cow's teat with a sharp fingernail, she may kick. Wash your hands as well.

Check and clean the udder. The cow's udder is a complex structure that needs good care. Before you start milking, check the udder for problems or injuries. Then check for abnormal milk by squirting a little into a small bowl. If the udder and the milk are fine, wash the udder with a sanitary solution (obtained from your vet or a dairy supplier) mixed with warm water. Remove any dirt on the teats so it won't get into the milk. Washing the udder with warm water also stimulates the cow to relax and let down her milk. Use a clean paper towel to wash and dry the udder and teats before you milk.

How to Milk

Make yourself comfortable on a stool beside the cow's udder. Hold a clean empty seamless stainless steel bucket between your legs. Start with two teats — the front teats or the back ones, or two on the same side. Hold one teat in each hand and squeeze one at a time, squeezing the milk down and out through the teat opening. Begin the squeeze at the top of the teat, with your index finger and thumb grip. Finish the squeeze with the lower fingers. By applying pressure with your thumb and index finger, you keep the milk from going back up the teat, so when you squeeze with the rest of your fingers, the milk comes down and out through the hole.

Aim the stream of milk into the bucket; don't let it squirt off to one side. When you become good at milking, you can easily direct the stream and even aim a squirt toward a barn cat waiting to catch the milk in its mouth.

After each squirt, release your grip. More milk will flow down into the teat. Keep up a nice rhythm by alternating squirts. When the first two teats are soft and flat and you can't get any more milk from them, milk out the other two quarters. If one quarter seems to have

When you get good at milking you can easily aim a squirt in the direction of a barn cat waiting to catch the milk in its mouth.

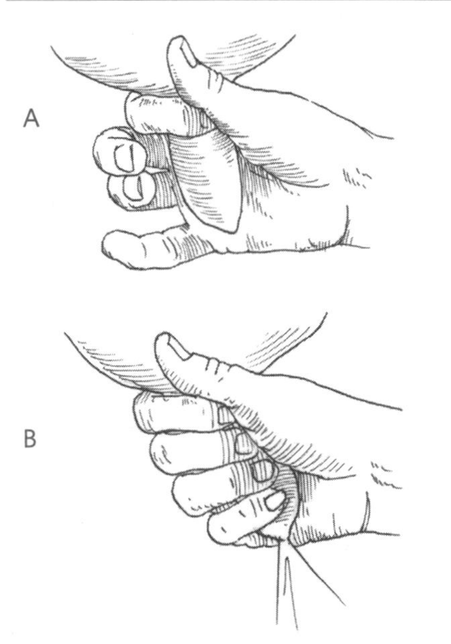

To milk, apply pressure with your thumb and index finger (A) while squeezing with your lower fingers (B).

The Cow's in Charge

Cows can let down or hold up their milk. No milk is in the teat until the cow lets it down, which she does by relaxing muscles that keep the teat canal closed. When a cow wants her calf to nurse, or thinks it is milking time, she lets the milk flow down from the udder into the teat. If a cow doesn't want to let down her milk, you won't get much milk out of her. Washing and massaging may encourage a stubborn cow to release her milk.

more milk than another, perhaps one of your hands is not yet as strong as the other and you aren't getting quite as much milk out with each squeeze.

Some Things to Consider

Before you begin milking, you should be aware of the following.

Tips for First Milking

Sometimes a cow's udder is so sore right after calving that she doesn't want you to touch it. She may kick at her calf as you try to help it nurse, or kick at you when you try to milk. In this case, press your head firmly into her flank as you milk to prevent her from swinging that leg forward. Try for the area right in front of the stifle joint. If you press hard every time you feel her tensing up to kick, she won't be able to kick well. If a cow is nervous the first few times you milk her, talk softly to her or hum a little to reassure her and keep her relaxed.

Sometimes a cow will kick if her teats get sore and cracked, as they may do in cold weather. If they get too cracked, they will bleed. To prevent chapping on teats, rub a little ointment (such as Bag Balm) on the teats after you finish each milking.

Sore Muscles

Once you get used to the proper squeezing motion and rhythm, milking is easy. But milking uses muscles in your forearms and hands you may not have used much, and they will get tired. Your arms and hands may ache afterward. The more often you milk, the stronger your arms and your grip will get.

Easy Milkers, Hard Milkers

Some cows are easy to milk. They have teats that are easy to hold on to, and they let their milk down freely. The milk almost flows from the udder into your bucket. Other cows are harder to milk. More effort is needed to squeeze out the milk, and milking takes longer. A cow with short teats may be a difficult milker because getting a good hold is more difficult. More time is needed to milk if you must squeeze with just one or two fingers instead of with your whole hand.

Mastitis

Bacteria sometimes enter the udder through the teat canal. In the warm environment inside the udder they multiply rapidly and cause infection, resulting in mastitis. The cow's body sends white blood cells to fight the infection. The infected quarter is swollen and often feels hot. If a quarter has abnormal milk — if it is lumpy, watery, or bloody or has any other abnormality — or heat and swelling, treat the cow for mastitis. A cow with mastitis may show no other symptoms.

You can test milk for mastitis by using the California Mastitis Testing kit, which measures the amount of somatic cells in the milk. Somatic cells are white blood cells and mammary cells that have been damaged by infection. The somatic cell count in the milk will be high as long as the quarter is infected. A test kit comes with complete instructions and is available from most dairy supply companies.

Your vet can provide antibiotic preparations to treat mastitis or you can buy antibiotic tubes at your farm store. Before giving the medication, thoroughly milk out the quarter. To administer the medication, gently insert a syringe tube with a long plastic tip into the teat and squirt the antibiotic up into the quarter. Gently massage the treated quarter to help spread the antibiotic to all parts. Follow directions on the label for proper use and the number of days of treatment. Call your vet if you notice any of the following symptoms:

- The mastitis does not clear rapidly; you may need to have a milk sample cultured to find out which organism is the culprit and what type of antibiotics it is most sensitive to.
- The cow shows signs of systemic illness, such as being off feed, acting dully, or having a fever.

Not Safe for Humans

Milk from an infected cow should not be consumed by humans. Even after the infection is cleared up, the milk can't be used until it contains no more antibiotic. The label on the medication tells how long the antibiotic persists in the cow's milk. While the cow has mastitis, keep milking her regularly to hasten her recovery. In most cases, the milk from a recovering cow can be fed to calves. Consult your vet for guidance.

Taking Care of Milk

Take care to keep milk clean. Use a clean bucket, a clean strainer, and clean storage containers. After each use, rinse all equipment in warm water to remove the milk fat. Then scrub everything with hot water and dishwashing soap, using a stiff plastic brush. The brush will clean the equipment much better than a dishcloth. Always use a plastic brush or plastic scouring pad rather than metal; metal leaves scratches on the surface of the equipment where bacteria can cling. Rinse everything thoroughly in clean water.

If you wish, you can pasteurize your milk to make sure it is perfectly safe to drink. (The milk you buy at the grocery store is pasteurized.) Pasteurized milk is heated to a specific temperature, kept there for a short time, and then quickly cooled. A home pasteurizing unit can be purchased from a dairy supply catalog. It consists of a metal container with a heating element in the bottom. Fill the container with water, set a gallon (4 L) of milk in a covered metal pail into the water, and then plug in the unit. The water heats to the desired temperature for the proper time, and a buzzer sounds when it's done. (See page 208 for instructions for pasteurizing milk without an electric unit; cow milk can be pasteurized in the same manner as goat milk.)

Can you save money by milking your own cow? That depends on how much you pay for the cow — though the expense is spread over her lifetime and also is part of the cost of getting a calf each year to sell or raise for beef — the price of feed (which varies considerably with type, season, and location), how much milk she produces, and how much you are currently paying for your family's supply of milk.

During the summer months your cow may milk adequately on good pasture with just a little grain fed at milking time. A family milk cow on good pasture often does fine on 2 to 3 pounds (1 to 1.5 kg) of grain per milking. During winter months when grazing is sparse or nonexistent, however, she will need good-quality hay — preferably legume hay while she is still lactating. Your hay costs can be minimized if you dry her up during winter, giving her at least two months of rest before her next calf. She won't need as much quantity or quality of hay while she is dry. If you plan her calving date for spring, she will have green grass (your cheapest feed) during her peak lactation. By purchasing a young cow, you will have her milk and calves for a dozen years or more, making the purchase price a good investment.

Milking Schedule

A cow should be milked twice a day, 12 hours apart. Not sticking to this schedule can be harmful to the cow.

Rebreeding Your Cow

To keep the milk flowing in the future, your cow will need to be rebred about three months after she gives birth to her calf. Keep track of her periods of heat when she starts cycling again, so you can have her bred at the proper time. It may be several weeks before she starts having heat cycles.

A few weeks or months after breeding, have your veterinarian check her for pregnancy if she hasn't returned to heat. It helps to know if she is actually pregnant — sometimes a cow will lose the pregnancy early on but won't return to heat — and when to expect her calf to arrive the next year, so you can plan to dry her up on time before her next calving. If she is not pregnant, you must try to get her rebred.

Signs of Heat

A cow can be bred only when she comes into heat, or estrus. If she is to be bred by artificial insemination (AI), you must be able to determine when she comes into heat, and then have a technician insert a capsule of semen into her uterus at the proper time. Determining when your cow is in heat can be difficult if no other cattle are around. Signs of heat include increased restlessness, pacing the fence or bawling, or a mucous discharge from the vulva. However, not all cows show obvious signs. One clue is that milk production usually decreases temporarily on the day the cow is in heat.

If the cow is living with other cattle, telling when she comes into heat is

Keep a Record of Heat Cycles

When your cow starts cycling, keep a record of her heat periods, which are usually three weeks apart, so you'll know when she should be bred. If you keep good records, you'll be less likely to miss the period during which you want her bred in order for her to calve at the best time of the year for your climate.

Parts of a Cow's Reproductive System

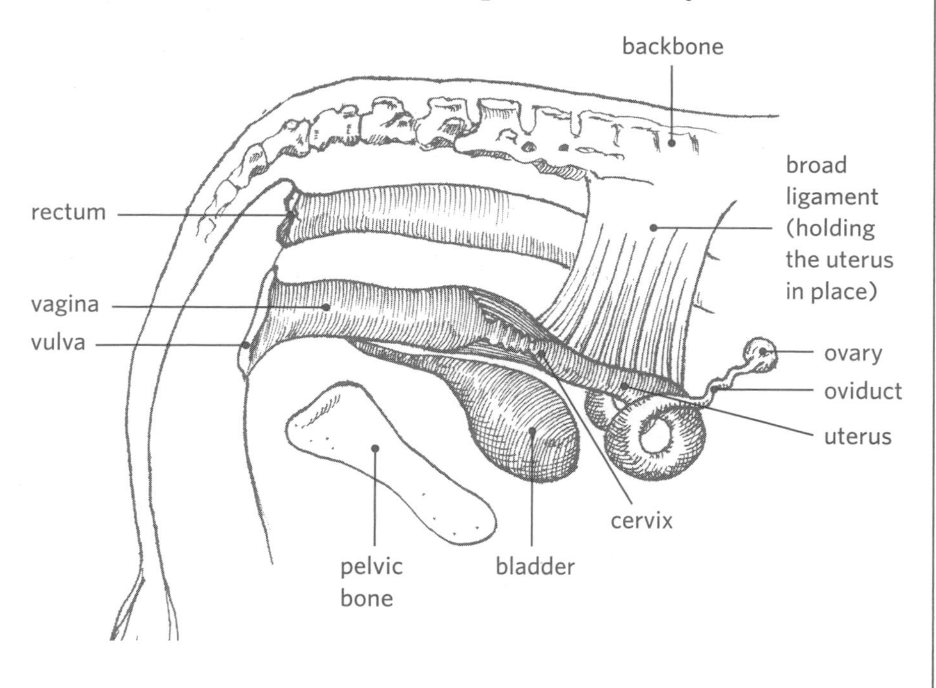

easier. The other cattle will mount her, or she will mount them even if no bull is present. The hair over her tail and hips may be ruffled from this activity. The easiest way to tell is to put her with another cow, heifer, or steer for a short time. If you don't have any other cattle, take her where she can be left with a bull for one to three weeks until she is bred. The bull will know when she is in heat.

Selecting a Sire

Breed your cow to the best bull available. If your cow is a registered purebred and you want a purebred calf that you can register, she must be bred to a registered bull of the same breed. When a dairy cow is mated to a good bull of her breed and her calf is a heifer, the calf will be worth more.

The nice thing about breeding your cow by AI is that you can select an outstanding bull from anywhere in the United States. You have your pick of the best bulls in the breed. These bulls are kept in central locations called bull studs. Get a catalog from one of the major bull studs. Your AI technician can obtain one for you.

If your cow is a crossbred or if you want to raise a crossbred calf, choose a bull of a different breed or a crossbred bull. A crossbred beef-dairy calf makes a good beef calf if you plan to have it butchered.

If you bought your cow from a local cattleman, you might ask him if he would consider putting her with a bull at his place, and what he would charge. Or you may have a neighbor with a bull who is amenable to breeding him with your cow. Ask the bull's owner to keep track of the breeding date so you'll have it for your records. With this information, you can predict your cow's calving date the next spring.

Duration of Pregnancy

The duration of gestation is about 285 days, but a cow may calve as much as 9 days before or after her due date. Most cows calve within 3 or 4 days of their due dates.

Breeding by Artificial Insemination (AI)

Using AI, a large number of cows can be bred to one bull. The bull's semen is collected, divided into many small portions, and put into tubes called straws. The straws are stored in liquid nitrogen, which keeps them at the cold temperature of –320°F (–195°C). The frozen straws can be shipped anywhere, using semen tanks that keep them frozen.

If you can tell when your cow is having heat cycles, you can have her bred by AI. Talk with your local AI technician about ordering semen from a bull of your chosen breed. Several breeding services collect semen from champion bulls all across the United States. Some ranchers and most dairymen use AI instead of buying bulls.

The price of semen varies. Some bulls, especially the most popular champions in their breed, are expensive. You don't need the most expensive semen. Choose the sire ahead of time and make arrangements with the AI technician so you can purchase the semen and have your cow inseminated at the proper time.

Watch your cow closely to tell when she comes into heat. She will probably be in heat for 12 to 18 hours. Try to spend at least 30 minutes twice a day, morning and evening, watching her for signs of heat. (You may not need to spend this much time if she gives obvious clues.) When you see that she is in heat, call the AI technician.

While your cow is restrained in a chute, the semen is inserted into her uterus through the vagina. With good luck, she will settle, or become pregnant. If she does not conceive, she will return to heat 17 to 25 days later and can be bred again.

Size of Fetus during Pregnancy

At 2 months, the fetus is the size of a mouse. By 5 months, it's the size of a large cat. After the fifth month, the fetus will be large enough that you may be able to see or feel it kicking. The cow's right side may ripple and move. As you lean your head against the cow's right flank during milking, you may feel a bump from a small foot. After the fifth month the fetus grows rapidly, becoming calf-size by 9 months.

Make Sure She's Pregnant

Once your cow is bred, watch her closely for the next few weeks, especially during the time she would have her next heat period. If she does not come into heat at that time, she's probably pregnant. To make sure, have your vet check her for pregnancy two or three months after the breeding.

Vaccinations during Pregnancy

Your cow should be on a vaccination schedule in which she receives booster shots for certain diseases once or twice a year. Some vaccines can be given during pregnancy, but others should not. Talk with your vet about the vaccinations your cow needs.

Also ask the vet about a vaccination to help protect your cow's calf against scours (infectious diarrhea). If your pens or pastures have held baby calves and have been contaminated with calf diarrhea, vaccinating your cow against scours will create antibodies against many of the diseases that cause scours. She will pass the antibodies on to her calf when he nurses.

Managing the Dry Cow

To get ready for the new calf and new milk production, the cow's body needs a rest from making milk. Allow her to go dry for a couple of months before her next calving. The length of the dry period varies depending on the cow's age and condition. She needs at least 45 days of rest to be able to produce a lot of milk during her next lactation and to make enough antibodies in her colostrum for the next calf. A six- or seven-week dry period is adequate for the average cow. But young cows calving for the second time and high-producing cows generally need eight weeks (56 to 60 days).

To dry up a cow, simply stop milking her. The transition should be abrupt; don't try to ease the cow into it by partial milking. Cows are designed to stop producing milk when the udder is full and tight, which is what happens under natural conditions when a calf is weaned or dies. The pressure in a cow's udder signals her body to stop making milk. She will be uncomfortable at first, but after a few days the pressure will ease. Her body gradually resorbs the milk left in her udder.

To help a cow dry up, reduce her feed, especially grain. Eliminating grain helps the cow's body adjust to not making milk. Pasture or hay should provide enough nutrition for your cow to go through the dry period without becoming fat. She will need grain only if she needs to gain weight. If she is in good condition at calving time, she will produce milk well. If she is thin, she won't be able to milk as well as she should, and she'll have trouble rebreeding on schedule.

Check your cow's udder closely while she is drying up. After the last milking, treat each quarter with an antibiotic recommended by your vet to help prevent mastitis. Watch for heat or swelling in the udder.

Vaccinate Before Rebreeding

Before rebreeding your cow, vaccinate her against leptospirosis, infectious bovine rhinotracheitis, bovine virus diarrhea, and other diseases as recommended by your veterinarian. Vaccinate at least three weeks before rebreeding. Some of the live virus vaccines (such as infectious bovine rhinotracheitis or bovine virus diarrhea) may cause abortions in a cow if given while she is pregnant. Also the cow needs time to build immunities against these diseases before she becomes pregnant, since some of these diseases can cause abortions.

Get Ready for Calving

As calving time approaches, your cow will get a large belly and become clumsy. She should be in a safe place where she won't slip on ice or get stuck in a ditch. Make sure you have a good place for your cow to calve. A shed in her pen or pasture will work if the weather is cold, wet, or windy. If she is confined in a pen or a barn, make sure she has clean bedding. A calf that is born in an unclean place may get an infection.

If your cow is calving in summer and the weather is nice, she can calve at pasture if you check her often. Be sure the pasture is safe and clean, and covered with grass rather than dirt or mud. She should have shade if the weather is hot, no gullies or ditches to get stuck in, and strong fences she can't crawl through when she becomes restless during early labor.

If your cow will have her calf in the barn, put lime onto the barn floor before covering it with new bedding. The lime not only helps disinfect the floor but also makes a nonslippery base. The floor must provide good footing so the calf will be able to stand up and the cow won't injure herself. She'll be getting up and down during labor, and you don't want her to slip and injure her legs or damage her udder.

The stall should be large and roomy. If the stall is too small, the cow may lie too close to the wall, and the calf may get jammed into the wall as it emerges.

Make sure the bedding is clean. Never use wet sawdust, moldy straw, or any damp, moldy, dusty material. Many cases of mastitis are caused by dirty bedding. Wet or dirty bedding containing mold or manure will have germs that can invade the uterus or udder of the calving cow or infect the calf's navel.

Things to Have on Hand at Calving Time

- Halter and rope in case you need to tie the cow
- Tamed iodine (Betadine) or chlorhexidine solution in a small, widemouthed jar, for dipping the calf's navel
- Towels for drying the calf
- Bottle and lamb nipple, in case you need to feed the calf
- Obstetric chains or short, small-diameter (½ inch [1 cm]), smooth nylon rope with a loop at each end, in case you need to pull the calf
- Disposable obstetric gloves (from your vet) and lubricant (obstetric "soap") in a squeeze bottle
- Flashlight for checking on your cow at night

Early labor may last two or three hours (a heifer having her first calf may experience early labor for four to six hours or even longer).

Signs the Cow Will Soon Calve

As your cow approaches calving, her udder gets full. The vulva becomes large and flabby; the muscles are relaxing so they can stretch when the calf comes through. The area between the cow's tailhead and pinbones becomes loose and sunken. These changes may start several weeks or just a few days before she calves. Her teats may fill with milk, or milk may leak from the teats. Every cow is a little different, so be alert, observant, and ready. If your cow is with other cattle, put her into a separate pen so they can't bother her when she calves.

Stages of the Birth Process

The process of calving has two stages. Knowing what occurs at each stage will help you decide when or if your cow needs help.

First Stage: Early Labor

The signs of early labor are restlessness and mild discomfort. The cow has a few early uterine contractions as the uterus prepares to push the calf out. She may kick at her belly or switch her tail.

Contractions become more frequent and more intense as labor progresses. The contractions of early labor usually help turn the calf toward the birth canal.

Early labor may last two or three hours (a heifer having her first calf may experience early labor for four to six hours or even longer). She will get restless and may pace the fence. If she is at pasture, she may go into the bushes or a secluded corner.

Second Stage: Active Labor

When the cervix is fully open and the calf or the water sac — which often precedes the calf — starts into the birth canal, active labor has begun. The birth should take place in 30 minutes to 2 hours.

The water sac is dark and purplish. When it breaks, dark yellow fluid rushes out. If the sac breaks before it comes out, all you'll see is fluid pouring from the vulva. The water sac should not be confused with the amnion, a white sac full of thick, clear fluid. The amnion protects the calf while it is in the uterus.

Active labor is more intense than early labor. The cow has strong abdominal contractions. The entrance of the calf into the birth canal stimulates hard straining. Each contraction forces the calf farther along. The calf's feet soon appear at the vulva. Although the calf can safely remain in this position for a couple of hours, it is best if it is born within one hour. Give the cow time to stretch her tissues, however; helping her out by pulling on the calf too soon may injure her.

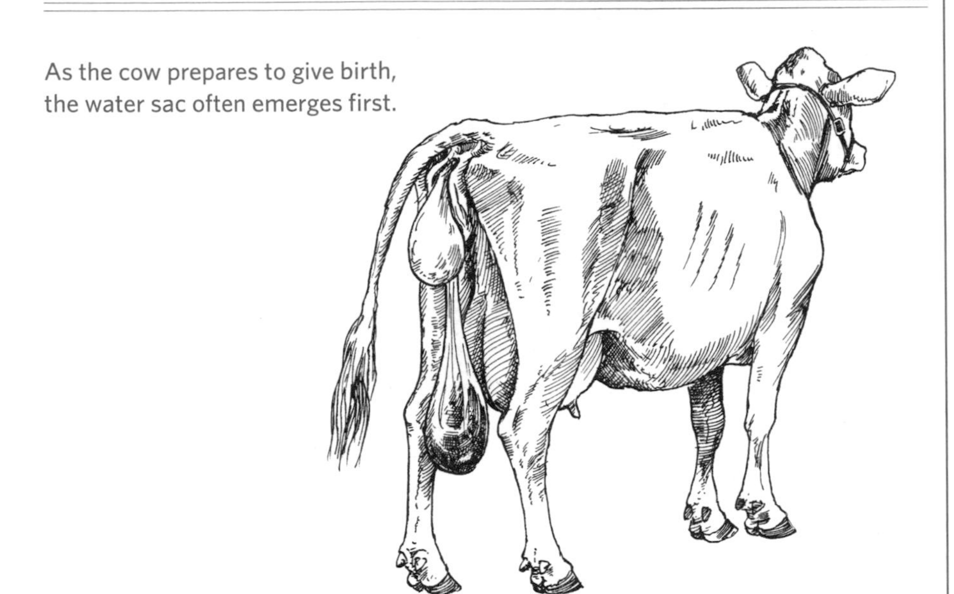

As the cow prepares to give birth, the water sac often emerges first.

Your cow can get up and down a lot in early labor, but once she starts straining hard, she will probably stay down. Make sure she doesn't lie with her hindquarters against the fence or the stall wall.

Passing the calf's head may take a while, but as long as the cow is making progress, you won't need to help. After the head emerges, the rest of the calf usually comes easily. Fluid will flow from the calf's mouth and nostrils as its rib cage is squeezed through the cow's pelvis. This fluid was in the calf's air passages while it floated around in the uterus, and it comes out now so it can start to breathe.

When the Calf Is Born

After the calf is born, the cow may rest for a few minutes; labor is hard, and she may be tired. But the calf must begin breathing immediately. If it doesn't, or if the sac over its head does not break and is still full of fluids, you must quickly help. Pull the sac away from its nose. Clear the fluid away and make sure the calf starts breathing. The cow will probably turn to see her new calf. She should sniff at it and then start to lick it.

Shedding the Afterbirth

When the cow gets up after calving, a lot of red tissue will be hanging from her vulva. This mass is the placenta, which surrounded the calf and is attached it to the uterus with buttons — dark red, dollar-sized attachments spaced over the uterus. The afterbirth may take 30 minutes to a few hours to completely detach from her uterus and come out. Never pull on the afterbirth while it is still hanging from the cow. If she doesn't shed the afterbirth for many hours, call the vet. Cows eat their afterbirth so it won't attract predators; watch for the afterbirth to detach, and remove it from your cow's pen right away so she won't choke on it.

If the cow takes longer than 10 hours to shed the afterbirth, she may develop a uterine infection. Keep a close watch for pus discharge or illness, which are signs of infection. If your cow won't eat or develops a fever, she'll need immediate treatment. Call the veterinarian.

Helping a Cow Calve

You may need to help with the birth. The calf may be positioned wrong in the uterus so it cannot enter the birth canal or come through it. The calf may be a bit too big to pass through easily, or maybe your cow has twins. A normal calf should be born within one hour of the start of second-stage labor. If a cow is too long at labor and nothing is happening, have her checked by your veterinarian or another experienced person. Definitely call for help if you see only one foot or hind feet coming out.

Your help will most often be needed for birthing twins, because they may be tangled together, and bull calves, because they are bigger than heifer calves.

Be There to Help

Many calving problems can be corrected if someone is there to help. When a cow goes into labor, check on her frequently to make sure the birth is progressing normally. All too often, assistance is given only after the cow or the calf is in critical condition. Be on hand so you can give or get help quickly.

Checking inside the Cow

During a problem birth, a careful examination inside the cow may be necessary. If your cow is gentle, you can check her. First, tie her up so that she can't move around. If she's lying down and won't get up, check her where she is. Work as cleanly as possible to avoid introducing infection into her. If possible, use a disposable long-sleeved plastic glove that covers your whole arm. If nothing has yet appeared at the vulva, reach into the birth canal to see whether you can feel two feet. If you feel just one foot, or some other abnormality, you'll know why the birth is not progressing.

If no feet have come into the birth canal, reach farther in and examine the cervix. If it is not opening yet, you are interfering too soon. A cervix that is completely open will be 6 to 7 inches (15.5 to 18 cm) wide, and you can easily reach into the uterus.

If the calf is not positioned correctly, the first part of it you touch may be its head, tail, foot, or some other part of its body. In that case, you will need immediate help to reposition the calf so it can be born.

Pulling a Calf

Often, the only problem is that the calf is a little too big and needs a pull. But don't pull on a calf unless it is in perfect position to come out. If the feet have been showing for an hour and you've felt inside the vulva to make sure the nose is right there and the head is advancing properly, pull the calf.

If the calf's nose is showing and the cow's straining starts to push the head out, you can wait. But if she isn't making progress after the feet have been showing for one hour, you should help her. First, feel inside the birth canal to see if the head has room to pass through the cow's pelvic opening. If you cannot get your fingers between the top of the calf's head and the top of the birth canal, the opening may be too small. If that's the case, call a vet.

If you think the head can come through, go ahead and pull on the calf's legs. Having two people working as a team to pull the calf makes the job easier. Pull alternately on one leg and then the other to ease the calf through the pelvis one shoulder at a time.

The calf has to come out in an arc. When its feet emerge from the vulva, pull straight out. But after the head comes out, pull slightly downward, more toward the cow's hocks, as its body arches up over the pelvis and then down. If you watch a normal birth, you'll notice that the calf curves around toward the cow's hind legs as she is lying there and it slides out.

In a difficult birth, one person can pull on the front legs with obstetric chains or ropes secured around the calf's legs above the fetlock joints, so they won't injure the joints or the feet. At the same time, the other person stretches the cow's vulva (see drawing below). Stretching the vulva helps the head come through more easily. One person pulls while the other stands beside the cow if she's up, or sits beside her hips if she's down, facing to the rear. If you are the one doing the stretching, put your fingers between the calf's head and the cow's vulva, pulling and stretching the vulva each time the cow strains. You and your partner should pull and stretch the vulva only when she strains, and rest while she rests. Don't pull when she is not straining.

Pulling a Calf

In a difficult birth, you may need to help out by stretching the vulva and pulling the calf using obstetric chains or ropes looped around the calf's legs above the fetlock joints. Pull only when the cow strains. Do not pull when she is not straining.

When in Doubt, Get Help

Try to recognize problems early and get help before the cow or the calf is in serious trouble. When you are in doubt about a situation or unsure of your ability to handle it, call the vet or an experienced cattleman or dairyman.

Hip Lock

Sometimes in a hard birth, you get the calf partly out, only to have it stop at the hips. Don't panic. Remember the calf has to come up and over the pelvic bones in an arc. As the calf's body comes out, start pulling downward, toward the cow's hind legs. To avoid hurting the calf's ribs, get it out far enough that its rib cage is free before you pull hard downward. Once its rib cage is out, the calf can start to breathe if the umbilical cord pinches off.

If the cow is standing, pull straight downward and underneath her, pulling the calf between her hind legs, which raises the calf's hips higher, to where the pelvic opening is the widest. If the cow is lying down, pull the calf between her hind legs, toward her belly.

Backward and Breech

A calf coming backward, with hind feet protruding from the vulva, has its heels up. Front legs have the toes pointing down. If the bottoms of the feet are up, the calf may be backward. Before you assume that, reach inside the birth canal to see whether you feel knees (front legs) or hocks (hind legs). The calf can be rotated just a little sideways or upside down. If the calf is backward, call the vet quickly to help you. He will use a calf puller to get the calf out before it suffocates.

A breech calf is positioned backward, but the legs do not enter the birth canal; he is trying to come rump first. The cow may not start second-stage labor at all. Nothing is in the birth canal to stimulate hard straining, and she seems to be too long in early labor. If you wait too long before checking, the placenta will eventually detach and the calf will die. If you check inside her, all you'll feel is the calf's rump or tail. Call the vet.

Leg Turned Back

Sometimes one of the calf's front legs will be turned back. One foot will appear but not the other, or sometimes the head and one front foot will show. If you can detect this problem early, before the head is pushed out far, you can push the calf back into the uterus, rearrange it, and get the other leg unbent and coming properly. Otherwise, get help immediately.

Get the Calf Breathing

After a difficult delivery, make sure the calf starts breathing as soon as possible. Stimulate it to breathe by sticking a clean piece of straw or hay up one nostril to make it sneeze and cough. If you get no response, give the calf artificial respiration. If it is still alive, you can feel its heartbeat near the rib cage, on the left side, behind its front leg. Blow a full breath of air into one nostril, holding the other nostril and its mouth closed with your hand. Blow until you see the calf's chest rise, then let the air come back out on its own. Blow in another breath, and keep breathing for the calf until it regains consciousness and starts breathing on its own.

Some calves are still encased in the amnion and its fluids after sliding out. If the sac doesn't break, the calf will die because it cannot breathe. This is a very good reason to be on hand when your cow calves.

Once the calf is born, the cow should get up soon and start licking it. The licking stimulates the calf's circulation and encourages it to try to get up and nurse. If possible, let the cow lick the calf dry. If the weather is cold and the cow is not licking, rub the calf with clean towels.

Disinfect the Navel Stump and Medicate

Navel ill is a serious infection that can kill or cripple a calf. Bacteria that enter through the navel may create an abscess in the navel area or may get into the bloodstream and cause a general infection called septicemia, which can be fatal. Or the bacteria may settle in the joints. Even with diligent treatment, it can be difficult to save a calf once it gets navel ill.

Disinfect the calf's navel stump as soon as the umbilical cord is broken and you have made sure the calf is breathing. Have a widemouthed plastic container (such as a yogurt container) ready ahead of time, containing ½ inch (1.5 cm) of tamed iodine or chlorhexidine. Immerse the navel stump in the iodine by holding the jar tightly against the calf's belly and making sure the navel cord is thoroughly soaked in the iodine.

The iodine not only kills germs but also acts as an astringent, shrinking the tissues and helping the navel stump dry

To disinfect a newborn's navel stump, use a small jar containing tamed iodine (Betadine) or chlorhexidine (Nolvasan) solution.

Calf Birthing Positions

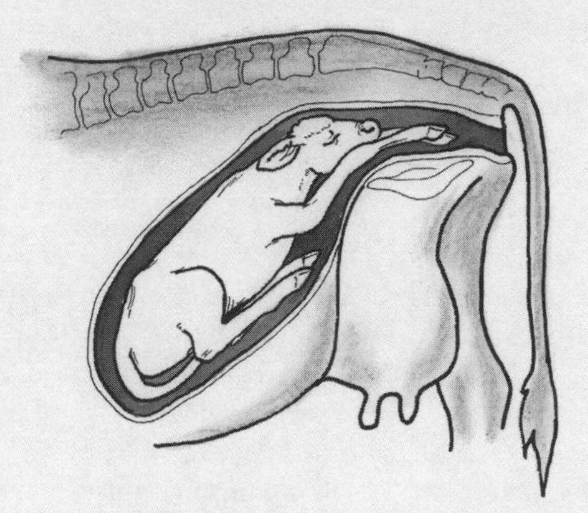

Normal birth position.

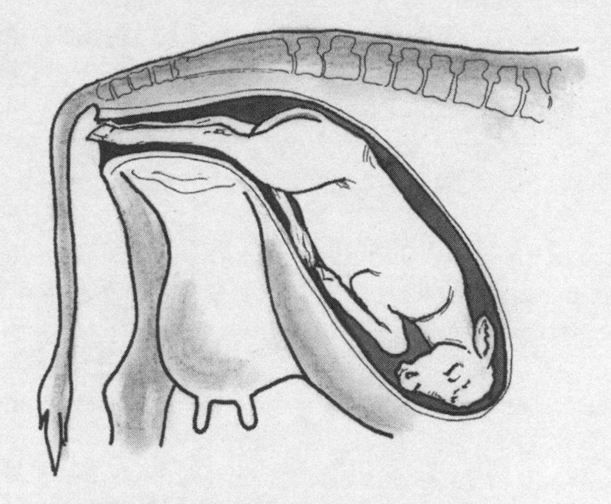

Posterior presentation. Birth is usually too slow to result in a live calf unless assisted.

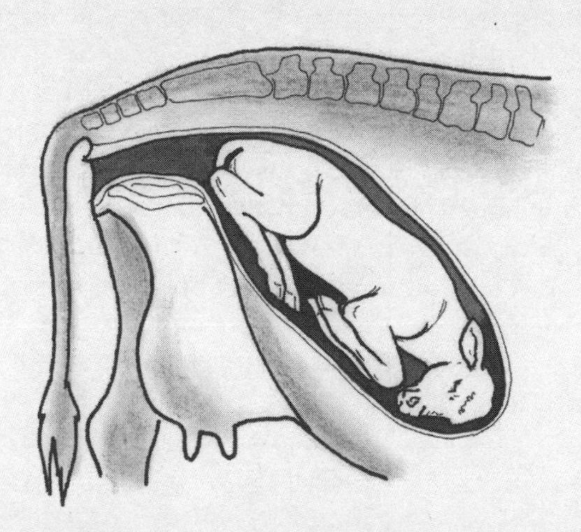

Breech. The calf must be pushed forward far enough so each hind leg can be tightly flexed at the hock and brought into the birth canal.

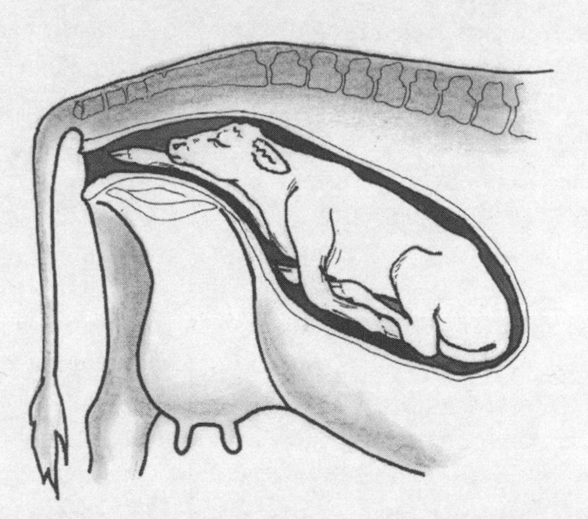

One front leg turned back. The calf must be pushed back and the leg brought into the birth canal.

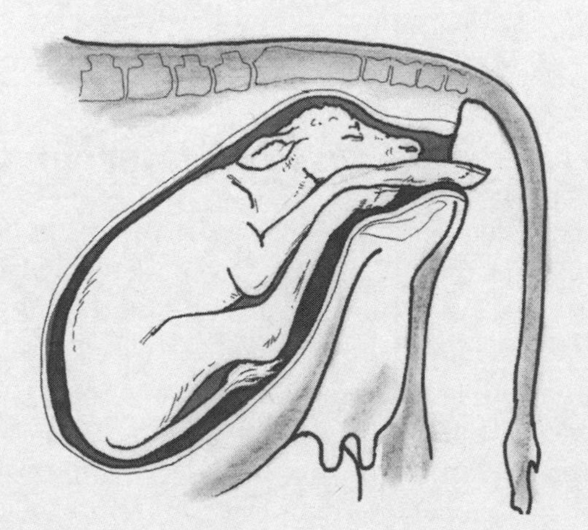

All four feet coming into the birth canal. The calf must be pushed back far enough to push the hind legs back over the pelvic rim and into the uterus.

up quickly and seal off so that bacteria cannot enter the calf. Don't touch the navel cord with your hands unless they are really clean. You need to touch the navel cord only if it drags on the ground when the calf is standing up. In that case, cut it with clean, sharp scissors; leave a 3-inch (7.5 cm) stump and be careful not to pull on it. Pulling or jerking on the cord can injure the calf internally. As soon as you cut the cord, immediately soak the navel stump in iodine.

The navel stump of a baby heifer usually dries up after just one application of iodine. But a bull calf's stump may take longer. Baby bulls urinate close to the navel, and if they urinate while lying down, as they often do, the navel stump gets wet repeatedly and doesn't dry up quickly. While it is still wet, bacteria can enter it. Reapply the iodine a few hours later, and again if necessary, to dry up the stump within the first two hours after birth.

Also give the new calf an injection of vitamin A, if recommended by your vet, and any necessary medications. In some areas, newborn calves need selenium to prevent white muscle disease. You may need to vaccinate newborns against enterotoxemia (a highly fatal gut infection) or tetanus. Discuss your calves' vaccination needs with your vet ahead of time.

Make Sure the Calf Nurses

Colostrum, the cow's first milk, has twice the calories of ordinary milk, with rich, creamy fat that is easily digested and high in energy. Colostrum helps the calf pass its first bowel movements. Most important to the calf are the antibodies in the colostrum that provide immunity to disease. This temporary immunity lasts several weeks, until its immune system starts making its own antibodies. A calf that gets no colostrum or that doesn't nurse until it is several hours old may develop scours or pneumonia.

The best time for the calf to absorb antibodies is during the first two hours (and preferably the first 30 minutes) after birth. Make sure the calf is up and nursing within one hour of birth. If it can't stand up on its own within that crucial time, help the calf, or milk a little out of your cow's udder and feed it to the calf with a bottle. Before the calf first nurses, wash the cow's udder and teats well with warm water.

If you weren't there when the calf was born, don't assume that it nursed well just because it is with the cow. To make sure it gets an adequate amount of colostrum, try to get the calf to nurse as soon as you arrive on the scene.

Without help, most calves will eventually manage to nurse, but some won't, because they have a hard time getting onto teats. This problem is especially likely if the cow has a low udder or the teats are full and big. If too much time passes before a calf nurses, it will not absorb enough antibodies because its gut lining loses the ability to absorb colostral antibodies after the first couple of hours of life.

Colostrum and Transitional Milk

A calf can be fed for several days with the colostrum you milk from your cow. But people cannot drink the milk until it no longer has colostrum in it. You must milk the cow at least twice a day to get the remaining colostrum out of her udder and hasten the production of regular milk. The milk will be ready to be used by people in four to seven days.

True colostrum, the undiluted first milk, is obtained only from the first milking or nursing. The calf needs this first milk immediately after it is born. Later milkings produce transitional milk, a mixture of colostrum and regular milk. It is mostly colostrum at first but becomes more and more diluted by regular milk with time, until no more colostrum is left in the udder.

The colostrum is thicker and richer than regular milk and is usually yellow and sticky. It is waxy when cold. One indication that milk still has colostrum is that it will not easily go through a strainer.

If a cow's colostrum is not good — if it is bloody or she has mastitis — her newborn calf should be fed colostrum from another cow. A 90-pound (41 kg) calf, such as a Holstein, needs about 3 quarts (3 L) of colostrum at its first feeding right after birth. A 50-pound calf (23 kg) should receive 1½ quarts (1.5 L).

If a cow has a lot of colostrum when she calves, you can milk a quart (1 L) from her while her own calf is nursing and store it in a plastic container or milk carton. This extra colostrum can be saved for emergencies, such as when a cow has poor colostrum or dies during birth. Extra colostrum will keep for up to one week in the refrigerator and for several years in the freezer.

Heat Stress

If the weather is hot when your cow calves, make sure she and the newborn are in the shade. Heat stress can reduce the immunity the calf receives from colostrum, making it more susceptible to diseases.

If a Calf Can't Nurse

If the newborn calf has trouble nursing or is too weak to stand and nurse after a difficult birth, milk colostrum from its mother to feed it. You'll need about a quart (1 L) to get it started. Pour the colostrum into a small-necked bottle (one that a lamb nipple will fit) and feed it to the calf. This bit of colostrum

Learn to Use an Esophageal Feeder

An esophageal feeder is a handy way to get fluids into a calf that is too weak or sick to nurse. The esophageal feeder is a container attached to a tube or stainless steel probe that goes down the calf's throat and into the stomach. Get an esophageal feeder from your vet, and ask the vet or other experienced person to show you how to put the feeding tube down a calf's throat.

will usually give the calf strength to get up and nurse. If the calf is too weak to suck a bottle for its first nursing you will need to use an esophageal feeder.

Share the Milk

If your dairy cow gives more milk than your family can use, let her raise her own calf and one extra, while you take part of the milk. Let the calves nurse on one side while you milk the other.

By raising a calf for beef in addition to the calf your cow produces, you can save money on milk replacer by feeding the beef calf the extra colostrum and transitional milk from your cow. Every time a cow calves, her milk for the first five days can be used to feed any calf or to mix with milk replacer. Don't let the 8 to 25 gallons (30 to 95 L) of colostrum your cow produces go to waste; that amount will feed calves for quite a while.

Many combinations can be used to feed calves, including mixes of whole milk and milk replacer, or sour colostrum and milk replacer. Make any changes in a calf's diet gradually to avoid digestive upsets. Don't suddenly switch a calf from milk replacer to sour colostrum; mix the two. Avoid changing the feed of a young calf during its first two weeks of life. After that, gradual changes in diet won't affect the calf much.

If the Cow Won't Cooperate

Sometimes a cow, particularly a first-calf heifer, is slow to mother her calf. Or sometimes a cow won't let a calf nurse because her udder is sore. If a cow kicks at her calf and it cannot nurse, tie her up or restrain her in a stanchion and tie one of her hind legs back so she can't kick. Leave enough slack in the rope so she can stand comfortably on that leg but cannot kick. You can then help the calf without having the cow strike out at it or you.

If the cow keeps kicking when the calf attempts to nurse, she may have a lot of swelling (called cake) in her udder, which makes nursing painful. You may have to put hobbles onto her hind legs for a few days so the calf can nurse without being kicked. It may take a few days, a week, or even longer, but eventually the cow will come to accept her calf. Once she starts showing some affection for it and stops kicking at it, the hobbles can be removed.

Some owners of backyard dairy cows let their calves nurse. Others leave the calf with its mother for only a few hours, then put it into a separate pen. A calf that is left with the cow more than 12 hours may bunt at her udder when it nurses. This bunting can bruise the mammary tissue, especially if the cow has a large, full udder or much swelling. Bruising of the mammary tissue can cause mastitis.

Cow tying prevents a cow from kicking her newborn calf when it tries to nurse.

You may have to put hobbles onto the hind legs of a mother cow if she continues to kick her calf.

A small three-sided shed like this one protects calves from the weather while keeping cows out.

Care of the Cow after Calving

Offer the cow feed and clean, lukewarm water to drink. She should eat and drink soon after calving. If she won't eat or drink or seems dull, she may have a problem; consult your vet. Take her temperature to see if she has a fever.

Most cows do fine after calving. Just watch your cow closely to make sure all is well. Give her as much good hay as she wants, but only a small amount of grain at first, until the swelling of her udder disappears. A high-producing cow fed a lot of grain just before or just after calving may have problems.

Milk your cow at least twice a day, even if the calf is still with her. The calf cannot drink all of her milk. Milking the cow will help reduce the pressure and swelling in her udder and will relieve a quarter (one of the four teats) the calf may have missed.

Natural Nursing

A commercial dairy takes calves away from their mothers for several reasons: so they can utilize the cows' milk; it can be cheaper to raise calves on milk replacer than on milk; it simplifies management; and a high-producing dairy cow's udder is large and delicate and may be easily injured and subsequently infected (develop mastitis) by a calf that bunts at the udder while nursing. A family milk cow, however, generally doesn't have the massive udder of a big-time dairy cow, and many owners of a family cow let their calves nurse. You can let your cow nurse her calf and still have milk for yourself, in one of two ways:

- Pen the calf away from the cow at night. Milk the cow in the morning, then let the calf back in with the cow to nurse during the day. However, when a calf lives with the cow, the cow may prefer to have the calf nurse rather than let you milk; this tractability issue varies from cow to cow.
- Keep the calf penned separately from the cow and put them together at milking time, letting the calf nurse one side of the udder (two quarters) while you milk your side. By this method, a cow that would rather nurse her calf than be milked will be more cooperative and let down her milk. You just have to make sure you can milk about as fast as the calf nurses, or it may try to rob your side when its side is empty. Take care not to let the calf slobber in your milk bucket.

Calf Management Procedures

Dehorning, removing extra teats on a heifer, and castrating a bull should be done while the calf is still young. For these procedures you may want your calf restrained and lying on the ground. To hold the calf still while it's lying on its side being castrated or having extra teats removed, kneel down and hold the calf's front leg (folded at the knee). Put gentle pressure on the neck with your knee so the calf cannot rise.

Removing Extra Teats

Most heifers have just four teats, but some are born with an extra one or two. An extra teat is of no use to a cow and may cause problems when she is being milked. If your calf has an extra teat, it should be removed.

As soon as the heifer is big enough for you to tell which teat is the extra one (generally at 1 to 3 weeks of age), the teat can be removed. The extra teat is easier to locate when she is lying down. Examine her udder closely. The four regular teats will be arranged symmetrically, with the two rear teats slightly closer together than the two front ones. An extra teat is usually smaller than the others and located close to the main teats. *Before removing a teat, make sure it is truly the extra one. You don't want to make a mistake and remove the wrong teat.*

Flanking a Calf

To get your calf to the ground easily and gently, without a big struggle, flank it. Stand close to the calf. Reach over its back and grab hold of the flank skin with one hand and the front leg (at its knee) with your other hand. Lift the calf off its feet, gently lowering its body to the ground.

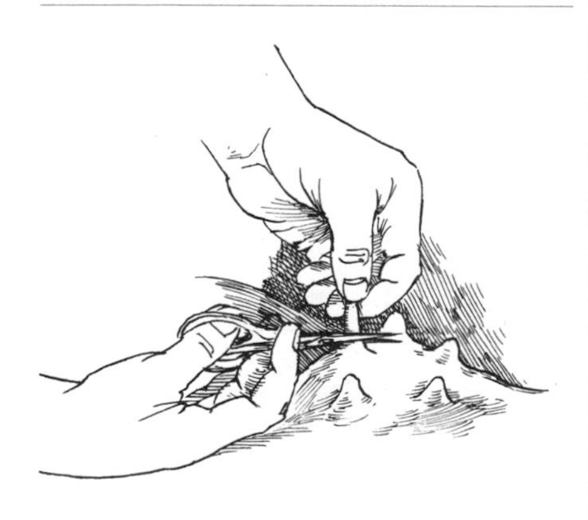

Snip off an extra teat with sharp, disinfected scissors.

If you are not sure which teat should be taken off, wait until the calf is older and the teat is more developed, then have your veterinarian do the removal. Removal of an extra teat in an older heifer should always be done by a vet because the wound will bleed more and may need stitches.

The small extra teat on a baby heifer is easy to remove. Flank the heifer (see box above) and put her onto the ground. Disinfect the teat to be removed and snip it off with sharp, disinfected scissors. Hold the scissors with the handle directed toward the front of the calf and the blades pointed toward her tail end. Make the cut lengthwise with the body. Afterward, dab the wound with tamed iodine.

Another method is to tie the heifer and remove the teat while she is standing, preferably with someone holding her back end so that she can't move around. Pull the extra teat down and snip it off cleanly where it meets the udder. Swab with disinfectant.

A young heifer rarely bleeds when an extra teat is removed. In the summertime, use fly repellent to keep flies away from the small wound.

Castrating

A bull calf should be castrated at an early age, the younger the better. All bull calves should be castrated, unless they will be used for breeding.

Castration is harder on a calf the older he gets. As the testicles grow, the blood vessels supplying them also become larger. Bleeding is always a risk when castration is done surgically, a procedure that should be performed by an experienced person.

Baby calves are easy to castrate by using an elastrator ring, which is a strong rubber ring that resembles a large Cheerio. To put the rubber ring onto a calf, he should be lying down on his side, with someone holding his head and front leg so he can't wiggle around while another person puts on the ring.

Using an elastrator tool, the ring is stretched and placed over the scrotum at the top of the scrotal sac, above the testicles. The tool is removed, leaving the rubber ring to constrict tightly, cutting off all circulation and feeling below it. The testicle tissue gradually dies, and the scrotal sac shrivels up and drops off a few weeks later, leaving a small raw spot that soon heals. Elastrator rings are the easiest and safest way to castrate baby calves, because they cause no bleeding.

Dehorning

If your calf has horns, they should be removed; cattle with horns may hurt one another or you. Dehorning should be done when the calf is young and the horns are small. Older calves or mature cattle with large horns may bleed a lot when their horns are removed, or they may get infections.

A caustic dehorning paste or stick can be used for a calf that's a few days old. An electric dehorner is often used on calves up to 3 months of age. The hot dehorning iron is held firmly against the head over the horn for long enough to kill the horn-growing cells at the base of the horn. Unless you are raising a lot of calves, you may prefer to have an experienced person who owns the appropriate appliance dehorn your calves for you.

Keeping Your Cattle Healthy

Maintaining proper housing and sanitary conditions, providing an adequate diet, and adhering to a vaccination program go a long way toward keeping your cattle healthy. The following measures will help you avoid having your calves or cow suffer needlessly, as well as avoid incurring the expense of treating an illness.

Vaccinations. For some illnesses, no vaccines are available. Most diseases, however, can be prevented with good care, prevention of exposure to infectious agents, and vaccination. Cattle need vaccinations against such diseases as blackleg, malignant edema, and brucellosis. Talk with your veterinarian about a vaccination program to protect your cattle from the diseases common in your area. A single family cow, or a calf or two raised for beef, may not be exposed to some diseases that are more likely to affect cattle in larger groups.

Keep cattle comfortable. To prevent illness and avoid stress, feed your cattle properly and make sure they have shelter from the elements in both cold and hot weather. A heat wave in summer may kill cattle if temperature and humidity get too high. During hot weather, use a fan in the stall, or hose down your animal periodically with a misty wet spray from a garden hose.

Sanitation. Good sanitation is important. It is easier to prevent infectious diseases than to cure them. If several animals share a pen or a barn, or if others have been there before, bacteria, parasites, and viruses are lurking. To help prevent disease and infections, thoroughly clean and disinfect your facilities before installing a new calf. Get rid of all old bedding and scrub the walls and floor with a good disinfectant; your veterinarian or dairyman can recommend one.

Keep pens clean and well bedded. A cow must not lie on dirty bedding or in mud and manure. Mastitis and blind quarters can result from bacteria entering the teat when a cow lies in dirty places. A blind quarter is a quarter of the udder that does not produce milk because it has been damaged by infection. If your cow gets an udder infection, consult your vet. Do not let this infection go untreated, or it may ruin your cow for milking.

Signs of Illness

Pay close attention to your cattle every day so you can tell whether they are feeling fine or getting sick. If you feed your animals twice a day, you can check for signs of illness at each feeding. Even cattle that receive good care may become sick. If you notice illness early and start treatment promptly,

Watch for Loners

If you have several cattle, one that is off by itself should be checked on. A sick animal often wants to be alone.

you should be able to catch most problems in time to help the animal recover quickly. Get to know your cattle so you can recognize the following signs.

Behavior and Appearance

A healthy animal is bright and alert and has a good appetite. It comes eagerly to its feed or grazes at regular times each day. A sick animal may spend a lot of time lying down. It may seem dull, and its ears may droop instead of standing up alert. It may stop chewing its cud because of pain, fever, or a digestive problem.

A healthy animal usually stretches when it first gets up and shows an interest in its surroundings. It responds with curiosity to sounds and movement and spends some time grooming itself. It walks with a free and easy gait. In contrast, a sick animal may show less interest in things around it. It may stand up slowly or with effort. All of its movements will be slow, and it lacks the spark of vitality and health shown by a normal animal. The more serious the illness, the more indifferent the animal will be to its surroundings, and the more reluctant to move.

Respiration Rate

A sick animal with a fever breathes fast. However, exercise or hot weather also make a healthy animal breathe faster. Cattle don't have many sweat glands. They cool themselves by panting.

Check each animal's respiration rate when it is standing quietly or lying down. Its respiration rate should be about 20 breaths per minute (10 to 30 is normal). An easy way to figure the respiration rate is to watch the animal's sides move in and out. Each in-and-out movement counts as one breath. Using the second hand on your watch, count the breaths for 15 seconds, then multiply by 4 to get the number of breaths per minute.

Temperature

The normal body temperature for a cow or calf is 101.5°F (38.6°C). A temperature of more than 102.5°F (39.2°C) signals a fever. Learn how to take your animal's temperature by using a rectal animal thermometer, which you can buy from your veterinarian. Tie a string to the ring end to avoid losing it in the animal's rectum.

To take your calf's temperature, restrain the animal in a chute. Digital thermometers are inexpensive, available at any drugstore, easy to read, and big enough to find quickly if you drop one. If using a mercury-in-glass thermometer, shake it until it reads less than 98°F (36.6°C). Moisten the thermometer with petroleum jelly so it will slide in easily. Gently lift the calf's tail and insert the thermometer into the rectum. Keep hold of the string, because the animal may poop out the thermometer, along with a bunch of manure. Leave the thermometer in for two minutes to get an accurate reading. When you take out the thermometer, wipe it with a tissue or paper towel so that you can read it. Disinfect the thermometer after each use.

A fever indicates that the cow or calf is fighting a bacterial or viral infection. If her temperature is 103.5°F (39.7°C) or above, call your veterinarian. Below-normal temperatures can also be a sign of something wrong; if she acts sick and has a low temperature, call your veterinarian.

On hot, humid days, a cow or calf's temperature can easily reach 103°F (39.4°C), even in the shade. If she's in the sun and panting, move her to a shady spot immediately. To cool her, keep pouring cool water down her back, from head to tail.

Normal Vital Signs

- Normal respiration is 10 to 30 breaths per minute, with an average of about 20.
- Normal body temperature is 101.5°F (38.6°C).

Other Health Signs

Another indication of the health of an animal is whether its eating habits are normal. Does it chew and swallow properly, or does swallowing seem painful? Is it drooling, dribbling food from its mouth, or having trouble with belching and chewing its cud?

The animal's bowel movements and urination can also indicate illness. With some digestive problems, the animal becomes constipated. If it has a gut blockage, the manure may become firm and dry, or it may not be able to excrete at all. Other problems in a sick animal may cause diarrhea. Manure should be moderately firm (not runny and watery) and brown or green.

If a bottle-fed or nursing calf has blood in its manure, call your vet. This is often the sign of a serious disease. Early treatment is important.

Urination may become difficult if the urinary tract is blocked, as when a steer develops a bladder stone. The steer may dribble only small amounts of urine, remain in the urinating position for a long time, kick at his belly in pain, or stand stretched. If he shows any of these signs, call your veterinarian.

Pay attention to abnormal posture. Resting a leg may mean lameness. Arching the back with all four legs bunched together is usually a sign of pain. A bloated animal may stand with its front legs uphill to make belching gas easier. A sick animal may lie with its head tucked around toward its flank.

Common Health Problems

If your animal becomes sick, recognizing the symptoms early and knowing what to do can make all the difference. A disease can have an infectious or noninfectious cause. Noninfectious diseases include bloat, poisoning, founder, and injury. Infectious diseases are those caused by microorganisms or parasites. The most common microorganisms are bacteria, viruses, and fungi. Infectious diseases can be contagious, such as pinkeye, or noncontagious, such as infection of a wound or abscess

Acidosis

A large increase in an animal's grain ration can cause overproduction of lactic acid in the rumen, resulting in acidosis — too much acid in the animal's body.

Symptoms/effects: If acidosis is not promptly treated, the rumen will stop working. The animal may get fever, diarrhea, or founder (see page 317), and it may die.

Prevention: Acidosis occurs most often when the grain ration is increased too rapidly. It can also happen if something interferes with the regular feeding schedule, causing an animal to overeat. For example, if a calf's water has manure in it, the calf won't drink it; then, because it's thirsty, the calf may quit eating. After the calf's water tub is cleaned and it drinks again, the calf may load up on feed.

To prevent acidosis, stick to your feeding schedule. Split the daily grain ration and feed twice a day, so that an animal doesn't eat a large amount of grain at once. When increasing the grain ration, do it gradually, over a couple of weeks.

Treatment: Fast action may be needed to save your cow or calf or to prevent founder. Call your vet immediately if your animal shows symptoms of acidosis.

Blackleg and Other Clostridial Diseases

Blackleg is one of several serious diseases caused by clostridia, a group of bacteria that live in the soil. This family of diseases includes tetanus, red water, malignant edema, and enterotoxemia.

Symptoms/effects: Cattle become sick suddenly and usually die.

Prevention: A good vaccine that prevents blackleg is available. Your calf should be vaccinated at about 2 months of age and receive a second dose of vaccine around weaning time.

Vaccines are available that protect against several of the other clostridial diseases. Check with your vet to see which vaccine you should use for your calf.

Some clostridial diseases are a problem only in calves. Once calves are vaccinated, they have lifelong immunity. Others, such as red water, can be deadly at any age. If red water is a problem in your area (as it is in the Northwest), all cattle must be vaccinated every six months.

Bloat

Bloat is a digestive problem that is often caused by highly fermentable feeds. Harmful bacteria that create gas when they multiply can also cause bloat. If too much gas builds up in the rumen, burping may not get rid of it, especially if the gas is frothy. The tight rumen eventually puts so much pressure on the lungs the animal can't breathe, and pressure on large internal blood vessels interferes with blood circulation.

Symptoms/effects: When viewed from behind, an animal with bloat looks puffed up on its left side, where the rumen is located. As bloat gets worse, both sides puff up, and the animal has trouble breathing.

Prevention: Avoid giving feeds that promote bloat, such as lush alfalfa pasture, rich alfalfa hay, too much grain, or finely chopped hay or grain.

Treatment: Severe bloat must be relieved quickly. Your vet may pass a tube into the animal's stomach to release the gas. If the bloat is frothy, it won't come out easily, and the vet will pour mineral oil through the tube to break up the foam.

Cattle that bloat often can be fed Bloat Guard, which contains an antifoaming agent. It comes in block form. Give Bloat Guard in place of a salt block, and let your cattle use it for several days before you start them on fermentable feed.

To treat a small calf for bloat, you can use an esophageal feeder to administer Therabloat Drench Concentrate mixed with water.

Brucellosis

Brucellosis is also called Bang's disease. Although it is not common in North American cattle, it is still taken seriously since it can be transmitted to people — causing fever and headache — through contact with an infected animal or infected meat or milk.

Symptoms/effects: Brucellosis causes abortion in cows.

Prevention: In areas where brucellosis may be a risk, all heifers must be vaccinated between 2 and 10 months of age. A cow that has been vaccinated will have a small metal tag in her ear with a number on it; the same number will be tattooed in her ear. If you live in an area that requires vaccination, never buy a cow that has not been vaccinated. A cow cannot be vaccinated past calfhood, because if tested for brucellosis she may be a reactor (test positively).

The requirement to vaccinate heifers has nearly eradicated this disease in cattle in the United States. In some states, especially those in the greater Yellowstone area, heifers must still be vaccinated because the disease is spread by infected wildlife, such as elk and bison. Since the disease occasionally appears in other areas, vaccination requirements change from time to time. Your county Extension agent can tell you if cows in your area are required to be vaccinated. Steers do not need this immunization.

Coccidiosis

Coccidiosis is a disease of calves caused by protozoa — tiny one-celled animals that damage the intestinal lining. The protozoa are transmitted in the manure of sick calves and carrier animals. Carriers are not sick but have some protozoans living in their intestines. A calf may pick up the coccidia by eating contaminated feed or water or licking itself after lying on dirty ground or bedding.

Symptoms/effects: Coccidiosis causes severe diarrhea. The loose, watery manure often contains blood.

Treatment: Have a veterinarian examine a calf that starts passing really loose manure, and especially if it has blood in it or the calf strains after passing the loose bowel movements. The vet can recommend an appropriate treatment.

Prevention: Deccox medicated crumbles mixed with feed, or Deccox medicated powder mixed with milk, can help prevent coccidiosis. You can also have the feed mill add Deccox to your calf grain, or you can use milk replacer that already contains Deccox.

Be Careful with Medications

When using any medications or vaccines, always read the labels and follow the directions for dosage, administration, and other factors. If you don't understand the directions, ask your vet for clarification. Store medicines properly. Some need to be refrigerated, whereas others shouldn't be. Some need to be kept out of the sunshine. Some need to be shaken well before use.

Diphtheria

An infection of the throat and mouth, diphtheria is caused by the same bacteria that cause foot rot. Injury to tissues of the mouth or throat can let the bacteria gain entrance. Cattle are most susceptible through 2 years of age. Emerging teeth or injuries caused by coarse feed or sharp seeds can open the way for infection.

Symptoms/effects: Calves with diphtheria have fever. They may act dull and uninterested in eating. The calf may have a cough and will drool because it has a hard time swallowing. It may have swelling of the cheek tissues, and its breath may smell bad. Swelling at the back of the throat can block the windpipe and make breathing difficult. The calf may die of infection or of obstruction of the air passages unless treated quickly.

Treatment: Diphtheria can be serious. If the symptoms are present, call your vet immediately. Proper antibiotic therapy is important.

Flies

Flies bite and suck blood and annoy and irritate cattle.

Symptoms/effects: Cattle trying to escape flies may spend all their time in the shade. They therefore may not graze or eat as much as they should.

Treatment: Horn flies and face flies can often be controlled by using insecticide ear tags. While in the animal's ear, the tag continuously releases insecticide as it rubs against the hair. The animal rubs the insecticide over its body as it reaches around and scratches itself. For a milking cow, use only insecticides approved for dairies.

Foot Rot

Foot rot is caused by bacteria that live in the soil. Cattle may get the infection if they have a break in the skin on a foot and walk through wet, muddy areas.

Symptoms/effects: Foot rot causes swelling, heat, and pain in the foot, resulting in severe lameness. The swelling and lameness come on suddenly; the animal may be fine one day and lame the next. The foot may be too sore to walk on.

Treatment: Foot rot heals quickly if it is treated early with appropriate antibiotics. Consult your veterinarian.

Founder

An animal can develop founder if fed too much grain or its ration is changed suddenly.

Symptoms/effects: The attachments between the hoof wall and the sole of the foot become sore and may separate, resulting in malformed hooves and severe lameness. Acidosis (see page 316) is the main cause of founder.

Treatment: Founder is a serious emergency. Contact your vet immediately.

Grubs and Heel Flies

Cattle grubs, also called warbles, sometimes appear under the skin on the backs of cattle in late winter or early spring. The grub is the maggot stage of the heel fly. This fly lays its eggs on the lower part of the legs of cattle during the warm days of early summer. The grub travels through the body to the animal's back.

Symptoms/effects: Grubs look and feel like marbles under the hide. An animal bothered by heel flies may run wildly with its tail in the air and may even crash into fences. It will look for shade and try to stand in a water hole to escape the flies.

Treatment: If grubs and heel flies are a problem in your area, consult your vet about the best way to get rid of them. Pour-on products are available that treat lice and grubs at the same time.

Hardware Disease

Hardware disease is also called traumatic reticulopericarditis. Cattle are indiscriminate eaters and will swallow bits of wire and other objects as they eat. Sharp objects that penetrate the reticulum can cause peritonitis, which may be life threatening, or extend into the heart cavity, killing the cow.

Symptoms/effects: An affected cow goes off feed, shows signs of pain, may have a fever, and may have difficulty breathing.

Prevention: Maintain a safe environment free from wires and other metallic objects. A cow kept in a barn or small enclosure can be given a cow magnet by mouth to try to bind metal objects and keep them in the reticulum.

If your pasture fence is old and broken but you can't replace it just yet, use a metal detector along the fence line to pick up stray staples or pieces of barbed or woven wire that a cow could accidentally ingest.

Treatment: Antibiotics can be used but in a severe case will not be effective. Surgery may be considered for a valuable animal.

Internal Parasites

Internal worms commonly infest cattle, especially young animals that have not developed resistance to them. These parasites are most often a problem in cattle on pasture.

Symptoms/effects: A calf with worms won't gain weight fast and may lose weight. It may have a rough hair coat, a poor appetite, diarrhea, or a cough. To tell whether your cattle have worms, have your vet examine a sample of manure.

Prevention and treatment: Medication is available to eradicate worms. However, cattle raised in a clean place will probably not get internal parasites. A calf housed alone in a small pen or hutch that was clean before the calf went in won't have to be dewormed. Worms become a problem once a calf goes out on pasture with other calves or where other cattle have been.

Johne's Disease

Johne's disease is also called paratuberculosis. It is caused by a bacterium, *Mycobacterium johne*, also known as *Mycobacterium paratuberculosis*. Johne's is a reportable disease: When it is found in an animal, others in the herd must be tested by skin injection or blood or fecal culture. Affected herds are usually quarantined and regularly examined by state regulatory authorities until they are deemed healthy. These measures have drastically reduced the incidence of Johne's disease in cattle.

Store Medical Supplies Safely

Make sure all livestock medications and treatments (such as iodine, insecticides, and antibiotics) and needles and syringes are kept in a safe place, out of the reach of children.

Symptoms/effects: Johne's disease causes chronic diarrhea and wasting (severe loss of condition) in adult cattle.

Prevention: Johne's disease has no known treatment and no vaccine. Affected animals must be destroyed. The best way to avoid this disease is to purchase animals from a herd that is free of Johne's disease.

Leptospirosis

This bacterial disease is spread by mice, rats, and other rodents; wildlife; and infected domestic animals, such as pigs, dogs, and other cattle. Cattle can get lepto from contaminated feed or water.

Symptoms/effects: Leptospirosis is a mild disease in cattle, but it can have serious side effects, such as abortions in pregnant cows.

Prevention: Heifers and cows should be vaccinated against leptospirosis at least once a year. Some veterinarians recommend vaccination every six months.

Lice

Lice multiply swiftly in cool or cold weather and can build up in large numbers on cattle.

Symptoms/effects: Calves that are itchy, rubbing out the hair over the neck and shoulders, may have lice.

Treatment: Delouse all cattle in late fall and again in late winter. Talk with your veterinarian about a good lice control program. Your veterinarian, county Extension agent, or local farm supply store can recommend a product for getting rid of lice. When treating a dairy cow, use only products approved for dairies. Wear protective clothing, goggles, and rubber gloves when you apply the product. Read the label thoroughly before use. If it is a powder, do not breathe the dust. Apply it on a calm day with no breeze. Easier to use than a powder is a pour-on product that needs to be applied only along the animal's back. The treatment will kill the lice but not the louse eggs. It must be repeated at least once to kill lice that have not yet hatched.

For the more organically minded, dusting your cattle with diatomaceous earth will help control lice.

Lumpy Jaw

Lumpy jaw can occur after a calf eats hay or grass with sharp seeds in it. Foxtail or downy brome (cheat grass) seed pods have sharp stickers that can get caught in the mouth and poke into the cheek tissue. A sticker that pokes in deeply may open the way for bacteria to cause infection, and the wound may become an abscess.

Symptoms/effects: An abscess in a calf's mouth may get as large as a tennis ball, causing the cheek to bulge prominently.

Treatment: The abscess must be punctured and drained. Abscesses sometimes break and drain on their own, but they heal faster and with less scar tissue if they are lanced and drained. A veterinarian or other experienced person should treat an abscess. Severe cases may require antibiotic therapy.

Navel Hernia/Navel Abscess

In some cases, the abdominal wall at the navel area does not close up properly after the birth of the calf, causing a bulge at the navel.

Symptoms/effects: In a hernia, soft tissue passes back and forth through the hole in the abdominal wall. If the swelling at the navel is firm, an abscess is the cause.

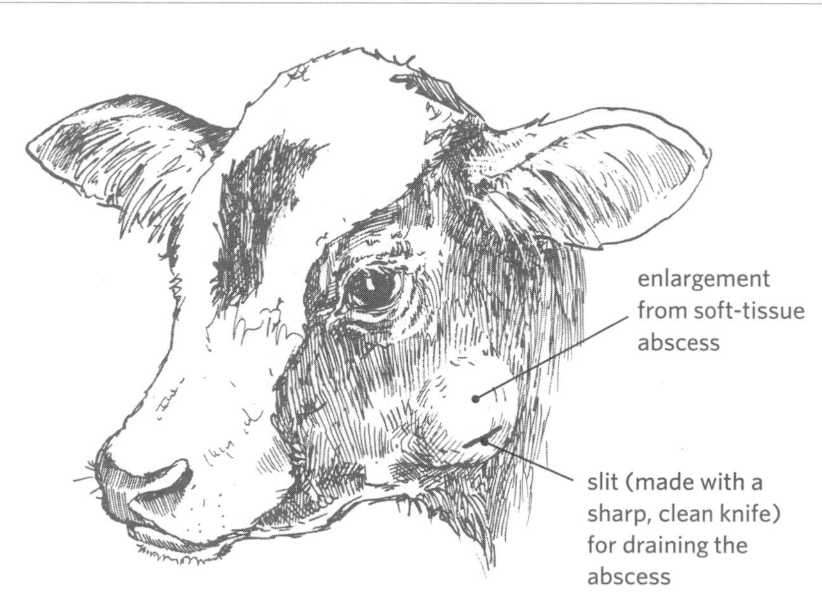

A cheek abscess may need to be lanced to return the cow to good health.

A small hernia may go away as the calf grows. However, a large hernia is serious, because a loop of intestine may come through it and strangulate, causing a portion of it to die and kill the calf. A bacterial abscess can spread through the body, causing severe illness and serious joint infections.

Treatment: If your calf has a swelling at the navel, have a veterinarian look at it. An abscess can be treated by lancing, draining, and flushing with antibiotics. If a hernia is present, the vet can tell you whether it will get better on its own or whether stitches are needed to close up the hole.

Pinkeye

Pinkeye usually appears in summer, because the bacteria that cause it are spread by face flies. Pinkeye often occurs when flies, dust, sunlight, or tall grass irritates eyes. Flies carry the bacteria from one animal to another. Pinkeye is contagious, and it may not be a problem if an animal lives alone, unless other cattle are within the distance face flies travel, which can be several miles.

Symptoms/effects: An animal with pinkeye will hold the eye shut, because it is sensitive to light. The eye will water. After a day or two, a white spot will appear on the cornea at the front of the eyeball. As the disease worsens, the spot grows larger and the cornea becomes cloudy and blue.

Treatment: A mild case of pinkeye may clear up without treatment, but a serious case can cause blindness. Pinkeye should be treated promptly with an antibiotic powder or spray, which your vet can recommend. Restrain the animal so you can squirt the medication right into its eye. Because the calf's tears will wash medications out of the eye, administer the treatment at least twice a day.

Early detection and diligent treatment should clear up pinkeye within a few days. If you can, bring the cow or calf inside, out of the bright sunlight, to make her more comfortable. A long-acting injection of the antibiotic oxytetracycline can also help clear up the infection.

Pneumonia

Pneumonia is the second most common killer of young calves. If you spot the warning signs early and start treatment quickly, pneumonia will be a lot easier to clear up.

Symptoms/effects: Pneumonia can be mild or swift and deadly. An animal coming down with pneumonia usually stops eating and lies around or stands humped up, looking depressed and dull. Its ears may droop. Respiration may be fast or labored and grunting. The calf may cough or may have a snotty or runny nose.

Causes: Pneumonia can be caused by viruses or bacteria. The germs that cause pneumonia are relatively plentiful in the environment. They make a calf sick only if its immunity is poor or its resistance is lowered by stress. Cattle of all ages can develop pneumonia, but young calves 2 weeks to 2 months of age are especially susceptible.

A newborn calf that doesn't nurse soon enough or get enough colostrum (see page 310) will not receive the antibodies needed for adequate immunity. A young calf that has been weakened by a bad case of scours may come down with pneumonia. Severe stress can also make cattle of any age susceptible to pneumonia. Stressful conditions include overcrowding, wet and cold weather, sudden changes in weather from one extreme to another, a long truck haul, or bad weather during weaning. Poor ventilation in sheds and barns can also lead to pneumonia.

Treatment: Confine the animal and take its temperature. If it's greater than 102.5°F (39.2°C), call your veterinarian. A temperature greater than 104°F (40°C) is serious. Antibiotics should be given immediately, even if the illness is caused by a virus. Antibiotics will fight secondary bacterial invaders, which are the killers. Your vet may leave medication with you, along with instructions for treatment. A sick calf will need good basic care, which includes shelter to keep it warm and dry, and sufficient fluids, especially if it isn't eating or nursing enough. Fever can cause dehydration.

For a serious case of pneumonia, try to reduce the pain and fever so the animal will feel better and start eating again. Dissolve two aspirin tablets in a little warm water and use a syringe to squirt the mixture into the back of the calf's mouth. For a larger animal, other medications can be used; consult your veterinarian.

Don't stop treatment too soon. You may think you can stop giving medication because the animal seems to be getting better and its fever is down. But if the symptoms return, the animal will be twice as hard to treat. Give antibiotics for at least two full days after symptoms have disappeared and the body temperature is normal again. A case of pneumonia usually requires at least one full week of care.

Vaccinate your cattle against the most common viral diseases, such as infectious bovine rhinotracheitis, bovine virus diarrhea, and parainfluenza type 3. These diseases often progress to pneumonia because the virus weakens the animal and allows bacteria to move into the lungs. Discuss a vaccination schedule with your veterinarian.

Ringworm

This fungal infection occurs most commonly in winter. It is spread from calf to calf directly or by calves rubbing against the same posts.

Don't Mix Medications with Milk

Do not mix electrolytes with milk, especially if your electrolyte mix contains sodium bicarbonate. Mixing medications with milk or milk replacer can prevent curd formation and can worsen diarrhea. Wait two or three hours after milk feeding before giving the fluid with electrolytes. Space the fluid treatments between the regular feedings.

Symptoms/effects: Ringworm causes hair to fall out in 1- to 2-inch- (2.5 to 5 cm) wide circles. The exposed skin is crusty or scaly.

Treatment: Ringworm generally clears up on its own in spring and may not need to be treated. Ringworm is contagious to other cattle and to people; use gloves or wash thoroughly after handling affected animals.

Scours

Scours, or diarrhea, is the most common disease of young calves and causes the most deaths. Scours may not be a problem in a big weaned beef calf, but it could be life threatening to a young dairy calf. Viral scours tends to affect calves during the first 2 weeks of life, whereas bacterial and parasitic scours can strike from birth up to about 3 months or even up to 1 year. The youngest calves are usually most susceptible and most adversely affected.

Symptoms/effects: The manure is runny and watery. Severe infections, such as those caused by salmonella bacteria, may cause rapid death, sometimes even before scouring is seen.

Prevention: The key to preventing scours is good management — a healthy cow, an uncontaminated area for calving, clean bedding for the cow, and adequate colostrum for the calf soon after it is born. Vaccinating cows before calving can prevent some but not all types of scours. Following are further preventive measures:

- Protection from weather. A newborn calf needs protection from bad weather.
- Keep it clean. Wash the teats of any cow you have to assist in calving, clean all your equipment between calves (especially bottles and nipples, and your esophageal feeder tube), and move the calf to a clean pasture as soon as it is up and about.
- Isolate a sick calf. Never put a sick calf into the same barn stall or pen in which a cow will calve. Shelter a sick calf in a different shed, and keep a pregnant and calving cow separate from a cow that has already calved.
- Provide timely colostrum. Make sure every calf gets colostrum as soon as possible after birth, no more than two hours later. Don't thaw frozen colostrum in a microwave or get it too hot; excessive heat destroys the antibodies. Put the container in hot water to let it thaw. It should feel comfortably warm but not hot on your skin before you feed it to a calf.

Causes: Calf scours is a complex problem. Diarrhea can be caused by many things, including bacteria, viruses, and protozoan parasites. Often more than one agent is involved. Overfeeding a calf or using a poor-quality milk replacer can also lead to scours. Poor nutrition and a dirty, wet environment can make calves more susceptible to infections that cause scours. Get help from your veterinarian to determine the cause of diarrhea and how to treat it.

Homemade Electrolyte Mix

To make one dose of electrolyte mix for a scouring calf, mix ½ teaspoon (2.5 mL) of regular table salt (sodium chloride) and ¼ teaspoon (1.25 mL) of Lite salt (sodium chloride and potassium chloride) into 1 quart (1 L) of warm water for a small calf or 2 quarts (2 L) for a large one.

If your calf is critically ill, add ½ teaspoon (2.5 mL) of baking soda (sodium bicarbonate). If the calf is weak, add 1 tablespoon (15 mL) of powdered sugar to the mix. Add a liquid antibiotic, such as neomycin sulfate solution (sometimes called Biosul — you can purchase it from your vet). Use the proper amount for the size of the calf by following the directions on the label.

To help soothe the gut and slow down the diarrhea, you can add 2 to 4 ounces (¼ to ½ cup [59 to 118 mL]) of Kaopectate to this fluid mix. Or, instead of Kaopectate, you can use a human adult dose of Pepto-Bismol.

Treatment: Treatment includes giving the animal electrolyte fluids and administering antibiotics in liquid or pill form. (For information on homemade electrolytes, see above).

To treat a calf at the first hint of sickness, watch it closely. Many clues besides a messy hind end can help you spot trouble before a calf becomes critically ill. Often a calf will act a little dull before it shows diarrhea. It may quit nursing or lie down off by itself. Feeding time is a good time to check on a calf and observe it for signs of illness. Also check the mother's udder. If a cow has a full udder or is only partly nursed out, take a closer look at her calf. The first sign that a calf doesn't feel well may be that it stops nursing.

If you catch scours early enough, you may be able to halt the infection before the calf needs fluids and electrolytes. Neomycin sulfate solution can be put into a syringe, to squirt into the back of the calf's mouth. Use 1 mL per 40 pounds, which means about 2 to 3 mL for a young calf.

A calf with diarrhea must be given fluids. In the early stages, while the calf is still strong, you can give electrolyte fluids with a nursing bottle if the calf will drink it. Oral fluids (by mouth or into the stomach) are adequate and effective because the calf's gut can still absorb them. However, if a sick calf refuses to drink, fluids will have to be administered by stomach tube or esophageal feeder. A dehydrated calf should be given the fluid-electrolyte mix every six to eight hours, or until the calf feels well enough to nurse from a bottle or nipple bucket again.

As disease progresses, more gut lining is destroyed. The calf becomes weaker, unable to absorb fluid from the digestive tract, and more dehydrated. A calf that becomes this ill requires intravenous fluids given by a veterinarian.

For treating bacterial diarrhea, antibiotic pills are not as effective as a liquid antibiotic. Liquid antibiotic can be squirted into the back of the calf's mouth or added to a fluid-and-electrolyte mix.

If the antibiotic recommended by your vet is available only in pill form, crush the pills or dissolve them and give them in a liquid. Add enough water so the crushed pills can be added to your fluid mix or squirted into the back of the mouth with a syringe. The liquid antibiotic is needed only once a day, but make sure the calf is nursing or getting fluid three or four times a day to avoid dehydration.

If you give antibiotic to a calf for more than three days, it may kill off the good bacteria as well as the bad. You may then have to give the calf a pill or paste containing the proper rumen bacteria (obtained from your vet) after it recovers, to restore proper digestive function.

Milk need not be withheld from a scouring calf. Rather, the calf should be encouraged to nurse at its regular feedings to keep up its fluid and energy levels. If the calf is too sick to nurse or won't nurse enough, give it regular feedings by an esophageal tube. Administer medicated fluids and electrolytes between feedings.

It is preferable for a calf to nurse instead of taking milk through an esophageal tube. The act of nursing activates a reflex that causes milk to go directly to the abomasum, bypassing the rumen. When milk is given by tubing, this reflex does not occur, and milk enters the rumen, which could cause irritation (rumenitis). Electrolytes and water, however, are not a problem.

Selenium Deficiency

Selenium deficiency is also called white muscle disease and nutritional myopathy. All cattle need adequate selenium in their diets. Many areas of the world are selenium deficient. Some areas, on the other hand, have an excess of selenium in the soils, and forages and grains produced in those areas may contain toxic levels of selenium.

Symptoms/effects: A newborn calf that is selenium deficient may be too weak to stand or to nurse. An older calf may die suddenly from heart failure, especially after exercise or handling. Heifers and cows may abort.

Prevention: When needed, provide selenium in trace mineral supplements. Selenium salt blocks are not adequate in severely deficient areas. Give a selenium injection to pregnant animals a few months before calving. Newborn calves may benefit from a selenium injection. Talk with your veterinarian or Extension agent about the need for selenium supplementation in your area.

Treatment: Weak calves may respond to selenium injection, but often the damage is severe and they still die.

Observing Withdrawal Periods

Whenever you administer any drug to your cow or calf, determine the drug's appropriate withdrawal period — the amount of time that must pass while the drug metabolizes from the animal's system. A withdrawal period has been established for each drug to ensure that any remaining drug residue has been reduced to an acceptable level, at or below which the animal's milk or meat is considered safe for humans to eat.

Warts

Warts are skin growths caused by a virus. They affect calves and yearlings more than adult cattle, since mature cattle have usually developed some resistance to the virus. Warts are unsightly but clear up on their own after a few months.

Giving Injections

Many medications and most vaccines are given to cattle by injection with a syringe and needle. A calf or cow should always be restrained before you give it an injection or oral medication. A cow should be restrained in a chute. A calf can be confined in a chute or tied up and pushed against a fence; if it is merely tied to the fence, the calf may kick you. A tied calf can't swing its head away or hit you with its head while trying to avoid an oral medicine. Never stand behind or beside a calf unless it is restrained so it cannot move.

Safety Precaution

After giving an injection, discard the syringe and needle if they are disposable. If they are reusable, boil them before the next use.

Injections are given in one of these three ways:

Intramuscularly: deep into a big muscle; most injections are given this way

Subcutaneously: under the skin, between the skin and the muscle; used to slow the absorption of vaccines and some medications

Intravenously: directly into a large vein; used to speed up a medication's absorption

Intravenous injections can be dangerous to the animal and should be given only by a veterinarian. But you can learn to give intramuscular or subcutaneous injections. Have an experienced person show you how to fill a syringe, measure a proper dose, and give an injection.

Intramuscular Injections

If you're raising cattle for beef, give injections into the neck muscle to avoid damage to the best cuts of meat, which are in the rump and buttocks. Sometimes an injection causes a local reaction and a knot in the muscle, or even a small abscess. It's better to have this damage occur in the neck muscle, where it can be trimmed out more easily during butchering.

The injection should be given in the thickest muscle of the neck. Make sure the area where you will put the needle is clean, free of mud or manure, or the

A balling gun is a long-handled tool that lets you give your calf or cow a pill without having your fingers bitten.

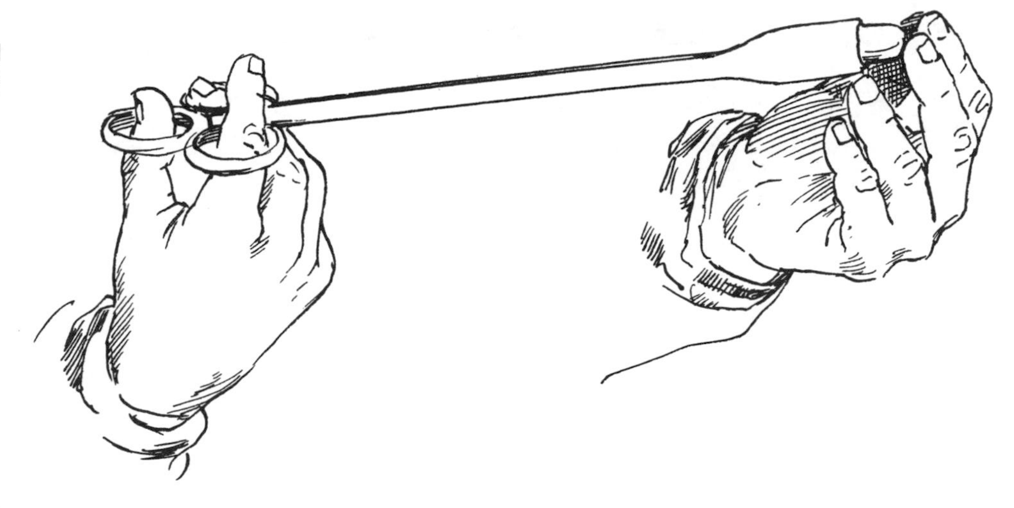

needle will take bacteria into the muscle with it. Wet skin and hair increase the risk that bacteria will enter the muscle with the needle.

Detach the needle from the syringe. Before putting in the needle, press the area firmly for a moment to desensitize the skin so the animal will not jump as much when jabbed. Insert the needle with a forceful thrust so it breaks the skin easily.

A new, sharp needle goes in with minimal effort and causes minimal pain for the calf. If the calf jumps, wait until it settles down again before you attach the syringe and give the injection. If the needle starts to ooze blood, you've hit a vein. Take the needle out and try again in a different spot; never give an injection into a vein.

Subcutaneous Injection

To give a subcutaneous injection, lift a fold of skin on the shoulder or neck, where the skin is loosest, and slip the needle in. Aim it sideways so it goes under the skin you have pulled up but not into the muscle. With a little practice, you'll find that subcutaneous injections are easy to give.

Giving Oral Medications

Oral medications are given by mouth and are absorbed through the lining of the stomach or intestines. They come in one of two forms.

- Pills can be given with a balling gun, which is a long-handled tool that holds a pill as you insert it toward the back of the animal's mouth. When you press the plunger, the balling gun pushes the pill out of its slot, forcing the animal to swallow it. This handy tool keeps your fingers from being bitten.
- Liquids might be liquid medications, liquid antibiotics, or pills dissolved in water. They are easy to administer using a big syringe without the needle. Special syringes known as dose syringes have a long nozzle end designed for giving liquids by mouth. Fill the syringe with the proper dose, stick the syringe into the corner of the animal's mouth, and slowly squirt the medication into the back of the mouth. If the dose is large, give the medication a little at a time, allowing the animal to swallow each portion before you squirt in the rest. Keep the animal's head tipped up so that the medication cannot run back out.

Don't Get Bitten!

Be careful when examining the inside of an animal's mouth or giving pills. A calf has no top teeth in front but can still hurt your fingers with the molars if it bites down. Position your fingers in the space between the lower molars and lower incisors (front teeth).

Glossary

Abdomen. On a honey bee, the third section of the body, containing the intestine (ventriculus), reproductive and execretory systems, and sting apparatus.

abomasum. Fourth stomach or true stomach of the ruminant animal, in which enzymatic digestion occurs.

abscess. Boil; localized collection of pus.

acidosis. Severe digestive upset from change in rumen bacteria.

acute infection. An infection or disease that has rapid onset and pronounced signs and symptoms.

African honey bee. The highly aggressive southern African honey bee *Apis mellifera scutellata*, imported by Brazil in 1956.

Africanized honey bee. A hybrid cross between an African honey bee and a European honey bee, resulting in a bee that is more aggressive than the European bee.

afterbirth. Placental tissue attached to the uterus during gestation and expelled after the birth.

air cell. Air space usually found in the large end of an egg.

albumen. The white of the egg.

anemia. A deficiency in the oxygen-carrying capacity of blood, possibly caused by loss of blood or by certain disease conditions.

anestrus. The nonbreeding season; the state (for females) of being not in heat.

artificial insemination (AI). The process in which a technician puts semen from a male animal into the uterus of a female animal to create pregnancy.

aviary netting. Fencing woven in a honeycomb pattern with ½-inch (1.25 cm) openings.

Balling gun. A device used to administer a bolus (a large pill).

Bang's disease. *See* brucellosis.

bantam. A diminutive chicken about one-fourth the size of a regular chicken, some breeds of which are distinct while others are miniatures of large breeds.

barrow. A castrated male pig. He is the meat animal that is the basis of the pork industry, although gilts do produce leaner carcasses.

beak. The upper and lower mandibles of chickens and turkeys.

bean. A hard protuberance on the upper mandible of waterfowl.

beard. Feathers bunched under the beaks of some chicken breeds, such as the Antwerp Belgian, the Faverolle, and the Houdan; also coarse hairlike bristles growing from a turkey's chest; also a clump of long hairs growing under a goat's chin.

bedding. Straw, wood shavings, shredded paper, or any other material used to cover the floor of an animal pen to absorb moisture and manure. Also called *litter*.

bee space. The ¼- to ⅜-inch (0.7–1 cm) space conserved by the honey bee to provide passageway through the comb, and the basis for development of the moveable-frame hive on which beekeeping is based.

bevy. A flock of ducks.

bill. The upper and lower mandibles of waterfowl.

bleaching. The fading of color from the beak, shanks, and vent of a yellow-skinned laying hen.

bloat. An excessive accumulation of gas in a ruminant's rumen and reticulum, resulting in distension.

blood spot. Blood in an egg caused by a rupture of small blood vessels, usually at the time of ovulation.

bloom. The moist, protective coating on a freshly laid egg that dries so fast you rarely see it; also, peak condition in exhibition poultry.

blowout. Vent damage caused by laying an oversize egg.

boar. An intact male hog kept for breeding.

body capacity. The internal dimensions of an animal's body.

bolus. A large pill for animals; also, regurgitated food that has been chewed (cud) by a ruminant.

bovine. Pertaining to or derived from cattle.

bovine virus diarrhea (BVD). A viral disease that can cause abortion, diseased calves, or suppression of the immune system.

break up. To discourage a female bird from being broody.

breech. The buttocks; a birth in which the fetus is presented rear first.

breed. A group of animals with the same ancestry and characteristics.

breeder ration. A nutritious feed used to boost the reproductive ability of breeding animals.

brisket. The front of a cow above the legs.

broiler. A young chicken grown for its tender meat. Also called a *fryer*.

broken. A color pattern in which blotches of color appear on a white background.

broken mouth. Having lost teeth.

brood. In poultry: to set on a nest of eggs until they hatch. Also, the resulting hatchlings, collectively. In bees: collectively, the immature stages of the honey bee, including eggs, larvae, and pupae.

brood chamber. A section of the hive used for brood rearing of honey bees.

brooder. A mechanical device used to imitate the warmth and protection a mother bird gives her chicks.

broody hen. A setting hen.

browse. Bushy or woody plants; to eat such plants.

brucellosis. A bacterial disease that causes abortion.

buck. A mature male goat or rabbit.

buck rag. A cloth rubbed onto a male goat and imbued with his odor and kept in a closed container until it is exposed to a doe to observe her reaction to help determine if she's in heat.

buckling. A young male goat or rabbit.

bull. An uncastrated male bovine of any age.

bummer. A lamb that has to be bottle-fed because it is either an orphan or a lamb whose mother doesn't produce enough milk for multiple lambs.

bunny. A affectionate term for rabbit. Also a baby rabbit; also called a *kit*.

burdizzo. A castrating device that crushes the spermatic cords to render a male animal sterile.

butcher hog. A hog being readied for or sold on the slaughter market. Also called a *market hog*.

Calf. A young bovine of either sex, less than a year old.

California Mastitis Test (CMT). A do-it-yourself kit to determine if a female milking animal has mastitis.

calve. To give birth to a calf.

candle. To determine the interior quality of an egg by shining a light through it.

cannibalism. The bad habit some chickens and turkeys have of eating one another's flesh or feathers.

capped brood. The portion of the honey bee brood that has been covered with wax and houses the pupae.

caprine. Pertaining to or derived from a goat.

caprine arthritis encephalitis (CAE). A serious and widespread type of viral arthritis.

carrier. An animal that carries a disease but doesn't show signs of it.

caruncle. A fleshy nonfeathered outgrowth around the eyes and head, and sometimes neck, of certain species of fowl.

castrate. To remove the testicles of a male animal to make him permanently incapable of breeding.

cervix. The opening (usually sealed) between the uterus and the vagina, which widens to allow an animal to give birth.

chalazae. White, twisted, ropelike structures that anchor the egg yolk in the center of the egg by their attachment to the layers of thick albumen.

chevon. Goat meat.

chicken wire. Fencing woven in a honeycomb pattern with 1-inch (2.5 cm) openings.

clean legged. Description of a chicken having no feathers growing down the shanks.

cloaca. The cavity just inside a fowl's vent, into which the intestinal and genitourinary tracts empty.

closed face. In sheep, having heavy wool about the eyes and cheeks.

Closed face

clostridial diseases. Diseases caused by *Clostridia* bacteria (including tetanus and enterotoxemia) that produce powerful toxins causing sudden illness.

clutch. A batch of eggs that are hatched together, either in a nest or in an incubator.

coccidiosis. An intestinal disease, caused by protozoa, that usually causes diarrhea.

coccidiostat. A drug used to prevent coccidiosis.

cock. A male chicken; also called a *rooster*.

cockerel. A male bird under 12 months of age.

colic. An abdominal condition of mammals generally characterized by severe pain.

colostrum. The first milk from a female animal that has just given birth, which contains antibodies that give the newborn animal temporary protection against certain diseases.

comb. The fleshy prominence on the top of the head of fowl. Also, to remove short fibers from wool and leave long fibers laid out straight and parallel. Also, collectively the wax cells of a honey bee's nest, which are constructed back to back into a solid slab and usually surrounded by a wooden frame.

concentrate. Feed — consisting of grains and oil meals — that is low in fiber and high in food value.

condition. Degree of health.

conformation. An animal's overall shape and other physical attributes.

congenital. A birth defect.

coop. The house or cage in which poultry live.

cover. The fat a meat animal lays on beneath the skin as it approaches market weight.

cow. A bovine female that has had one or more calves.

creep feeder. An enclosed feeder for supplementing the ration of young animals that excludes larger animals.

Creep feeder

creep-feeding. Providing extra feed (such as grain) to young animals that are still nursing their mothers.

crest. The elongated feathers on the head of some breeds of duck and chicken.

crop. An enlargement of the gullet of fowl where food is stored and prepared for digestion.

crossbred. The offspring of two different breeds.

cross-fencing. A fence used to subdivide a pasture into smaller paddocks.

crotch. To trim wool or hair from around a sheep's tail and udder.

cryptosporidiosis. Diarrhea in young animals caused by protozoa; may also cause diarrhea in humans.

cud. In ruminant animals, a wad of food burped up from the rumen to be rechewed.

cull. To remove a substandard animal from the herd.

cycling. Heat cycles in a nonpregnant female.

Dam. The female parent.

dehorn. To remove the horns of an animal that was not disbudded soon after birth.

dewclaw. A horny structure on the lower leg above the hoof.

dewlap. Loose skin under the neck.

diphtheria. A bacterial disease in the mouth or the throat.

disbud. To remove the horn buds from a young animal to prevent horn growth.

disbudding iron. A tool, usually electric, that is heated to burn the horn buds from young animals.

dished face. Having a concave nose, such as that of the Saanen goat.

dock. To cut off the tail; also the remaining portion of the tail that has been docked.

doe. A female goat or rabbit.

doeling. A young female goat or rabbit.

down. The furlike covering of newly hatched poultry. Also the fluffy bottom part of a chicken or turkey feather. Also, the inner layer of soft, light feathers on waterfowl.

drake. A male duck.

drake feather. One of three curly feathers on a drake's tail.

drakelet. A young male duck.

drench. To give liquid medication.

dress. To clean meat in preparation for cooking.

drove. A herd or group of pigs.

dry. Not producing milk.

dry period. The time when a female animal is not producing milk.

drylot. A large lot used to maintain animals that is devoid of vegetation due to foot traffic and numbers present.

duck. Any member of the family Anatidae and specifically a female.

ducklet. A young female duck.

duckling. A baby duck.

Ear canker. A scabby condition inside a rabbit's ear caused by mites.

edema. Swelling due to excess accumulation of fluid in tissue spaces.

egg tooth. A small, horny protuberance attached to upper mandible of a hatching bird's break or bill that it uses to help break open the shell, then falls off several days after hatching.

elastrator. The tool used to apply elastrator rings.

elastrator rings. Castrating rings, resembling rubber bands, applied to the scrotum so it will atrophy and fall off.

electrolytes. Important body salts, including sodium, potassium, calcium, and magnesium, that must be replaced as a result of dehydration.

endotoxic shock. Shock caused by body systems shutting down in reaction to bacterial poisons.

endotoxin. The poison created when bacteria multiply in the body.

enteritis. Inflammation of the intestine.

enterotoxemia. A bacterial gut infection caused by *Clostridium perfringens*, usually resulting in death; also called pulpy kidney disease and overeating disease.

Escherichia coli. A type of bacterium that has more than 100 strains, some of which cause serious infection.

esophageal feeder. A tube put down an animal's throat to force-feed fluids from a feeder bag.

estrous cycle. The time and physiological events that take place in one heat period.

estrus. The period when a female animal is in heat and will willingly mate with a male animal.

ewe. A mature female sheep.

Face. To trim wool from around the face of closed-face sheep.

feather legged. Having feathers growing down the shanks.

feather out. To grow a full set of plumage.

feed additive. Anything added to a ration, including preservatives, growth promotants, and medications.

feeder pig. A young pig — most often between 40 and 70 pounds (18.1 and 31.8 kg) — produced by one farmer and sold to another for feeding out to market weight.

fetus. An animal in the uterus or within an eggshell prior to its birth.

finish. To mature and fatten enough to butcher (to reach butchering condition).

flight feathers. The primary feathers on the wing of fowl; sometimes used to denote the primaries and secondaries.

flight zone. The proximity you can get to an animal before it flees.

flock. A group of chickens, turkeys, or sheep.

flush. To feed a female more generously than usual two to three weeks before breeding in order to stimulate the onset of heat and improve the chances of conception.

foot rot. An infection in the foot of a hooved animal, causing severe lameness.

forage. The hay and/or grassy portion of an animal's diet.

forced-air incubator. A mechanical device for hatching fertile eggs that has a fan to circulate warm air.

foundation. Thin sheets of beeswax (sometimes plastic) embossed with hexagonal cells, and the template used by honey bees to construct the cells that becomes the comb.

founder. Inflammation of the hooves.

fowl. A term applied collectively to chickens, ducks, geese, and the like, or the market class designation for old laying birds.

frame size. The measure of hip height, used to determine skeletal size.

free choice. Feed that is available to be eaten at all times.

free range. Poultry allowed to range and forage at will.

freshen. To give birth and begin to produce milk.

frizzle. Having feathers that curl rather than lie flat.

fryer. *See* broiler.

full feeding. Allowing an animal to consume all the feed it desires daily; also called *free choice*.

Gaggle. A flock of geese.

gander. A male goose.

germinal disc. In an egg, the fertility spot from which an embryo grows.

gestation. The time between breeding and birth.

gilt. A female hog that has yet to bear young.

gizzard. The muscular stomach of fowl that contains grit for grinding food.

goose. The female goose, as distinguished from the gander.

gosling. A young goose of either sex.

grade. Unregistered; not purebred.

graft. To have an adult female accept and mother a young animal that isn't her own.

green. Young waterfowl that have gone into the first molt.

grit. The hard, insoluble materials eaten by fowl and used by the gizzard to grind up food.

grow-out period. The time it takes for an animal to grow from weaning to harvest.

gut. The digestive tract.

Hackle. The long feathers on a rooster's neck and saddle.

halter. A rope or leather headgear used to control or lead an animal.

hardware disease. Peritonitis (infection in the abdomen) caused by a sharp foreign object penetrating the gut wall.

hatch. To come out of the egg; also, a group of birds that come out of their shells at roughly the same time.

hatchability. The percentage of fertilized eggs that hatch under incubation.

hatchling. A bird that has just hatched.

Hatchlings

hay. Dried forage.

heat. *See* estrus.

heifer. A young female bovine that has not calved.

hemoglobin. The compound in red blood cells that carries oxygen.

hen. A female chicken more than 12 months of age.

hen feathered. In a cock, having round feathers on the hackle and saddle.

herd. A group of goats or cattle.

hock. The large joint halfway up the hind leg.

hog. A swine weighing more than 120 pounds (54.4 kg).

hog forty. A 40 percent crude-protein supplement used widely to formulate swine rations.

hoof rot. *See* foot rot.

hopper. A food container that is filled from the top and dispenses from the bottom and is used for free-choice feeding of grain, grit, and other supplements.

horn bud. A small bump from which a horn grows.

hurdle. A short, solid gate used when handling and herding hogs and sheep.

hutch. A rabbit cage.

Incubate. To keep eggs warm and properly humid until they hatch.

incubation period. The number of days required for eggs to hatch once they are warmed to incubation temperature.

incubator. A mechanical device for hatching fertile eggs.

infectious bovine rhinotracheitis (IBR). A respiratory disease caused by a virus; also called red nose.

international unit (IU). A standard unit of potency of a biologic agent such as a vitamin or antibiotic.

intramuscular (IM). Into a muscle.

intravenous (IV). Into a vein.

iodine. A chemical used for disinfecting.

J-clip. A J-shaped metal clip used in hutch construction. Special pliers are required for application.

Johne's disease. A wasting, often fatal form of enteritis.

jug. A small pen large enough for just one ewe and her offspring.

junior. *See* green.

Ked. An external parasite that affects sheep.

keel. The breastbone or sternum of fowl.

kemp. Straight, brittle, chalky white mohair fiber.

ketosis. An overaccumulation of ketones in the body as a result of abnormal fat metabolism.

kid. A goat under 1 year of age; also, in goats, to give birth.

kindle. To give birth to a litter of rabbits.

kit. A baby rabbit.

knob. A rounded protuberance appearing at the base of the bill (between the eyes) of some species of goose.

Lactation. The period in which an animal is producing milk; the secretion or formation of milk.

lamb. A newborn or immature sheep, typically under 1 year of age.

lay ration. Feed that is formulated to stimulate egg production.

legume. A plant belonging to the pea family (such as alfalfa and clover).

leptospirosis. A bacterial disease that can cause abortion.

letdown. A physiological process that allows milk to be removed from the udder.

lice. Tiny external parasites on the skin that cause discomfort by either biting or sucking.

limit feeding. Restricting of the amount of energy feedstuffs an animal receives daily, often done to keep broilers from growing too fast or to keep brood sows from becoming excessively fat and having farrowing difficulties.

listeriosis. A bacterial disease that can cause abortion.

litter. Collectively, the offspring of a rabbit from a single birth. *See also* bedding.

liver flukes. Parasites that infest snails and spend part of their life cycle in cattle, damaging the liver and making the host more susceptible to red water.

lumpy jaw. An abscess in the mouth caused by infection.

Maintenance ration. A feed used for adult animals that are not in production.

malocclusion. An abnormal coming together of teeth.

mammary tissue. Milk-producing tissue in the udder.

mandible. The upper or lower bony portion of a bird's beak or bill.

mange. A skin disease caused by mites that feed on the skin.

marbled. Having flecks of fat interspersed in muscle.

mash. A mixture of finely ground grains.

mastitis. Infection and inflammation in the udder.

mature. Old enough to reproduce.

meconium. The dark, sticky first bowel movement of a newborn animal.

milking bench (or stand). A raised platform, usually with a seat for the milker and a stanchion for the goat's neck, that a goat stands upon to be milked.

mites. Tiny parasites that feed on skin, causing mange or scabies.

molt. To shed old feathers, fur, or hair and grow a new coat.

mount. To rear up over the back of an animal, as a bull does a cow when breeding.

mutton. Meat from a mature or aged sheep over 1 year old.

Nectar. Sweet sugar solution secreted by flowers that is the raw material for honey.

needle teeth. Also called wolf teeth, these are the two large teeth on each side of the upper nad lower jaw that are present at birth. If not clipped off, they can cut the sow's udder or other pigs, and infection will result.

nematode. Roundworm.

nest box. A place for fowl to lay eggs or rabbits to give birth.

Nolvasan. An all-purpose disinfectant.

nucleus hive (nuc). A starter unit consisting of a nucleus colony (three to five frames of bees), usually installed into a standard ten-frame hive, that becomes the foundation for a new full-size colony.

Off feed. Not eating as much as normal.

omasum. One of the four stomach compartments in the ruminant animal.

open. Not pregnant.

open brood. Brood of honey bees that has yet to be capped with wax, containing eggs and larvae.

open face. In sheep, not having much wool around the eyes and cheeks.

Open face

overconditioned. Overfed or fat.

oviduct. The long, glandular tube of female fowl, leading from the ovary to the cloaca, in which egg formation takes place.

Package bees. A screened container holding a specific number of adult honey bees, a caged queen, and a can of sugar syrup sold to start new colonies.

parainfluenza (P13). A viral respiratory agent that by itself causes a mild disease but in combination with bacterial infection can be severe.

parasite. An organism that lives in or on an animal.

parturition. The birth process.

pastern. The area between the hoof and the fetlock joint, corresponding to the human ankle.

pasting. Loose droppings sticking to the anal area.

pathogen. A harmful invasive microorganism, such as a bacterium or virus.

peck order. The heirarchy of status or social ranking.

pedigree. The list of an animal's forebears.

peritonitis. An infection in the abdominal cavity.

pH. The measure of acidity or alkalinity; on a scale of 1 to 14, 7 is neutral, 1 is most acid, and 14 is most alkaline.

pheromone. A chemical substance secreted by organisms that elicits specific behavior in others of its species, such as queen pheromones in honey bees that keep worker bees from laying eggs.

pig. A young swine.

pigeon-toed. Having toes turning inward instead of pointing straight ahead.

pinfeathers. New feathers just emerging from the skin.

pinion. The tip of a wing. Also, to cut off the tip of a wing to prevent flight.

pinkeye. A contagious eye infection spread by face flies.

pip. The hole a chick makes in its egg's shell when it is ready to hatch; also, the act of making the hole.

placenta. The afterbirth, which is attached to the uterus during pregnancy as a buffer and lifeline for the developing animal.

plumage. All of a bird's feathers, collectively.

poll. The top of the head.

polled. Born without horns; naturally hornless.

pollen. Dustlike grains produced in the anthers of flowers as the male element of their reproductive system, collected and stored by honey bees as the protein part of their diet.

post legged. Having hind legs that are too straight, with not enough angle in the hocks and stifles.

poult. A young turkey of either gender.

preen. To clean and organize feathers with the beak or bill.

primary feather. One of the long feathers at the end of a wing.

prolapse. Protrusion of an inverted organ such as rectum, vagina, or uterus.

propolis. A resin collected by honey bees from certain trees and used for filling cracks and crevices in the comb, as a comb reinforcement, and to cover and isolate objectionable materials bees cannot physically remove from the hive.

protozoa. One-celled animals, some of which cause disease.

puberty. The age at which an animal matures sexually and can reproduce.

pullet. A female chicken less than 1 year of age.

purebred. An animal whose ancestry can be traced back to the establishment of a breed through the records of a registry association.

Quarantine. To keep an animal isolated from other animals to prevent the spread of infections.

quarter. One of a cow's four teats.

quill. A primary feather.

Ram. A mature male sheep.

ram lamb. An immature male sheep.

ration. The combination of all feed consumed in a day.

raw milk. Milk as it comes from an animal; unpasteurized milk.

red water. A deadly bacterial disease of cattle caused by *Clostridium haemolyticum*.

registered. Description of an animal with birth and ancestry recorded by a registry association.

relative humidity. The percentage of moisture saturation in the air.

rennet. An enzyme used to curdle milk and make cheese.

reticulum. The second of the four stomach compartments in the ruminant animal.

ringworm. A fungal infection causing scaly patches of skin.

roaster. A young chicken of either sex, usually 3 to 5 months of age, that has tender meat, soft, pliable, smooth-textured skin, and a breastbone cartilage somewhat less flexible than that of the broiler-fryer.

roost. A perch on which fowl rest or sleep. Also, to rest on a roost.

rooster. A male chicken; also called a *cock*.

rotational grazing. The use of various pastures in sequence to give each one a chance to regrow before grazing it again.

roughage. Feed that is high in fiber and low in energy, such as hay or pasture.

royal jelly. A nutritious food produced by worker bees and fed exclusively to larvae that will develop into queens. Also called bee milk.

rumen. The largest stomach compartment in the ruminant animal, in which roughage is digested with the aid of microorganisms in a fermentation process.

ruminant. An animal that chews its cud and has four stomach compartments.

Saltpeter. Sodium nitrate; a curing product.

Satin. A breed of rabbit with transparent hair shafts that create an extremely lustrous coat.

scabies. A skin disease caused by a certain type of mite.

scales. The small, hard, overlapping plates covering the shanks and toes of fowl.

scours. Persistent diarrhea in a young animal.

scrapie. A usually fatal disease of the nervous system of sheep and goats.

scratch. The habit chickens have of scraping their claws against the ground to dig up tasty things to eat; also, any grain fed to chickens.

scurs. Horny tissue or rudimentary horns attached to the skin rather than to the skull.

selenium. A mineral needed in small amounts in the diet (too much is poisonous).

self-feeding. Allowing an animal to meet its own nutritional needs from a feeder that stores large amounts of feedstuffs so the animal has unlimited access to them. Also called free-choice feeding or full feed.

set. To keep eggs warm so they will hatch.

settle. To become pregnant.

sex. To sort by gender.

shank. The part of a fowl's leg between the claw and the first joint.

shoat. A hog from the time of weaning until it reaches 120 pounds (54.4 kg).

sickle hocked. Having too much angle in the hind legs (weak construction).

silage. Feed cut and stored green and preserved by fermentation.

sire. The male parent; also to father.

snood. The fleshy appendage on the head of a turkey.

snuffles. A highly contagious respiratory disease of rabbits marked by nasal discharge.

sore hocks. Ulcerated footpads in rabbits.

sow. A female hog that has borne young.

splayfooted. Having toes that turn out.

spurs. The sharp points on the shanks of roosters and tom turkeys.

stag. A late-castrated steer or hog, or improperly castrated steer that still shows masculine characteristics; also a mature rooster.

stanchion. A framework consisting of vertical bars used to secure an animal in a stall or at a feed trough.

standard. The description of an ideal specimen for its breed.

standing heat. The time during heat when the female animal allows the male animal to mount and breed.

started. Having survived the first few critical days or weeks of life and begun to develop.

starter. A ration formulated for newly hatched fowl. Also a second-stage feed for young pigs, given from the age of 14 days until they reach 40 to 50 pounds (18.1 to 22.7 kg) in weight.

steer. A male bovine after castration.

stifle. The large joint high on an animal's hind leg by the flank, equivalent to the human knee.

straight run. Description of newly hatched fowl that have not been sorted by gender. Also called unsexed or as hatched.

straw. Dried plant matter (usually oat, wheat, or barley leaves and stems) used as bedding; also, the glass tube semen is stored in for artificial insemination.

strip. To remove the last of the milk from an udder.

subcutaneous (SQ). Under the skin.

supplement. A feed additive that supplies something missing in the diet, such as additional protein, vitamins, minerals. Also to feed such an additive.

swarm. A portion of a honey bee colony, including the queen, that leaves its home to establish a new one elsewhere. Also the act of such a group leaving the colony.

swine. A generic term roughly equal to (but a little fancier than) hog.

Tag. A lock of wool contaminated by dung and dirt. Also to cut away such dung locks.

tallow. The extracted fat from sheep and cattle.

tattoo. The permanent identification of animals produced by placing indelible ink under the skin; to apply a tattoo.

tom. A male turkey.

top-dress. To place an additive or treatment on top of an animal's regular ration for consumption at the same time.

topline. The top of an animal running along the back.

toxemia. A condition in which bacterial toxins invade the bloodstream and poison the body.

trace minerals. Minerals needed in the diet in tiny amounts.

trichomoniasis. A disease caused by protozoa.

trio. Two females and one male of the same breed and variety.

tuft. A puff of feathers on top of a goose's head.

Udder. Mammary glands and teats.

udder wash. A dilute chemical solution, usually an iodine compound, for washing udders before milking.

Vaccine. A fluid containing killed or modified live germs, injected into the body to stimulate production of antibodies and immunity.

vent. The external opening from the cloaca of fowl, through which it emits eggs and droppings.

vermifuge. Any chemical substance administered to an animal to kill internal parasitic worms.

vibriosis. A venereal disease of cattle that causes early abortion.

vulva. The external opening of the vagina.

Warts. Skin growths caused by a virus.

wattle. A small, fleshy appendage that dangles under the chin of some fowl species.

wean. To separate a young animal from its mother or stop feeding it milk.

wether. A castrated male sheep or goat.

whey. The liquid remaining when curd is removed from curdled milk as part of the process of making cheese.

white muscle disease. Nutritional muscular distrophy, a fatal condition in calves and goats, caused by selenium deficiency, in which heart muscle fibers are replaced with connective tissue.

withdrawal period. The amount of time that must elapse for a drug to be eliminated from an animal's body before it is butchered so the meat or milk contains no residue.

wolf teeth. *See* needle teeth.

wool block. An illness in rabbits caused by swallowed fur forming a blockage in the digestive tract.

Yearling. A male or female cow between 1 and 2 years of age.

Resources

Note: Many periodicals, organizations, and suppliers pertain to livestock raising; the lists below are just a sampling of the major sources. Many more can be found online. In addition, numerous organizations are dedicated to specific breeds; a quick Internet search will reveal them.

For more information or help with a specific problem or question, contact your county Extension agent, state Extension service, state agriculture education and FFA program, state Department of Agriculture, specialists at land grant colleges, or your local library. For updated addresses of organizations and publications, consult a recent edition of the *Directory of American Agriculture* (published by Agricultural Resources & Communications, Inc., 785-456-9705, www.agresources.com), available through your library.

GENERAL RESOURCES

Extension Service

For more information about the livestock you are interested in and programs in your area, contact the Cooperative Extension Service in your state. This program is affiliated with each of the nation's land-grant universities and the U.S. Department of Agriculture and can provide information on a wide range of topics.

To find your nearest Extension office, contact:

Cooperative Extension System
National Institute of Food and Agriculture (NIFA)
202-720-4423
www.csrees.usda.gov/Extension

Books

Angora Goats the Northern Way, 4th ed., by Susan Black Drummond. Stony Lonesome Farm, 1993.

The Angora Goat: Its History, Management and Diseases, 2nd ed., by Stephanie Mitcham Sexton and Allison Mitcham. Crane Creek Publications, 1999.

The Backyard Homestead, edited by Carleen Madigan. Storey Publishing, 2009.

Basic Butchering of Livestock & Game, rev ed., by John J. Mettler Jr. Storey Publishing, 2003.

The Cattle Health Handbook, by Heather Smith Thomas. Storey Publishing, 2008.

Chicken Coops, by Judy Pangman. Storey Publishing, 2006.

The Chicken Health Handbook, by Gail Damerow. Storey Publishing, 1994.

Chicken Tractor, by Andy Lee and Patricia Foreman. Good Earth Publications, 2000.

Day Range Poultry, by Andy Lee and Patricia Foreman. Good Earth Publications, 2002.

Earth Ponds, 2nd ed., by Tim Matson. Countryman Press, 1991.

The Fairest Fowl: Portraits of Championship Chickens, photography by Tamara Staples, essay by Ira Glass. Chronicle Books, 2001.

Fences for Pasture & Garden, by Gail Damerow. Storey Publishing, 1992.

For the Love of Poultry, A Backyard Poultry Anthology, the First Year – 2006, by the editors of *Backyard Poultry*. Countryside Publications, 2010.

Goat Health Handbook, by Thomas R. Thedford. Winrock International, 1983.

Greener Pastures on Your Side of the Fence, 4th ed., by Bill Murphy. Arriba Publishing, 1998.

A Guide to Canning, Freezing, Curing & Smoking Meat, Fish & Game, rev ed., by Wilbur F. Eastman Jr. Storey Publishing, 2002.

Home Cheese Making, 3rd ed., by Ricki Carroll. Storey Publishing, 2002.

The Home Creamery, by Kathy Farrell-Kingsley. Storey Publishing, 2008.

Home Sausage Making, 3rd ed., by Susan Mahnke Peery and Charles G. Reavis. Storey Publishing, 2003.

How to Build Animal Housing, by Carol Ekarius. Storey Publishing, 2004.

Keeping Livestock Healthy, 4th ed., by N. Bruce Haynes. Storey Publishing, 2001.

Raising Rare Breeds: Heritage Poultry Breeds Conservancy Guide, edited by Valda Blake and Daniel Price-Jones. Joywind Farm Rare Breeds Conservancy, 1994.

Renovating Barns, Sheds & Outbuildings, by Nick Engler. Storey Publishing, 2001.

Storey's Guide to Raising Chickens, 3rd ed., by Gail Damerow. Storey Publishing, 2010.

Storey's Illustrated Guide to Poultry Breeds, by Carol Ekarius. Storey Publishing, 2007.

Storey's Illustrated Breed Guide to Sheep, Goats, Cattle, and Pigs, by Carol Ekarius. Storey Publishing, 2008.

Magazines

Acres USA Magazine
800-355-5313
www.acresusa.com

BackHome
800-992-2546
www.backhomemagazine.com

Countryside & Small Stock Journal
800-551-5691
www.countrysidemag.com

Farming Magazine
800-915-0042
www.farmingmagazine.net

Grit
Ogden Publications, Inc.
866-803-7096
www.grit.com

Hobby Farms
BowTie, Inc.
800-627-6157
www.hobbyfarms.com

Mother Earth News
Ogden Publications, Inc.
800-234-3368
www.motherearthnews.com

Small Farmer's Journal
800-876-2893
www.smallfarmersjournal.com

Small Farm Today
800-633-2535
www.smallfarmtoday.com

The Stockman Grass Farmer
800-748-9808
http://stockmangrassfarmer.net

Organizations

American Livestock Breeds Conservancy
919-542-5704
www.albc-usa.org

ATTR: National Sustainable Agriculture Information Service
800-346-9140
www.attra.org

Rare Breeds Canada
514-453-1062
www.rarebreedscanada.ca

The Samuel Roberts Noble Foundation
580-223-5810
www.noble.org

Slow Food USA
877-756-9366
www.slowfoodusa.org

Suppliers

American Livestock Supply, Inc.
800-356-0700
www.americanlivestock.com

Animal Health Express, Inc.
800-533-8115
www.animalhealthexpress.com

Dominion Veterinary Laboratories, Inc.
800-465-7122
www.domvet.com

Eldon's Jerky and Sausage Supply
800-352-9453
www.eldonsausage.com

Farmstead Health Supply
919-643-0300
www.farmsteadhealth.com

Infratherm, Inc.
715-469-3280
www.sweeterheater.com

Jeffers Livestock
800-533-3377
www.jefferslivestock.com

Kencove Farm Fence, Inc.
800- 536-2683
www.kencove.com

Lambert Vet Supply
800-344-6337
www.lambertvetsupply.com

LEM Products
877-336-5895
www.lemproducts.com

Nasco
800-558-9595
www.eNASCO.com

Nite Guard, LLC
800-328-6647
www.niteguard.com

Omaha Vaccine Company
CSR Company Inc.
800-367-4444
www.omahavaccine.com

PBS Animal Health
800-321-0235
www.pbsanimalhealth.com

Pipestone Veterinary Clinic
800-658-2523
www.pipevet.com
Pipestone Vet Clinic is probably the best-known sheep vet clinic in the country.

PolyDome
800-328-7659
www.polydome.com
Polyethylene huts and hutches for use in raising farm animals

Premier1 Supplies
800-282-6631
www.premier1supplies.com

Randall Burkey Company
800-531-1097
www.randallburkey.com

The Sausage Maker, Inc.
888-490-8525
www.sausagemaker.com

Valley Vet Supply
800-419-9524
www.valleyvet.com

Western Ranch Supply
800-548-7270
www.westernranchsupply.com

POULTRY

Magazines

Backyard Poultry
800-551-5691
www.backyardpoultrymag.com
A bimonthly magazine devoted to topics related to raising poultry on a small scale.

Feather Fancier Newspaper
519-542-6859
www.featherfancier.on.ca
Monthly paper covering Canadian exhibition poultry.

Poultry Press
765-827-0932
www.poultrypress.com
Monthly newspaper and source for clubs, shows, and suppliers of birds and equipment.

Organizations

American Bantam Association
973-383-8633
www.bantamclub.com

American Poultry Association
724-729-3459
www.amerpoultryassn.com

National Poultry Improvement Plan
Animal and Plant Health Inspection Service
www.aphis.usda.gov/animal_health/animal_dis_spec/poultry

National Turkey Federation
202-898-0100
www.eatturkey.com

Society for the Preservation of Poultry Antiquities
570-837-3157
www.feathersite.com/Poultry/SPPA/SPPA.html

Suppliers

Cutler's Pheasant, Poultry, and Bee Supplies
810-633-9450
www.cutlersupply.com

Egganic Industries
800-783-6344
www.henspa.com

EggCartons.com
888-852-5340
www.eggcartons.com

GQF Manufacturing Company
912-236-0651
www.gqfmfg.com

Klubertanz Equipment Co., Inc.
800-237-3899
www.klubertanz.com

My Pet Chicken
888-460-1529
mypetchicken.com

NatureForm Hatchery Systems
904-358-0355
www.natureform.com

Smith Poultry & Game Bird Supply
913-879-2587
www.poultrysupplies.com

Twin City Poultry Supplies
614-595-8608
www.twincitypoultrysupplies.com

Wine Country Coops
707-829-8405
www.winecountrycoops.com

Breeders and Hatcheries

This guide is provided for your convenience and not as an endorsement of individual breeders.

Cackle Hatchery
417-532-4581
www.cacklehatchery.com

Eagle Nest Poultry
419-562-1993
www.eaglenestpoultry.com

Hoffman Hatchery Inc.
717-365-3694
www.hoffmanhatchery.com

Holderread Waterfowl Farm & Preservation Center
541-929-5338
www.holderreadfarm.com

Hoover's Hatchery
800-247-7014
www.hoovershatchery.com

Ideal Poultry Breeding Farms, Inc.
254-697-6677
www.idealpoultry.com

Marti's Poultry Farm
660-647-3156
www.martipoultry.com

Metzer Farms
800-424-7755
www.metzerfarms.com

Meyer Hatchery
888-568-9755
www.meyerhatchery.com

Mt. Healthy Hatcheries
800-451-5603
www.mthealthy.com

Murray McMurray Hatchery
800-456-3280
www.mcmurrayhatchery.com

Privett's Hatchery, Inc.
877-774-8388
www.privetthatchery.com

Ridgway Hatchery
800-323-3825
www.ridgwayhatchery.com

Sand Hill Preservation Center
563-246-2299
www.sandhillpreservation.com

Strombergs' Chicks & Gamebirds Unlimited
800-720-1134
www.strombergschickens.com

Sun Ray Chicks Hatchery
319-636-2244
www.sunrayhatchery.com

Townline Poultry Farm
888-685-0040
www.townlinehatchery.com

Welp, Inc.
800-458-4473
www.welphatchery.com

Websites

BackYardChickens.com
www.backyardchickens.com

TheCityChicken.com
http://home.centurytel.net/thecitychicken

Community Chickens
Ogden Publications, Inc.
www.communitychickens.com

FeatherSite
www.feathersite.com

poultryOne
http://poultryone.com

ThePoultrySite.com
www.thepoultrysite.com

Poultry Small Flock Information for County Extension Personnel
University of Arkansas Cooperative Extension Service
www.aragriculture.org/poultry/small_flock_information.htm

Small Flock Management of Poultry
MSUcares.com
http://msucares.com/poultry/management

RABBITS

Organizations and Publications

The American Rabbit Breeders Association publishes an excellent bimonthly magazine, *Domestic Rabbits*, that comes free with ARBA membership. Additionally, for each recognized breed, a newsletter and a guidebook is produced by the specialty club devoted to their particular breed and sent to club members. Contact information for these breed-specialty clubs and for hundreds of state, regional, and local breeder associations, many of which also publish newsletters and maintain websites, can be found in the ARBA *Yearbook*, which is included in ARBA membership.

American Rabbit Breeders Association, Inc.
309-664-7500
www.arba.net

Rabbit and Equipment Suppliers

Your area farm supply and feed dealers are likely sources of prefabricated rabbit hutches, feeding and watering equipment, and other items such as welded wire mesh and the clips, rings, and pliers used for hutch assembly. You might also find hutches and accessories at pet stores, but prices there are apt to be much higher than those at the farm stores.

Bass Equipment Company
800-798-0150
www.bassequipment.com

KW Cages
800-447-2243
www.kwcages.com

Quality Cage Company
888-762-2336
www.qualitycage.com

Rabbit Mart
GregRobert Enterprises
866-331-1920
www.rabbitmart.com

Websites

DebMark Rabbit Education Resource
www.debmark.com/rabbits

Rabbits Online
www.rabbitsonline.net

Rudolph's Rabbit Ranch
www.rudolphsrabbitranch.com

BEES

Periodicals

American Bee Journal
217-847-3324
www.americanbeejournal.com

Bee Culture
Root Candles
800-289-7668
www.beeculture.com

Organizations

Apimondia: International Federation of Beekeepers' Associations
http://beekeeping.com/apimondia

Apimondia Foundation
www.apimondiafoundation.org

Suppliers (Bees and Equipment)

Betterbee
800-632-3379
www.betterbee.com

Blue Sky Bee Supply
877-529-9299
www.blueskybeesupply.com

Brushy Mountain Bee Farm
800-233-7929
www.brushymountainbeefarm.com

Cook & Beals, Inc.
308-745-0154
www.cooknbeals.com

Cowen Manufacturing, Inc.
800-257-2894
www.cowenmfg.com

Dadant and Sons, Inc.
888-922-1293
www.dadant.com

Draper's Super Bee Apiaries, Inc.
800-233-4273
www.draperbee.com

Mann Lake Ltd.
800-880-7694
www.mannlakeltd.com

Maxant Industries
978-772-0576
www.maxantindustries.com

Miller Bee Supply
888-848-5184
www.millerbeesupply.com

Pierco Beekeeping Equipment
800-233-2662
www.pierco.net

Rossman Apiaries Inc.
800-333-7677
www.gabees.com

Walter T. Kelley Co.
800-233-2899
www.kelleybees.com

Western Bee Supplies, Inc.
800-548-8440
www.westernbee.com

Websites

APIS Information Resource Center at Squidoo
www.squidoo.com/Apis

Bee Health
UMass Extension
www.extension.org/bee health

Beekeeping: The Beekeeper's Home Pages
www.badbeekeeping.com

Bees-Online
www.bees-online.com

Beesource
www.beesource.com

GOATS

Magazines

Dairy Goat Journal
800-551-5691
www.dairygoatjournal.com

The Goat Magazine
325-653-5438
www.goatmagazine.info

Ruminations
978-827-1305
www.smallfarmgoat.com

United Caprine News
817-297-3411
www.unitedcaprinenews.com

Organizations

American Dairy Goat Association
828-286-3801
www.adga.org

Canadian Meat Goat Association
306-598-4322
www.canadianmeatgoat.com

International Goat Association
501-454-1641
www.iga-goatworld.org

Miniature Dairy Goat Association
509-591-4256
www.miniaturedairygoats.com

The Miniature Goat Registry
619-669-9978
www.tmgronline.org

National Miniature Goat Association
NminiaturegoatA@comcast.net
www.nmga.net

Suppliers

Caprine Supply
800-646-7736
www.caprinesupply.com

Hamby Dairy Supply
800-306-8937
http://hambydairysupply.com

Hoegger Supply Co.
800-221-4628
www.hoeggergoatsupply.com

Websites

Bar None Meat Goats
www.barnonemeatgoats.com/menu.html

The Biology of the Goat
www.goatbiology.com

The Boer & Meat Goat Information Center
www.boergoats.com

Clear Creek Farms
www.motesclearcreekfarms.com

Fias Co Farm
http://fiascofarm.com

South African Boer Goats
www.sa-boergoats.com

SHEEP

Magazines

The Banner Sheep Magazine
309-785-5058
www.bannersheepmagazine.com

Sheep!
800-551-5691
www.sheepmagazine.com

Sheep Canada
888-241-5124
www.sheepcanada.com

Organizations

American Polypay Sheep Association
641-942-6402
www.polypay.org

American Romney Breeders Association
secretary@americanromney.org
www.americanromney.org

Continental Dorset Club
cdcdorset@cox.net
www.dorsets.homestead.com

Katahdin Hair Sheep International
479-444-8441
www.katahdins.org

National Tunis Sheep Registry Inc.
641-942-6402
www.tunissheep.org

United Suffolk Sheep Association
435-563-6105
www.u-s-s-a.org

Suppliers

Premier1 Supplies
800-282-6631
www.premier1supplies.com

Sheepman Supply Co.
800-331-9122
www.sheepman.com

Vittetoe, Inc.
800-848-8386
www.vittetoe.com
These folks specialize in equipment for showing, but they also carry a variety of other products for shepherds.

Website

Sheep 101
www.sheep101.info

PIGS

Organizations

Certified Pedigree Swine
309-691-0151
www.cpsswine.com
This is a unified organization of the Chester White Swine Record Association, National Spotted Swine Records, and Poland China Record Association.

National Swine Registry
765-463-3594
www.nationalswine.com
A consolidation of the American Yorkshire Club, the Hampshire Swine Registry, the American Landrace Association, and the United Duroc Swine Registry.

Websites

ThePigSite
www.thepigsite.com

American Association of Swine Veterinarians
www.aasp.org

BEEF AND DAIRY COWS

Magazines

American Red Angus Magazine
Red Angus Association of America
940-387-3502
http://redangus.org

American Salers Magazine
American Salers Association
http://salersusa.org

Angus Journal
800-821-5478
www.angusjournal.com

Brangus Journal
International Brangus Breeders Association
210-696-8231
www.int-brangus.org

Cattle Today
205-932-8000
www.cattletoday.com

Charolais Journal
American-International Charolais Association
816-464-5977
www.charolaisusa.com/journalhome.html

Gelbvieh World
American Gelbvieh Association
303-465-2333
www.gelbvieh.org/maingw.html

Hereford World
American Hereford Association
816-842-3757
www.herefordworld.org

The Register
American Simmental Association
406-587-2778
www.simmental.org

Texas Longhorn Trails
Texas Longhorn Breeders Association of America
817-625-6241
www.tlbaa.org

Western Cowman
916-362-2697
www.westerncowman.com

Dairy Suppliers

Dairy Connection Inc.
608-242-9030
www.dairyconnection.com

Family Milk Cow Dairy Supply Store
816-449-1314
www.familymilkcow.com

Hamby Dairy Supply
800-306-8937
http://hambydairysupply.com

Jeffers Livestock
800-533-3377
www.jefferslivestock.com

Websites

FamilyCow.net
http://familycow.net

Keeping a Family Cow
http://familycow.proboards.com

Small Beef Cattle Farm Information and Resources
www.small-beef-cattle.com

Illustration Credits

Cathy Baker: 287, 288 (overhand, double half-hitch, square knots), 293

Sarah Brill: 238 top

Bethany Caskey: 14 (all except top row second from left), 19, 29, 31, 42 left, 43, 89

Jeffrey Domm: 286

Brigita Fuhrmann: 262 right

Carol Jessup: 112, 116, 117, 119, 122 top right, 123 left, 131, 132, 136, 137, 139, 141 top, 182, 183 top, 186 top, 207, 208, 216, 217, 225, 228, 231, 232, 234, 240, 244, 289 top, 325, 330

Kimberlee Knauf: 91

© Elayne Sears: 14 (top row second from left), 16 (all except middle row right), 18, 20–22, 25, 27, 30, 33–36, 40, 41, 42 right, 45, 51, 52 (all except top left), 53 (top left, right), 55 bottom, 56–58, 67, 70, 71, 72 top, 73 (center left, right), 74 (center right, right), 75 (top left, top center), 76, 77 (all except right), 78 (center left, right), 80–83, 86, 87, 90 top, 93, 95–97, 101–103, 107, 108 (all except top left), 110, 113 (top, middle), 115, 121 top, 123 right, 127, 129, 133, 141 bottom, 145–168,172–174, 175 bottom left, 176, 179, 181, 183 bottom, 185, 186 bottom, 187, 191, 194, 195, 197–205, 211, 213, 224 (all except top left, middle right), 227, 229, 233, 235–238, 242, 245, 249, 250 (top left, top center), 251 (top left, bottom), 252 bottom, 255, 257, 258 bottom. 260 left, 261 (left, bottom right), 264 right, 266, 273, 275–277, 279–281, 285, 288 (bowline, quick-release knots), 289 bottom, 290, 292, 300, 306, 308, 309, 311–313, 319, 323, 326, 328

© Casey Alexander Southard: full-color section

© Elara Tanguy: 7–9, 13, 16 (middle row right), 39, 52 top left, 53 bottom left, 55 top, 61–63, 72 bottom, 73 (left, center right), 74 (left, center left), 75 (all except top left, top center), 77 right, 78 (left, center right), 90 bottom, 94, 108 top left, 109, 113 bottom, 122 middle, 130, 138, 141, 175 top left, 177, 178, 184, 189, 193, 224 (top left, middle right), 250 (top right, bottom left, bottom right), 251 top right, 252 top row all, 253, 259 bottom, 260 right, 264 left, 268

Becky Turner: 262 left, 261 top

Index

Page references in *italics* indicate illustrations; page references in **bold** indicate charts.

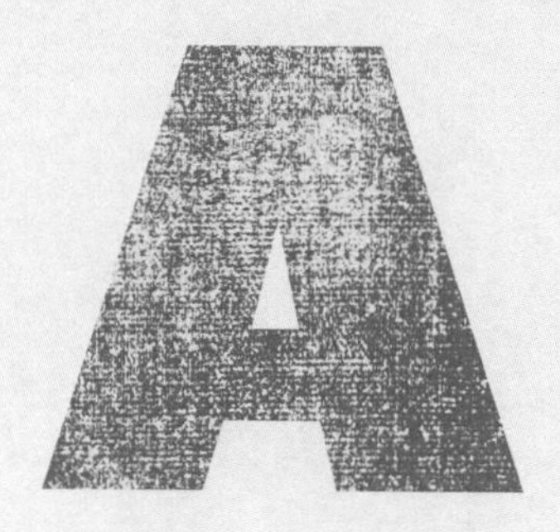

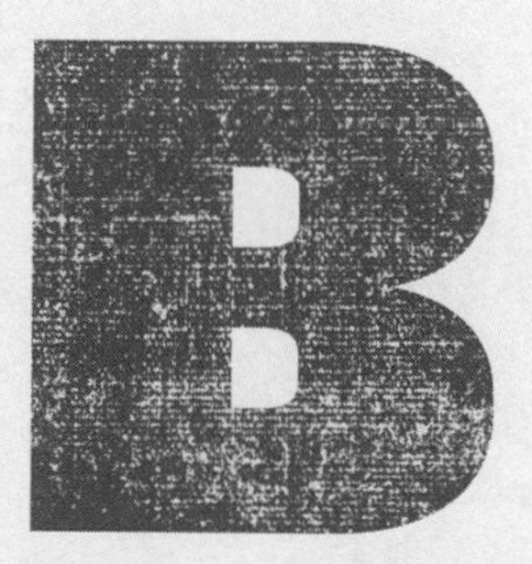

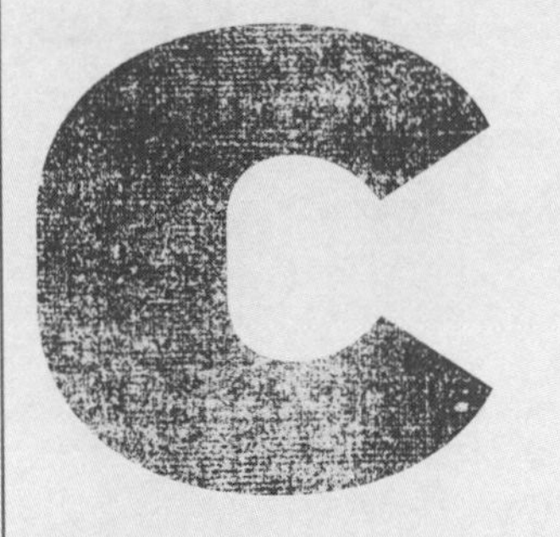

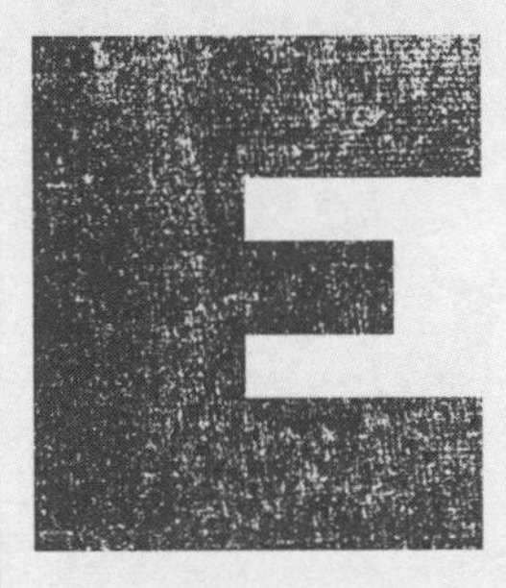

H

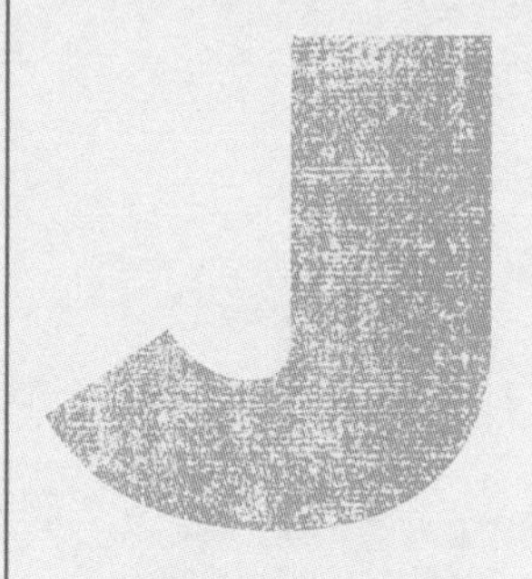

M

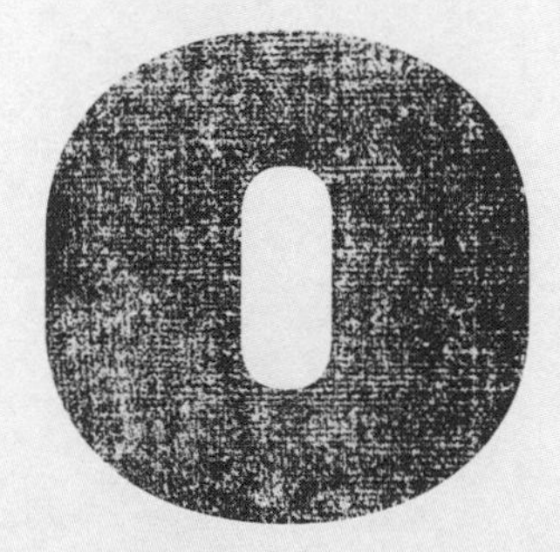

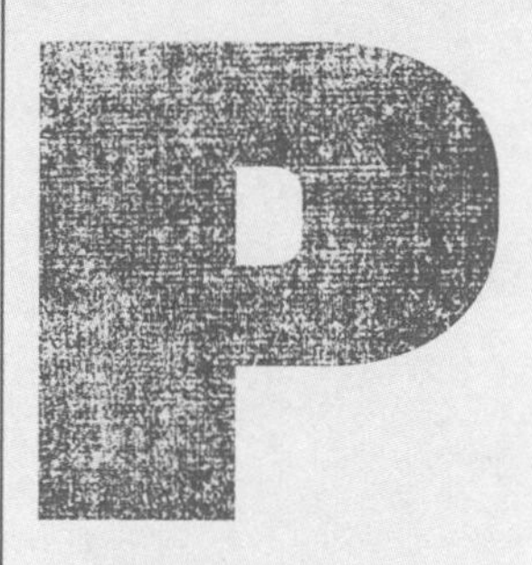

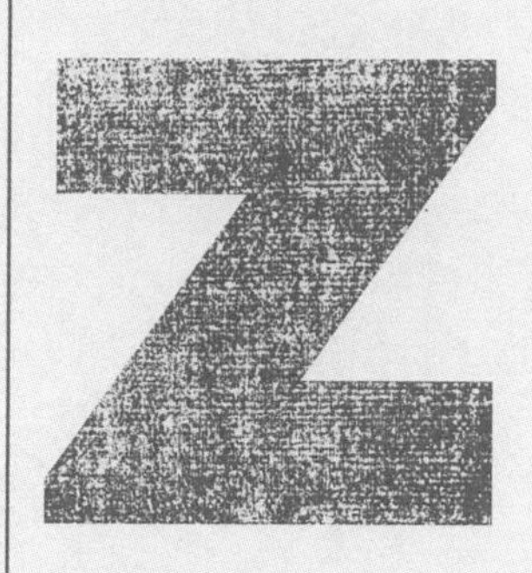

Other Storey Titles You Will Enjoy

The Backyard Homestead, edited by Carleen Madigan.

A complete guide to growing and raising the most local food available anywhere — from one's own backyard.
368 pages. Paper. ISBN 978-1-60342-138-6.

Small-Scale Livestock Farming, by Carol Ekarius.

A natural, organic approach to livestock management to produce healthier animals, reduce feed and health care costs, and maximize profit.
224 pages. Paper. ISBN 978-1-58017-162-5.

Starting & Running Your Own Small Farm Business, by Sarah Beth Aubrey.

A business-savvy reference that covers everything from writing a business plan and applying for loans to marketing your farm-fresh goods.
176 pages. Paper. ISBN 978-1-58017-697-2.

Storey's Guide to Raising Series.

Everything you need to know to keep your livestock and your profits healthy.
All new editions of *Beef Cattle*, *Sheep*, *Pigs*, *Dairy Goats*, *Meat Goats*, *Chickens*, *Ducks*, *Turkeys*, *Poultry*, *Rabbits*, *Raising Horses*, *Training Horses*, and *Llamas*. New additions to the series: *Miniature Livestock* and *Keeping Bees*.
Paper and hardcover. Learn more about each title by visiting *www.storey.com*.

Storey's Illustrated Breed Guide to Sheep, Goats, Cattle, and Pigs, by Carol Ekarius.

A comprehensive, colorful, and captivating in-depth guide to North America's common and heritage breeds.
320 pages. Paper. ISBN 978-1-60342-036-5.
Hardcover with jacket. ISBN 978-1-60342-037-2.

Storey's Illustrated Guide to Poultry Breeds, by Carol Ekarius.

A definitive presentation of more than 120 barnyard fowl, complete with full-color photographs and detailed descriptions.
288 pages. Paper. ISBN 978-1-58017-667-5.
Hardcover with jacket. ISBN 978-1-58017-668-2.

These and other books from Storey Publishing are available wherever quality books are sold or by calling 1-800-441-5700.
Visit us at *www.storey.com*.